BirdLife Conservation Series No. 5

KEY AREAS FOR THREATENED BIRDS IN THE NEOTROPICS

David C. Wege and Adrian J. Long

TED PARKER

John Maier

Ted Parker will be remembered by countless people and in as many different ways. At BirdLife he is most closely identified with *Threatened birds of the Americas,* which he co-authored and which forms the cornerstone of *Key Areas*. However, during his involvement in the project, and after, he left us with far more than just a portion of his quite astonishing knowledge. Ted's energy and enthusiasm for all aspects of birds and their conservation rubbed off on everyone who spent time with him.

The fact that Ted was a friend and an inspiration to so many is obvious from the numerous articles devoted to his memory. However, perhaps the most moving salute to the influence he exerted on Neotropical ornithology and conservation is the fact that two special memorial publications (a double issue of BirdLife's journal *Bird Conservation International*, and one of the American Ornithologists' Union's series of *Ornithological Monographs*) have been filled to capacity with papers contributed by admiring co-workers, colleagues and acquaintances. This is a measure of the goodwill he generated and the void he left behind.

Key Areas is a plea for action. Had Ted still been alive today he would have added his voice in support of the conservation that is so urgently needed if the threatened birds, for which these Key Areas are so important, are to be saved. We owe it to Ted to ensure this plea is answered.

Distributed in the Americas by Smithsonian Institution Press, Washington, DC

BirdLife International is a U.K. registered charity

ISBN 1 56098 529 1

Series editor Duncan Brooks
Design CBA (Cambridge) and Duncan Brooks
Layout, text preparation and indexing Duncan Brooks, Michelle Hines, Regina Pfaff

Text set in Times (9/11 pt) and Optima

Imageset, printed and bound in Great Britain by The Burlington Press (Cambridge) Ltd.

Cover illustration Seven-coloured Tanager *Tangara fastuosa*, by Peter Hayman. Endemic to north-east Brazil, this threatened species occurs in six Key Areas but is exposed to the twin threats of habitat loss and heavy exploitation for trade.

The BirdLife International Partnership in the Neotropics

■

Partners and Partners Designate

ARGENTINA

Asociación Ornitológica del Plata
25 de Mayo 749 - 2º Piso
1002 Buenos Aires
Argentina
tel./fax +54 1 3128958

CHILE

Unión de Ornitólogos de Chile (UNORCH)
Casilla 572-11
Santiago, Chile
tel. +56 2 2712865 ext. 257
fax +56 2 2727363
email ecofisio@abello.seci.uchile.cl

BELIZE

Belize Audubon Society
PO Box 1001
Belize City
Belize
tel. +501 2 77369
fax +501 2 34985

ECUADOR

Fundación Ornitológia del Ecuador (CECIA)
Av. de los Shyris 20-30 y la Tierra
PO Box 17-17-906, Quito, Ecuador
tel./fax +593 2 468876
email cecia@cipa.ecx.ec

BOLIVIA

Fundación Armonía
Box 3081
Santa Cruz de la Sierra
Bolivia
tel. +591 3 522919
fax +591 3 324971

PARAGUAY

Fundación Moisés Bertoni
Av. Rodriguez de Francia 770
C.C. 714, Asunción, Paraguay
tel. +595 21 444253
fax +595 21 440239
email moises@fmb.py

VENEZUELA

Sociedad Conservacionista Audubon de Venezuela
Apartado 80450, Caracas 1080-A
Venezuela
tel. +58 2 9932525/922812
fax +58 2 910716
email crodner@conicit.ve

■

Representatives

EL SALVADOR

Asociación Audubon de El Salvador
la Calle Poniente
Condominio Monte Maria
Edificio A 2-2. S.S., El Salvador
tel. +503 2980811
fax +503 2981634
+503 2749180

MEXICO

CIPAMEX
Depto de Zoología, Instituto de Biología
UNAM AP 70-153
México DF 04510, Mexico
tel. +52 5 6225703
fax +52 5 5500164
email escalant@servidor.unam.mx

HONDURAS

Sherry Thorne
Apdo 30289 Toncontín
Tegucigalpa
Honduras
tel. +504 341869

URUGUAY

GUPECA
Casilla de Correo 6955
Montevideo, Uruguay
tel./fax +598 2959678
email gual@fcien.edu.uy

CONTENTS

FOREWORD

by Enrique H. Bucher
Professor of Zoology, University of Córdoba, Argentina
Council Member of BirdLife International

ACCORDING to the Convention on Biological Diversity that emerged from the 1992 Earth Summit conference in Rio de Janeiro, the protection of the biological diversity of this planet is vital to the continued health and welfare of humanity, and immediate national action and international cooperation are required for the protection of ecosystems and for the conservation of biodiversity and its various components. It is recommended that all countries prepare an inventory of their natural and social science data that are relevant to sustainable development, and that the development of comprehensive national strategies for the conservation of biological diversity be made a priority with the widest possible global participation.

Key Areas for threatened birds in the Neotropics represents a significant contribution within the priorities and recommendations of the convention, both in terms of the approach followed and the product delivered. This timely book summarizes the results of an international effort that involves the dedication and commitment of the authors as well as the long-term effort of all those ornithologists who have contributed to our understanding of the Neotropical region. The book provides, in the first place, a synthesis of current information on threatened bird species and, secondly, a practical and objective way of highlighting areas for these species that deserve priority consideration for conservation.

This book builds on and is a useful complement to another extremely valuable volume produced by BirdLife International, namely *Threatened birds of the Americas: the ICBP/IUCN Red Data Book* (Collar *et al.* 1992). Together, they represent an invaluable tool for researchers, conservationists and decision-makers interested in developing national bird conservation strategies for the Neotropics.

Key Areas is based on objective evidence about the shrinking distributions or populations of a considerable number of species of the Neotropical avifauna. As birds are generally good indicators of environmental conditions, the observed problems of the most threatened should be considered the 'tip of the iceberg' of a much more extended process of degradation, which is also being observed in other taxa of plants and animals. These threatened birds should be seen as 'warning lights' which may play an important role in convincing decision-makers and the general public of the rapidly increasing threats in a continent where the perception of environmental problems is incipient. Even today, large sectors of Latin American society think of their countries as immensely rich in natural resources just waiting to be exploited. Such widespread views of unlimited richness are encouraged by the present worldwide attitude towards free commerce and open economies that stimulates an unrestricted race for the exploitation of all natural resources in order to promote rapid economic growth.

Unfortunately, ignorance of the fragility and limits of the planet's biological resources discourages the control and monitoring necessary to ensure the sustainability of their use. As resource overexploitation results in short periods of economic boom followed by degradation and loss of productivity, the general economic situation deteriorates. An unfortunate association between poverty and degradation of natural resources develops, which easily becomes self-perpetuating.

Hopefully, availability of well-documented evidence and priorities will result not only in better management decisions but may also help to convince decision-makers, politicians and the whole Latin American society of the urgent need to include environmental threats in our social and political agenda. *Key Areas for threatened birds in the Neotropics* is a valuable step in the right direction.

ACKNOWLEDGEMENTS

KEY AREAS has benefited immeasurably from a wealth of new information supplied by the many selfless correspondents listed below. For useful comments, adding new data, reviewing accounts or entire national inventories, and generally helping us make this publication as comprehensive and up-to-date as possible we extend our sincere thanks to E. I. Abadie, W. Adsett, A. Aleixo, J. C. Almada, G. I. Andrade, the late T. A. Andrews, M. C. Arizmendi, C. Balchin, G. Bencke, P. J. Benstead, K. S. Berg, B. J. Best, M. Bettinelli, J. Blincow, R. Boçom, P. Boesman, A. Bosso, R. C. Brace, T. M. Brooks, E. H. Bucher, C. Bushell, S. Butchart, L. Carlos, O. Carrillo, M. Catsis, J. C. Chebez, R. P. Clay, J. Curson, D. Dacol, S. Davis, C. Downing, A. Drewitt, G. Engblom, E. Enkerlin, P. B. Escalante, E. Z. Esquivel, C. Estades, J. Estudillo, J. Farthing, D. Finch, C. S. Fontana, B. C. Forrester, I. Franke, F. M. Gabelli, M. Galetti, D. Gallegos-Luque, J. Goerck, D. Gómez, L. P. Gonzaga, A. Di Giacomo, G. Green, N. Guedes, F. E. Hayes, A. J. Hesse, S. N. G. Howell, R. Innes, L. Jammes, M. Johnston, M. G. Kelsey, G. Kirwan, N. Krabbe, C. Ladd, F. R. Lambert, M. Lammertink, M. Lentino, A. Lewis, E. Londoño, B. López Lanús, J. C. Lowen, E. de Lucca, A. Luy, A. Madroño N., P. Martuscelli, J. L. Mateo, S. Mayer, J. Mazar, B. Milá, A. G. Navarro, R. Naveen, A. J. Negret, F. Olmos, Y. Oniki, J. F. Ornelas, J. F. Pacheco, J. Pearce-Higgins, M. Pearman, S. Peris, A. T. Peterson, D. M. Pullan, M. Reid, J. V. Remsen, L. M. Renjifo, K. Renton, M. Retamosa, R. S. Ridgely, M. B. Robbins, S. K. Robinson, C. A. Rodríguez-Yáñez, J. C. Rudolf, P. Salaman, M. Santos, T. S. Schulenberg, C. Seger, J. M. C. da Silva, H. Gómez de Silva, C. Sharpe, G. Speight, G. Spinks, F. G. Stiles, J. Tobias, H. Torres, E. P. Toyne, P. L. Tubaro, S. J. Tyler, S. Umpiérrez, D. Uribe, F. Uribe, N. Varty, B. P. Walker, J. Wall, B. M. Whitney, A. Whittaker, M. Whittingham, E. O. Willis, R. S. R. Williams, R. G. Wilson, and offer our apologies to anyone whom we may have inadvertently overlooked.

Key Areas relies extensively on the wealth of data compiled and published in *Threatened birds of the Americas* (Collar *et al.* 1992), and it is to the over 550 correspondents, representing a network of expertise throughout the Americas, that contributed to this cornerstone publication that we would like to offer our thanks once more. In particular, and often overlooked by the conservation community, is the enormous contribution to our knowledge and understanding made by museum collections, both in terms of the raw data from the specimens themselves, but also the experience and knowledge so generously (and selflessly) imparted by their various curators, taxonomists and ornithologists; this information alone forms the backbone of *Threatened birds of the Americas*, and to the over 60 museums (and associated people) acknowledged in Collar *et al.* (1992) we again extend our gratitude.

Also deserving special mention and thanks are Mark Pearman for reviewing and adding to most of the national inventories, Jonathan Rhind, Simon Blyth and Gillian Bunting from the World Conservation Monitoring Centre (WCMC) for their invaluable assistance with the provision of map coverages, and Jon Fjeldså for producing at short notice the evocative line-drawings that adorn the beginning of each national inventory.

The production of any publication is almost invariably a team effort, and this one is no exception. At the BirdLife Secretariat, for his substantial and invaluable editorial guidance, encouragement and friendship, we give our wholehearted thanks and gratitude to Nigel Collar. We are also greatly indebted to the publications team at BirdLife: Duncan Brooks improved the book enormously through his editorial excellence and layout design, capably and meticulously assisted by Regina Pfaff and Michelle Hines. Invaluable support, advice and assistance was also given by many of our colleagues at BirdLife, among whom we thank Beatriz Torres, Alison Stattersfield, Sue Squire, Andrew Rayner, Sandra Loor-Vela, Martin Kelsey, Jenifer Fletcher, Henk van Dijkhuizen, Mike Crosby and Colin Bibby. Amongst our supportive friends and relatives, David Wege wishes particularly to thank Nicola Wege for her patience and understanding during the long days, evenings and weekends of work that have typified the last year of this project, and likewise Adrian Long thanks Melanie Heath for her continued encouragement, and for her help in the final stages.

BirdLife International gratefully acknowledges the support of Shell International Petroleum Company Limited

KEY AREAS IN CONTEXT

The beauty and genius of a work of art may be reconceived, though its first material expression be destroyed; a vanished harmony may yet again inspire the composer; but when the last individual of a race of living beings breathes no more, another heaven and another earth must pass before such a one can be again.

C. William Beebe (1906) *The bird: its form and function.*

THE NEW WORLD, and in particular its Neotropical region, has long been recognized as supporting a disproportionately high level of biological diversity. Of the world's 9,700 bird species (Sibley and Monroe 1990, 1993), we compute from a variety of sources that no fewer than 4,130 (43%) occur in the New World (although this constitutes only 29% of the planet's land area), with 3,600 (39%) present in the non-Caribbean Neotropics (a mere 16% of the planet's land area). However, some 327 species of this rich and diverse New World avifauna were, in 1992, considered at risk of extinction (Collar *et al.* 1992), and with 290 (89%) of these being restricted to the Neotropics there is clearly an urgent need for conservation priority-setting within this latter region.

A multiplicity of economic, social and political factors, especially in the twentieth century, has brought about profound changes to the Neotropical environment. The forests are being felled, the grasslands cultivated, wetlands drained and pristine habitats increasingly fragmented and disturbed, as a consequence of which a number of species have already been lost forever. In the past hundred years the Neotropics has seen the last of such birds as Colombian Grebe *Podiceps andinus*, Glaucous Macaw *Anodorhynchus glaucus* and Slender-billed Grackle *Quiscalus palustris*, with others such as Magdalena Tinamou *Crypturellus saltuarius*, Brazilian Merganser *Mergus octosetaceus* and Pale-headed Brush-finch *Atlapetes pallidiceps* (see Appendix 3) possibly gone already, or critically close to extinction. Moreover, birds such as Alagoas Curassow *Mitu mitu*, Socorro Dove *Zenaida graysoni* and Spix's Macaw *Cyanopsitta spixii* are all extinct in the wild or effectively so, and survive only by virtue of their captive populations.

With such losses, which are mirrored in other groups of organisms and in other regions of the world, it is timely that the conservation of nature has at last entered the global agenda of the community of nations, the formal recognition of which was given when the Convention on Biological Diversity came into force in December 1993. Arising from this (and explicit within the convention articles) is the need for clear communication to decision-makers—at local, national and global levels—of the priorities which most urgently deserve our attention, given the constraints of limited time, funds and other resources. This presents a conservation challenge which lies, as Collar *et al.* (1994) put it, in simply defining biodiversity, deciding what strategies are most appropriate to which of its components, and setting priorities accordingly—a challenge that we believe is met, at least in part, by *Key Areas for threatened birds in the Neotropics*.

PRIORITY-SETTING FOR CONSERVATION

BirdLife International has long been committed to priority-setting activities for *in-situ* conservation. Using birds (which have been shown to be effective indicators for biodiversity conservation: see Box 1) as a focus, BirdLife's rationale has been to identify and document the potential losses from the global stock of species, and, by establishing sites of sympatric occurrence of threatened (or locally endemic) forms as the units of concern (a policy espoused in, e.g., Collar and Stuart 1988, Myers 1988,

Box 1. Birds as indicators for biodiversity conservation.

Birds make a unique contribution to the conservation of biodiversity in a number of ways.

- Birds have dispersed to, and diversified in, all terrestrial regions of the world and virtually all habitat types and altitudinal zones, and can therefore be used as global measures of diversity and richness.
- Birds are the only major group for which the taxonomy and geographical distribution of individual species are sufficiently well documented to permit a comprehensive and rigorous review and analysis at the global level (see Box 3).
- Birds have an enormous public and scientific following and, as they are well covered in field guides and relatively easy to identify, large amounts of data have been generated.
- Birds generate strong public interest and are thus an excellent subject for conservation education and advocacy.

1990, ICBP 1992), identify the sites where those potential losses can be saved in cost-efficient aggregations. This rationale is explicit within the requirements of the Convention on Biological Diversity, which charges contracting nations to identify those ecosystems, habitats, species and communities important for the conservation and sustainable use of natural resources, with an emphasis on the need for *in-situ* conservation.

BirdLife has traditionally documented those elements of the global avifauna that are most fragile through Red Data Books and site-based analyses, initially in the form of *Key forests for threatened birds in Africa* (Collar and Stuart 1988), but more recently as 'Endemic Bird Areas' and 'Important Bird Areas' (see Box 3). *Key Areas for threatened birds in the Neotropics* is a further contribution to BirdLife's commitment to conserving the world's birds. 'Key Areas' are the most important places *currently known* for the conservation of the 290 globally threatened birds in the Neotropics (see Box 2). As these areas have been identified using threatened birds, the majority of which are considered such owing to habitat loss, they represent some of the most precarious or damaged ecosystems and habitats within the region. By presenting *Key Areas* in an easy-to-use format, we aim to make accessible information that is useful to all biologists, planners, managers and decision-makers involved in Neotropical conservation issues and thereby offer a positive contribution to the preparation of national strategies and action plans in the Neotropics.

Key Areas and Red Data Books

Red Data Books present a synthesis, involving often widely scattered sources of data on threatened birds in a species-by-species format (Box 3). Extensive data on the threatened birds of the Neotropics were assembled in *Threatened birds of the Americas* in 1992 (Collar *et al.* 1992). Within this publication there was some cross-referencing (between species) of the obvious sites of sympatry and an attempt to match species localities to formally protected areas, but this was not done in any systematic way. A clear need therefore was to make a comprehensive assessment of the site-based conservation priorities for threatened birds—and *Key Areas for threatened birds in the Neotropics* is a response to this.

Key Areas and Endemic Bird Areas

Endemic Bird Areas (EBAs) are generally quite extensive, and of prime importance for targeting the conservation of restricted-range, and therefore potentially vulnerable, bird species concentrated within them (see Box 3). As 78% of threatened birds in the Neotropics have restricted ranges, many Key Areas are located within EBAs. These same areas will most likely be some of the most endangered (and therefore important) within the EBA and will almost certainly support many other commoner, but potentially vulnerable, restricted-range birds.

Box 2. Key Areas for threatened birds in the Neotropics.

The Key Area concept

- Key Areas are the most important places currently known for globally threatened bird species in the Neotropics.
- Key Areas are important for the conservation of 280 threatened birds in the region.
- Key Areas have been identified following a simple set of pragmatic criteria.
- Key Areas often harbour aggregations of threatened bird species.
- Key Areas represent some of the most precarious or damaged ecosystems and habitats in the region.

***Key Areas*: the book**

- *Key Areas* is a species-driven conservation tool presenting site-based priorities for threatened birds.
- *Key Areas* uses *Threatened birds of the Americas* as the primary information source but incorporates a large body of new data.
- *Key Areas* satisfies a major requirement of the Convention on Biological Diversity.

Box 3. Priority-setting mechanisms for bird conservation.

Red Data Books

- *Red Data Books document information relevant to the conservation of species considered to be at risk of extinction.*

Red Data Books (RDBs) were instituted by IUCN in the mid-1960s to identify and document threatened species and thereby help prevent global extinctions. RDBs provide a format in which information relevant to the conservation of threatened species is compiled, published and later updated. Thus the guiding principle behind all RDBs is to search out, analyse and include material on the distribution, population and ecology of threatened species, and document the causes of and cures for their endangerment, in the belief that the best decisions over the needs of threatened species can only be made with the fullest review of their status. The publication of RDBs such as *Threatened birds of Africa* (Collar and Stuart 1985), *Threatened birds of the Americas* (Collar *et al.* 1992) and the global summary *Birds to watch 2* (Collar *et al.* 1994) has long been part of the BirdLife programme, and they provide the official source for birds on the IUCN Red List (e.g. Groombridge 1993).

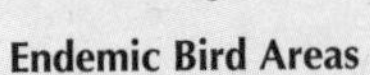

Endemic Bird Areas

- *Endemic Bird Areas support at least two restricted-range bird species, and thus represent areas where global extinctions are likely to occur unless the integrity of habitat can be guaranteed.*

Endemic Bird Areas (EBAs) have been identified by mapping the overlapping distributions of birds which have had, in historical times, breeding ranges of less than 50,000 km². An area to which two or more of these restricted-range birds are totally confined in close geographic proximity is designated as an EBA (up to 60 species have been recorded within an EBA). Patterns of endemism for birds often reflect the overall patterns in other plant and animal groups (Thirgood and Heath 1994, Balmford and Long in press: see Box 2). EBAs are thus areas where global extinctions would be caused by the wholesale modification or destruction of the habitat, and are therefore of prime importance from a strategic conservation perspective, although the conservation of restricted-range birds within an EBA will usually require the identification of smaller, more defined areas.

EBAs can be large areas (70% of them are greater than 10,000 km²) and are often islands (or island groups), or distinct biogeographic regions defined by a particular habitat-type, altitudinal range or climatic zone. Thus topographical features such as isolated mountains (or mountain ranges), river valleys and islands are often sufficiently biogeographically distinct to constitute EBAs. Over 2,600 species or 27% of the world's birds have restricted ranges, and are accommodated within just 221 EBAs: these same areas harbour 75% of the world's threatened bird species. EBAs were first documented in *Putting biodiversity on the map: priority areas for global conservation* (ICBP 1992), with a more detailed presentation to be published in *A global directory of Endemic Bird Areas* (Stattersfield *et al.* in prep.).

Important Bird Areas

- *The identification and protection of Important Bird Areas is critical for the long-term viability of populations of species whose conservation requires a site-based approach.*

Important Bird Areas (IBAs), as defined and identified by BirdLife, are of global importance for threatened species, species with restricted ranges, biome-restricted species and species exposed to danger by virtue of their congregatory habits. They exist as generally small, actual or potential protected areas, or areas which can be appropriately managed for wildlife conservation, providing (seasonally or in combination with other sites) the requirements of the birds for which they are important. The function of the IBA programme is to identify and protect, at the national level, a global network of sites critical for the conservation of those species for which a site-based approach is appropriate. IBAs are identified nationally using the criteria detailed in Appendix 5, and are now central to BirdLife policy and programme activities. IBA inventories have been completed by the BirdLife Partnership for all countries in Europe (Grimmett and Jones 1989) and the Middle East (Evans 1994), and individual country programmes are now also under way in Africa, Asia and the Americas.

Key Areas and Important Bird Areas

Key Areas contribute directly towards a network of Important Bird Areas (IBAs) for the Neotropics. BirdLife's IBA programme seeks to form a global network (albeit a minimum set) of sites necessary for the conservation of bird species for which a site-based approach is appropriate (Box 3). IBAs are identified under a variety of criteria (listed in Appendix 5), the first of which concerns their complement of threatened birds, so the Key Area analysis presented here will form a cornerstone of the IBA research programme in the Neotropics (which commences in 1996). Key Areas will commonly meet other IBA criteria (i.e. those concerning restricted-range species, biome restricted assemblages or congregations) on the basis of their bird species complements, and every Key Area will thus have to be reviewed as part of each national IBA initiative. New data on species will be added and supplementary information incorporated on geography, protection status, habitats, land use and threats.

■

KEY AREAS: THE BOOK

THIS BOOK treats what we have loosely termed the mainland or non-Caribbean Neotropics (and associated islands), and thus includes the threatened species that occur in Mexico, Central and South America, but also (where such species are present) what Collar *et al.* (1992) termed the 'Neotropical Pacific' islands, i.e. those belonging to Mexico, Ecuador and Chile, plus the Caribbean islands belonging to Colombia and Venezuela. The majority of the Caribbean islands have thus been omitted from this book, but will be covered in detail as part of the proposed IBA programme (see Box 3 in '*Key Areas* in context', p. 9).

Key Areas treats 290 threatened species, including Nava's Wren *Hylorchilus navai*, which has recently been recognized as a species distinct from another threatened bird, Sumichrast's Wren *H. sumichrasti*. Although *Threatened birds of the Americas* documented only in abbreviated form the threatened birds on the Neotropical Pacific islands and the Nearctic–Neotropical migrants that winter in the area under review, we have endeavoured to be as comprehensive as possible and identify Key Areas for each of these species. The exception is Eskimo Curlew *Numenius borealis*, for which no wintering areas—or indeed breeding areas—are currently known.

KEY AREAS IN THE MAKING

Following a simple set of criteria and using a database of localities for individual threatened species in conjunction with a Geographic Information System, potential Key Areas were identified and, through a process of consultation, review and literature searches, new data were incorporated to complete and update the analysis.

Criteria

The aim of this book is to ensure that wherever possible each threatened species is adequately represented within Key Areas. To achieve this aim, a simple set of criteria was devised (Box 1), such that wherever possible *the three most important areas from which a threatened bird is known were selected as Key Areas*. A pragmatic approach has been taken to the application of these criteria. Where the information about a species was poor we selected each locality (within reason) from which it had been recorded, and documented the inadequacy in the text introducing each country's Site Inventory under 'Old records and little-known birds'. Also, where there was no distinction to be made between areas, or

Box 1. Criteria for the identification of Key Areas.

Each threatened species must (where possible) be represented within at least three Key Areas (with seasonal migrants represented in each region to or from which they move). Areas are selected on the basis of the following guidelines.

- The area should support an extant population of the species in question;

or, if this is not known,

- The area should be where the species has been most recently recorded;

or, where there are no such records,

- The area is one in which there is good reason to suspect the species' continued existence.

If a choice presents itself, preference is given to Key Areas that:

- Are the larger or more intact;

or

- Are already protected;

or

- Ensure the entire range of a species is represented;

or

- Ensure that a species is represented in more than one country.

As protected areas are often the best places in which to implement conservation actions, an attempt has been made to identify those protected areas harbouring two or more threatened species.

where a species is known to be nomadic or widely dispersed (e.g. some parrot species, lowland forest raptors), more than three Key Areas have been chosen to ensure adequate coverage.

Data sources

The cornerstone of *Key Areas* is *Threatened birds of the Americas: the ICBP/IUCN Red Data Book* (Collar *et al.* 1992). In that study, every effort was made to make the dataset as complete as possible: all traceable references (both published and unpublished) containing information relevant to the conservation of the species under review were consulted (the bibliography lists 2,600 works); over 550 individuals are acknowledged for their contributions of knowledge and opinion; and as many as 60 museums in the Americas and Europe were contacted or visited for their unpublished specimen data, the exceptional value of which is demonstrated by the fact that it yielded 'new' material for virtually every species treated in the book. Thus, the data presented in *Threatened birds of the Americas* are a complete reflection of available knowledge up to 1992, and the information presented therein was used almost without modification for the present analysis.

Our knowledge of threatened birds is, however, constantly increasing, and the status of many areas in which they are known to occur is steadily deteriorating. To keep up-to-date with these changes (which often affect our perceptions of priorities) we have relied on information from concerned reserve managers, field biologists, ornithologists, birdwatchers, etc. (i.e. both the informal BirdLife network and the formal BirdLife Partnership: see p. 3), and, through a process of targeted consultation and general literature review, we have endeavoured to present as much up-to-date and corrected information as possible. All new information compiled during this project, whether corrections, information previously overlooked or post-Red Data Book updates, is referenced to source within the appropriate Key Area account, with the individual contributors and references used for each country documented under 'Data sources' within the appropriate national inventory.

The list of threatened species presented in *Threatened birds of the Americas* has recently been superseded by a new analysis. *Birds to watch 2* (Collar *et al.* 1994) covers, in abbreviated form, all of the world's threatened bird species, and represents an entirely new evaluation using the new IUCN criteria for assigning threat status and category (IUCN SSC 1994: see Appendix 4, p. 305). Inevitably there are a number of changes between the 1992 Americas listing and the 1994 global one, but as *Birds to watch 2* is an abbreviated RDB, there is not sufficient detail compiled for any of these new species to enable the extension of the Key Areas analysis to include them; thus, for the sake of consistent and authoritative results, all (and only the) species documented as threatened within *Threatened birds of the Americas* have been used for the present analysis, although these new species are listed and discussed under 'Recent changes to the threatened list' within the introductory section of each national inventory, with a full list of threatened species for the region (from *Birds to watch 2*) presented in Appendix 3 (p. 300).

Data storage

This book is effectively the product of a number of different databases: distributional information, in the form of geo-referenced point-locality data derived from *Threatened birds of the Americas*, is stored for all threatened Neotropical species in a species–locality database originally developed for the BirdLife Biodiversity Project (details of which are given in ICBP 1992 and Crosby 1994). Various attributes of these threatened birds are coded and stored in a database that contains a single record for each species (see Collar *et al.* 1994 and in press); the Key Area accounts are largely an output from a database that contains a single (geo-referenced) record for each area covered; and the references have been stored (and coded) in a bibliographic database.

Data analysis

Applying the criteria outlined above, three (or more) 'key localities' were selected for each species and marked within the species–locality database, and these localities were mapped using a geographic information system. By overlaying these key localities with all other threatened bird localities, each threatened bird recorded within 5 km of a key locality was selected and marked within the database. This allowed the production of a list of threatened species (and localities) to be found in the immediate vicinity of each key locality for a particular threatened species. After some rationalization (to take into account the geography and status of an area), these species and localities were amalgamated into the Key Area. However, if these (and indeed other species) fell inside the boundary of a protected area, the entire protected area with all the threatened species to be found within it was designated as the Key Area. Through a process of review, we have endeavoured to ensure that the areas identified are sensible at the local level, especially with regard to the continued presence of the appropriate habitat.

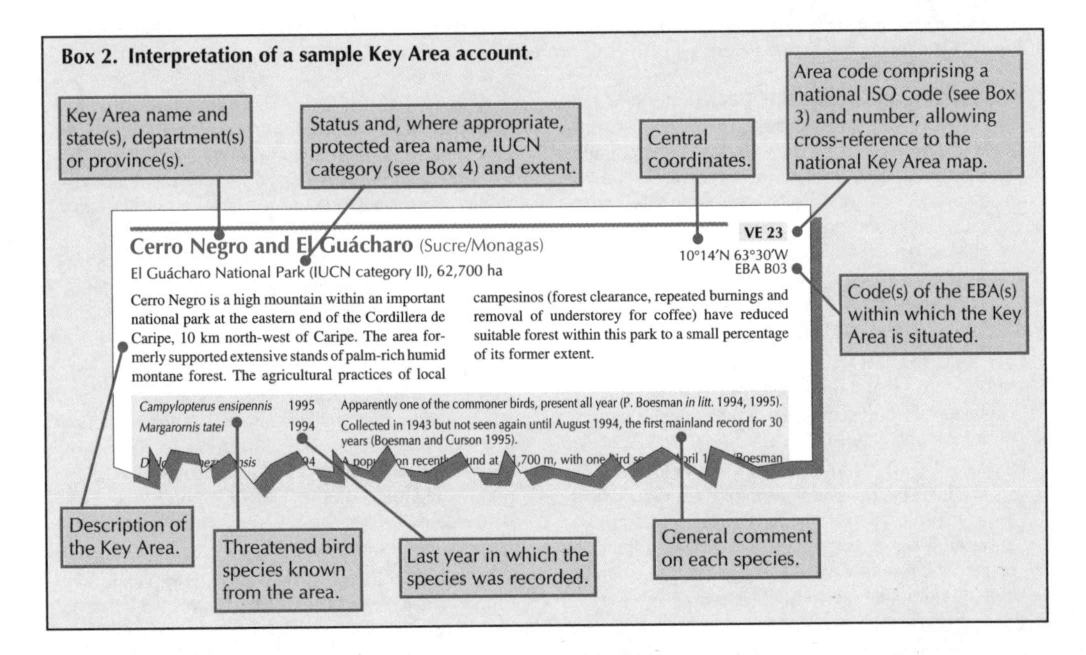

KEY AREAS PRESENTED

For each country featured in this analysis, an inventory is presented—although owing to the paucity of threatened species, and thus Key Areas, in two regions it has been necessary to present groups of countries in single chapters: with just six globally threatened species between them, the north Central American countries of Belize, Guatemala, Honduras, El Salvador and Nicaragua are treated together, and with only four threatened birds between them, the countries of Guyana, Surinam and French Guiana are likewise dealt with collectively. Each national (or regional group) inventory is divided into an introductory section and a Site Inventory.

Introductory section

Each introductory section outlines general details about the avifauna of the country (or groups of countries), but also, under the following headings, various aspects of the threatened birds and Key Areas at the national level.

Threatened birds indicates attributes of the threatened component of the avifauna, including general ecological preferences, threats faced and, owing to the close relationship between threatened and restricted-range bird species (see Box 3 in '*Key Areas* in context', p. 9), the species' distributions in terms of occurrence within the various Endemic Bird Areas (EBAs).

Key Areas outlines the representation of threatened species within Key Areas.

Key Area protection shows the extent to which Key Areas have formal protected status and the coverage of threatened species within such areas.

Recent changes to the threatened list indicates which species have been added and which dropped from the national threatened species list in Collar *et al.* (1994) (see Appendix 3), and what effect these changes could have on the Key Area analysis.

Old records and little-known birds is an account of those species within the country which are known particularly poorly.

Outlook highlights particularly important areas, species or general conservation needs as they relate to Key Areas.

Site Inventory

Each Site Inventory comprises a national Key Area map, and accounts for each Key Area (Box 2). The fact of a Key Area's presence within an EBA is flagged in the Key Area account, and Appendix 2

Box 3. ISO codes for Central and South American countries.

The codes are established by ISO (International Organization for Standardization).

AR	Argentina	HN	Honduras
BO	Bolivia	MX	Mexico
BR	Brazil	NI	Nicaragua
BZ	Belize	PA	Panama
CL	Chile	PE	Peru
CO	Colombia	PY	Paraguay
CR	Costa Rica	SR	Surinam
EC	Ecuador	SV	El Salvador
GF	French Guiana	UY	Uruguay
GT	Guatemala	VE	Venezuela
GY	Guyana		

Box 4. Categories and management objectives of protected areas (from IUCN 1992).

I SCIENTIFIC RESERVE/STRICT NATURE RESERVE To protect nature and maintain natural processes in an undisturbed state in order to have ecologically representative examples of the natural environment available for scientific study, environmental monitoring, education and the maintenance of genetic resources in a dynamic and evolutionary state.

II NATIONAL PARK To protect natural and scenic areas of national or international significance for scientific, educational and recreational use.

III NATURAL MONUMENT/NATURAL LANDMARK To protect and preserve nationally significant natural features because of their special interest or unique characteristics.

IV MANAGED NATURE RESERVE/WILDLIFE SANCTUARY To assure the natural conditions necessary to protect nationally significant species, groups of species, biotic communities or physical features of the environment where these require specific human manipulation for their perpetuation.

V PROTECTED LANDSCAPE OR SEASCAPE To maintain nationally significant natural landscapes which are characteristic of the harmonious interaction of man and land while providing opportunities for public enjoyment through recreation and tourism within the normal lifestyle and economic activity of these areas.

VI RESOURCE RESERVE To protect the natural resources of the area for future use and prevent or contain development activities that could affect the resource pending the establishment of objectives which are based on appropriate knowledge and planning.

VII NATURAL BIOTIC AREA/ANTHROPOLOGICAL RESERVE To allow the way of life of societies living in harmony with the environment to continue undisturbed by modern technology.

VIII MULTIPLE-USE MANAGEMENT AREA/MANAGED RESOURCE AREA To provide for the sustained production of water, timber, wildlife, pasture and outdoor recreation, with the conservation of nature primarily orientated to the support of economic activities (although specific zones may also be designed within these areas to achieve specific conservation objectives).

IX BIOSPHERE RESERVE To provide protection to unique areas or wetlands, but also satisfy a range of objectives such as research, monitoring, training and demonstration, as well as conservation. Sites are designated as part of an international scientific programme, the UNESCO Man and the Biosphere Programme, and thus in most cases the human component is vital to the functioning of the Biosphere Reserve.

lists the Key Areas to be found within the boundaries of each EBA. Maps of the EBAs associated with each country are provided in the introductory sections of the national inventories.

The information in each Key Area account comes largely from *Threatened birds of the Americas*, and for the sake of brevity and clarity of presentation the use of this source should be considered implicit within the text. Information on protected area names, IUCN category, size and area boundaries comes almost exclusively from the *Protected areas of the world* (IUCN 1992), and for South American countries the central coordinates, area names and the framework for the area descriptions were often verified using the relevant (and invaluable) ornithological gazetteers (see Paynter, Paynter and Traylor, and Stephens and Traylor under 'References', p. 26): the use of these references should similarly be considered implicit within each area account. All other sources of information have been cited in the usual way. In all other aspects, Key Areas follows the protocols (e.g. for species taxonomy) as set out in the introduction to *Threatened birds of the Americas*, and readers should refer to that book for further details.

OVERVIEW

APPROXIMATELY 3,600 species of bird occur in the non-Caribbean Neotropics. Among them, c.988 (27%) have restricted ranges (Stattersfield *et al.* in prep.), over 600 (17%) are national endemics (see the introductory sections of the various national inventories) and 290 (8%) were listed as threatened in *Threatened birds of the Americas*. This current book documents 596 Key Areas which are the most important areas currently known for the conservation of these 290 threatened birds, over 25% of all such birds in the world.

THREATENED BIRDS

With 290 (8%) of the Neotropical bird species considered at risk of extinction, the search for clear patterns and causes of endangerment can produce insights into why particular birds are threatened and therefore which species may become threatened in the future, but also which regions, countries, Endemic Bird Areas and habitats are of greatest importance for threatened birds and their conservation. The identification of such large-scale trends, some of which are described below, has the potential to influence large-scale conservation or development policy, and may therefore have profound implications for conservation at all levels.

Threats

- *The primary threat to birds in the Americas is habitat loss, although the effects of trade and hunting are also important.*

Throughout the Americas the primary threat to birds is the destruction and disturbance (or alteration) of the habitats on which their existence depends. Almost 75% of the threatened species are regarded as such owing in part to loss of habitat, and for almost 48% of threatened birds in the New World this factor alone is the main threat (Collar *et al.* in press). Over 8% of birds are threatened solely because of their restricted ranges. Hunting is an important component of the threat profile for 15% of New World species, and trade is important for 11%—but hardly ever is either of these factors the sole danger for a threatened species (Collar *et al.* in press). Thus, trade features only once as the sole threat, nearly always otherwise occurring alongside habitat loss, and hunting in combination only with habitat loss accounts for over 5% of all threatened birds (Collar *et al.* in press).

In an analysis of threatened species representation in the various families concerned, Collar *et al.* (in press) show that only two families varied significantly from expected levels of endangerment. Parrots are the most strongly represented, with 28% (39) of all New World species (of which there are about 140) threatened owing to the potent combination of habitat loss and trade; similarly, 26% of the cracids (13 threatened species in a global or New World total of 50) are in danger from the combination of habitat loss and hunting (Collar *et al.* in press) (Figure 1).

Habitats

- *Over half of all threatened birds in the Americas rely on wet forest habitats, with dry forest and grasslands also featuring highly.*

Discounting islands, 39% of threatened birds in the Americas are tropical lowland species (0–500 m), 20% are submontane (500–2,000 m) and 19% are highland temperate forms (2,000 m and above), with the altitude preferences of the remainder less clearly defined (Collar *et al.* in press). Within these various zones, the three main habitat types for the threatened birds are, perhaps not surprisingly, wet forest, dry forest and grassland (Collar *et al.* in press) (Figure 2).

Wet forest

Almost 55% of threatened species are confined to the broad wet forest habitat type (Collar *et al.* in press). The Atlantic coastal forests of (primarily) Brazil immediately stand out as of critical importance (Figure 3), with 15–24 threatened species in some of the 1° grid squares in Bahia, Espírito Santo, Rio de Janeiro, Minas Gerais and São Paulo. This is empha-

Figure 1. The distribution of Key Areas that support threatened parrot and cracid species.

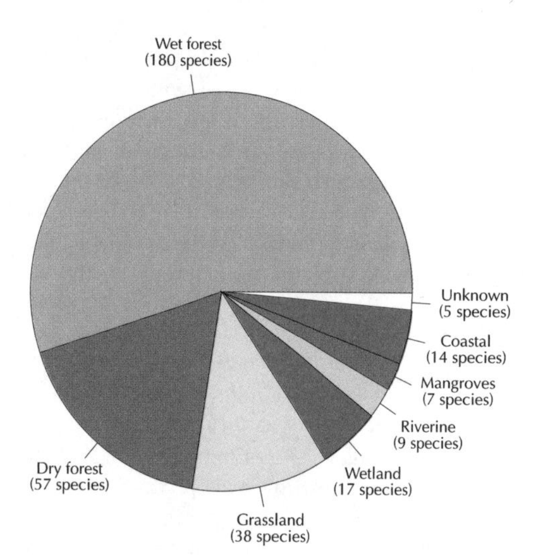

Figure 2. Key habitats for threatened birds in the Americas.

sized by the fact that the majority of Key Areas supporting populations of 10 or more threatened species (Box 2) are to be found in these Brazilian states. Also highlighted are the northern Andes (and adjacent lowlands), particularly in western Colombia (and exemplified at the site level by Key Area CO 53), Ecuador (e.g. EC 11, 32) and northern Peru (e.g. PE 06, 07, 16). In comparison with all other habitat types, these wet forest areas support two to four times as many threatened species.

Dry forest

Over 17% of threatened birds rely on dry forest habitats, which include deciduous forest, dry scrub and caatinga (Collar *et al.* in press). With 3–6 species occurring in each of the most important 1° grid squares, Figure 3 clearly shows the priority that should be given to the Tumbesian region of south-west Ecuador (including Key Areas EC 19, 20, 39, 43) and north-west Peru (e.g. PE 01), and also the Río Marañón (e.g. PE 21, 33). There is a thin spread of threatened species in the caatinga region of Brazil: especially important are the dry forests in the central Brazilian states of Bahia, Minas Gerais and Goiás (e.g. BR 048, 050, 064, 065), which have been devastated by clearance for agriculture and charcoal production (Collar *et al.* in press).

Grassland

With almost 12% of threatened birds confined to grasslands (which include páramo, campo limpo, campo sujo, savanna and open cerrado), this habitat harbours over twice as many threatened species as any of the other non-forest habitat types (Collar *et al.* in press). The Andean páramo/puna and Patagonian grasslands hold a number of these birds (11 species occur in páramo/ puna and montane scrub), but especially important and threatened are the grasslands of southern Brazil and northern Argentina (Figure 3). As a result of agricultural development, nearly all the species endemic or near-endemic to the open vegetation of central Brazil have suffered drastic declines, and a few may even have gone completely from large parts of their former range (Collar *et al.* in

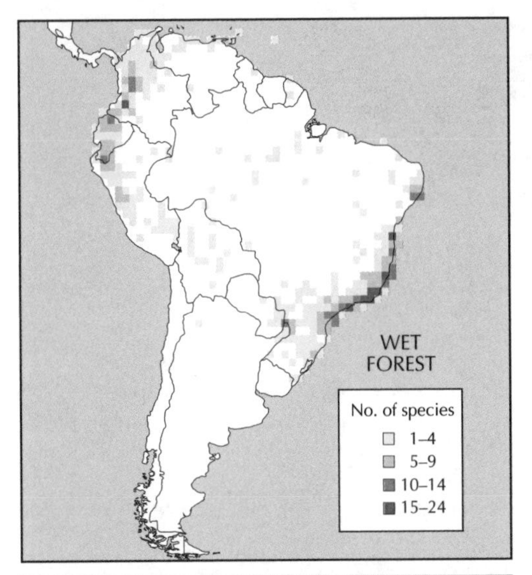

Figure 3. Density distributions of three groups of threatened species, each restricted mainly to one of three major habitat types (plotted on the basis of 1° squares).

press). Most of the remaining campo and cerrado habitat specialists of the region would also be considered threatened but for the fact that they retain reasonably healthy populations in the largely pristine grassland of north-central Bolivia (Collar *et al.* in press). The wet 'Mesopotamia' grasslands of Entre Ríos and Corrientes provinces in Argentina, holding 13 threatened species, have suffered overgrazing and uncontrolled annual burning and are inadequately protected (Collar *et al.* in press).

Distribution

- *The highest concentrations of threatened species in the Neotropics are in the Atlantic coastal forests of Brazil, south-west Ecuador and north-west Peru, and in western Colombia.*

Of the 327 threatened birds of the Americas documented in Collar *et al.* (1992), 29 (9%) are found within Central America, and 260 (almost 80%) in South America, with a large number in the Andes (35% of the total) and the relatively restricted regions of the Atlantic forest (17%) and cis-Andean (Pacific and Caribbean) lowlands (11%); the islands of the eastern Pacific (5%) also register levels out of proportion to their land areas (Collar *et al.* in press).

In a ranking of countries by number of threatened bird species (Figure 4), Brazil comes top with 97 species, of which 65 are endemic; Peru is next, with 64 threatened species, 30 being endemic; with 57 threatened species, Colombia comes very close to Peru in overall importance, and has two species more that are endemic (32). Although Ecuador and Argentina lie fourth and fifth in importance, they share a high proportion of their threatened species with neighbouring countries, and in terms of complementarity of species, Mexico, although sixth in terms of total numbers, moves up to fourth: indeed, there is almost no overlap in species between any of these four top countries (Collar *et al.* in press).

Figures 5–6 show the density distribution of all threatened species in Central and South America, and again highlight the critical importance of the Atlantic coastal forests of (mainly) Brazil, but also the northern Andes and adjacent lowlands, and in particular the central East Andes, Central and West Andes of Colombia, the northern Ecuadorean Andes, the Pacific slope of Colombia and Ecuador (the Chocó), the Andes of southern Ecuador and northern Peru, and the central Peruvian Andes. The impor-

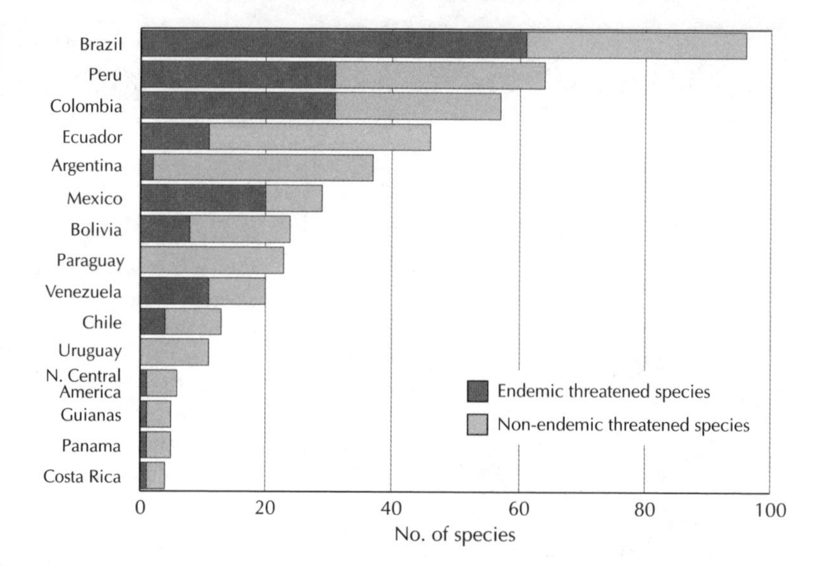

Figure 4. Numbers of threatened bird species in each country, and the proportions which are country endemics (see 'Key Areas presented'). Precise numbers are given in the introductory section to each national inventory.

tance of these areas cannot be overemphasized.

There are 1,009 restricted-range bird species in the New World, contained within 79 distinct EBAs (Stattersfield *et al.* in prep.). Of these species, 255 (25%) are threatened, but more significantly, perhaps, these 255 species represent 78% of the New World's threatened birds, emphasizing the close relationship between range-size and extinction probability. In addition, a further 52 threatened species with ranges greater than 50,000 km^2 also occur in EBAs (Bibby 1994), so that altogether no fewer than 93% of the species covered in *Threatened birds of the Americas* are embraced by EBAs.

Within the non-Caribbean Neotropics, there are some striking examples of highly threatened EBAs (in terms of the number of threatened restricted-range bird species), most notable among which are the East Andes of Colombia (EBA B10), subtropical inter-Andean Colombia (B12), the Chocó and Pacific-slope Andes (B14), Tumbesian western Ecuador and Peru (B20), the high Peruvian Andes (B27), the Alagoan Atlantic slope (B47) and the south-east Brazilian lowland and foothills (B52). Each of these EBAs is home to over 10 threatened species, and they are all among the most critically important target regions for bird conservation in the Neotropics (see Appendix 2, p. 298, for a list of Key Areas within the various EBAs, also the maps within the introductory section of each national inventory).

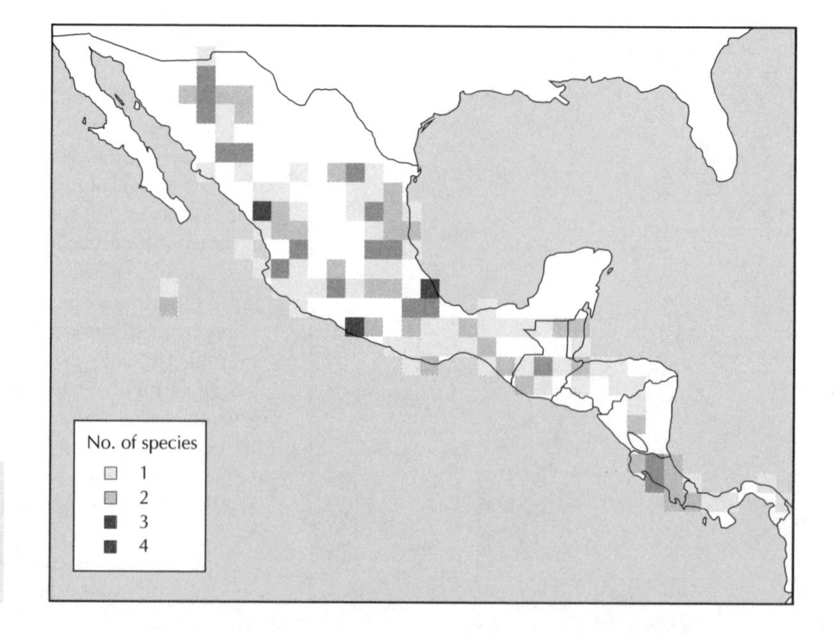

Figure 5. Density distribution of all threatened species in Central America, plotted on the basis of 1° squares.

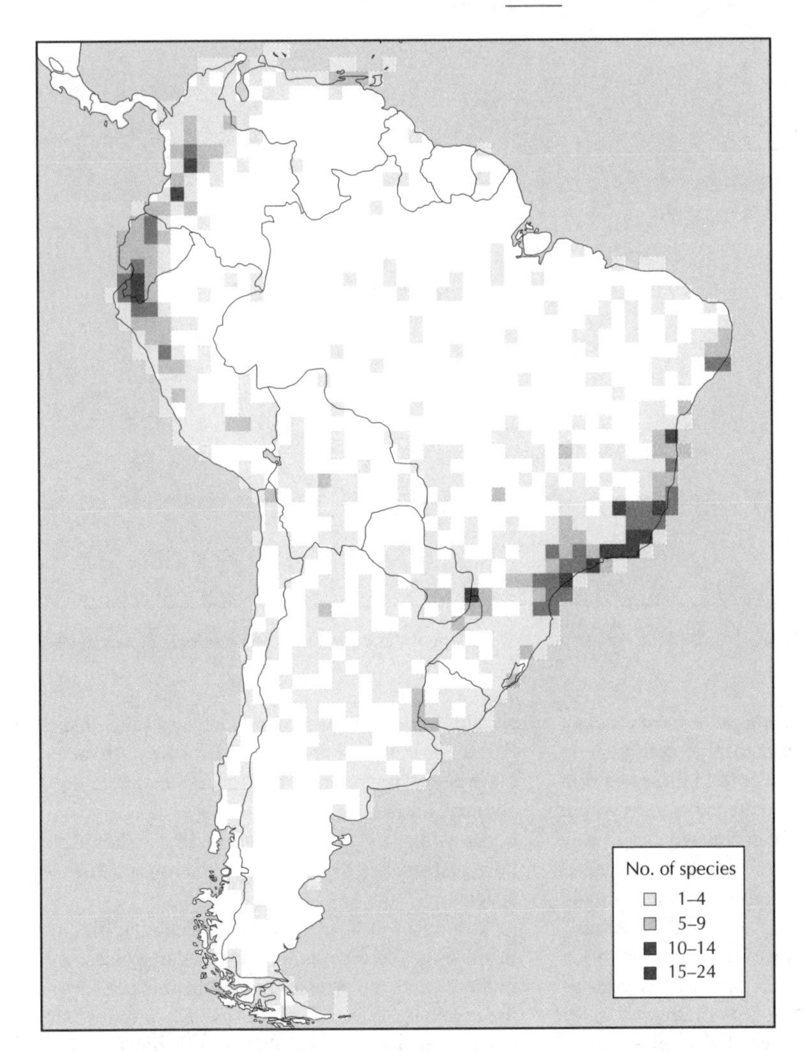

Figure 6. Density distribution of all threatened species in South America, plotted on the basis of 1° squares.

KEY AREAS

- *596 Key Areas for threatened birds in the Neotropics have been identified in this analysis.*

By following a simple set of criteria (see Box 1 in '*Key Areas*: the book', p. 11), and identifying the most important sites currently known for each species, the initial total of c.7,000 individual localities from which these same species are known to occur (or have occurred) has been honed to a list of 596 Key Areas (Figure 7) which would, if adequately protected, help ensure the conservation of 280 (97%) of the threatened species in the region. As these areas have been identified using current knowledge, the inventory should not be seen as static or exclusive: important new populations and areas may yet be found, and Key Areas defined in the present analysis may in time be found no longer to support the threatened species for which they were identified.

As there is a close relationship between the number of threatened species and the number of Key Areas (compare Figures 4 and 10), the top four countries for numbers of Key Areas are, unsurprisingly (in descending order), Brazil, Peru, Colombia and Ecuador. Similarly, the characteristic habitats of the Key Areas reflect those preferred by the threatened species: 65% of Key Areas support wet forest, 23% dry forest and 19% grassland, with 4% lake and marsh vegetation and 3% mangroves. Over 20% of areas support two or more of these broad habitat types (for more detail of which, see 'Habitats', above). The distribution of Key Areas supporting each of the three main habitat types is shown in Figure 9.

Areas

- *Over half of all Key Areas harbour two or more threatened bird species, with 18 Key Areas each having more than 10 species.*

Of the 596 Key Areas identified for threatened birds, 337 (56%) are important for two or more (up to 18) threatened species. These areas with 'cost-effective'

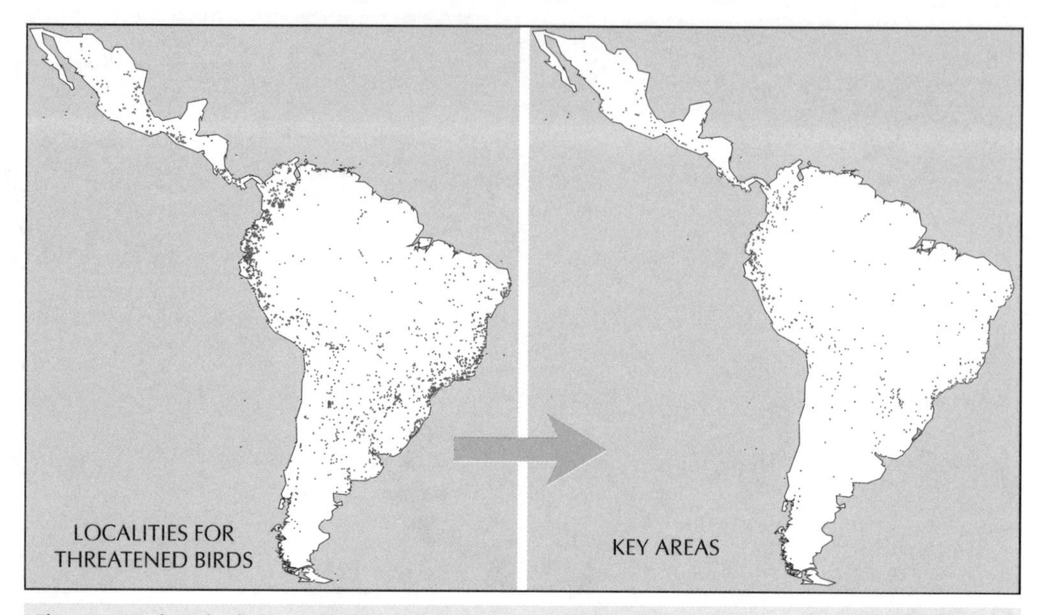

Figure 7. Localities for threatened birds in the Neotropics and the Key Areas which have been selected from them.

aggregations of species are perhaps the most efficient areas currently known on which to target conservation initiatives, whether protection or field surveys. From a frequency distribution of threatened species within Key Areas (Figure 8) it can be calculated that 77 Key Areas (13%) each harbour five or more threatened species (these areas are listed in Box 2), and should perhaps be considered among the 'top' Key Areas.

In a number of cases, however, the conservation of threatened species within Key Areas can be achieved only by considering a set of areas (e.g. for seasonal migrants), and, except for species known from a single Key Area, each national inventory should be regarded as a complementary suite of site-based conservation priorities. Many Key Areas are of primary importance for species endemic to a country, known from just one area or not known to occur within any protected area, and these should not be overshadowed by areas with numerous threatened species.

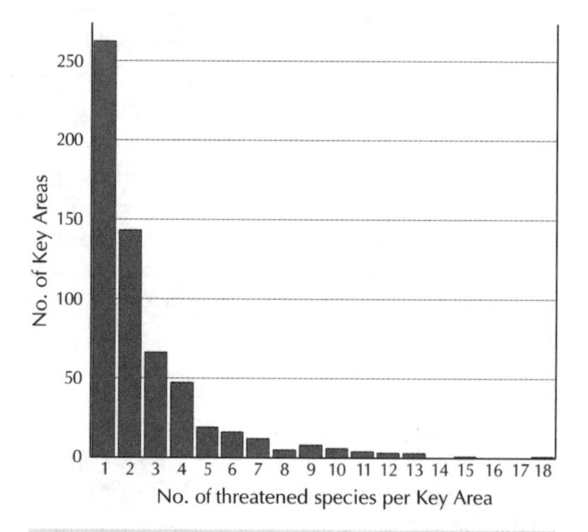

Figure 8. Frequency distribution of numbers of threatened species within Key Areas.

A number of Key Areas have been identified on the basis of old records for a particular species, and in these cases the primary requirement is to confirm the continued presence of the species, or indeed the habitat, in the area. In other cases, however, the bird(s) may have been regularly and recently recorded within a particular area, so that we can readily indicate the need for site conservation if the birds are to persist there.

Species

- *Represented in just one Key Area each are 47 threatened species—areas for these species should be high priorities for conservation.*

Known only from one Key Area are 47 threatened species, which are thus totally reliant for their survival on the integrity of habitat within their respective Key Areas, suggesting that in these cases targeted single-species site-based conservation is a necessity (these Key Areas are included in Box 2). For only 15 of these 47 Key Areas is there only one species present, and thus most often, even when the focus is single-species conservation, other threatened birds will benefit. There are many more species that are known only from two Key Areas, or for which the primary population is within just one area, and such

Figure 9. Key Areas supporting threatened species characteristic of wet forest, dry forest and grassland.

species should also be carefully considered in conservation planning, as they too may warrant special attention. These species, and the Key Areas in which they can be found, are mentioned in the introductory sections of the national inventories, often being listed in the table of 'top' Key Areas. The Key Areas from which each species is known are listed in Appendix 1 (p. 290).

The Key Area analysis is best suited to species for which site-based conservation is most appropriate; less so for some seasonal migrants and widely dispersed or nomadic species, in compensation for which we have endeavoured to ensure that such species are liberally represented. However, for some birds it has not been possible to identify any Key Areas, and not represented within this analysis are 10 species as follows: *Charadrius melodus*, *Numenius borealis*, *Anodorhynchus glaucus*, *Popelairia letitiae*, *Eriocnemis godini*, *Myrmotherula fluminensis*, *Pithys castanea*, *Calyptura cristata*, *Sporophila melanops* and *Conothraupis mesoleuca* (see also Box 1). Most of these birds are extremely poorly known; *Anodorhynchus glaucus* is now considered extinct, and the little-known but dispersed winter range (primarily outside the region) of *Charadrius melodus* is not well suited to the Key Area approach.

KEY AREA PROTECTION

- *Fewer than 50% of Key Areas have any form of protected status, demonstrating a clear need for further conservation action in the region.*

About 44% (262) of the Key Areas currently have some form of protected status, 22% (132) as national parks, strict nature reserves or Biosphere Reserves (IUCN categories I, II and IX: see Box 4 in '*Key Areas*: the book', p. 14). Many of the unprotected Key Areas urgently require attention in the form of appropriate conservation measures if the populations of their threatened species are to survive. A simple national comparison of the proportion of Key Areas under protection highlights some countries within which the development of the protected-area system should be seen as a priority (Figure 10). Critically important countries with fewer than 50% of their Key Areas under any form of protection include Peru, Colombia, Ecuador, Mexico and Bolivia. However, this can disguise the fact that in countries such as Bolivia the few protected Key Areas often embrace enormous tracts of habitat. Peru, with the second highest number of threatened species and Key Areas, has the lowest percentage (12%) of Key Areas pro-

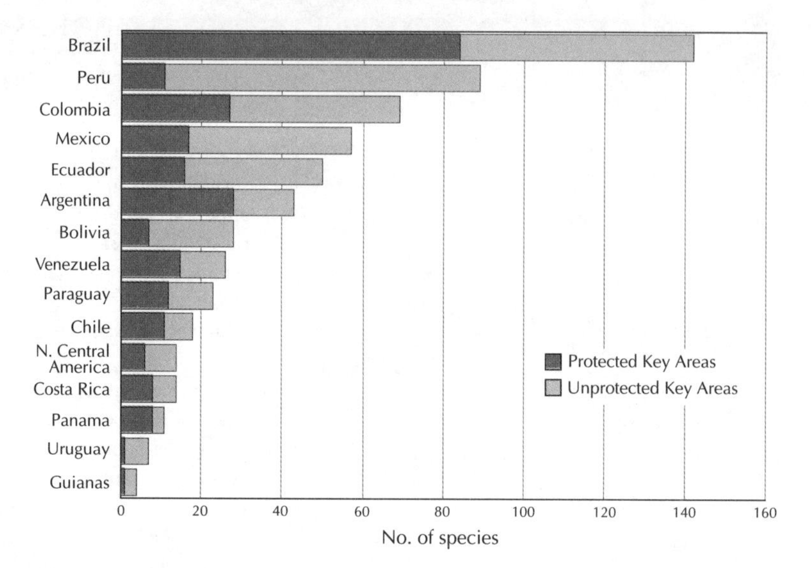

Figure 10. Numbers of Key Areas in each country, and the proportions with some form of protected status (see 'Coverage'). Precise numbers are given in the introductory section to each national inventory.

tected, and therefore has the most pressing need for the development of a more extensive protected-area system.

Legal measures do not always guarantee protection, but they do demonstrate to the government, and to society in general, the official recognition of the importance of a site, although this importance is most often recognized in terms of landscape and scenic value or an area's potential for educational, economic or sustainable natural resource use (see Box 4 on p. 14). The Key Area analysis has shown that many of these protected areas also have immense importance for the conservation of threatened birds and therefore for biodiversity.

Many protected areas remain under threat, and in many, habitat degradation, uncontrolled hunting, etc., continue unchecked, suggesting that effective management is still required of activities undertaken within them. Threatened bird species that are represented within the protected-area system are mentioned in the introductory section of each national inventory. The representation of threatened species within protected areas has sometimes been determined solely by historic occurrence, which should not therefore be interpreted as an indication of present-day guaranteed safety. In the first instance, as mentioned above, many protected areas remain under threat, but perhaps more importantly the species may no longer be present in the area. The need for targeted fieldwork in this latter case is clear.

RECENT CHANGES TO THE THREATENED LIST

- *A recent revision of the threatened species list using new criteria has increased the number of threatened birds in the region from 290 to 326.*

Threatened birds of the Americas listed 289 species as being at risk of extinction within the Neotropics (one of which, *Hylorchilus sumichrasti*, is now considered to best split into two species, both being included in the Key Area analysis). This list has recently been superseded by that presented in *Birds to watch 2* (see '*Key Areas* and Red Data Books', p. 8), which documents 326 species for the region (see Appendix 3, p. 300). These differences (which amount to 25% between the two listings) have been brought about by the reassessment and relegation of 18 species, two of which (*Podiceps andinus* and *Anodorhynchus glaucus*) are now considered extinct, and the addition of 55 species. Of these 55, nine (*Geotrygon carrikeri*, *Cypseloides storeri*, *Heliangelus zusii*, *Grallaria kaestneri*, *Doliornis remseni*, *Onychorhynchus swainsoni*, *Phylloscartes kronei*, Chocó Vireo *Vireo* sp., *Hylorchilus navai*) are species described as new to science or are a result of taxonomic revisions since the 1992 listing. Most of the remaining additions were listed as Near Threatened in *Threatened birds of the Americas*, with the new criteria (see Appendix 4, p. 305) and/or new data having led to their upgrading. The changes to the threatened list as they relate to each country are detailed within the introductory section of each national inventory.

Unfortunately, insufficient detail was compiled for any of these additions to the threatened species list to allow for their inclusion within the Key Area analysis, and although we have endeavoured to assess the impact they would have on the identification of Key Areas, there is an urgent need for the compilation of information relevant to their conservation. These species should be considered in any future priority-setting initiatives.

OLD RECORDS AND LITTLE-KNOWN BIRDS

- *Many threatened species are still so poorly known that their conservation status is extremely difficult to assess—information on these species is urgently required.*

There is still much to learn about threatened (and indeed all) birds in the Neotropics. At the national level, many species are known only from old records, or by insubstantial information (and these are listed within the introductory section of each national inventory), but in combination with our knowledge from neighbouring countries we can still establish a working understanding of their status and distribution. However, some of these birds are truly little known even at a regional level, and in the hope that simply naming them may stimulate interest or generate new information, they are listed in Box 1.

With a near-continuous flow of new information concerning threatened birds and the areas in which they occur, our perceptions and understanding of their status are constantly changing. The maintenance of up-to-date information on the status of threatened birds or Key Areas is essential for a continuing review process and for the formulation of clear and authoritative conservation priorities. It is hoped that the format of the information presented in each national inventory will stimulate the submission of new information either to the relevant BirdLife Partner (see p. 3) or other national conservation organization, or direct to the BirdLife Secretariat.

OUTLOOK

Key Areas is a priority-setting initiative. Comprehensive and up-to-date information on threatened birds in the Neotropics has been used to identify the most important areas for their conservation. However, simply identifying the priorities is just the first and often easiest step. The 596 Key Areas documented in this book present a major challenge in that each warrants conservation attention if the species for which they are so important are to survive in the long term.

Box 1. Some of the most poorly known threatened birds in the Neotropics.

Black Tinamou *Tinamus osgoodi*
Magdalena Tinamou *Crypturellus saltuarius*
Kalinowski's Tinamou *Nothoprocta kalinowskii*
Guadalupe Storm-petrel *Oceanodroma macrodactyla*
Blue-billed Curassow *Crax alberti*
Austral Rail *Rallus antarcticus*
Speckled Crake *Coturnicops notatus*
Eskimo Curlew *Numenius borealis*
Blue-eyed Ground-dove *Columbina cyanopis*
Purple-winged Ground-dove *Claravis godefrida*
Cayenne Nightjar *Caprimulgus maculosus*
Coppery Thorntail *Popelairia letitiae*
Chestnut-bellied Hummingbird *Amazilia castaneiventris*
Táchira Emerald *Amazilia distans*
Turquoise-throated Puffleg *Eriocnemis godini*
Imperial Woodpecker *Campephilus imperialis*
Orinoco Softtail *Thripophaga cherriei*
Recurve-billed Bushbird *Clytoctantes alixii*
Rondônia Bushbird *Clytoctantes atrogularis*
Rio de Janeiro Antwren *Myrmotherula fluminensis*
Fringe-backed Fire-eye *Pyriglena atra*
White-masked Antbird *Pithys castanea*
Moustached Antpitta *Grallaria alleni*
Táchira Antpitta *Grallaria chthonia*
Brown-banded Antpitta *Grallaria milleri*
Kinglet Cotinga *Calyptura cristata*
Hooded Seedeater *Sporophila melanops*
Pale-headed Brush-finch *Atlapetes pallidiceps*
Cone-billed Tanager *Conothraupis mesoleuca*
Cherry-throated Tanager *Nemosia rourei*
Baudó Oropendola *Psarocolius cassini*
Selva Cacique *Cacicus koepckeae*

The need for action

- *The primary action required for most Key Areas is the guaranteed integrity of the habitats they support.*

The Key Areas which should perhaps be considered the highest priorities for attention are listed in Box 2. In an analysis of recommendations made in *Threatened birds of the Americas*, Bibby (1994) showed how central is site conservation to the future of all the New World's threatened species: over 70% of the individual recommendations made (there were 1,045 for the 327 species) concern a set of four site-specific proposals (namely, to find new sites, to protect known sites, to manage them and to enumerate populations within them); for only five species do the recommendations contain no site-oriented actions. Thus, at the site level, the primary action required for Key Areas is the protection of those currently unprotected, but also increased protection

Box 2. Top Key Areas for threatened birds in the Neotropics: those which hold populations of five or more threatened species (the number is given with each area), and those for threatened national endemics known from just one area.

Code	Area	No.
Argentina		
AR 15	Iguazú	12
AR 16	Arroyo Urugua-í	9
AR 28	Colonia Carlos Pellegrini	7
AR 31	Gualeguaychú	6
Bolivia		
BO 03	Trinidad	2
BO 13	Nevado Illampu	1
Brazil		
BR 008	Borba	1
BR 012	Serra do Cachimbo	2
BR 014	Cachoeira Nazaré	1
BR 016	Serra das Araras	2
BR 035	Pedra (Serra) Branca and Fazenda Bananeira	13
BR 036	Pedra Talhada	11
BR 037	São Miguel dos Campos	5
BR 039	Curaçá	1
BR 040	Canudos	1
BR 043	Santo Amaro	1
BR 051	Serra do Ouricana	9
BR 055	Porto Seguro	7
BR 056	Monte Pascoal	11
BR 061	Emas	7
BR 062	Brasília	8
BR 064	Brejo do Amparo	3
BR 066	Serra do Cipó	4
BR 067	Caratinga	7
BR 068	Rio Doce	11
BR 074	Córrego Grande (Fazenda São Joaquim)	9
BR 076	Sooretama	13
BR 077	Linhares	9
BR 078	Augusto Ruschi (Nova Lombardia)	10
BR 079	Jatibocas	1
BR 084	Desengano	18
BR 086	Itatiaia	13
BR 087	Fazenda União	5
BR 088	Serra dos Órgãos	10
BR 090	Serra do Tinguá	6
BR 092	Tijuca	7
BR 094	Angra dos Reis	5
BR 100	Serra da Bocaína	6
BR 102	Ubatuba	9
BR 103	Bairro de Corcovado	10
BR 106	Boracéia and Bertioga	11
BR 107	Ilha São Sebastião	5
BR 108	Carlos Botelho	6
BR 109	Fazenda Intervales	12
BR 110	Alto Ribeiro	6
BR 111	Foz do Rio Ribeira	5
BR 112	Juréia	8
BR 113	Iguape	8
BR 114	Jacupiranga	10
BR 115	Ilhas Comprida and Cananéia	7
BR 116	Ilha do Cardoso	15
BR 120	Guaragueçaba	6
BR 126	Iguaçu	6
BR 128	Salto do Piraí	5
Chile		
CL 17	Islas Robinson Crusoe and Santa Clara	3
CL 18	Isla Alejandro Selkirk	1
Colombia		
CO 01	Isla de San Andrés	1
CO 05	Ayacucho	1
CO 09	Puerto Valdivia	5
CO 25	Ucumarí	7
CO 26	Alto Quindío–Laguneta	7
CO 28	Río Toche	6
CO 34	San Gil	1
CO 42	Laguna de Pedropalo	5
CO 52	Los Farallones de Cali	5
CO 53	Munchique	10
CO 55	Santa Helena–Cerro Munchique	2
CO 56	Puracé	6
CO 59	Cueva de los Guácharos	7
CO 63	Isla de Tumaco and Isla Boca Grande	1
CO 65	Río Ñambi	5
Ecuador		
EC 07	Mindo	6
EC 11	Río Palenque	7
EC 12	Cayambe–Coca	6
EC 19	Machalilla	10
EC 20	Cerro Blanco, Guayaquil	7
EC 22	Sangay	5
EC 25	Manta Real	6
EC 27	Cajas and Río Mazan	3
EC 32	Buenaventura	8
EC 37	Alamor	9
EC 38	Celica	8
EC 41	Catacocha	6
EC 43	Tambo Negro	6
EC 47	Podocarpus	9
EC 50	Islas Galápagos	5
Mexico		
MX 01	Isla de Guadalupe	2
MX 29	Socorro Island	4
MX 43	Atoyac de Alvarez–Teotepec road	3
Paraguay		
PY 06	Mbaracayú	9
PY 08	Estancia La Fortuna	5
PY 09	Estancia Itabó	7
Peru		
PE 01	Tumbes National Forest	11
PE 06	Cerro Chinguela	5
PE 07	San José de Lourdes	5
PE 16	Northern Cordillera de Colán	6
PE 17	Southern Cordillera de Colán	5
PE 23	Jesús del Monte	4
PE 34	Hacienda Tulpo	2
PE 44	Huascarán	6
PE 52	Cerros del Sira	2
PE 54	Balta	1
PE 59	Lago de Junín	3
PE 71	Bosque Ampay	2
Venezuela		
VE 02	El Tamá	5
VE 23	Cerro Negro and El Guácharo	5
VE 25	Península de Paria	5
VE 26	Río Capuana	1

and management for areas already having protected status.

- *Targeted fieldwork is needed in many Key Areas.*

Many Key Areas urgently need attention in the form of fieldwork to confirm the continued presence or assess the status of the relevant threatened species or appropriate vegetation. For some Key Areas such as those identified on the basis of old records of species, fieldwork may need to be more exploratory, targeting remnant patches of habitat in the vicinity of the Key Area. In the Neotropics, while we know more about the distribution of birds than any other comparable group of organisms, there remains a general paucity of information, although our knowledge of threatened birds and the areas in which they occur is steadily increasing. Therefore, an aim of *Key Areas* is to provide a format and forum through which to solicit additional information not only on the individual areas and threatened birds within them, but also on recently discovered areas or populations. Of particular importance is the need for additional information concerning the little-known species listed in Box 1, but also the compilation of data for those birds recently added to the threatened species list (see 'Recent changes to the threatened list', above).

Saving species

- *Priority-setting must not be seen as an end in itself: priorities are identified in order to be acted upon and thereby save species.*

No single priority-setting exercise is comprehensive. *Key Areas* has *not* documented every area likely to be important for bird conservation—either those already known of, or the large areas in the Neotropics that are ornithologically unknown, but potentially of critical importance. The identification of, and targeted fieldwork in, these areas (such as Conservation International's 'Rapid Assessment Program': e.g. Foster *et al.* 1994) must be encouraged and supported if we are to advance the priority-setting process, although there is an urgent need for *more* biologists to become involved in the survey and assessment of biologically important areas.

However, while exploratory fieldwork is clearly important, and further priority-setting may ultimately be desirable (e.g. as part of BirdLife's Important Bird Areas programme), priorities have already been identified and our quest for perfect answers must not obstruct the need to initiate conservation action immediately. What else needs to be known, for example, about Black-hooded Antwren *Formicivora erythronotos* before we start action to conserve the last remaining population confined to one small, unprotected site?

The threatened bird species of the Neotropics have already been documented in detail, with recommendations made for their conservation, and this book now identifies Key Areas on which to target such action. With such a comprehensive suite of priorities established, and given the current pace of habitat loss, there must be no excuse to delay the implementation of the action that is so obviously necessary if species are to be saved from extinction.

REFERENCES

References listed here are those cited within the preceding three introductory chapters and the appendices. For references cited in the national inventories, see separate lists within each.

BALMFORD, A. AND LONG, A. J. (in press) Across-country analyses of biodiversity congruence and current conservation effort in the tropics. *Conserv. Biol.*

BIBBY, C. J. (1994) Recent past and future extinctions in birds. *Phil. Trans. R. Soc. Lond.* B 344: 35–40.

COLLAR, N. J. AND STUART, S. N. (1985) *Threatened birds of Africa and related islands: the ICBP/ IUCN Red Data Book*, 1. Cambridge, U.K.: International Council for Bird Preservation, and International Union for Conservation of Nature and Natural Resources.

COLLAR, N. J. AND STUART, S. N. (1988) *Key forests for threatened birds in Africa*. Cambridge, U.K.: International Council for Bird Preservation (Monogr. 3).

COLLAR, N. J., GONZAGA, L. P., KRABBE, N., MADROÑO NIETO, A., NARANJO, L. G., PARKER, T. A. AND WEGE, D. C. (1992) *Threatened birds of the Americas: the ICBP/ IUCN Red Data Book* (Third edition, part 2). Cambridge, U.K.: International Council for Bird Preservation.

COLLAR N. J., CROSBY, M. J. AND STATTERSFIELD, A. J. (1994) *Birds to watch 2: the world list of threatened birds*. Cambridge, U.K.: BirdLife International (BirdLife Conservation Series no. 4).

COLLAR, N. J., WEGE, D. C. AND LONG, A. J. (in press) Patterns and causes of endangerment in the New World avifauna. In J. V. Remsen, ed. *Natural history and conservation of Neotropical birds*. American Ornithologists' Union (Orn. Monogr.).

CROSBY, M. J. (1994) Mapping the distributions of restricted-range birds to identify global conservation priorities. Pp.145–154 in R. I. Miller, ed. *Mapping the diversity of nature*. London: Chapman and Hall.

EVANS, M. I. (1994) *Important Bird Areas in the Middle East*. Cambridge, U.K.: BirdLife International (BirdLife Conservation Series no. 2).

FOSTER, R. B., CARR, J. L. AND FORSYTH, A. B., EDS. (1994) *The Tambopata–Candamo Reserved Zone of southeastern Perú: a biological assessment (Rapid Assessment Program)*. Washington, D.C.: Conservation International.

GRIMMETT, R. F. A. AND JONES, T. A. (1989) *Important Bird Areas in Europe*. Cambridge, U.K.: International Council for Bird Preservation (Techn. Publ. 9).

GROOMBRIDGE, B. (1993) *1994 IUCN Red List of threatened animals*. Gland, Switzerland, and Cambridge, U.K.: International Union for Conservation of Nature and Natural Resources.

ICBP (1992) *Putting biodiversity on the map: priority areas for global conservation*. Cambridge, U.K.: International Council for Bird Preservation.

IUCN (1992) *Protected areas of the world: a review of national systems. Volume 4: Nearctic and Neotropical.* Gland, Switzerland and Cambridge, U.K.: International Union for Conservation of Nature and Natural Resources.

IUCN SSC (IUCN SPECIES SURVIVAL COMMISSION) (1994) *IUCN Red List Categories, as approved by the 40th meeting of the IUCN Council Gland, Switzerland.* Gland, Switzerland: IUCN The World Conservation Union.

MACE, G. AND STUART, S. (1994) Draft IUCN Red List categories. *Species* 21–22: 13–24.

MYERS, N. (1988) Threatened biotas: 'hotspots' in tropical forests. *Environmentalist* 8: 187–208.

MYERS, N. (1990) The biodiversity challenge: expanded hot-spots analysis. *Environmentalist* 10: 243–256.

PAYNTER, R. A. (1982) *Ornithological gazetteer of Venezuela.* Cambridge, Mass.: Museum of Comparative Zoology.

PAYNTER, R. A. (1985) *Ornithological gazetteer of Argentina*. Cambridge, Mass.: Museum of Comparative Zoology.

PAYNTER, R. A. (1988) *Ornithological gazetteer of Chile.* Cambridge, Mass.: Museum of Comparative Zoology.

PAYNTER, R. A. (1989) *Ornithological gazetteer of Paraguay*. Second edition. Cambridge, Mass.: Museum of Comparative Zoology.

PAYNTER, R. A. (1992) *Ornithological gazetteer of Bolivia.* Second edition. Cambridge, Mass.: Museum of Comparative Zoology.

PAYNTER, R. A. (1993) *Ornithological gazetteer of Ecuador.* Second edition. Cambridge, Mass.: Museum of Comparative Zoology.

PAYNTER, R. A. (1994) *Ornithological gazetteer of Uruguay*. Second edition. Cambridge, Mass.: Museum of Comparative Zoology.

PAYNTER, R. A. AND TRAYLOR, M. A. (1981) *Ornithological gazetteer of Colombia.* Cambridge, Mass.: Museum of Comparative Zoology.

PAYNTER, R. A. AND TRAYLOR, M. A. (1991) *Ornithological gazetteer of Brazil.* Cambridge, Mass.: Museum of Comparative Zoology.

ROSE, P. M. AND SCOTT, D. A. (1994) *Waterfowl population estimates*. Slimbridge, U.K.: International Waterfowl and Wetlands Research Bureau (IWRB Spec. Publ. 29).

SIBLEY, C. G. AND MONROE, B. L. (1990) *Distribution and taxonomy of birds of the world.* New Haven: Yale University Press.

SIBLEY, C. G. AND MONROE, B. L. (1993) *A supplement to distribution and taxonomy of birds of the world.* New Haven: Yale University Press.

STATTERSFIELD, A. J., CROSBY, M. J., LONG, A. J. AND WEGE, D. C. (in prep.) *A global directory of Endemic Bird Areas*. Cambridge, U.K.: BirdLife International (BirdLife Conservation Series).

STEPHENS, L. AND TRAYLOR, M. A. (1983) *Ornithological gazetteer of Peru*. Cambridge, Mass.: Museum of Comparative Zoology.

STEPHENS, L. AND TRAYLOR, M. A. (1985) *Ornithological gazetteer of the Guianas.* Cambridge, Mass.: Museum of Comparative Zoology.

THIRGOOD, S. J. AND HEATH, M. F. (1994) Global patterns of endemism and the conservation of biodiversity. Pp.207–227 in P. L. Forey, C. J. Humphries and R. I. Vane-Wright, eds. *Systematics and conservation evaluation*. Oxford: Clarendon Press (Systematics Association Special Volume 50).

NATIONAL INVENTORIES

ARGENTINA

Rufous-throated Dipper *Cinclus schulzi*

ARGENTINA harbours between 992 and 1,025 species of resident and migratory bird (depending on which taxonomic treatment is followed), of which 62 (6%) have restricted ranges (Stattersfield *et al.* in prep.) and 37 (4%) are threatened (Collar *et al.* 1992). This analysis has identified 43 Key Areas for the threatened birds in Argentina (see '*Key Areas*: the book', p. 11, for criteria).

THREATENED BIRDS

Of the 37 Argentine species which Collar *et al.* (1992) considered to be at risk of extinction, two are entirely confined to the country (see Table 1). The majority of the species (31 or 83%) are, at least in part, in the lowlands below 500 m, with eight (21%) in the submontane zone, and four (11%) in the High Andes temperate zone (Collar *et al.* in press). These threatened birds rely primarily on grasslands (15 species, 41%) and wet forest (11, 30%), with a small but significant number (five, 14%) from wetlands. Only Brazil has more threatened species confined to Neotropical grassland. The main threat is habitat loss affecting 26 (70%) of the species, but also significant are trade (10, 27%) and hunting (five, 13%). The distributions of threatened birds and their relationship to Endemic Bird Areas are shown in Figure 1.

KEY AREAS

The 43 Key Areas that have been identified in Argentina would, if adequately protected, help to ensure the conservation of 34 (92%) of its threatened species, including all those that are confined to the country (always accepting that important new populations and areas may yet be found). Of these areas, 26 are important for two or more (up to 12) threatened species, and these are therefore perhaps the most efficient areas currently known in which to conserve Argentina's threatened birds (see 'Outlook', below); the 12 areas which each harbour four or more threatened species (see Table 3) together represent potential sanctuaries for 29 threatened species, 78% of the total number. However, as vital as these areas are for the conservation of Argentine threatened species, they should not detract from the significance of the remaining Key Areas, as these support five species whose habitat would otherwise be unrepresented.

From Tables 1 and 3 it can be seen that six species

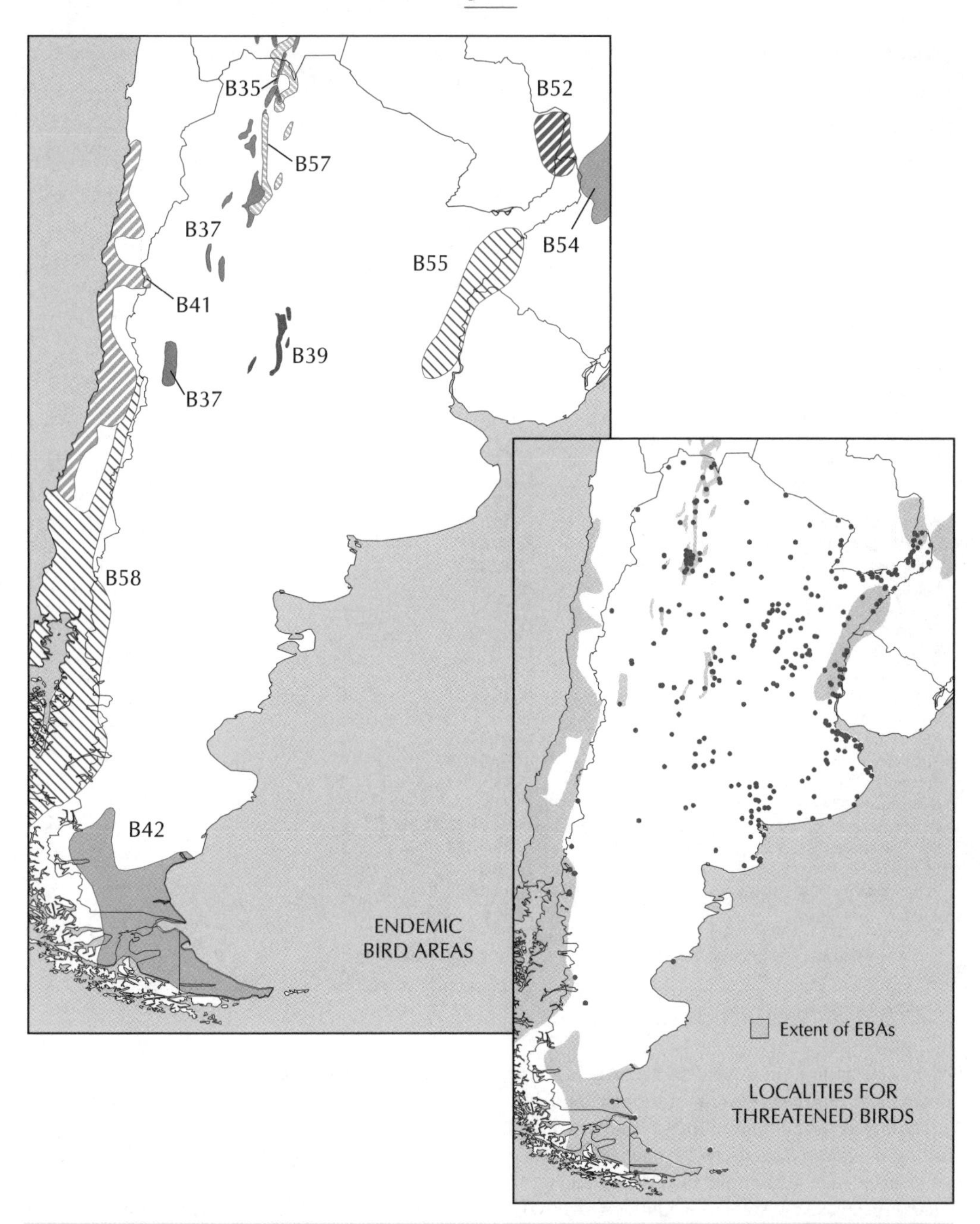

Figure 1. The localities where threatened birds have been recorded in Argentina and their relationship to Endemic Bird Areas (EBAs).

- Threatened birds are distributed throughout most regions of the country with, however, notable concentrations in the wet grasslands of Entre Ríos and Corrientes (EBA B55 in part), the forests of Misiones (EBAs B52, B54) and the northern Andes (EBAs B35, B37, B57). Of the threatened birds in Argentina 14 (38%) have restricted ranges, 11 of which occur together in various combinations within EBAs, which are listed below (figures are numbers of these species in each EBA).

B35 Bolivian Andes (1)
B37 North Argentine Andes (1)
B39 Argentine cordilleras (0)
B41 Central Chile (0)
B42 Tierra del Fuego and the Falklands (0)
B52 South-east Brazilian lowland and foothills (4)
B54 South-east Brazilian *Araucaria* forest (2)
B55 Entre Ríos wet grasslands (2)
B57 Boliviano-Tucuman yungas (1)
B58 Valdivian forests of central Chile and Argentina (0)

Table 1. Coverage of threatened species by Key Area. Areas in bold currently have some form of protected status.

	Key Areas occupied	No. of Key Areas protected	Total nos. of Key Areas	
			Argentina	Neotropics
Dwarf Tinamou *Taoniscus nanus*	—	0	0	3
Brazilian Merganser *Mergus octosetaceus*	**15,16**,19	2	3	7
Crowned Eagle *Harpyhaliaetus coronatus*	**03,10,12,28,37,38**	6	6	18
Black-fronted Piping-guan *Pipile jacutinga*	**15,16**,19,**22**	3	4	29
Austral Rail *Rallus antarcticus*	41	0	1	2
Dot-winged Crake *Porzana spiloptera*	**11,32,33**,34	3	4	4
Speckled Crake *Coturnicops notatus*	**11**,27,**32**	2	3	8
Horned Coot *Fulica cornuta*	**01,02,07**	3	3	6
Eskimo Curlew *Numenius borealis*	—	0	0	0
Olrog's Gull *Larus atlanticus*	**11,33**,35,**36**,43	3	5	5
Purple-winged Ground-dove *Claravis godefrida*	**15,16**	2	2	9
Glaucous Macaw *Anodorhynchus glaucus*	—	0	0	0
Red-spectacled Amazon *Amazona pretrei*	**16,21,26**	3	3	14
Vinaceous Amazon *Amazona vinacea*	**15,18,20,21**	4	4	35
Blue-bellied Parrot *Triclaria malachitacea*	**15,16**	2	2	17
Sickle-winged Nightjar *Eleothreptus anomalus*	**14,15,24,26,30**,31	5	6	10
Helmeted Woodpecker *Dryocopus galeatus*	**15,16**,17,**22**,23	3	5	15
Austral Canastero *Asthenes anthoides*	**38,39,42**	3	3	5
White-bearded Antshrike *Biatas nigropectus*	**15,16,20**	3	3	8
Black-capped Manakin *Piprites pileatus*	**20**	1	1	5
Dinelli's Doradito *Pseudocolopteryx dinellianus*	08,**11**,13	1	3	5
São Paulo Tyrannulet *Phylloscartes paulistus*	**15**,23	1	2	17
Russet-winged Spadebill *Platyrinchus leucoryphus*	**15**	1	1	19
White-tailed Shrike-tyrant *Agriornis andicola*	**07**	1	1	11
Strange-tailed Tyrant *Yetapa risora*	**14,16,24**,27,**28**	4	5	9
Ochre-breasted Pipit *Anthus nattereri*	25	0	1	6
Rufous-throated Dipper *Cinclus schulzi*	**03,04,05,07**	4	4	6
Tucumán Mountain-finch *Poospiza baeri* [E]	**03,04,07**,09	3	4	4
Temminck's Seedeater *Sporophila falcirostris*	**15,16**	2	2	14
Buffy-throated Seedeater *Sporophila frontalis*	**15,22**	2	2	22
Rufous-rumped Seedeater *Sporophila hypochroma*	**14,24,26,28**,31,**32**	5	6	16
Marsh Seedeater *Sporophila palustris*	25,**26**,27,**28**,29,31,**32**	3	7	15
Entre Ríos Seedeater *Sporophila zelichi* [E]	**28,30**,31	2	3	3
Yellow Cardinal *Gubernatrix cristata*	**12,14,28**,29,**30**,31,34, **37**,40	5	9	11
Rufous-bellied Saltator *Saltator rufiventris*	**03**,06	1	2	8
Saffron-cowled Blackbird *Xanthopsar flavus*	**14,24**,25,27,**28**,31	3	6	12
Pampas Meadowlark *Sturnella militaris*	34	0	1	4

[E] Endemic to Argentina

each occur in just one Argentine Key Area (*Rallus antarcticus*, *Piprites pileatus*, *Agriornis andicola*, *Platyrinchus leucoryphus*, *Anthus nattereri* and *Sturnella militaris*), and are thus reliant for their survival in Argentina on the integrity of habitat in their respective Key Areas (see 'Outlook', below).

There are a number of threatened birds found in only one or two Key Areas in Misiones province whose main range lies within the Atlantic coastal forests of south-east Brazil. Most of these species (*Claravis godefrida*, *Triclaria malachitacea*, *Biatas nigropectus*, *Phylloscartes paulistus*, *Platyrinchus leucoryphus*, *Sporophila falcirostris* and *S. frontalis*) are known in Argentina from only a handful of records.

Not represented within the Argentine Key Area analysis are three of the country's threatened species, as follows.

Taoniscus nanus is known in Argentina from just two specimens collected at the turn of the century in either Formosa or Chaco provinces.

Numenius borealis formerly wintered in the southern half of Argentina but even by the 1870s was considered very rare. The only possible sighting in recent times was of four birds reported on Mar Chiquita (AR 11) in October 1990.

Anodorhynchus glaucus was known locally from northern Argentina in the 1800s but there have been no sightings in the twentieth century and it is now considered extinct (Collar *et al.* 1994).

KEY AREA PROTECTION

Encouragingly, 27 (63%) of Argentina's Key Areas currently have some form of protected status (i.e. as national or provincial parks, Biosphere, strict nature, scientific or private reserves); 22 (51% of the total) are strict nature reserves, national parks or Biosphere Reserves (IUCN categories I, II or IX). The 16 unprotected Key Areas (37% of the total) require attention in the form of appropriate conservation measures if the populations of their threatened species are to survive. However, even the formally protected areas remain under threat and, in many, habitat degradation continues unchecked; effective management is thus required of activities undertaken within them.

With 37% of the Key Areas lacking any form of protection, three Argentine threatened species (*Rallus antarcticus*, *Anthus nattereri* and *Sturnella militaris*) are not currently recorded from any protected Key Area; a further seven Argentine threatened species are known from only one protected Key Area (see Table 1).

RECENT CHANGES TO THE THREATENED LIST

With the publication of Collar *et al.* (1994), three of the 37 threatened species (*Anodorhynchus glaucus*, *Eleothreptus anomalus* and *Sporophila hypochroma*) were dropped from the Argentine threatened list, and seven were added—Andean Flamingo *Phoenicopterus andinus*, Puna Flamingo *P. jamesi*, Blue-winged Macaw *Ara maracana*, Military Macaw *A. militaris*, Chestnut Canastero *Asthenes steinbachi*, Black-and-white Monjita *Heteroxolmis dominicana* and Black-masked Finch *Coryphaspiza melanotis*—but these have not been included within the site inventory (see '*Key Areas*: the book' p. 12).

These changes represent a 14% difference in the composition of the threatened species list; all of the species added (except *Ara militaris*) were listed as Near Threatened in Collar *et al.* (1992), and the new criteria and/or new data having promoted their upgrading. These additions to the list will not have a major distributional impact on the Key Area analysis, as most of the species are known to occur sympatrically with at least one other threatened bird.

Two of the most important sites for *Phoenicopterus andinus* and *P. jamesi* are Laguna de Pozuelos (AR 01) and Laguna Pululos (AR 02). Both the macaws are now very rare in Argentina (Chebez 1994); *Ara maracana* may even be extinct (Nores and Yzurieta 1994), but the last records of it were from Key Areas in Misiones and the most recent records of *A. militaris* were from Baritú National Park (AR 03) in 1991 (Nores and Yzurieta 1994, J. C. Chebez *in litt.* 1995). The Key Areas already identified support the most important populations of *Heteroxolmis dominicana* (AR 15, 24, 26, 27, 28, 30, 31). *Asthenes steinbachi* occurs within Pozuelos Biosphere Reserve (AR 01), but other areas known for the species such as Sierra de Narvaez Strict Nature Reserve and El Leoncito Strict Nature Reserve (J. C. Chebez *in litt.* 1995) are not included in this analysis. There are very few records of *Coryphaspiza melanotis* from Argentina, but one of the most recent is a probable sighting from Mbaracayú National Park (AR 26) (J. C. Chebez *in litt.* 1995).

OLD RECORDS AND LITTLE-KNOWN BIRDS

Our knowledge of threatened birds in Argentina has improved significantly in recent years through survey work targeted at threatened species (e.g. Tubaro and Gabelli 1993, Chebez 1994), habitats holding

Table 2. Top Key Areas in Argentina. Those with the area number in bold currently have some form of protected status.

Key Area		No. of threatened spp.	Comments
07	Sierra de Aconquija	4	These mountains are a stronghold for *Poospiza baeri* as well as three other threatened species.
11	Laguna Mar Chiquita and Bañados del Río Dulce	4	A major stronghold for *Pseudocolopteryx dinellianus* and still likely to be an important area for two threatened Rallidae.
15	Iguazú	12	This protected area is a stronghold for many species of the Atlantic coastal forests.
16	Arroyo Urugua-í	11	This nature reserve supports a large number of Atlantic forest threatened species.
25	San Carlos	3	Grasslands supporting the most important population currently known of *Anthus nattereri*, and probably a significant population of *Xanthopsar flavus*.
28	Colonia Carlos Pellegrini	7	Some of the best remaining grasslands in northern Argentina with key populations of several threatened species.
31	Gualeguaychú	6	A key breeding area for three *Sporophila* and three other threatened species.
34	Bahia Blanca	3	The principal remaining breeding area for *Sturnella militaris*.
36	Islas Jabalí and Gama	1	The main breeding area for *Larus atlanticus*.

Table 3. Matrix of threatened species by Key Area.

	01	02	03	04	05	06	07	08	09	10	11	12	13	14	15	16	17	18	19	20	21	22	23	24	25	26	27
Taoniscus nanus	–	–	–	–	–	–	–	–	–	–	–	–	–	–	–	–	–	–	–	–	–	–	–	–	–	–	–
Mergus octosetaceus	–	–	–	–	–	–	–	–	–	–	–	–	–	–	●	●	–	–	●	–	–	–	–	–	–	–	–
Harpyhaliaetus coronatus	–	–	●	–	–	–	–	–	–	●	–	●	–	–	–	–	–	–	–	–	–	–	–	–	–	–	–
Pipile jacutinga	–	–	–	–	–	–	–	–	–	–	–	–	–	–	●	●	–	–	●	–	–	●	–	–	–	–	–
Rallus antarcticus	–	–	–	–	–	–	–	–	–	–	–	–	–	–	–	–	–	–	–	–	–	–	–	–	–	–	–
Porzana spiloptera	–	–	–	–	–	–	–	–	–	–	●	–	–	–	–	–	–	–	–	–	–	–	–	–	–	–	–
Coturnicops notatus	–	–	–	–	–	–	–	–	–	–	●	–	–	–	–	–	–	–	–	–	–	–	–	–	–	–	●
Fulica cornuta	●	●	–	–	–	–	●	–	–	–	–	–	–	–	–	–	–	–	–	–	–	–	–	–	–	–	–
Numenius borealis	–	–	–	–	–	–	–	–	–	–	–	–	–	–	–	–	–	–	–	–	–	–	–	–	–	–	–
Larus atlanticus	–	–	–	–	–	–	–	–	–	–	●	–	–	–	–	–	–	–	–	–	–	–	–	–	–	–	–
Claravis godefrida	–	–	–	–	–	–	–	–	–	–	–	–	–	–	●	●	–	–	–	–	–	–	–	–	–	–	–
Anodorhynchus glaucus	–	–	–	–	–	–	–	–	–	–	–	–	–	–	–	–	–	–	–	–	–	–	–	–	–	–	–
Amazona pretrei	–	–	–	–	–	–	–	–	–	–	–	–	–	–	–	●	–	–	–	–	●	–	–	–	–	●	–
Amazona vinacea	–	–	–	–	–	–	–	–	–	–	–	–	–	–	●	–	–	●	–	●	●	–	–	–	–	–	–
Triclaria malachitacea	–	–	–	–	–	–	–	–	–	–	–	–	–	–	●	●	–	–	–	–	–	–	–	–	–	–	–
Eleothreptus anomalus	–	–	–	–	–	–	–	–	–	–	–	–	–	●	●	–	–	–	–	–	–	–	–	●	–	●	–
Dryocopus galeatus	–	–	–	–	–	–	–	–	–	–	–	–	–	–	●	●	●	–	–	–	–	●	●	–	–	–	–
Asthenes anthoides	–	–	–	–	–	–	–	–	–	–	–	–	–	–	–	–	–	–	–	–	–	–	–	–	–	–	–
Biatas nigropectus	–	–	–	–	–	–	–	–	–	–	–	–	–	–	●	●	–	–	–	●	–	–	–	–	–	–	–
Piprites pileatus	–	–	–	–	–	–	–	–	–	–	–	–	–	–	–	–	–	–	–	●	–	–	–	–	–	–	–
Pseudocolopteryx dinellianus	–	–	–	–	–	–	–	●	–	–	●	–	●	–	–	–	–	–	–	–	–	–	–	–	–	–	–
Phylloscartes paulistus	–	–	–	–	–	–	–	–	–	–	–	–	–	–	●	–	–	–	–	–	–	–	●	–	–	–	–
Platyrinchus leucoryphus	–	–	–	–	–	–	–	–	–	–	–	–	–	–	●	–	–	–	–	–	–	–	–	–	–	–	–
Agriornis andicola	–	–	–	–	–	–	●	–	–	–	–	–	–	–	–	–	–	–	–	–	–	–	–	–	–	–	–
Yetapa risora	–	–	–	–	–	–	–	–	–	–	–	–	–	●	–	●	–	–	–	–	–	–	–	●	–	–	●
Anthus nattereri	–	–	–	–	–	–	–	–	–	–	–	–	–	–	–	–	–	–	–	–	–	–	–	–	●	–	–
Cinclus schulzi	–	–	●	●	●	–	●	–	–	–	–	–	–	–	–	–	–	–	–	–	–	–	–	–	–	–	–
Sporophila falcirostris	–	–	–	–	–	–	–	–	–	–	–	–	–	–	●	●	–	–	–	–	–	–	–	–	–	–	–
Poospiza baeri	–	–	●	●	–	–	●	–	●	–	–	–	–	–	–	–	–	–	–	–	–	–	–	–	–	–	–
Sporophila frontalis	–	–	–	–	–	–	–	–	–	–	–	–	–	–	●	–	–	–	–	–	–	●	–	–	–	–	–
Sporophila hypochroma	–	–	–	–	–	–	–	–	–	–	–	–	–	●	–	–	–	–	–	–	–	–	–	●	–	●	–
Sporophila palustris	–	–	–	–	–	–	–	–	–	–	–	–	–	–	–	–	–	–	–	–	–	–	–	–	●	●	●
Sporophila zelichi	–	–	–	–	–	–	–	–	–	–	–	–	–	–	–	–	–	–	–	–	–	–	–	–	–	–	–
Gubernatrix cristata	–	–	–	–	–	–	–	–	–	–	–	●	–	●	–	–	–	–	–	–	–	–	–	–	–	–	–
Saltator rufiventris	–	–	●	–	–	●	–	–	–	–	–	–	–	–	–	–	–	–	–	–	–	–	–	–	–	–	–
Xanthopsar flavus	–	–	–	–	–	–	–	–	–	–	–	–	–	●	–	–	–	–	–	–	–	–	–	●	●	–	●
Sturnella militaris	–	–	–	–	–	–	–	–	–	–	–	–	–	–	–	–	–	–	–	–	–	–	–	–	–	–	–
No. of species	1	1	4	2	1	1	4	1	1	1	4	2	1	5	12	9	1	1	2	3	2	3	2	4	3	4	4

concentrations of threatened species (e.g. Pearman and Abadie in press) and inventories for protected areas (e.g. Di Giacomo 1995, Di Giacomo *et al.* 1995). Most species are little-known somewhere within their range, but recent records from other areas mean that we are able to make informed recommendations for their conservation. The exceptions are those species known only from old specimen records or chance observations (from which there can be no guarantees of a current viable population), and in Argentina these include the following.

Mergus octosetaceus was formerly distributed throughout the shallow, fast-flowing rivers of Misiones province. A survey of 376 km of seemingly suitable rivers within Misiones in July–September 1993 resulted in the observation of a single bird and the conclusion that there is little prospect for the species' long-term survival in Argentina (Benstead *et al.* 1993, 1994).

Piprites pileatus is known in Argentina from only a single specimen collected in 1959 (AR 20).

Agriornis andicola is known in Argentina mainly from Sierra de Aconquija (AR 07), though it has not been seen there since 1952.

OUTLOOK

This analysis has identified 43 Key Areas, each of which is of major importance for the conservation of threatened birds in Argentina. However, nine Key Areas stand out as top priorities (Table 2), and together they have supported populations of 20 (74%) of the Argentine threatened birds. These and the remaining Key Areas (which are individually important for many threatened species: see 'Key Areas' and 'Key Area protection', above) all need some degree of conservation action if the populations of their threatened birds are to remain viable. For many

Table 3 (cont.)

28	29	30	31	32	33	34	35	36	37	38	39	40	41	42	43	No. of areas
–	–	–	–	–		–	–	–	–	–	–	–	–	–	–	0
–	–	–	–	–	–	–	–	–	–	–	–	–	–	–	–	3
●	–	–	–	–	–	–	–	–	●	●	–	–	–	–	–	6
–	–	–	–	–	–	–	–	–	–	–	–	–	–	–	–	4
–	–	–	–	–	–	–	–	–	–	–	–	–	●	–	–	1
–	–	–	–	●	●	●	–	–	–	–	–	–	–	–	–	4
–	–	–	–	●	–	–	–	–	–	–	–	–	–	–	–	3
–	–	–	–	–	–	–	–	–	–	–	–	–	–	–	–	3
–	–	–	–	–	–	–	–	–	–	–	–	–	–	–	–	0
–	–	–	–	–	●	–	●	●	–	–	–	–	–	–	●	5
–	–	–	–	–	–	–	–	–	–	–	–	–	–	–	–	2
–	–	–	–	–	–	–	–	–	–	–	–	–	–	–	–	0
–	–	–	–	–	–	–	–	–	–	–	–	–	–	–	–	3
–	–	–	–	–	–	–	–	–	–	–	–	–	–	–	–	4
–	–	–	–	–	–	–	–	–	–	–	–	–	–	–	–	2
–	–	●	●	–	–	–	–	–	–	–	–	–	–	–	–	6
–	–	–	–	–	–	–	–	–	–	–	–	–	–	–	–	5
–	–	–	–	–	–	–	–	–	–	●	●	–	–	●	–	3
–	–	–	–	–	–	–	–	–	–	–	–	–	–	–	–	3
–	–	–	–	–	–	–	–	–	–	–	–	–	–	–	–	1
–	–	–	–	–	–	–	–	–	–	–	–	–	–	–	–	3
–	–	–	–	–	–	–	–	–	–	–	–	–	–	–	–	2
–	–	–	–	–	–	–	–	–	–	–	–	–	–	–	–	1
–	–	–	–	–	–	–	–	–	–	–	–	–	–	–	–	1
●	–	–	–	–	–	–	–	–	–	–	–	–	–	–	–	5
–	–	–	–	–	–	–	–	–	–	–	–	–	–	–	–	1
–	–	–	–	–	–	–	–	–	–	–	–	–	–	–	–	4
–	–	–	–	–	–	–	–	–	–	–	–	–	–	–	–	2
–	–	–	–	–	–	–	–	–	–	–	–	–	–	–	–	4
–	–	–	–	–	–	–	–	–	–	–	–	–	–	–	–	2
●	–	–	●	●	–	–	–	–	–	–	–	–	–	–	–	6
●	●	–	●	●	–	–	–	–	–	–	–	–	–	–	–	7
●	–	●	●	–	–	–	–	–	–	–	–	–	–	–	–	3
●	●	●	●	–	–	●	–	–	●	–	–	●	–	–	–	9
–	–	–	–	–	–	–	–	–	–	–	–	–	–	–	–	2
●	–	–	●	–	–	–	–	–	–	–	–	–	–	–	–	6
–	–	–	–	–	–	●	–	–	–	–	–	–	–	–	–	1
7	2	3	6	4	2	3	1	1	2	2	1	1	1	1	1	

areas there is a requirement to confirm the continued existence or assess the population of a threatened species, and for others, especially those areas based on old records, the need is for locating (and defining the extent of) suitable habitat in the vicinity of the Key Area.

DATA SOURCES

The above introductory text and the Site Inventory were compiled from information supplied by E. I. Abadie, P. J. Benstead, M. Bettinelli, A. Bosso, J. C. Chebez, F. M. Gabelli, D. Gallegos-Luque, D. Gómez, A. Di Giacomo, F. R. Lambert, B. López Lanús, E. de Lucca, J. Mazar, M. Pearman, S. Peris, R. S. Ridgely, H. Torres, P. L. Tubaro and S. J. Tyler, as well as from the following references.

BENSTEAD, P. J., HEARN, R. D., JEFFS, C. J. S., CALLAGHAN, D. A., CALO, J., GIL, G., JOHNSON, A. E. AND STAGI NEDELCOFF, A. R. (1993) 'Pato Serrucho 93': an expedition to assess the current status of the Brazilian Merganser *Mergus octosetaceus* in north-east Argentina. Unpublished report.

BENSTEAD, P. J., HEARN, R. D., JEFFS, C. J. S., CALLAGHAN, D. A., CALO, J., GIL, G., JOHNSON, A. E. AND STAGI NEDELCOFF, A. R. (1994) A recent sighting of Brazilian Merganser *Mergus octosetaceus* in Misiones Province, Argentina. *Cotinga* 2: 35–36.

CHEBEZ, J. C. (1994) *Los que se van*. Buenos Aires: Albatros.

COLLAR, N. J., CROSBY, M. J. AND STATTERSFIELD, A. J. (1994) *Birds to watch 2: the world list of threatened birds*. Cambridge, U.K.: BirdLife International (BirdLife Conservation Series no. 4).

COLLAR, N. J., GONZAGA, L. P., KRABBE, N., MADROÑO NIETO, A., NARANJO, L. G., PARKER, T. A. AND WEGE, D. C. (1992) *Threatened birds of the Americas: the ICBP/IUCN Red Data Book*. Cambridge, U.K.: International Council for Bird Preservation.

COLLAR, N. J., WEGE, D. C. AND LONG, A. J. (in press) Patterns and causes of endangerment in the New World avifauna. In J. V. Remsen, ed. *Natural history and conservation of Neotropical birds*. American Ornithologists' Union (Orn. Monogr.).

DI GIACOMO, A. G. (1995) An ornithological survey of Argentine national parks. *Cotinga* 3: 69.

DI GIACOMO, A., LÓPEZ LANÚS, B. AND DI GIACOMO, A. (1995) *Inventario ornitológico del Parque Nacional Baritú (Salta, Argentina)*. Buenos Aires: Literature of Latin America (LOLA Special Monogr. 8).

GIL, G., HAENE, E. AND CHEBEZ, J. C. (1995) Notas sobre la avifauna de Sierra de las Quijadas. *Nuestras Aves* 31: 26–28.

HALLOY, S., GONZÁLEZ, J. A. AND GRAU, A. (1994) *Proyecto de creación del Parque Nacional Aconquija (Tucumán, Argentina): report no. 4*. Tucumán: Fundacíon Miguel Lillo (Nature Conservation Series no. 9).

IUCN (1992) *Protected areas of the world: a review of national systems, 4: Nearctic and Neotropical*. Gland, Switzerland and Cambridge, U.K.: International Union for Conservation of Nature and Natural Resources.

DE LUCCA, E. R. (1993) El Aguila Coronada. *Nuestras Aves* 11: 14–17.

NAVAS, J. R. (1962) Reciente hallazgo de *Rallus limicola antarcticus* King (Aves, Rallidae). *Neotrópica* 8: 73–76.

NORES, M. AND YZURIETA, D. (1994) The status of Argentine parrots. *Bird Conserv. Internatn.* 4: 313–328.

PAYNTER, R. A. (1985) *Ornithological gazetteer of Argentina*. Cambridge, Mass.: Museum of Comparative Zoology.

PEARMAN, M. AND ABADIE, E. I. (in press) Mesopotamia grasslands and wetlands survey, 1991–1993: conservation of threatened birds and habitat in north-east Argentina.

SCOTT, D. A. AND CARBONELL, M. (1986) *A directory of Neotropical wetlands*. Cambridge and Slimbridge, U.K.: International Union for Conservation of Nature and Natural Resources and International Waterfowl Research Bureau.

STATTERSFIELD, A. J., CROSBY, M. J., LONG, A. J. AND WEGE, D. C. (in prep.) *Global directory of Endemic Bird Areas*. Cambridge, U.K.: BirdLife International (BirdLife Conservation Series).

TUBARO, P. L. AND GABELLI, F. M. (1993) Distribución actual y selección de habitat del Pecho Colorado Mediano *Sturnella defilippi*: informe final de proyecto. Unpublished report.
TYLER, S. J. (1994) The Yungas of Argentina: in search of Rufous-throated Dipper *Cinclus schulzi*. *Cotinga* 2: 38–40.
VEBLEN, T. T., MERMOZ, M., MARTIN, C. AND KITZBERGER, T. (1992) Ecological impacts of introduced animals in Nahuel Huapi National Park, Argentina. *Conserv. Biol.* 6: 71–83.

SITE INVENTORY

Laguna de Pozuelos (Jujuy)

AR 01
22°29′S 66°00′W

Pozuelos Biosphere Reserve (IUCN category IX), 364,000 ha
Laguna Pozuelos National Monument (IUCN category III), 28,000 ha

This large permanent lake lies at 3,500 m in the High Andes, c.50 km south-west of La Quiaca in northern Jujuy. Listed as a Ramsar site, it has little vegetation and large areas of mud are exposed during the dry season (Scott and Carbonell 1986).

Fulica cornuta	1992	Three or four pairs nested in January 1987. About 28 nests were reported from a small lake 4 km west of Laguna de Pozuelos in November 1984, and c.80 birds have been recorded from La Lagunilla, apparently close to Laguna de Pozuelos.

Lagunas Pululos and Vilama (Jujuy)

AR 02
22°35′S 66°44′W

Altoandina de la Chinchilla Provincial Reserve (IUCN category II), 119,730 ha

These small lakes, of c.100 ha, are located at c.4,400 m in the High Andes, c.110 km west of Abra Pampa, close to the Bolivian border in north-east Jujuy.

Fulica cornuta	1980s	About 40 pairs nested in October 1982 (photograph in Chebez 1994), although similar numbers were not found in the late 1980s.

Calilegua (Jujuy)

AR 03
23°35′S 64°54′W
EBAs B37, B57

Calilegua National Park (IUCN category II), 76,000 ha

This national park lies extends in altitude from 600 m to 3,000 m or more in the Serranía de Calilegua east of Valle Grande and north-west of Ledesma (Libertador General San Martín) in eastern Jujuy. The park comprises second-growth on the lower slopes, subtropical montane forest with limited areas of *Podocarpus* up to the treeline (at c.1,700 m) and a shrub zone and Andean grassland above this.

Harpyhaliaetus coronatus	1992	Three records since 1986 (A. Di Giacomo *in litt.* 1995).
Cinclus schulzi	1993	One record, close to the northern limit of the park (A. Di Giacomo *in litt.* 1995).
Poospiza baeri	1993	One record: seen just outside the western boundary of the park (A. Di Giacomo *in litt.* 1995).
Saltator rufiventris	1993	Observed three times since 1988, always on Cerro Hermoso (A. Di Giacomo *in litt.* 1995).

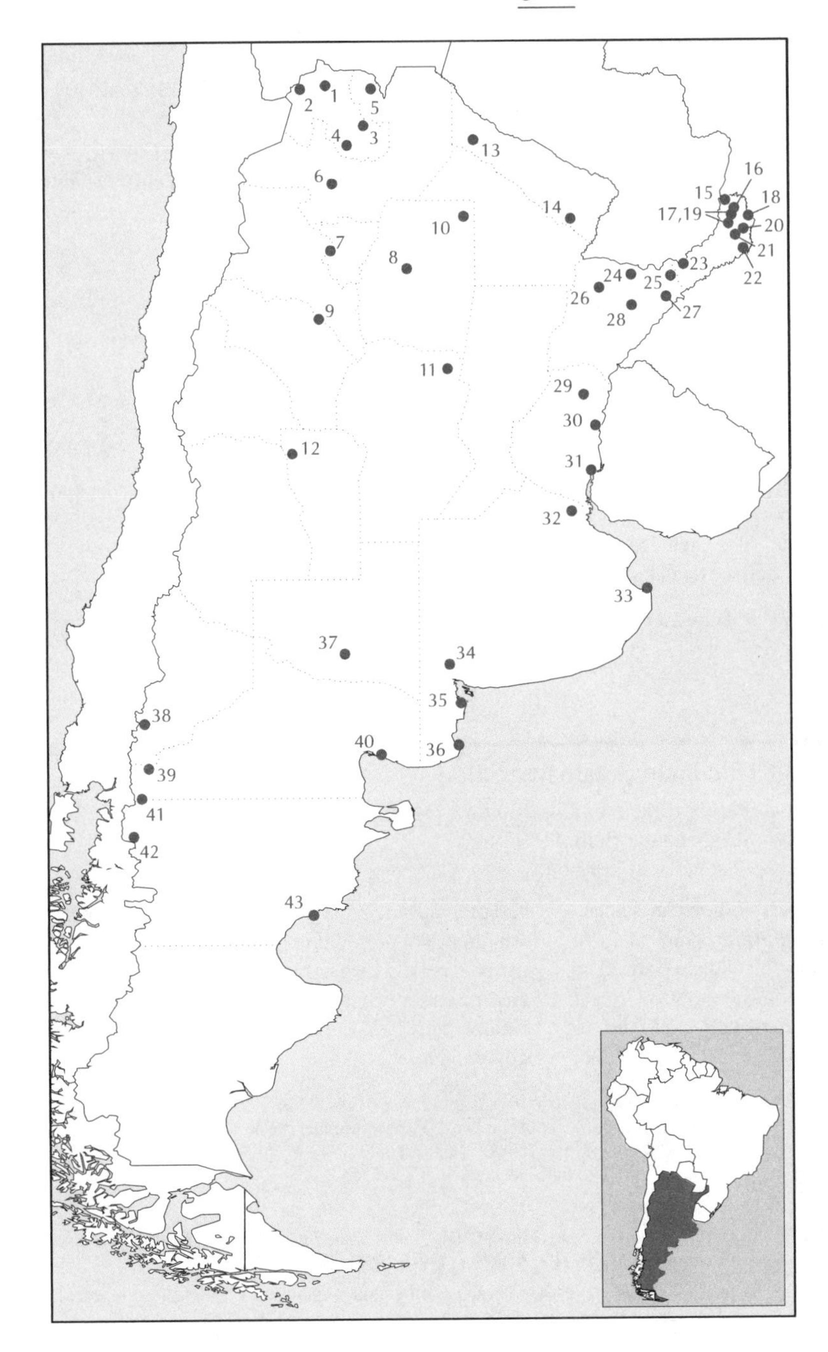

Figure 2. Key Areas in Argentina.

01 Laguna de Pozuelos
02 Lagunas Pululos and Vilama
03 Calilegua
04 Río Yala
05 Baritú
06 Cuesta del Obispo
07 Sierra de Aconquija
08 Bañados de Figueroa
09 Sierra del Manchao
10 Copo
11 Laguna Mar Chiquita and Bañados del Río Dulce
12 Sierra de las Quijadas
13 Bañado La Estrella
14 El Bagual
15 Iguazú
16 Arroyo Urugua-í
17 Sierra Morena
18 San Antonio
19 Arroyo Piray Miní
20 Tobuna
21 San Pedro
22 Yabotí
23 Campo San Juan
24 San Juan de Poriahú
25 San Carlos
26 Mburucuyá
27 Caza Pava
28 Colonia Carlos Pellegrini
29 Selva de Montiel
30 El Palmar
31 Gualeguaychú
32 Otamendi
33 Punta Rasa
34 Bahia Blanca
35 Isla Brightman
36 Islas Jabalí and Gama
37 Lihuel Calel
38 Lanín
39 Nahuel Haupi
40 Laguna del Monte
41 El Bolsón
42 Los Alerces
43 Isla Vernaci

AR 04

Río Yala (Jujuy)

24°07'S 65°23'W
EBAs B37, B57

Portrero de Yala Provincial Park (IUCN category II), 4,292 ha

This area refers to the Río Yala valley on the eastern slope of the Andes around Termas de Reyes, c.20 km north-west of San Salvador de Jujuy in east-central Jujuy. The area supports subtropical montane forest.

Cinclus schulzi	1993	Several sightings in 1993 included a pair along the Río Yala at c.1,700 m (Tyler 1994, P. J. Benstead *in litt.* 1994).
Poospiza baeri	1950	One record: specimen taken near Río la Quesera, apparently within the Yala valley.

Baritú (Salta)

AR 05

Baritú National Park (IUCN category II), 72,439 ha

22°35′S 64°40′W
EBAs B37, B57

This national park is located in the Andes of extreme northern Salta, with part of its north-western limits lying along the Río Bermejo, which marks the border with Bolivia. The park includes an extensive tract of subtropical montane forest which has only recently been visited by ornithologists.

Cinclus schulzi	1992	Common along Ríos Baritú and Lipeo (A. Di Giacomo *in litt.* 1995).

Cuesta del Obispo (Salta)

AR 06

Los Cardones National Park (proposed), 70,000 ha

25°10′S 65°50′W
EBA B37

A mountain ridge reaching 3,500 m and located 30 km east of Cachi and 35 km west of Rosario de Lerma in south-central Salta. The area has been proposed as a national park, and avifaunal surveys have recorded 112 species including an unsubstantiated record of *Agriornis andicola* (J. C. Chebez *in litt.* 1995).

Saltator rufiventris	1992	Recorded in 1990, 1991 and 1992 (D. Gallegos-Luque *in litt.* 1995).

Sierra de Aconquija (Tucumán/Catamarca/Salta)

AR 07

Campo de los Alisos National Park (proposed), 250,000 ha
La Florida Provincial Reserve (IUCN category II), 9,892 ha
Sosa Provincial Reserve (IUCN category II), 890 ha

27°00′S 65°53′W
EBAs B37, B57

The Sierra de Aconquija is east of the main Andean Cordillera in western Tucumán and eastern Catamarca, with a small area in southern Salta. The area comprises moist subtropical montane forest (c.700–1,800 m), some agricultural areas up to c.2,100 m (e.g. around the village of Tafí del Valle) and prepuna and puna grasslands at higher altitudes. An important proposal to gazette the entire area as a national park, which has included detailed feasibility studies and a proposed boundary for the park (see Halloy *et al.* 1994), is currently being considered by government.

Fulica cornuta	1984	At Laguna Cerritos, Laguna Los Patos (both on Cerro Muñoz at 26°46′S 65°51′W) and Laguna La Manga (15 km south-west of Cerro Muñoz), totals of 135 were present in March 1982 and 98 in September 1982, but only 10–12 in April and October 1984; at La Lagunilla, c.80 birds have been recorded, undated.
Agriornis andicola	1952	A number of old specimen records.
Cinclus schulzi	1994	During surveys around Tafí del Valle, birds were found on only the less polluted tributaries of the main Río Angostura (Tyler 1994).
Poospiza baeri	1993	Most often seen in vegetated, steep-sided gullies and ravines at 2,600–2,850 m, especially around El Infiernillo; some nests found in recent years (S. Peris *in litt.* 1993).

Bañados de Figueroa (Santiago del Estero)

AR 08

Unprotected

27°30′S 63°40′

These marshlands are located along the Río Salado, immediately north of Villa Figueroa and 80 km north-east of Santiago de Estero city in central Santiago de Estero. The marshes and lagoons support periodically flooded rushy and grassy vegetation with some shrubbery.

Pseudocolopteryx dinellianus	1990s	Often recorded, but relatively low numbers.

AR 09

Sierra del Manchao (Catamarca)

28°52′S 66°14′W
EBAs B37, B57

Unprotected

This small mountain range is a southerly extension of the larger Sierra de Ambato in the south-east of Catamarca, and the area presumably holds habitats similar to those in the nearby Sierra del Aconquija (AR 07).

Poospiza baeri	1980s	A number of specimens from the 1980s.

AR 10

Copo (Santiago del Estero)

26°05′S 62°00′W

Copo Scientific Reserve (IUCN category I), 114,250 ha

This protected area is located around Monte Quemado in extreme north-east Santiago del Estero, with its northern and eastern limits lying on the border with Chaco province. The area represents one of the best examples of protected dry Chaco in Argentina (J. C. Chebez *in litt.* 1995).

Harpyhaliaetus coronatus	1990s	Population present (D. Gómez *per* J. C. Chebez *in litt.* 1995).

AR 11

Laguna Mar Chiquita and Bañados del Río Dulce (Córdoba/Santiago del Estero)

30°15′S 62°30′W

Bañados del Río Dulce and Laguna de Mar Chiquita Natural Park (IUCN category V), 50,000 ha

This vast system of riverine marshes comprises permanent and seasonal shallow, fresh to brackish lakes and marshes and seasonally flooded grasslands along the Río Dulce and other rivers flowing into Laguna Mar Chiquita, a very large saline lake in south-east Santiago del Estero and adjacent north-east Córdoba (Scott and Carbonell 1986). There is an unsubstantiated report of four Eskimo Curlews *Numenius borealis* from Laguna Mar Chiquita in October 1990.

Coturnicops notatus	1973	One along the Río Dulce.
Porzana spiloptera	1974	Sightings along the Río Dulce in 1973 (8–10 birds) and 1974 suggest that the species was not uncommon.
Larus atlanticus	1991	Winter visitor to Laguna Mar Chiquita; the 1991 sighting involved 20 birds.
Pseudocolopteryx dinellianus	1991	The major global stronghold, with records from areas in both Córdoba and Santiago del Estero provinces.

AR 12

Sierra de las Quijadas (San Luis)

32°33′S 67°02′W

Sierra de las Quijadas National Park (IUCN category II), 150,000 ha

This recently decreed national park is found 100 km north of San Luis city in northern San Luis. The climate is arid, the terrain is hilly with small canyons, and the vegetation consists mainly of monte woodland.

Harpyhaliaetus coronatus	1990	Seen several times in the reserve during 1990 (Gil *et al.* 1995). This area was noted as a priority for the species by de Lucca (1993).
Gubernatrix cristata	1990s	Recorded recently.

Bañado La Estrella (Formosa)

AR 13

Unprotected

23°45'S 62°00'W

This extensive area of marshes in central and northern Formosa, south of the Río Pilcomayo, is considered one of the most important wetlands in Argentina, and a project investigating the creation of a nature reserve is under way (J. C. Chebez *in litt.* 1995).

Pseudocolopteryx dinellianus	c.1994	Recorded recently; no information on previous status (J. Mazar *per* J. C. Chebez *in litt.* 1995).

El Bagual (Formosa)

AR 14

El Bagual Ecological Reserve (IUCN category I), 6,000 ha

26°10'S 58°56'W

This reserve is 70 km west of Formosa City in Laishi department, south-east Formosa, and consists of seasonally inundated grasslands. There is a warden, who is currently studying the bird populations (A. Di Giacomo *in litt.* 1995).

Eleothreptus anomalus	1992	One record; not found during recent fieldwork (A. Di Giacomo *in litt.* 1995).
Yetapa risora	1995	An important breeding population that is currently being studied (A. Di Giacomo *in litt.* 1995).
Sporophila hypochroma	1995	Probable sightings of a male and 2–3 females (A. Di Giacomo *in litt.* 1995).
Gubernatrix cristata	1990s	Past records but not found during recent fieldwork (A. Di Giacomo *in litt.* 1995).
Xanthopsar flavus	1990s	Listed for the reserve but not recorded during recent fieldwork (A. Di Giacomo *in litt.* 1995).

Iguazú (Misiones)

AR 15

Iguazú National Park (IUCN category I), 53,609 ha; (IUCN category VIII), 7,675 ha
Iguazú National Park World Heritage Site (IUCN category X), 55,000 ha

25°41'S 54°26'W
EBA B52

This national park is centred on the famous Iguazú falls sited close to the junction of the Argentina, Brazil and Paraguay borders. The main habitat is humid evergreen subtropical forest. This protected area, in combination with the Iguaçu National Park (BR 126) in Brazil to the north and the Urugua-í Provincial Park (AR 16) to the south-east, forms one of the most extensive tracts of forest within the south-western portion of the Atlantic coastal forests. This area supports more threatened species than any other in Argentina, and, for several, it is the only known locality in the country.

Mergus octosetaceus	1978	Irregularly recorded between 1942 and 1978.
Pipile jacutinga	1993	A number of historical records, and recent observations along several rivers (Benstead *et al.* 1993). A stronghold for the species in Argentina.
Claravis godefrida	1991	A handful of records, the last being a sighting in May 1991 (F. R. Lambert *in litt.* 1993).
Triclaria malachitacea	1983	One seen with a group of Reddish-bellied Parrots *Pyrrhura frontalis*.
Amazona vinacea	1983	Probably now extinct in the park.
Eleothreptus anomalus	undated	One record, probably of a vagrant.
Dryocopus galeatus	1993	Most recent sightings have been from the Macuco trail and the Bernabé Méndez trail.
Biatus nigropectus	1990s	One record (Chebez 1994).
Phylloscartes paulistus	1993	A handful of records.
Platyrinchus leucoryphus	1989	One mist-netted in 1978 and a bird seen in 1989 are the only Argentine records.
Sporophila falcirostris	1990	A number of records, with breeding proven in August 1988.
Sporophila frontalis	1978	One mist-netted in 1978 is the only Argentine record.

Arroyo Urugua-í (Misiones)

AR 16

25°54′S 54°10′W
EBA B52

Urugua-í Natural Reserve (IUCN category IV), 84,000 ha

This large reserve embraces the middle and upper reaches of the Río Urugua-í (a major tributary of the upper Río Paraná) and is located in north-east Misiones east of but adjacent to the lake created by the Urugua-í dam, c.50 km east of Esperanza and 50 km north-west of Bernardo de Irigoyen. The northern boundary of this Key Area lies adjacent to Iguazú National Park (AR 15), and together these reserves form one of the largest tracts of humid evergreen subtropical forest remaining in Argentina. Much of the area has been under timber production so the quality of the forest varies greatly, with some areas of scrub and introduced tree species (Benstead *et al.* 1993).

Species	Year	Notes
Mergus octosetaceus	1988	No records since 1988 in spite of specific searches made by boat in 1993 (Benstead *et al.* 1993).
Pipile jacutinga	1993	A number of sightings indicate that the park is one of the main strongholds for the species in Argentina (Benstead *et al.* 1993).
Claravis godefrida	1957	One collected in 1957.
Triclaria malachitacea	1986	One bird seen c.30 km west of Bernardo de Irigoyen.
Amazona pretrei	1980	One seen near Arroyo Urugua-í in 1980, but a survey in 1993 failed to find the species (Benstead *et al.* 1993).
Dryocopus galeatus	1995	Recently observed on the edge of the reserve at Caá-Porá private nature reserve (B. López-Lanús *per* J. C. Chebez *in litt.* 1995).
Biatas nigropectus	1993	A number of historical records and specimens up to the 1960s, and a recent sighting of a female or immature bird at 26°13′S 53°50′W (Benstead *et al.* 1993).
Yetapa risora	1961	Known from historical records, but with no suitable habitat left the species is probably extinct in this part of its range (M. Pearman *in litt.* 1995).
Sporophila falcirostris	1958	A number of specimens collected in 1958.

Sierra Morena (Misiones)

AR 17

26°05′S 54°15′W
EBA B52

Unprotected

This site lies in the catchment area of the Arroyos Urugua-í and Aguaray-Guazú in extreme south-east Misiones. It supports a sizeable tract of humid forest (J. C. Chebez *in litt.* 1995).

Species	Year	Notes
Dryocopus galeatus	1994	Recorded recently; no information on previous status (J. C. Chebez *in litt.* 1995)

San Antonio (Misiones)

AR 18

26°07′S 53°45′W
EBA B52

San Antonio Strict Nature Reserve (IUCN category I), 600 ha

The area lies by the Río San Antonio, close to the Brazilian border in General Manuel Belgrano department, north-east Misiones. The habitat of the general area is upland forest (at 700 m) with stands of *Araucaria angustifolia*, but large areas have been lost to, or modified by, agriculture (Benstead *et al.* 1993). Since 1990 a small wardened nature reserve has been established (J. C. Chebez *in litt.* 1995).

Species	Year	Notes
Amazona vinacea	1993	Recent sightings of 5–7 birds along the Intercontinental Highway near Bernardo de Irigoyen, but, without habitat protection, long-term survival of this small population is unlikely (Benstead *et al.* 1993).

Arroyo Piray Miní (Misiones)

AR 19

Unprotected

26°19′S 54°20′W
EBA B52

This area embraces the mid- to lower reaches of the Arroyo Piray Miní (a tributary of the upper Río Paraná), which flows westwards in north-central Misiones. The main habitat is humid evergreen subtropical forest, but much of it has been cleared or heavily degraded by man's activities, primarily logging and agriculture (Benstead *et al.* 1993).

Pipile jacutinga	1993	Recorded in 1991 and 1993 (E. de Lucca *per* J. C. Chebez *in litt.* 1995).
Mergus octosetaceus	1993	The most recent record for Argentina, but from a degraded section of the river, and which probably could not hold a viable population (Benstead *et al.* 1993, 1994).

Tobuna (Missiones)

AR 20

Cruce Caballero Provincial Reserve (IUCN category II), 432 ha
Piñalitos Private Reserve (IUCN category I), 3,796 ha

26°28′S 53°54′W
EBAs B52,B54

The area is located c.30 km north-east of San Pedro in north-east Misiones and contains humid evergreen subtropical forest, with smaller areas of upland forest supporting stands of *Araucaria angustifolia*. Much of the forest has now been lost (J. C. Chebez *in litt.* 1995).

Amazona vinacea	1950s	Common in the 1950s and still likely to be present (J. C. Chebez *in litt.* 1995).
Biatas nigropectus	1953	Four old specimens.
Piprites pileatus	1959	One specimen, the only record for Argentina.

San Pedro (Misiones)

AR 21

Araucaria Provincial Park (IUCN category II), 92 ha

26°38′S 54°08′W
EBA B54

San Pedro is in the interior of north-central Misiones, the main habitat in the area being humid forest and relict patches of upland forest with *Araucaria angustifolia*.

Amazona pretrei	1995	Three seen in a group of c.100 *A. vinacea* in March 1995 (H. Torres *per* J. C. Chebez *in litt.* 1995).
Amazona vinacea	1995	A group of c.100 seen in March 1995 (H. Torres *per* J. C. Chebez *in litt.* 1995) and roosts within Araucaria Provincial Park (J. C. Chebez *in litt.* 1995).

Yabotí (Misiones)

AR 22

Yabotí Multiple Use Reserve (IUCN category VIII), 223,000 ha

27°00′S 53°55′W
EBA B52

This humid forest lies 50 km south-west of San Pedro in east-central Misiones. It forms one of the largest remnant tracts of humid forest in San Pedro department and is contiguous with Turvo State Park, which lies across the border in Brazil (J. C. Chebez *in litt.* 1995).

Pipile jacutinga	c.1993	Recorded recently; no information on previous status (J. C. Chebez *in litt.* 1995).
Dryocopus galeatus	c.1993	Recorded recently; no information on previous status (J. C. Chebez *in litt.* 1995).
Sporophila frontalis	c.1993	Recorded recently; no information on previous status (J. C. Chebez *in litt.* 1995).

Campo San Juan (Misiones)

AR 23
27°25'S 55°40'W
EBA B52

Unprotected

Campo San Juan is close to Posados near the Río Paraná in extreme south-west Misiones close to the border with Paraguay. It comprises a wide array of habitats with humid subtropical forest surrounded by grasslands, low-lying scrub and *Urunday* woodland in an area of only c.5,000 ha (J. C. Chebez *in litt.* 1995).

Dryocopus galeatus	1990s	Recorded recently; no information on previous status (J. C. Chebez *in litt.* 1995).
Phylloscartes paulistus	1990s	Recorded recently; no information on previous status (J. C. Chebez *in litt.* 1995).

San Juan de Poriahú (Corrientes)

AR 24
27°42'S 57°11'W

San Juan de Poriahú Private Reserve (IUCN category I, 4,199 ha, and IUCN category IV, 10,199 ha)

Located on the north-west borders of the Iberá grassland system, this estancia comprises extensive humid grassland and marsh with pools and small woodlots on slightly elevated land (M. Pearman *in litt.* 1995). This is a working estancia under a land management agreement with Fundación Vida Silvestre Argentina (M. Pearman *in litt.* 1995).

Eleothreptus anomalus	1989	One record, which requires confirmation (M. Pearman *in litt.* 1995).
Yetapa risora	1990s	A small but stable resident population (M. Pearman *in litt.* 1995).
Sporophila hypochroma	1990s	A fairly common summer visitor, confirmed breeding (M. Pearman *in litt.* 1995).
Xanthopsar flavus	1990s	One record (Chebez 1994).

San Carlos (Corrientes)

AR 25
27°44'S 56°02'W
EBA B58

Unprotected

This area comprises several large privately owned estancias including San Juan Bautista, San Miguelito and Los Alamos located 45 km south-west of Posadas in north-west Corrientes. The habitats consist of savannas, rich in grasses (particularly *Aristida jubata*), marshes in the low-lying terrain, isolated pockets of transitional woodland and extensive gallery forest along the Río Aguapey. The area is partially modified with *Pinus* and *Eucalyptus* plantations and soya and sunflower crops (Pearman and Abadie in press). However, cattle-raising remains the long-standing and principal land use on the grasslands and the populations of the threatened species (below) depends on sensitive management, especially a slow rotation of grazing pasture (M. Pearman *in litt.* 1995).

Anthus nattereri	1993	A density of 9 birds/km^2 (and confirmed breeding) was found in the best grasslands of Estancias San Juan Bautista and San Miguelito in 1993 (M. Pearman *in litt.* 1995). This is one of the only sites currently known for the species.
Sporophila palustris	1993	A male recorded in a marsh bordering the Río Aguapey.
Xanthopsar flavus	1993	Mobile, scattered flocks of nine, c.70 and over 305 were encountered in January 1993, the latter being the largest ever recorded (M. Pearman *in litt.* 1995). As the species is nomadic, it is not known whether these flocks represent a resident population (M. Pearman *in litt.* 1995).

Mburucuyá (Corrientes)

AR 26

Mburucuyá National Park (IUCN category II), 15,060 ha

28°03′S 58°07′W

Located in north-central Corrientes province, this area is principally forested with transitional woodland with palms *Copernicia alba* and *Butia yatay* and chaco-type woodland (dominated by *Prosopis* and *Aspidosperma*), among which are inundated marshlands and small areas of dry natural grassland (J. C. Chebez *in litt.* 1995, M. Pearman *in litt.* 1995). Until recently the area had been intensively overgrazed.

Species	Year	Notes
Amazona pretrei	1980s	One record, presumably of a wandering bird.
Eleothreptus anomalus	1989	Two records.
Sporophila palustris	1991	A male in March 1991; may have been a bird on passage.
Sporophila hypochroma	1990s	Fairly common breeder; at least 14 males found in recent surveys (M. Pearman *in litt.* 1995).

Caza Pava (Corrientes)

AR 27

Unprotected

28°18′S 56°10′W
EBA B55

Located on the extreme eastern border of the Esteros de Iberá, this area comprises cattle-grazing pastures which are bisected by low-lying marshes, most notably the Bañado Mora Cué and Estero Mbérity. The estero is bordered by the extensive Monte Mbérity (dense monte and transitional woodland running north–south), paddyfields and a limited area of soya crops.

Species	Year	Notes
Coturnicops notatus	1991	A single record from 1991 is the only record for Corrientes province, but the species' status remains unknown; fieldwork and a crake-trapping programme in 1992 did not detect it (M. Pearman *in litt.* 1995).
Yetapa risora	1992	A small but long-standing population at Estancia La Casualidad where natural grasslands dominate (M. Pearman *in litt.* 1995).
Sporophila palustris	1992	At least eight males (six apparently paired) encountered in recent surveys (M. Pearman *in litt.* 1995). Cagebird trappers were seen targeting this species (M. Pearman *in litt.* 1995).
Xanthopsar flavus	1992	A number of sightings; the largest flock of c.105–135 birds was at a roost near Santo Tomé in mid-May 1991.

Colonia Carlos Pellegrini (Corrientes)

AR 28

Iberá Natural Reserve (IUCN category IV), 1,200,000 ha

28°32′S 57°10′W
EBA B55

This area is defined as land within 50 km of Colonia Carlos Pellegrini (M. Pearman *in litt.* 1995) in north-central Corrientes, near the south-east shore of Laguna Iberá. The area is dominated by humid open campos with tall natural grassland and isolated monte woodlands. It holds some of the best remaining grasslands and marshes in Argentina, and is the stronghold for several threatened species. The reserve consists mainly of the open water of Laguna Iberá and protects little of the surrounding grassland and marsh.

Species	Year	Notes
Harpyhaliaetus coronatus	1992	A pair and immature were located in a remote estancia in 1992, and breeding and long-standing presence were confirmed by the estancia owner (M. Pearman *in litt.* 1995).
Yetapa risora	1992	The largest known population of the species, estimated at no more than 23,000 individuals in the Iberá grassland system (c.11,450 km^2) with the centre of abundance in a 50-km radius of Colonia Carlos Pellegrini (Pearman and Abadie in press).
Sporophila hypochroma	1993	Up to 15 males (some paired) recorded in surveys, 1992–1993 (M. Pearman *in litt.* 1995).
Sporophila palustris	1993	Up to 11 males (some paired) recorded in recent surveys (Pearman and Abadie in press).
Sporophila zelichi	1993	A pair caught at Estancia Rincón del Socorro (7 km south of Colonia Carlos Pellegrini) in February 1993 is the first record outside Entre Ríos province but probably refers to passage birds as none was located at the height of the breeding season (M. Pearman *in litt.* 1995).
Gubernatrix cristata	1992	One collected 1972; a pair holding territory in November 1992 (M. Pearman *in litt.* 1995).
Xanthopsar flavus	1993	Two pairs seen north of Colonia Carlos Pellegrini (M. Pearman *in litt.* 1995).

Selva de Montiel (Entre Ríos)

AR 29
30°58′S 58°35′W
EBA B55

Unprotected

This area lies next to the Río Gualeguay immediately south-east of Federal in north-east Entre Ríos. The area comprises one of the largest existing areas of monte woodland (dominated by *Prosopis* and *Aspidosperma*) in this province (M. Pearman *in litt.* 1995).

Sporophila palustris	1987	Found at Arroyo Feliciano in January 1987 (M. Bettinelli and A. Bosso *per* J. C. Chebez *in litt.* 1995).
Gubernatrix cristata	1993	Probably one of the last strongholds in the north of the species' range (Pearman and Abadie in press).

El Palmar (Entre Ríos)

AR 30
31°49′S 58°15′W
EBA B55

El Palmar National Park (IUCN category II), 8,500 ha

This national park is south-east of Ubajay in Colón department, east-central Entre Ríos. The area is dominated by the last extensive tract of yatay palm *Butia yatay*, but also supports natural grassland, monte and gallery forest along the Río Uruguay (M. Pearman *in litt.* 1995).

Eleothreptus anomalus	1985	Recorded along the Arroyo Palmar.
Sporophila zelichi	c.1982	One record, of a wintering bird in May. No other records of any threatened *Sporophila* in spite of surveys in 1992 (M. Pearman *in litt.* 1995).
Gubernatrix cristata	1981	A family group of five near Arroyo Los Loros is the only record from the park and the species may already be extinct (or close to extinction) in the region (M. Pearman *in litt.* 1995).

Gualeguaychú (Entre Ríos)

AR 31
33°03′S 58°23′W
EBA B55

Unprotected

This area comprises Puerta Boca and Estancias Centella, San Luis and Ñandubaysal, all located within 7–10 km of Gualeguaychú in east-central Entre Ríos. Natural grasslands alternate with zones of monte, gallery forest along the rivers and streams (Arroyo Sauzal in Estancia Ñandubaysal, Arroyo Capilla in Estancias San Luis and Ñandubaysal, and Arroyo Las Piedras) and low-lying marshes dominated by *Panicum* grass.

Eleothreptus anomalus	1992	A number of sightings of at least two birds near the Río Gualeguaychú at Puerto Boca. Looked for, without success, in 1993 (Pearman and Abadie in press).
Gubernatrix cristata	1993	Reported by workers at Estancia Centella.
Sporophila hypochroma	1992	Fewer records than for *S. palustris*. First seen in December 1991 (first record for Entre Ríos), since when it has apparently increased in numbers: two males and accompanying female at Arroyo Sauzal in December 1992, a male at Arroyo Capilla in Estancia Ñandubaysal in December 1992 and a singing male mist-netted at Puerto Boca also in December 1992 (Pearman and Abadie in press, M. Pearman *in litt.* 1995).
Sporophila palustris	1994	One seen at Puerto Boca in December 1994 (J. C. Chebez *in litt.* 1995) and up to eight birds and a nest found in 1991–1992 (M. Pearman *in litt.* 1995). Additionally 18, including six pairs holding territory (plus one nest), were found at Arroyo Capilla, three birds at Arroyo Las Piedras, and seven at Estancia Ñandubaysal, all in December 1992 (Pearman and Abadie in press, M. Pearman *in litt.* 1995).
Sporophila zelichi	1992	A pair and dependent juvenile were recorded at Puerto Boca in December 1991–January 1992, and a paired male and female were found at Arroyo Sauzal in December 1992; both records possibly refer to the same birds (Pearman and Abadie in press, M. Pearman *in litt.* 1995).
Xanthopsar flavus	1992	Recorded throughout the year with flocks typically of c.50 birds, but one of 230 at Estancia San Luis in April 1992 (Pearman and Abadie in press, M. Pearman *in litt.* 1995).

Otamendi (Buenos Aires)

AR 32

34°10′S 58°57′W

Otamendi Strict Nature Reserve (IUCN category I), 2,632 ha

This reserve lies along the Río Luján (which enters Río de la Plata) in north-east Buenos Aires. The habitat is mainly open marshland and wet grasslands with some gallery forest.

Porzana spiloptera	1990s	A number of sightings, especially in the Río Luján marshes where two were seen in c.4 ha in 1989 and 1991.
Coturnicops notatus	1991	One record, of a bird seen only in flight, so confirmation of presence is desirable (M. Pearman *in litt.* 1995).
Sporophila hypochroma	1993	The recent records from this area apparently represent a range extension (M. Pearman *in litt.* 1995).
Sporophila palustris	1993	Three records, all of a male at the same locality, so the possibility of an escaped cagebird cannot be excluded as trappers are known in the area and birds are sometimes released if they do not sing (M. Pearman *in litt.* 1995).

Punta Rasa (Buenos Aires)

AR 33

36°17′S 56°47′W

Punta Rasa Biological Station (IUCN category VI), area unknown

Punta Rasa is located on Cabo San Antonio, which is a prominent cape to the south-east of the Río de la Plata estuary, c.275 km south-east of Buenos Aires. The area has large areas of saltmarsh.

Porzana spiloptera	1989	A number of records in the late 1980s.
Larus atlanticus	1988	Non-breeding birds recorded almost annually since the 1970s, the largest single count being 182 in 1988.

Bahia Blanca (Buenos Aires)

AR 34

38°20′S 62°30′W

Unprotected

This area comprises a large portion of the open grassland pampas of south-west Buenos Aires province located inland from Bahia Blanca. The majority of this large area is now agricultural, with wheat, pasture and stubble, and only some remnants of the natural grassland, and monte woodland. From this region there are historical records of *Harpyhaliaetus coronatus*, *Rallus antarcticus*, *Numenius borealis*, *Yetapa risora* and *Xanthopsar flavus* (A. Di Giacomo *in litt.* 1995).

Porzana spiloptera	1990s	Recorded recently; no information on previous status (A. Di Giacomo *in litt.* 1995).
Gubernatrix cristata	1994	Breeds especially in the Bosques de Caldénon on the western edge of the area (A. Di Giacomo *in litt.* 1995).
Sturnella militaris	1993	A major breeding area: a census in November 1993 estimated 1,137 birds around Villa Iris, 3,825 around Nova Roma and 1,705 around Napostá (Tubaro and Gabelli 1993).

Isla Brightman (Buenos Aires)

AR 35

39°24′S 62°10′W

Unprotected

A flat, sandy island in the narrow bay formed between Península Verde and Punta Laberinto, c.70 km south-east of Bahía Blanca in southern Buenos Aires province.

Larus atlanticus	1990	A census in November 1990 estimated 280–350 pairs, a significant proportion of the world population of 1,112–1,366 pairs.

Islas Jabalí and Gama (Buenos Aires)

AR 36

40°33'S 62°15'W

Bahía San Blas (IUCN category I), 7,386 ha

These flat, sandy islands are located south of Bahía Anegada and north of Punta Rasa in southern Buenos Aires. Isla Jabalí (or Península Jabalí) lies just off the mainland, with Isla Gama c.8 km north of it.

Larus atlanticus	1990	A census in November 1990 estimated 279–339 pairs on Isla Gama and 147–179 on Isla Jabalí, a significant proportion of the world population.

Lihuel Calel (La Pampa)

AR 37

38°02'S 65°33'W

Lihuel Calel National Park (IUCN category II), 9,900 ha

This national park is located in the Serranías Lihuel Calel, a low range of hills in south-central La Pampa. The habitat of the entire area is monte woodland of relatively low stature.

Harpyhaliaetus coronatus	1991	Several records in 1991 including sightings of an adult and juvenile together, suggesting that this is a key area for the species (de Lucca 1993).
Gubernatrix cristata	1993	A family group of six seen in January 1993 (M. Pearman *in litt.* 1995). The reserve is thought to be a stronghold for the species, and so far appears not to have been affected by trapping (M. Pearman *in litt.* 1995).

Lanín (Neuquén)

AR 38

39°55'S 71°25'W

Lanín National Park (IUCN category II), 200,870 ha

This major national park, located west of Junín de los Andes near the Chilean border in south-west Neuquén, is dominated by Volcán Lanín and several large lakes, principally Lagos Huechulafquén and Epulafquen. The main habitat is grassland with patches of *Nothofagus* scrub and forests of *Araucaria araucana* and *Nothofagus dombeyi*.

Harpyhaliaetus coronatus	1980s?	Apparently accidental in this area.
Asthenes anthoides	1990s	Recorded regularly (M. Pearman *in litt.* 1995).

Nahuel Haupi (Neuquén/Río Negro)

AR 39

41°09'S 71°18'W

Nahuel Haupi National Park (IUCN category II), 475,781 ha

The national park is close to the border with Chile in the Andean Cordillera of south-west Neuquén and extreme western Río Negro. It includes numerous lakes including Lago Nahuel Huapi, 560 km^2 in area. There are many tall peaks within the park, notable among which is El Tronador at 3,554 m. Much of the park is forested with *Araucaria araucana* and *Nothofagus dombeyi* (Andean rainforest with trees as tall as 50 m), but higher up on the western slopes xeric woodland and bunchgrass of the Patagonian steppe predominates. Browsing by introduced deer and livestock is having a significant effect on the vegetation, and this poses a long-term threat to the continued existence of forest cover (Veblen *et al.* 1992).

Asthenes anthoides	1980s	Found to be 'not rare' near Bariloche.

Laguna del Monte (Río Negro)

AR 40

Unprotected

40°48′S 64°30′W

This private estancia is located midway between Viedma and San Antonio Oeste in north-central Río Negro. It has rather open, dry, chaco-type scrubland.

Gubernatrix cristata	1991	A regular site since the late 1970s.

El Bolsón (Río Negro)

AR 41

Unprotected

41°58′S 71°31′W

This area is located near the border with Chile in the Andean Cordillera of western Río Negro. Surveys to find *Rallus antarcticus* need to be carried out here and perhaps also in Nahuel Haupi National Park (AR 39) immediately to the north and Los Alerces National Park (AR 42) to the south.

Rallus antarcticus	1980s	A bird found freshly dead in 1959 (Navas 1962). The same collector who found the road-kill insisted that he recorded the species with some regularity in the 1980s (R. S. Ridgely *in litt.* 1995); there are still extensive marshes in the area.

Los Alerces (Chubut)

AR 42

Los Alerces National Park (IUCN category II), 186,730 ha

43°00′S 71°45′

Named after giant conifers which reach 60 m in height and 3 m in diameter, this national park is located west of Esquel city in the Andean Cordillera of north-west Chubut, close to the border with Chile. A principal feature is the many lakes, the largest of which is Lago Futalaufquen.

Asthenes anthoides	1990s	Apparently recorded regularly (M. Pearman *in litt.* 1994).

Isla Vernaci (Chubut)

AR 43

Unprotected

45°11′S 66°30′W

A cluster of small islands (each less than 10 ha) close to and north of Península Aristizábal, on the northern side of Golfo San Jorge in south-east Chubut. Isla Vernaci and Islote Galfrascoli (located 50 km east at 45°02′S 65°52′W) are the only known breeding sites for *Larus atlanticus* away from the main colonies in southern Buenos Aires Province (AR 35, 36).

Larus atlanticus	1990s	A census in November 1990 estimated 60–80 pairs.

■ BOLIVIA

Red-fronted Macaw *Ara rubrogenys*

BOLIVIA has one of the most species-rich avifaunas of any country in the world, with more than 1,300 species recorded; 16 are endemic to the country (Remsen and Traylor 1989, Armonía 1995, J. V. Remsen *in litt.* 1995), 68 (7%) have restricted ranges (Stattersfield *et al.* in prep.) and 24 (2%) are threatened (Collar *et al.* 1992). This analysis has identified 27 Key Areas for the threatened birds in Bolivia (see '*Key Areas*: the book', p. 11, for criteria).

THREATENED BIRDS

Of the 24 Bolivian species which Collar *et al.* (1992) considered at risk of extinction, eight are confined to the country (see Table 1). There are threatened species in most habitats and at most altitudes within Bolivia, the most important for numbers of such species being the savannas, the grassland and forest mosaic of the eastern lowlands, and the wet and dry forests of the East Andes. As with other Neotropical countries, habitat loss is the major reason for 46% of the species being listed as threatened, with trade and hunting being additional threats to cracids and parrots. The distributions of threatened birds and their relationship to Endemic Bird Areas are shown in Figure 1.

KEY AREAS

The 27 Bolivian Key Areas that have been identified would, if adequately protected, help ensure the conservation of 23 (85%) of the threatened species in the country (always accepting that important new populations and areas may yet be found), including eight of the nine threatened endemics. Of these areas, 15 are important for two or more (up to four) threatened species, and these are therefore perhaps the most efficient areas currently known in which to conserve Bolivia's threatened birds (see 'Outlook', below), together representing potential sanctuaries for 16 threatened species, two-thirds of the total number. However, vital as these sites are for the conservation of Bolivian threatened species, they should not detract from the significance of the other Key Areas, as these are important for the remaining four species (one of which is confined to the country) whose habitat would otherwise be unrepresented.

From Tables 1 and 2 it can be seen that six species occur in just one Bolivian Key Area, two of these

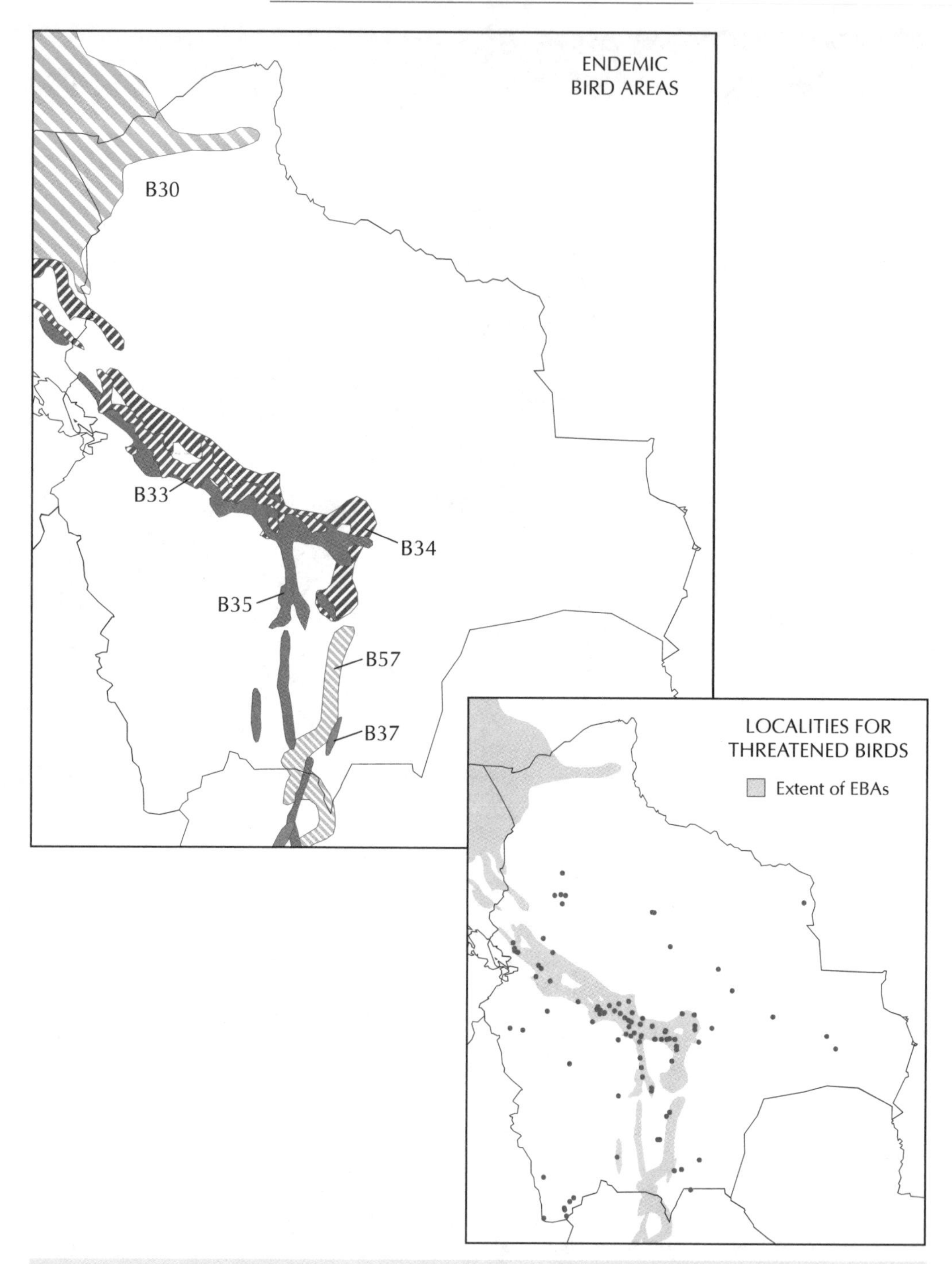

Figure 1. The localities where threatened birds have been recorded in Bolivia and their relationship to Endemic Bird Areas (EBAs).

- The most important regions for numbers of threatened birds are the humid and dry forests of the East Andes (EBAs B33, B34, B35, B37, B57), and the savanna, grassland and forest mosaic of the lowlands of Santa Cruz and El Beni in eastern Bolivia (no EBAs). As almost two-thirds of the threatened birds in Bolivia have restricted ranges, they occur together in various combinations within EBAs, which are listed below (figures are numbers of these species in each EBA).

B33 Upper Bolivian yungas (0)
B34 Lower Bolivian yungas (4)
B35 Bolivian Andes (6)
B37 North Argentine Andes (0)
B57 Boliviano-Tucuman yungas (1)

Table 1. Coverage of threatened species by Key Area. Areas in bold currently have some form of protected status.

	Key Areas occupied	No. of Key Areas protected	Total nos. of Key Areas	
			Bolivia	Neotropics
Crowned Eagle *Harpyhaliaetus coronatus*	**02**,11	1	2	18
Southern Helmeted Curassow *Pauxi unicornis*	**09**,17	1	2	4
Wattled Curassow *Crax globulosa*	—	0	0	4
Horned Coot *Fulica cornuta*	**25**	1	1	6
Hyacinth Macaw *Anodorhynchus hyacinthinus*	**05**,07	1	2	15
Blue-throated Macaw *Ara glaucogularis* [E]	03	0	1	1
Red-fronted Macaw *Ara rubrogenys* [E]	**09**,21,22,23	1	4	4
White-winged Nightjar *Caprimulgus candicans*	**02**	1	1	2
Coppery Thorntail *Popelairia letitiae* [E]	—	0	0	0
Royal Cinclodes *Cinclodes aricomae*	—	0	0	2
Berlepsch's Canastero *Asthenes berlepschi* [E]	13	0	1	1
Bolivian Recurvebill *Simoxenops striatus* [E]	**09**,14,17	1	3	3
Ashy Antwren *Myrmotherula grisea* [E]	**04**,**09**,12,14,17	2	5	5
Yellow-rumped Antwren *Terenura sharpei*	12,16	0	2	3
Ash-breasted Tit-tyrant *Anairetes alpinus*	15	0	1	7
Dinelli's Doradito *Pseudocolopteryx dinellianus*	—	0	0	5
Rufous-sided Pygmy-tyrant *Euscarthmus rufomarginatus*	01,**05**	1	2	9
White-tailed Shrike-tyrant *Agriornis andicola*	24	0	1	11
Rufous-throated Dipper *Cinclus schulzi*	26,**27**	1	2	6
Unicoloured Thrush *Turdus haplochrous* [E]	**02**,03,**04**,**05**	3	4	4
Cochabamba Mountain-finch *Poospiza garleppi* [E]	**18**,19,20	1	3	3
Rufous-rumped Seedeater *Sporophila hypochroma*	**02**,**05**,06,10	2	4	15
Black-and-tawny Seedeater *Sporophila nigrorufa*	**05**,06,08	1	3	4
Rufous-bellied Saltator *Saltator rufiventris*	**18**,19,20,24,26,**27**	2	6	8

[E] Endemic to Bolivia

species (*Ara glaucogularis* and *Asthenes berlepschi*) being endemic to the country and thus totally reliant for their survival on the integrity of habitat in their respective Key Areas (see 'Outlook', below). Four of the seven occur on their own within their respective Key Areas, emphasizing that, in these cases, targeted single-species site-based conservation is a necessity.

Three of the threatened Bolivian endemics to the lower yungas (*Pauxi unicornis*, *Simoxenops striatus* and *Myrmotherula grisea*), although known from a number of localities, have been recorded from only a single protected Key Area, Amboró National Park (BO 09). Not represented within the Bolivian Key Area analysis are four of the country's threatened species, as follows.

Crax globulosa is known from Bolivia by only three records from the lower Río Beni, with no recent sightings.

Popelairia letitiae is known just from three nineteenth-century specimens labelled simply Bolivia.

Cinclodes aricomae is known from a single specimen collected near Tilo Tilo in 1876.

Pseudocolopteryx dinellianus has been recorded only from Tarija province where two were collected in 1926.

KEY AREA PROTECTION

Only eight (30%) of Bolivia's Key Areas currently have some form of protected status, but four of these (16% of the total) are strict nature reserves, national parks or Biosphere Reserves (IUCN categories I, II or IX) and cover more than 2.25 million hectares. The 19 unprotected Key Areas (70% of the total) require attention in the form of appropriate conservation measures if the populations of their threatened species are to survive. However, even the formally protected areas remain under threat and, in many, habitat degradation continues unchecked; effective management of activities undertaken within them is thus required.

With so many of the Key Areas currently lacking any form of protection, the long-term survival of a number of threatened species must be in question. Six Bolivian threatened species are not currently recorded from any protected Key Area (*Anodorhynchus hyacinthinus*, *Ara glaucogularis*, *Asthenes berlepschi*, *Terenura sharpei*, *Agriornis andicola* and *Anairetes alpinus*), and three of these are endemic to the country (Table 1). A further six threatened Bolivian endemics or restricted-range endemics (*Pauxi unicornis*, *Ara rubrogenys*, *Simoxenops striatus*, *Cinclus schulzi*, *Poospiza garleppi* and *Sporophila nigrorufa*) are known from only one protected Key Area (Table 1) which in each case is

Table 2. Matrix of threatened species by Key Area.

	01	02	03	04	05	06	07	08	09	10	11	12	13	14	15	16	17	18	19	20	21	22	23	24	25	26	27	No. of areas
Harpyhaliaetus coronatus	–	●	–	–	–	–	–	–	–	–	●	–	–	–	–	–	–	–	–	–	–	–	–	–	–	–	–	2
Pauxi unicornis	–	–	–	–	–	–	–	–	●	–	–	–	–	–	–	–	●	–	–	–	–	–	–	–	–	–	–	2
Crax globulosa	–	–	–	–	–	–	–	–	–	–	–	–	–	–	–	–	–	–	–	–	–	–	–	–	–	–	–	0
Fulica cornuta	–	–	–	–	–	–	–	–	–	–	–	–	–	–	–	–	–	–	–	–	–	–	–	–	●	–	–	1
Anodorhynchus hyacinthinus	–	–	–	–	●	–	●	–	–	–	–	–	–	–	–	–	–	–	–	–	–	–	–	–	–	–	–	2
Ara glaucogularis	–	–	●	–	–	–	–	–	–	–	–	–	–	–	–	–	–	–	–	–	–	–	–	–	–	–	–	1
Ara rubrogenys	–	–	–	–	–	–	–	–	●	–	–	–	–	–	–	–	–	–	–	–	●	●	●	–	–	–	–	4
Caprimulgus candicans	–	●	–	–	–	–	–	–	–	–	–	–	–	–	–	–	–	–	–	–	–	–	–	–	–	–	–	1
Popelairia letitiae	–	–	–	–	–	–	–	–	–	–	–	–	–	–	–	–	–	–	–	–	–	–	–	–	–	–	–	0
Cinclodes aricomae	–	–	–	–	–	–	–	–	–	–	–	–	–	–	–	–	–	–	–	–	–	–	–	–	–	–	–	0
Asthenes berlepschi	–	–	–	–	–	–	–	–	–	–	–	–	●	–	–	–	–	–	–	–	–	–	–	–	–	–	–	1
Simoxenops striatus	–	–	–	–	–	–	–	–	●	–	–	–	–	●	–	–	●	–	–	–	–	–	–	–	–	–	–	3
Myrmotherula grisea	–	–	–	●	–	–	–	–	●	–	–	●	–	●	–	–	●	–	–	–	–	–	–	–	–	–	–	5
Terenura sharpei	–	–	–	–	–	–	–	–	–	–	–	●	–	–	–	●	–	–	–	–	–	–	–	–	–	–	–	2
Anairetes alpinus	–	–	–	–	–	–	–	–	–	–	–	–	–	–	●	–	–	–	–	–	–	–	–	–	–	–	–	1
Pseudocolopteryx dinellianus	–	–	–	–	–	–	–	–	–	–	–	–	–	–	–	–	–	–	–	–	–	–	–	–	–	–	–	0
Euscarthmus rufomarginatus	●	–	–	–	●	–	–	–	–	–	–	–	–	–	–	–	–	–	–	–	–	–	–	–	–	–	–	2
Agriornis andicola	–	–	–	–	–	–	–	–	–	–	–	–	–	–	–	–	–	–	–	–	–	–	–	●	–	–	–	1
Cinclus schulzi	–	–	–	–	–	–	–	–	–	–	–	–	–	–	–	–	–	–	–	–	–	–	–	–	–	●	●	2
Turdus haplochrous	–	●	●	●	●	–	–	–	–	–	–	–	–	–	–	–	–	–	–	–	–	–	–	–	–	–	–	4
Poospiza garleppi	–	–	–	–	–	–	–	–	–	–	–	–	–	–	–	–	–	●	●	●	–	–	–	–	–	–	–	3
Sporophila hypochroma	–	●	–	–	●	●	–	–	–	●	–	–	–	–	–	–	–	–	–	–	–	–	–	–	–	–	–	4
Sporophila nigrorufa	–	–	–	–	●	●	–	●	–	–	–	–	–	–	–	–	–	–	–	–	–	–	–	–	–	–	–	3
Saltator rufiventris	–	–	–	–	–	–	–	–	–	–	–	–	–	–	–	–	–	●	●	●	–	–	–	●	–	●	●	6
No. of species	1	4	2	2	5	2	1	1	4	1	1	2	1	2	1	1	3	2	2	2	1	1	1	2	1	2	2	

potentially of importance for the continued survival of the species in question (see 'Outlook', below).

RECENT CHANGES TO THE THREATENED LIST

With the publication of Collar *et al.* (1994), two of the 23 threatened species (*Turdus haplochrous* and *Sporophila hypochroma*) were dropped from the Bolivian threatened list, and six were added: Andean Flamingo *Phoenicopterus andinus*, Puna Flamingo *P. jamesi*, Military Macaw *Ara militaris*, Yellow-faced Amazon *Amazona xanthops*, Maquis Canastero *Asthenes heterura*, Black-masked Finch *Coryphaspiza melanotis* (but these have not been included in the Site Inventory; see '*Key Areas*: the book', p. 12).

Table 3. Top Key Areas in Bolivia. Those with the area number in bold currently have some form of protected status.

Key Area		No. of threatened spp.	Comments
02	Beni Biological Station	4	An important protected area with at least four threatened species recorded in recent years.
03	Trinidad	2	Virtually the entire range of *Ara glaucogularis* lies within this Key Area.
05	Noel Kempff Mercado	5	An important protected area, especially for *Sporophila nigrorufa*.
09	Amboró	4	The only protected area in Bolivia supporting *Pauxi unicornis* and also important for three threatened Bolivian endemics.
12	Serranía Bellavista	2	The only site where *Terenura sharpei* has been noted on more than one occasion.
13	Nevado Illampu	1	The entire range of *Asthenes berlepschi* is embraced by this Key Area.
18	Cerro Tunari and Liriuni	2	The only protected area supporting populations of *Poospiza garleppi* and also important for *Saltator rufiventris* (and *Asthenes heterura*).
25	Eduardo Avaroa	1	A stronghold for *Fulica cornuta*, with the largest concentration ever recorded (also important for the two flamingo species now added to the threatened list).
27	Tariquía	2	*Cinclus schulzi* occurs on the main rivers in higher densities than in any other Key Area.

Caprimulgus candicans, which was formerly known only from Brazil, has now been recorded from Bolivia (Davis and Flores 1994).

Although these changes represent a high percentage difference (39%), four of the six species added were previously listed as Near Threatened in Collar *et al.* (1992), with the new criteria and/or new data having promoted their upgrading.

The recent additions will have some distributional impact on the Key Area analysis, although the two flamingos and *Asthenes heterura* are covered within the existing Key Areas. Eduardo Avaroa National Faunal Reserve is the only regular breeding site for *Phoenicopterus jamesi* and in 1992–1993 supported up to 30,000 birds with 9,000 pairs breeding, as well as 1,000 pairs of *P. andinus* (Maier and Kelly 1994). *Asthenes heterura* is endemic to the arid, scrubby, *Polylepis*-dominated slopes in central Cochabamba and La Paz, and has been recorded at Cerro Tunari and Liriuni (BO 18). The status and distribution of *Amazona xanthops* and *Coryphaspiza melanotis* in Bolivia are still poorly understood and a careful revision of their current distributions would be needed to determine appropriate Key Areas for them.

OLD RECORDS AND LITTLE-KNOWN BIRDS

A number of threatened species in Bolivia suffer from a general paucity of recent records. Most species are little known somewhere within their range, but recent records from other areas mean that we are able to make informed recommendations for their conservation. The exceptions are those species known only from old specimen records such as *Crax globulosa*, *Popelairia letitiae* and *Cinclodes aricomae* for which no recent records exist.

OUTLOOK

This analysis has identified 27 Key Areas, each of which is of importance for the conservation of threatened birds in Bolivia. However, certain Key Areas stand out as top priorities (Table 3); they comprise those which each host populations of four threatened species or areas which are effectively the only site for a species confined to Bolivia, and the most important protected areas. These nine Key Areas represent populations of two-thirds of Bolivian threatened species. These and the remaining Key Areas (which are individually extremely important for many threatened Bolivian endemics: see 'Key Areas' and 'Key Area protection', above) all need some degree of conservation action if the populations of their threatened birds are to remain viable. For many areas, especially in the lower yungas, there is a requirement to confirm the continued existence or assess the populations of threatened species, and for others, especially those with no Key Areas, the need is to locate and define the extent of suitable habitat in the vicinity of past sightings.

DATA SOURCES

The above introductory text and the Site Inventory were compiled from information supplied by Armonía, J. Blincow, P. Boesman, R. C. Brace, S. Davis, A. J. Hesse, R. Innes, L. Jammes, S. Mayer, J. Pearce-Higgins, M. Pearman, J. V. Remsen, and S. J. Tyler, as well as from the following references.

ARMONÍA (1995) *Lista de las aves de Bolivia*. Santa Cruz de la Sierra: Armonía.

BRACE, R. C., HESSE, A. J. AND WHITE, A. G. (1995) The endemic macaws of Bolivia. *Cotinga* 3: 27–30.

CLARKE, R. O. S. AND DURAN PATIÑO, E. (1991) The Red-fronted Macaw (*Ara rubrogenys*) in Bolivia: distribution, abundance, biology and conservation. Buena Vista: Wildlife Conservation International and the International Council for Bird Preservation. Unpublished report.

COLLAR, N. J., GONZAGA, L. P., KRABBE, N., MADROÑO NIETO, A., NARANJO, L. G., PARKER, T. A. AND WEGE, D. C. (1992) *Threatened birds of the Americas: the ICBP/IUCN Red Data Book*. Cambridge, U.K.: International Council for Bird Preservation.

COLLAR, N. J., CROSBY, M. J. AND STATTERSFIELD, A. J. (1994) *Birds to watch 2: the world list of threatened birds*. Cambridge, U.K.: BirdLife International (BirdLife Conservation Series no. 4).

DAVIS, S. E. (1993) Seasonal status, relative abundance, and behavior of the birds of Concepción, Departamento Santa Cruz, Bolivia. *Fieldiana Zool.* n.s 71.

DAVIS, S. E. AND FLORES, E. (1994) First record of White-winged Nightjar *Caprimulgus candicans* for Bolivia. *Bull. Brit. Orn. Club* 114: 127–128.

HESSE, A. AND JAMMES, L. (1994) A preliminary assessment of the distribution of the Blue-throated Macaw *Ara glaucogularis*, Beni, Bolivia. Santa Cruz de la Sierra: Armonía. Unpublished report.

IUCN (1992) *Protected areas of the world: a review of national systems, 4: Nearctic and Neotropical*. Gland, Switzerland and Cambridge, U.K.: International Union for Conservation of Nature and Natural Resources.

JORDAN, O. C. AND MUNN, C. A. (1993) First observations of the Blue-throated Macaw in Bolivia. *Wilson Bull.* 105: 694–695.

KRATTER, K. W., SILLET, T. S., CHESSER, R. T., O'NEILL, J. P., PARKER, T. A. AND CASTILLO, A. (1993) Avifauna of a chaco locality in Bolivia. *Wilson Bull.* 105: 114–141.

LÓPEZ, N. E. (1992) Observaciones sobre la distribución de

psitacidos en el Departamento de Concepción, Paraguay. *Bol. Mus. Nac. Hist. Nat. Paraguay.* 11: 2–25.

MAIER, R. AND KELLY, A. (1994) Laguna Colorada and Eduardo Avaroa National Reserve, Bolivia. *Cotinga* 1: 36–41.

MAYER, S. (1992) Birds observed in and near the reserve of Tariquia, Depto. Tarija, Bolivia, in September/October 1992. Unpublished (birdwatching) report.

MAYER, S. (1994) Two birding sites near La Paz, Bolivia. Unpublished (birdwatching) report.

MAYER, S. (1995) Notes on the occurrence and natural history of Berlepsch's Canastero *Asthenes berlepschi. Cotinga* 3: 15–16.

MIRANDA, C. (1991) *Beni Biological Station Biosphere Reserve Management Plan.* La Paz: Artes gráficas.

PARKER, T. A., GENTRY, A. H., FOSTER, R. B., EMMONS, L. H. AND REMSEN, J. V. (1993) *The lowland dry forests of Santa Cruz, Bolivia: a global conservation priority (Rapid Assessment Program).* Washington, D.C.: Conservation International.

PAYNTER, R. A. (1992) *Ornithological gazeteer of Bolivia.* Second edition. Cambridge, Mass.: Harvard University.

PITTER, E. AND CHRISTIANSEN, M. B. (1995) Ecology, status and conservation of the Red-fronted Macaw *Ara rubrogenys. Bird Conserv. Internatn.* 5: 61–78.

REMSEN, J. V. AND TRAYLOR, M. A. (1989) *An annotated list of the birds of Bolivia.* Vermillion, South Dakota: Buteo Books.

STATTERSFIELD, A. J., CROSBY, M. J., LONG, A. J. AND WEGE, D. C. (in prep.) *Global directory of Endemic Bird Areas.* Cambridge, U.K.: BirdLife International (BirdLife Conservation Series).

TYLER, S. (in prep.) The Rufous-throated Dipper in Bolivia. *Cotinga.*

VÁSQUEZ, R. (1994) Las aves del Parque Nacional Noel Kaempff Mercado: lista de las familias, géneros y especies. Unpublished report.

WHITE, A., BRACE, R., DUFFIELD, G., HESSE, A., PAYNE, A. AND SPICK, S. (1993) Nottingham University Bolivia Project 1992: an ornithological survey of the Beni Biological Station. Unpublished report.

WHITE, A. G., BRACE, R. C. AND PAYNE, A. J. (1995) Additional records and notes on the Unicoloured Thrush *Turdus haplochrous*, a little known Bolivian endemic. *Bull. Brit. Orn. Club* 115: 29–33.

SITE INVENTORY

Riberalta (El Beni)

BO 01

11°00′S 65°50′W

Unprotected

This area comprises a long (at least 30 km) but narrow stretch of cerrado and natural grassland 35 km from Riberalta on the road to Guayamirím in extreme north-east El Beni.

Euscarthmus rufomarginatus	1994	The commonest bird (together with *Formicivora rufa*); 20 seen in only a small part of the area in April 1994 (S. Mayer *in litt.* 1994, *Cotinga* 1995, 3: 62).

Beni Biological Station (El Beni)

BO 02

14°38′S 66°18′W

Beni Biological Station (IUCN category I), 135,000 ha

This large reserve is located north-east of San Borja in south-west El Beni. The reserve mainly embraces humid forest and savanna, but other notable habitats include forested river islands, pampas grassland, swamps and lagoons.

Harpyhaliaetus coronatus	1992	Irregularly recorded; the last observation was of an immature in tall riverine forest 3 km south of Palma Flores (White *et al.* 1993).
Caprimulgus candicans	1987	One collected in September in dry open savanna near Estancia El Porvenir, representing a western range extension of c.1,500 km and the first record for Bolivia (Davis and Flores 1994).
Turdus haplochrous	1992	Population discovered in September in seasonally flooded forest along the south bank of Río Manaqui (White *et al.* 1993, 1995).
Sporophila hypochroma	1992	Apparently found throughout savanna areas of the reserve (Miranda 1991), though probably only a winter visitor. The last record was of an adult male at El Porvenir (White *et al.* 1993).

Figure 2. Key Areas in Bolivia.

01 Riberalta
02 Beni Biological Station
03 Trinidad
04 Isiboro Sécure
05 Noel Kempff Mercado
06 Concepción
07 San Fernando
08 San Ignacio de Velasco
09 Amboró
10 Santa Cruz de la Sierra
11 Estancia Perforación
12 Serranía Bellavista
13 Nevado Illampu
14 Santa Ana
15 Choquetanga valley
16 Chaparé
17 Bolívar and Palmar
18 Cerro Tunari and Liriuni
19 Quebrada Majón
20 Cerro Cheñua Sandra
21 Río Mizque valley
22 Lower Río Caine valley
23 Río Grande valley
24 Azurduy
25 Eduardo Avaroa
26 Tarija
27 Tariquía

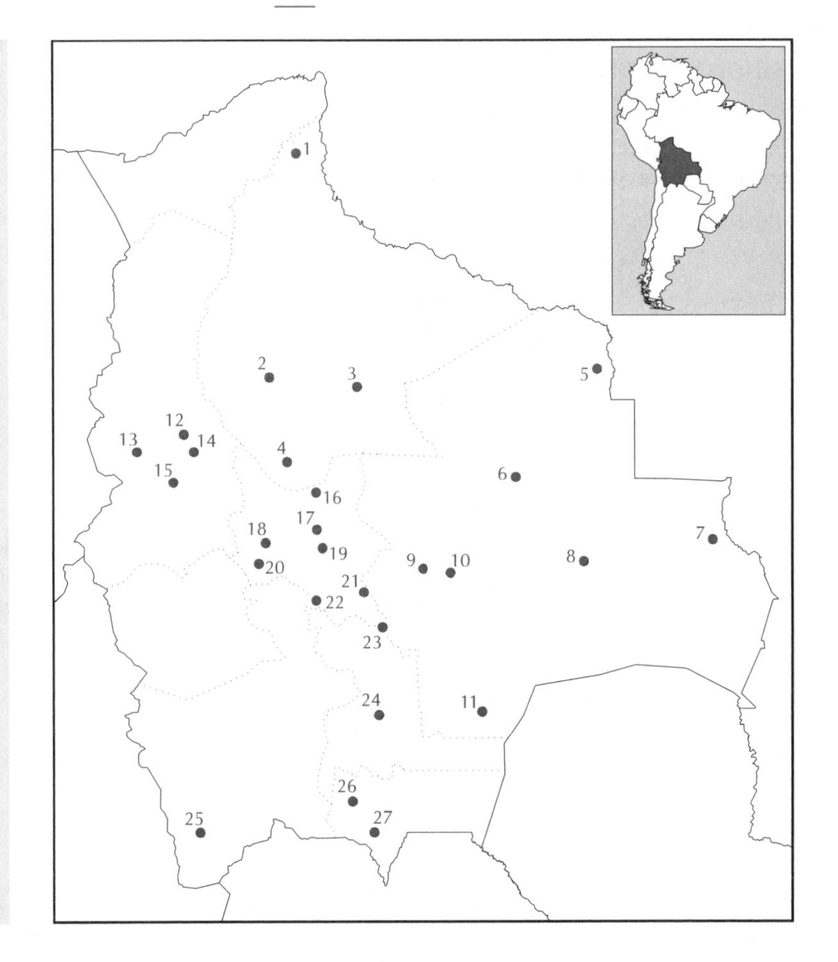

Trinidad (El Beni)

BO 03

14°47′S 64°47′W

Unprotected

The town lies on the right bank of the upper Río Mamoré in the humid lowlands of south-central El Beni. The main habitat is wooded and wetland savanna; especially important for the threatened macaw is seasonally flooded savanna with scattered islands of forest dominated by palms (*Scheelea princeps*, *Acrocomia totai*) and tall *Tabebuia impetiginusa* trees.

Ara glaucogularis	1994	Discovered here in 1992 after many years of its precise whereabouts being unknown to ornithologists (Jordan and Munn 1993). Surveys in 1993–1994 produced additional sites north-east and south of Trinidad and a population estimate of no more than 100 birds in c.15,000 km² (Hesse and Jammes 1994, Brace *et al.* 1995).
Turdus haplochrous	1984	Two birds taken 6 km south-east of Trinidad in semi-open woodland.

Isiboro Sécure (El Beni/La Paz)

BO 04

16°00'S 67°00'W
EBA B34

Isiboro Sécure National Park (IUCN category II), 1,100,000 ha

This national park, embracing humid forest and savanna, covers a considerable area of southern Beni and eastern La Paz between Río Mamoré and Río Beni.

Turdus haplochrous	1993	Two collected in the north-east section of the reserve, November 1993: one at Santa Rosa (Marbán province) and the other an hour by river from San Bernardo (Moxos province) towards Santa Rosa (S. Davis *in litt.* 1995).
Myrmotherula grisea	1981	One collected at 600 m in humid tropical forest, c.20 km upriver of Puerto Linares, Santa Ana de Huachi, in July 1981.

Noel Kempff Mercado (Santa Cruz)

BO 05

14°30'S 60°39'W

Noel Kempff Mercado National Park (IUCN category II), 914,000 ha

This extensive park in extreme north-east Santa Cruz is dominated by the Serranía de Huanchaca, a 150-km range of low mountains roughly paralleling the Brazilian border. The wide range of habitats includes tropical evergreen forest, gallery forest, pampas and seasonally flooded and savanna grassland.

Anodorhynchus hyacinthinus	c.1992	Sightings of c.5 birds near Río Itenes (record in López 1992), and listed for the park by Vásquez (1994).
Euscarthmus rufomarginatus	1989	Population found in June and October 1989 at two localities on the Serranía de Huanchaca plateau.
Turdus haplochrous	1989	One seen and tape-recorded in semi-deciduous forest near La Junta, which lies just outside the national park (White *et al.* 1995).
Sporophila hypochroma	1994	Regular at two sites at the base of the Serranía de Huanchaca. At Flor de Oro the species was regular (<10) in mixed-species flocks (with *S. nigrorufa* and *S. ruficollis*) of 30–60 birds in October–December 1994 (S. Davis *in litt.* 1995). At Los Fierros several males were found in mixed-species seedeater flocks in August–September 1994 (J. Pearce-Higgins *in litt.* 1995) but were not present in October or December 1994 (S. Davis *in litt.* 1995). Apart from a possible pair seen away from the flocks at Flor de Oro in October 1994, there is no evidence of breeding (S. Davis *in litt.* 1995), so the species probably only winters here.
Sporophila nigrorufa	1994	Regular at three sites at the base of the Serranía de Huanchaca. At Flor de Oro, found in several flocks in October–December 1994, the largest being a (single-species) group of 60–70 birds (S. Davis *in litt.* 1995); breeding begins in November and two nests were found in 1994 (S. Davis *in litt.* 1995). In the Los Fierros grasslands, seen in groups of fewer than five in March 1995 (F. Sagot *per* S. Davis *in litt.* 1995) and two adult males were among large mixed seedeater flocks in August 1994 (J. Pearce-Higgins *in litt.* 1995), but not during the breeding season (October–December) in 1993 or 1994 (S. Davis *in litt.* 1995). At Las Torres, a pair (male singing) was seen in inundated savanna in November 1994 (S. Davis *in litt.* 1995).

Concepción (Santa Cruz)

BO 06

16°15'S 62°04'W

Unprotected

The town is in the eastern lowlands of Bolivia, at 490 m, near the western edge of the Brazilian Shield, 140 km north-east of Santa Cruz de la Sierra in Ñuflo de Chávez province. The vegetation of the region is a mosaic of semi-deciduous forest, wooded savanna and savanna wetland (Davis 1993).

Sporophila hypochroma	1990s	Found in mixed *Sporophila* flocks, and, like the other species, is a non-breeding visitor (S. Davis *in litt.* 1995).
Sporophila nigrorufa	1987	A group of five males and two females/immatures in November 1986 and a male in April 1987; not seen in mixed-species *Sporophila* flocks (Davis 1993). These were the only records during 2½ years' fieldwork, so the species is probably only casual or a migrant (S. Davis *in litt.* 1995).

San Fernando (Santa Cruz)

BO 07

17°16′S 58°38′W

Unprotected

This cattle ranch is located in the northern pantanal along the Río San Fernando (Parker *et al.* 1993). Nearby habitats include cerrado, tall dry forest and gallery forest (Parker *et al.* 1993).

Anodorhynchus hyacinthinus	1991	A small but seemingly resident population (Parker *et al.* 1993).

San Ignacio de Velasco (Santa Cruz)

BO 08

17°37′S 60°53′W
EBA B35

Unprotected

The town lies at c.300 m, 30 km north-north-west of San José de Chiquitos in eastern Santa Cruz. The principal habitat is seasonally inundated savanna, wooded savanna and semi-deciduous forest.

Sporophila nigrorufa	1994	Two pairs held territories in inundated savanna by a road close to the town in December 1994 (S. Davis *in litt.* 1995).

Amboró (Santa Cruz)

BO 09

17°44′S 63°39′W
EBA B34

Amboró National Park (IUCN category II), 180,000 ha

The park is located in the foothills south of Buena Vista in central-west Santa Cruz, with its borders being Río Yapacaní to the west, Río Surutú to the east and latitude 17°51′ to the south. It holds substantial tracts of montane and lower montane rainforest and also temperate cactus woodland in arid intermontane valleys.

Pauxi unicornis	c.1992	Studied at 700 m on the upper Río Saguayo. This is the only site where it is seen regularly.
Ara rubrogenys	1980s	There is a population of c.8 pairs on the Río Zapillar. The species also breeds in small numbers in the Chañawaykho valley which lies just outside the park.
Simoxenops striatus	1989	Recorded at 700–800 m along the upper Río Saguayo in August 1989.
Myrmotherula grisea	1992	A few recorded along the upper Río Saguayo in March 1992; several seen at c.800 m in the Serranía Raigones, in the extreme south-east of Amboró.

Santa Cruz de la Sierra (Santa Cruz)

BO 10

17°48′S 63°10′W

Unprotected

The city is on the east bank of the Río Piray in the lowlands at 480 m in eastern Santa Cruz. Habitats in the area include semi-deciduous forest, wooded and open savanna, and savanna wetland.

Sporophila hypochroma	1995	Common in wetland and open savanna on the city outskirts (e.g. Viru Viru airport, Lomas de Arena and Palmasola), October–May; preliminary data indicate breeding November–March (S. Davis *in litt.* 1995).

Estancia Perforación (Santa Cruz)

BO 11

19°55′S 62°34′W

Unprotected

A cattle ranch c.130 km east of Charagua in the Chaco region of southern Santa Cruz (Parker *et al.* 1993). Dominant vegetation is low-lying dry woodland and extensively grazed chaco grassland.

Harpyhaliaetus coronatus	1991	An adult seen in remnant chaco grassland in June 1991 is one of the only Bolivian records (Kratter *et al.* 1993, Parker *et al.* 1993).

Serranía Bellavista (La Paz)

BO 12

15°33'S 67°46'W
EBA B34

Bella Vista Protection Forest Reserve (IUCN category VIII), 90,000 ha

This area lies in the eastern foothill yungas of the Bolivian Andes (1,100–1,600 m), located 35–47 km by road north of Caranavi in east-central La Paz. The areas falls within the humid upper tropical zone, receiving sufficient rainfall for montane forest and cloud-forest.

Myrmotherula grisea	1979	Two males and a female were mist-netted at 1,650 m in lower cloud-forest during June.
Terenura sharpei	1980	Noted daily in small numbers from 10 June to 2 July at 1,650 m in lower cloud-forest; also at 1,350 m (four sightings of pairs in July).

Nevado Illampu (La Paz)

BO 13

15°50'S 68°34'W
EBA B35

Unprotected

This area is the montane basin of the upper Mapiri valley at the northern end of the Cordillera Real, which forms the semi-arid slopes and inter-montane valleys of the Nevado Illampu. A number of sites near to the Nevado are included, namely: the area around Sorata, notably Río Challa Suyu (immediately north-east of the town) and Cotaña (15°49'S 68°37'W); Chilcani (c.15°44'S 68°40'W), in the San Cristóbal valley, north-west of the Nevado; Tacacoma (15°35'S 68°43'W), 30 km north-west of the Nevado.

Asthenes berlepschi	1992	Endemic to the semi-arid slopes and inter-montane valleys of Nevado Illampu. Several birds and three nests found in December 1991 at Río Challa Suyu at 2,800 m (Mayer 1995).

Santa Ana (La Paz)

BO 14

15°50'S 67°36'W
EBA B34

Unprotected

Santa Ana is located in the eastern foothills of the Bolivian Andes at 670 m in east-central La Paz, on the upper Río Coroico c.8 km south of Canevari.

Simoxenops striatus	1934	Known here only from specimens collected by Carriker in July and August, who considered it rare.
Myrmotherula grisea	1934	Specimens (the type-series) collected in July. This remains one of the only sites for the species.

Choquetanga valley (La Paz)

BO 15

16°20'S 67°57'W
EBA B35

Unprotected

The valley, immediately north of the paved highway between La Paz and Cotapata near the small village of Ponga, contains numerous patches (all less than 100 m^2) of low *Polylepis* and *Gynoxys* scrub among predominantly boulder and scree slopes.

Anairetes alpinus	1993	Two recent sightings (Mayer 1994), the first in Bolivia since 1935, but several observers failed to find it in 1994 (J. Blincow *in litt.* 1994).

Chaparé (Cochabamba)

BO 16

16°30′S 65°30′W
EBA B34

Unprotected

Chaparé is a province encompassing part of the yungas and lowlands in north-east Cochabamba. The humid upper tropical zone forest at 1,100–1,650 m is the key habitat and is threatened by accelerating clearance, especially for cultivation of coffee, coca, tea and citrus fruit.

Terenura sharpei	1979	One sight record.

Bolívar and Palmar (Cochabamba)

BO 17

17°06′S 65°29′W
EBA B34

Unprotected

This area is located in the Yungas de Cochabamba north-east of Cochabamba City. The area is an old collecting locality with dense, humid forest and much undergrowth, moss and epiphytes.

Pauxi unicornis	1937	The type-locality, which is described as 'in the hills above Bolívar, near Palmar'.
Simoxenops striatus	1935	Collected at 800 m.
Myrmotherula grisea	1937	A female collected at 800 m in July.

Cerro Tunari and Liriuni (Cochabamba)

BO 18

17°19′S 66°22′W
EBA B35

Tunari National Park (IUCN category VIII), 6,000 ha

This area is located above 3,000 m, c.10 km north of Quillacollo and 20-25 km north-east of Cochabamba City. There are some reasonable patches of *Polylepis* woodland which is the key habitat for the threatened species found here. A park ranger claimed that *Anairetes alpinus* occurs, but there is no confirmation.

Poospiza garleppi	1994	Recorded above Liriuni towards Cerro Tunari, and north of Quillacollo; quite common in the main valley north of Quillacollo, with up to 20 seen in December 1994 (P. Boesman *in litt.* 1995).
Saltator rufiventris	1994	Uncommon in the main valley north of Quillacollo, with up to six seen daily during a short visit in December 1994 (P. Boesman *in litt.* 1995). Past records include as many as 50 birds going to roost in Tunari National Park.

Quebrada Majón (Cochabamba)

BO 19

17°24′S 65°23′W
EBA B35

Unprotected

This small inter-Andean valley is located near Quehuiñapampa, 6.6 km by road beyond López Mendoza, and 98 km from Cochabamba along the Santa Cruz road. The area comprises steep, often rocky slopes with patchy *Polylepis* woodland especially in the more sheltered and wetter slopes facing south and south-west.

Poospiza garleppi	1984	Noted on 10 of 16 field days in May and August.
Saltator rufiventris	1984	Seen on 10 of 16 days field days in May and August, and thought fairly common.

Cerro Cheñua Sandra (Cochabamba)

BO 20

Unprotected

17°39′S 66°29′W
EBA B35

Located alongside Cerro Pararani, 70 km from Cochabamba on the Oruro road, this area holds watered ravines with scattered trees such as *Polylepis* and a variety of dense, thorny bushes providing ideal habitat for the two threatened species found there.

Poospiza garleppi	1980s	Two specimens.
Saltator rufiventris	1991	Two or three pairs found in a *Polylepis* woodland of c.4 ha at Cerro Cheñua Sandra in April 1987.

Río Mizque valley (Cochabamba)

BO 21

Unprotected

18°07′S 64°40′W
EBA B35

This area embraces the middle Mizque valley from Tin-Tin (18°01′S 64°25′W) to its confluence with the Río Grande (18°40′S 65°21′W) in south-east Cochabamba. The habitat holds both tropical and temperate cactus woodland, but much of the valley is used for agriculture. Clarke and Duran Patiño (1991) recommend that the lower and less populated part of the valley from Saipina to its confluence with the Río Grande (BO 23) be declared a wildlife sanctuary to protect the threatened parrot.

Ara rubrogenys	1991	Pitter and Christiansen (1995) found groups in December 1991 near Saipina and Vallegrande; local people reported it from several other localities.

Lower Río Caine valley (Cochabamba/Potosí)

BO 22

Unprotected

18°15′S 65°30′W
EBA B35

The semi-arid intermontane valley of the lower Río Caine from east of Torotoro (18°07′S 65°46′W) to its confluence with the Río Grande (18°24′S 65°21′W) forms this Key Area on the Cochabamba–Potosí border. There is excellent *Schinopsis* forest, but extensive areas adjacent to the river are under cultivation.

Ara rubrogenys	1994	A group of 80–100 has been studied for several years around Sucusuma, with 80 seen there in August 1993 (Brace *et al.* 1995, Pitter and Christiansen 1995).

Río Grande valley (Cochabamba/Chuquisaca)

BO 23

Unprotected

18°40′S 64°20′W
EBA B35

This area covers the Río Grande from north of Sucre to El Oro and holds temperate cactus woodland. Clarke and Duran Patiño (1991) suggest that this valley and the lower Río Mizque (BO 21) should be declared a wildlife sanctuary to protect the threatened parrot.

Ara rubrogenys	1991	Group of 35 seen at Puente Acre, and smaller groups seen further downriver (Pitter and Christiansen 1995).

Azurduy (Chuquisaca)

BO 24

Unprotected

20°06′S 64°25′W
EBA B35

The area is located in the Cordillera Central of the Bolivian Andes in the range running immediately north of the Pilcomayo valley in west-central Chuquisaca.

Agriornis andicola	1991	One seen in October at 2,700 m.
Saltator rufiventris	1991	A pair and a juvenile seen in October 1991.

Eduardo Avaroa (Potosí)

BO 25

Eduardo Avaroa National Faunal Reserve (IUCN category IV), 714,000 ha

22°00′S 67°30′W

This reserve covers an enormous area across the Bolivian altiplano of southern Potosí, its southern and western boundaries being the Bolivia–Chile frontier and its eastern boundary the Bolivia–Argentina frontier. The area includes a large number of mainly saline lakes, bogs and fast-flowing acidic rivers.

Fulica cornuta	1993	A stronghold in Bolivia, with important breeding and feeding grounds on several small lagoons in the reserve, namely Lagunas Khastor, Chojllas, Totoral, Catalcito and Pelada. Probably the largest concentration ever recorded anywhere was 2,800 at Laguna Pelada in November 1982.

Tarija (Tarija)

BO 26

Unprotected

21°30′S 64°52′W
EBA B57

Just west of Tarija town in western Tarija, and running south for c.80 km from Serranía Sama and Cerro Condor to Cerro Negro, is a prominent Andean ridge with many peaks of over 4,000 m. The climate is rather arid, but there is *Polylepis* and *Alnus* woodland on the eastern slope, especially along streams.

Cinclus schulzi	1994	Found on several streams in 1994, all of which are tributaries of the Guidalquivir (Tyler in prep.).
Saltator rufiventris	1994	Seen in several gorges during river surveys (Tyler in prep.).

Tariquía (Tarija)

BO 27

Tariquía National Reserve (IUCN category IV), 246,870 ha

22°00′S 64°30′W
EBA B57

A very large national park lying on the humid eastern slopes of the central valley of Tarija, located between the towns of Padcaya and Bermejo in southern Tarija, close to the border with Argentina. Moving down an elevational gradient the principal vegetation types are montane grassland, *Polylepis* woodland and broadleaf forest dominated by *Alnus*.

Cinclus schulzi	1992	Found on several streams in 1992, notably Ríos Escalera, Achirales and Lorayo, where it was common (Mayer 1992).
Saltator rufiventris	1992	A single record in a narrow, *Polylepis*-clad valley (Mayer 1992).

■ BRAZIL

Black-fronted Piping-guan *Pipile jacutinga*

BRAZIL harbours one of the most species-rich avifaunas of any country in the world, with 1,635 species recorded, of which 143 (9%) are migrants, 177 (11%) are endemic to the country (Sick 1993), 187 (12%) have restricted ranges (Stattersfield *et al.* in prep.) and 96 (6%) were listed as threatened by Collar *et al.* (1992; but see 'Recent changes to the threatened list', below). This analysis has identified 142 Key Areas for Brazil's threatened bird species (see '*Key Areas*: the book', p. 11, for criteria).

THREATENED BIRDS

Brazil holds much the highest number of threatened bird species of any country in the Americas: of the total of 96 such species, 61 (64%) are confined to the country (see Table 1). Thus nearly 19% of the threatened birds of the Americas are endemic to Brazil and almost 30% of all the continent's threatened birds occur there. They rely primarily on wet forests (58% of species), but also important are dry forests (17% of species) and grasslands (19%). Two-thirds of the species are confined to lowland, with a further 22% having lowland and submontane distributions. Habitat loss is the main reason for 83% of the species being listed as threatened and is the only reason for 55%, with trade and hunting being additional threats, especially to cracids and parrots. The distributions of threatened birds and their relationship to Endemic Bird Areas are shown in Figure 1.

KEY AREAS

The 142 Key Areas that have been identified in Brazil would, if adequately protected, help to ensure the conservation of 88 (92%) of the country's threatened species (always accepting that important new populations and areas may yet be found).

Of these areas, 89 are important for two or more (up to 18) threatened species, and these are therefore perhaps the most efficient areas currently known in which to conserve Brazil's threatened birds (see 'Outlook', below). A closer look at the distribution patterns reveals that the 39 Key Areas which each harbour five or more threatened species together represent potential sanctuaries for 59 threatened species (43% of the total) and that the 47 areas supporting four or more species together hold 67

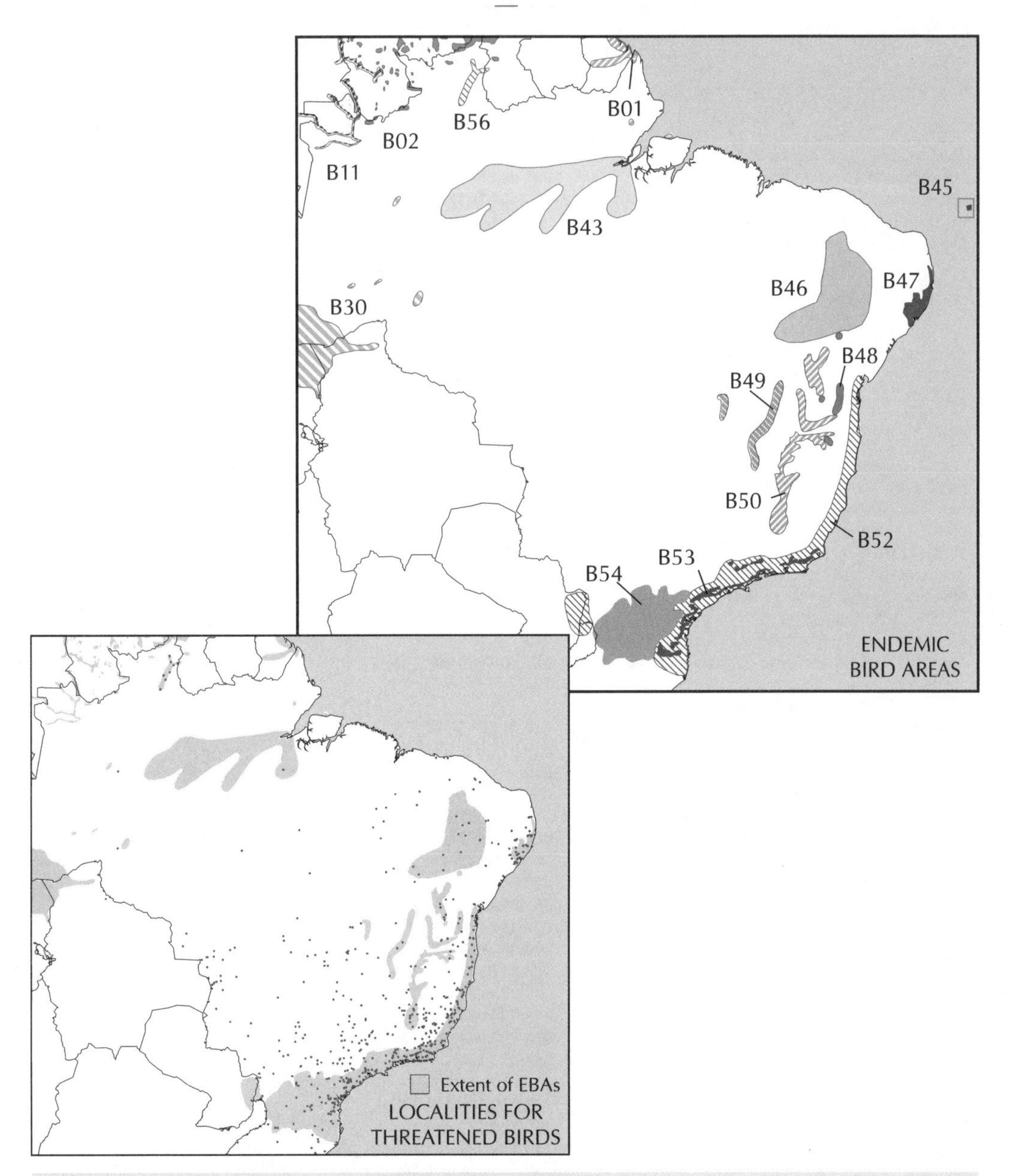

Figure 1. The localities where threatened birds have been recorded in Brazil and their relationship to Endemic Bird Areas (EBAs).

- The threatened species are not distributed evenly through Brazil: prominent concentrations are found in the Atlantic coastal forests (55 species; in EBAs B47, B52, B53), the interior dry forests and savannas (15 species; partly in B46, B48, B49, B50) and the inland wet grasslands (15 species; no EBAs), with very few threatened species occurring elsewhere; only six species are confined to the Amazon basin, for example. As almost two-thirds of Brazil's threatened birds have restricted ranges, they occur together in various combinations within EBAs, which are listed below (figures are numbers of these species in each EBA).

B11 Upper Rio Negro and Orinoco white-sand forest (0)
B19 Napo and upper Amazon lowlands (0)
B30 South-east Peruvian lowlands (0)
B43 Central Amazonian Brazil (2)
B45 Fernando de Noronha (0) *
B46 North-east Brazilian caatinga (2)
B47 Alagoan Atlantic slope (12)
B48 Bahian deciduous forests (2)
B49 Minas Gerais deciduous forests (2)
B50 Serra do Espinaço (1)
B52 South-east Brazilian lowland and foothills (29)
B53 South-east Brazilian mountains (1)
B54 South-east Brazilian *Araucaria* forest (2)
B56 Upper Rio Branco (2)

* See 'Recent changes to the threatened list'

Table 1. Coverage of threatened species by Key Area. Areas in bold currently have some form of protected status.

	Key Areas occupied	No. of Key Areas protected	Total nos. of Key Areas	
			Brazil	Neotropics
Lesser Nothura *Nothura minor* [E]	**061**,**062**,**096**,098,**104**	4	5	5
Dwarf Tinamou *Taoniscus nanus*	060,**062**,**104**	2	3	3
Brazilian Merganser *Mergus octosetaceus*	**058**,**061**,**072**	3	3	7
White-necked Hawk *Leucopternis lacernulata* [E]	**036**,**056**,**066**,**068**,**074**,**076**,**077**,**078**,**084**,**088**,**089**,**092**,**102**,**103**,**106**,**113**,**114**,**115**,**116**,**120**,**125**,128	21	22	22
Crowned Eagle *Harpyhaliaetus coronatus*	**058**,**061**,**062**,**090**,**100**	5	5	18
Chestnut-bellied Guan *Penelope ochrogaster* [E]	017,**018**,**057**	2	3	3
Black-fronted Piping-guan *Pipile jacutinga*	**056**,**068**,**074**,**086**,**088**,**100**,**105**,**106**,**107**,**108**,**109**,**110**,**112**,**114**,**116**,**120**,**126**	17	17	29
Alagoas Curassow *Mitu mitu* [E]	037	0	1	1
Red-billed Curassow *Crax blumenbachii* [E]	**053**,**056**,**068**,**074**,**076**,**077**	6	6	6
Wattled Curassow *Crax globulosa*	**006**,007	1	2	4
Rufous-faced Crake *Laterallus xenopterus*	**062**	1	1	3
Speckled Crake *Coturnicops notatus*	101	0	1	8
Olrog's Gull *Larus atlanticus*	—	0	0	5
Blue-eyed Ground-dove *Columbina cyanopis* [E]	**016**	1	1	1
Purple-winged Ground-dove *Claravis godefrida*	**078**,**086**,**088**,**103**,**106**,**126**	6	6	9
Glaucous Macaw *Anodorhynchus glaucus*	—	0	0	0
Hyacinth Macaw *Anodorhynchus hyacinthinus*	**010**,**011**,015,017,019,020,**023**,025,**057**,059	4	10	15
Lear's Macaw *Anodorhynchus leari* [E]	**040**	1	1	1
Spix's Macaw *Cyanopsitta spixii* [E]	039	0	1	1
Golden-capped Parakeet *Aratinga auricapilla* [E]	042,**044**,051,**056**,063,**067**,**068**,**072**	5	8	8
Golden Parakeet *Guaruba guarouba* [E]	**009**,**013**,**022**	3	3	3
Blue-chested Parakeet *Pyrrhura cruentata* [E]	**044**,051,**054**,**055**,**056**,**067**,**068**,**074**,**075**,**076**,**077**,**084**,087	11	13	13
Brown-backed Parrotlet *Touit melanonota* [E]	**084**,**086**,**088**,**092**,**095**,**103**,**111**,**113**,**115**,**116**	10	10	10
Golden-tailed Parrotlet *Touit surda* [E]	029,**035**,**036**,037,046,051,**054**,**055**,**056**,**074**,**076**,**078**,**084**,**086**,087,**088**,099,**106**,**107**,**109**,**114**,**116**	16	22	22
Red-tailed Amazon *Amazona brasiliensis* [E]	**111**,**112**,**113**,**114**,**115**,**116**,**120**,**123**,**124**	9	9	9
Red-spectacled Amazon *Amazona pretrei*	**132**,**133**,**134**,135,**136**,**137**,**138**,**140**,141,142	7	10	14
Red-browed Amazon *Amazona rhodocorytha* [E]	037,**055**,**056**,**068**,**073**,**074**,**075**,**076**,**077**,**078**,**084**,087,**100**	11	13	13
Vinaceous Amazon *Amazona vinacea*	063,064,**067**,**073**,**097**,099,**109**,**110**,**114**,117,**118**,119,121,127,129,**131**,**132**,**133**,**134**,135,**136**,**137**,**138**	14	23	35
Blue-bellied Parrot *Triclaria malachitacea*	**076**,**078**,**080**,**084**,**090**,**100**,**106**,**108**,**109**,**110**,**112**,**114**,**116**,**137**,139	14	15	17
White-winged Nightjar *Caprimulgus candicans*	**061**	1	1	2
Sickle-winged Nightjar *Eleothreptus anomalus*	**062**,098,**104**,**122**	3	4	10
Hook-billed Hermit *Glaucis dohrnii* [E]	**055**,**056**,**074**,**077**	4	4	4
Three-toed Jacamar *Jacamaralcyon tridactyla* [E]	**067**,**068**,081,082,083,085	2	6	6
Helmeted Woodpecker *Dryocopus galeatus*	**108**,**109**,**116**,**126**	4	4	15

cont.

Table 1. (cont.)

	Key Areas occupied	No. of Key Areas protected	Total nos. of Key Areas	
			Brazil	Neotropics
Moustached Woodcreeper *Xiphocolaptes falcirostris* [E]	025,**026**,047,063,064	10	50	50
Chestnut-throated Spinetail *Synallaxis cherriei*	**010**,**011**,015	2	3	4
Plain Spinetail *Synallaxis infuscata* [E]	**032**,**034**,**035**,**036**	4	4	4
Hoary-throated Spinetail *Synallaxis kollari*	**001**,002	1	2	3
Cipó Canastero *Asthenes luizae* [E]	**066**	1	1	1
Striated Softtail *Thripophaga macroura* [E]	046,051,**076**,**084**	2	4	4
Alagoas Foliage-gleaner *Philydor novaesi* [E]	**035**	1	1	1
Great Xenops *Megaxenops parnaguae* [E]	**024**,025,**028**,**033**,042, **044**,047,049,059,063	4	10	10
White-bearded Antshrike *Biatas nigropectus*	**084**,**086**,**088**,**109**,**126**	5	5	8
Rondônia Bushbird *Clytoctantes atrogularis* [E]	014	0	1	1
Plumbeous Antvireo *Dysithamnus plumbeus* [E]	**067**,**068**,**076**,**077**,**078**	5	5	5
Rio de Janeiro Antwren *Myrmotherula fluminensis* [E]	—	0	0	0
Alagoas Antwren *Myrmotherula snowi* [E]	**035**	1	1	1
Pectoral Antwren *Herpsilochmus pectoralis* [E]	021,**038**,041,**044**	2	4	4
Black-hooded Antwren *Formicivora erythronotos* [E]	094	0	1	1
Narrow-billed Antwren *Formicivora iheringi* [E]	048,050,065	0	3	3
Restinga Antwren *Formicivora littoralis* [E]	**091**,**093**	2	2	2
Orange-bellied Antwren *Terenura sicki* [E]	030,**035**,**036**	2	3	3
Rio Branco Antbird *Cercomacra carbonaria*	003,004	0	2	3
Fringe-backed Fire-eye *Pyriglena atra* [E]	043	0	1	1
Slender Antbird *Rhopornis ardesiaca* [E]	048,050	0	2	2
Scalloped Antbird *Myrmeciza ruficauda* [E]	**031**,**032**,**034**,**035**,**036**, **075**,**076**	7	7	7
Stresemann's Bristlefront *Merulaxis stresemanni* [E]	052	0	1	1
Brasília Tapaculo *Scytalopus novacapitalis* [E]	**062**,**066**,**071**,**072**	4	4	4
Bahia Tapaculo *Scytalopus psychopompus* [E]	045,052	0	2	2
Shrike-like Cotinga *Laniisoma elegans* [E]	051,**067**,**071**,**078**,**084**, **086**,087,**090**,**092**,**100**, **103**,**106**,**107**,**116**	12	14	14
Grey-winged Cotinga *Tijuca condita* [E]	**088**,**090**	2	2	2
Black-headed Berryeater *Carpornis melanocephalus* [E]	**035**,051,**055**,**056**,**074**, **076**,**077**,**080**,**106**,**108**, **109**,**111**,**112**,**113**,**114**, **116**,**120**,**123**	17	18	18
Buff-throated Purpletuft *Iodopleura pipra* [E]	029,**035**,051,**084**,**086**, **088**,094,**102**,**103**,**113**	7	10	10
Kinglet Calyptura *Calyptura cristata* [E]	—	0	0	0
Cinnamon-vented Piha *Lipaugus lanioides* [E]	051,**067**,**068**,**071**,**076**, **078**,**080**,**086**,**088**,**102**, **109**,**110**,**114**,**120**	13	14	14
Banded Cotinga *Cotinga maculata* [E]	**054**,**055**,**056**,**068**,**074**, **076**,**077**	7	7	7
White-winged Cotinga *Xipholena atropurpurea* [E]	029,**031**,**035**,**036**,037, **053**,**054**,**055**,**056**,**074**, **076**,**077**,**084**	11	13	13
Golden-crowned Manakin *Pipra vilasboasi* [E]	012	0	1	1
Black-capped Manakin *Piprites pileatus*	**086**,**097**,**100**,**137**	4	4	5
Rufous-sided Pygmy-tyrant *Euscarthmus rufomarginatus*	005,012,**016**,**062**,**069**, **096**	4	6	10
Long-tailed Tyrannulet *Phylloscartes ceciliae* [E]	**035**,**036**	2	2	2
São Paulo Tyrannulet *Phylloscartes paulistus*	**084**,**089**,**102**,**103**,**106**, **108**,**109**,**114**,**116**,**126**, 128	10	11	17
Minas Gerais Tyrannulet *Phylloscartes roquettei* [E]	064	0	1	1
Fork-tailed Pygmy-tyrant *Hemitriccus furcatus* [E]	051,**084**,**086**,**102**,**103**, **105**,**106**	6	7	7
Buff-breasted Tody-tyrant *Hemitriccus mirandae* [E]	**026**,027,**036**	2	3	3
Kaempfer's Tody-tyrant *Hemitriccus kaempferi* [E]	128,130	0	2	2
Buff-cheeked Tody-flycatcher *Todirostrum senex* [E]	008	0	1	1
Russet-winged Spadebill *Platyrinchus leucoryphus*	**078**,**084**,**092**,094,**102**, **103**,**108**,**109**,**110**,**112**, **116**,**120**,**125**,**126**,128	13	15	19

cont.

Table 1. (cont.)

	Key Areas occupied	No. of Key Areas protected	Total nos. of Key Areas: Brazil	Neotropics
Strange-tailed Tyrant *Yetapa risora*	—	0	0	9
Ochre-breasted Pipit *Anthus nattereri*	098,**134**,**138**	2	3	6
Cinereous Warbling-finch *Poospiza cinerea* [E]	**061**,**062**,**066**,**069**,**070**	5	5	5
Temminck's Seedeater *Sporophila falcirostris*	**053**,**084**,**086**,**090**,**102**, **103**,**111**,**112**,**113**,**115**, **116**	11	11	14
Buffy-throated Seedeater *Sporophila frontalis*	**078**,**084**,**086**,**088**,**089**, **092**,**102**,**103**,**105**,**106**, **107**,**109**,**111**,**112**,**113**, **114**,**115**,**116**,**123**,129	19	20	22
Rufous-rumped Seedeater *Sporophila hypochroma*	019,**061**	1	2	15
Hooded Seedeater *Sporophila melanops* [E]	—	0	0	0
Black-and-tawny Seedeater *Sporophila nigrorufa*	019	0	1	4
Marsh Seedeater *Sporophila palustris*	017,019,**061**	1	3	15
Yellow Cardinal *Gubernatrix cristata*	—	0	0	11
Cone-billed Tanager *Conothraupis mesoleuca* [E]	—	0	0	0
Cherry-throated Tanager *Nemosia rourei* [E]	079	1	1	1
Seven-coloured Tanager *Tangara fastuosa* [E]	**031**,**032**,**034**,**035**,**036**, 037	5	6	6
Black-backed Tanager *Tangara peruviana* [E]	**084**,**091**,**092**,**093**,094, **105**,**106**,**107**,**112**,**113**, **115**,**116**	11	12	12
Black-legged Dacnis *Dacnis nigripes* [E]	**084**,**086**,087,**088**,**090**, **092**,094,**102**,**109**,**110**, **115**,**116**,129	10	13	13
Saffron-cowled Blackbird *Xanthopsar flavus*	**137**,**138**	2	2	12
Pampas Meadowlark *Sturnella militaris*	128	0	1	4
Forbes's Blackbird *Curaeus forbesi* [E]	**035**,**036**,**068**	3	3	3
Yellow-faced Siskin *Carduelis yarrellii*	**024**,029,**031**,**033**,**034**, **035**,036	6	7	8

species (50% of the total). However, as vital as these areas are for the conservation of Brazilian threatened species, they should not detract from the significance of Key Areas supporting no more than three species, as these are of importance for the remaining 22 species, 17 of which are confined to the country.

The Atlantic coastal forests have the highest concentration of threatened birds in the Americas and many of the species are known from a large number of sites. There are 21 species confined to this biome and found there in 10 or more Key Areas, a notably higher number of Key Areas than for many other threatened birds in the Neotropics, probably reflecting a combination of factors peculiar to the Atlantic forests. Within the highly fragmented forests many of the Key Areas cover a small area (less than 1,000 ha) in which some of the threatened species occur seasonally or in only small numbers and are now very vulnerable to hunting or capture for trade. For instance, eight of the 21 species found in 10 or more Key Areas are parrots, all of which either occur at low density, are nomadic, or are captured for the bird trade; five are cotingas which occur only at low densities within each forest fragment; and other species include *Leucopternis lacernulata* (21 Key Areas), a low-density forest raptor, *Pipile jacutinga* (17 Key Areas) which has been hunted close to extinction in many of the forests, and two *Sporophila* species which are nomadic (following flowering bamboo) and so are present in many areas only infrequently.

There are many threatened species in Brazil which occur in only a handful of Key Areas and 19 species occur in just one Brazilian Key Area, 15 of which (*Mitu mitu*, *Columbina cyanopis*, *Anodorhynchus leari*, *Cyanopsitta spixii*, *Asthenes luizae*, *Philydor novaesi*, *Clytoctantes atrogularis*, *Formicivora erythronotos*, *Myrmotherula snowi*, *Pyriglena atra*, *Merulaxis stresemanni*, *Pipra vilasboasi*, *Phylloscartes roquettei*, *Todirostrum senex* and *Nemosia rourei*) are endemic to the country and thus totally reliant for their survival on the integrity of habitat in their respective Key Areas (see Tables 1 and 3–6, and 'Outlook', below). Of these species, five occur on their own within their respective Key Areas, emphasizing that in these cases targeted single-species site-based conservation is a necessity.

Not represented within the Brazilian Key Area analysis are nine of the country's threatened species, as follows.

Larus atlanticus is a rare winter visitor to coastal Rio Grande do Sul.

Anodorhynchus glaucus was found from Paraná state southwards but there have been no twentieth-century sightings, and the species is now considered extinct (Collar *et al.* 1994).

Myrmotherula fluminensis is known from just one individual netted in July 1982 in a highly degraded woodlot near Santo Aleixo, Majé, in Rio de Janeiro state. The discovery of this bird after seven years of research at the site was taken to indicate that it was a straggler there (Collar *et al.* 1992).

Calyptura cristata was apparently not uncommon in the nineteenth century within its tiny range in the foothills north of Rio de Janeiro, but has not been seen in the twentieth century.

Yetapa risora is an austral migrant in Brazil and known only from a handful of sightings, the last being in 1914.

Gubernatrix cristata is a scarce resident in south-east Rio Grande do Sul (Belton 1984–1985) but there is no information on its current status in Brazil.

Sporophila melanops is known from a single specimen taken near Registro do Araguaia in extreme west-central Goiás in October 1823.

Conothraupis mesoleuca is known from just one specimen collected in 1938, apparently from dry forest in Mato Grosso state.

KEY AREA PROTECTION

Some 85 (60%) of Brazil's Key Areas currently have some form of protected status (i.e. as national or state parks, ecological, biological, forestry or private reserves, or environmental protection or indigenous areas); 48 Key Areas (a third of the total) are biological reserves or national parks (IUCN categories I or II). The 57 unprotected Key Areas (40% of the total) require attention in the form of appropriate conservation measures if the populations of their threatened species are to survive. However, even the formally protected areas remain under threat and, in many, habitat degradation and uncontrolled hunting continue unchecked; effective management of activities undertaken within them is thus required.

With so many Key Areas lacking any form of protection, 16 threatened species are not currently known from any protected Key Area (Table 1), and 12 of these are endemic to Brazil: *Cyanopsitta spixii*, *Clytoctantes atrogularis*, *Formicivora erythronotos*, *Formicivora iheringi*, *Pyriglena atra*, *Rhopornis ardesiaca*, *Merulaxis stresemanni*, *Pipra vilasboasi*, *Hemitriccus kaempferi*, *Phylloscartes roquettei*, *Todirostrum senex* and *Nemosia rourei*. For *F. erythronotos* and *P. atra*, both known from single highly degraded woodlots, the benefits of establishing protected areas might be too late, with remedial actions such as habitat creation being their only hope.

RECENT CHANGES TO THE THREATENED LIST

With the publication of Collar *et al.* (1994), five of the 96 threatened species (*Eleothreptus anomalus*, *Synallaxis cherriei*, *Todirostrum senex*, *Poospiza cinerea* and *Sporophila hypochroma*) were dropped from the Brazilian threatened list, and the following 13 were added: Yellow-faced Amazon *Amazona xanthops*, Blue-winged Macaw *Ara maracana*, Tawny Piculet *Picumnus fulvescens*, Ochraceous Piculet *P. limae*, Red-shouldered Spinetail *Gyalophylax hellmayri*, Salvadori's Antwren *Myrmotherula minor*, Unicoloured Antwren *M. unicolor*, Band-tailed Antwren *M. urosticta*, Black-and-white Monjita *Heteroxolmis dominicana*, Atlantic Royal Flycatcher *Onychorhynchus swainsoni*, Restinga Tyrannulet *Phylloscartes kronei*, Black-masked Finch *Coryphaspiza melanotis* and Noronha Vireo *Vireo gracilirostris* (but these have not been included within the Site Inventory; see '*Key Areas*: the book', p. 12). Additionally, *Micrastur buckleyi* was listed as threatened in Collar *et al.* (1992) and Brazil was included in its range based on two specimens from Acre, but these have now been identified as *M. gilvicollis*.

These changes represent a 17% difference in the composition of the threatened species list, reflecting the large number of species listed as Near Threatened in Collar *et al.* (1992), for many of which the new criteria and/or new data have led to their upgrading. New data have shown, for example, that *Gyalo phylax hellmayri* and three species of *Myrmotherula* (*M. minor*, *M. unicolor* and *M. urosticta*) now occur in only a handful of localities within their formerly wider ranges (Whitney and Pacheco 1994, 1995). Taxonomic changes are reflected in two additions, namely the tentative treatment of *Onychorhynchus swainsoni* as a full species and the newly described *Phylloscartes kronei* (Willis and Oniki 1992).

The recent additions to the Brazilian threatened species list will have some distributional impact on the Key Area analysis. *Vireo gracilirostris* is endemic to Fernando de Noronha and does not overlap in distribution with other threatened species. Several species added to the threatened list have large ranges, such as *Amazona xanthops*, *Ara maracana*, *Heteroxolmis dominicana* and *Coryphaspiza melanotis*, and a careful revision of their current distributions would be needed to determine appropriate Key Areas for them. Indeed, it has been suggested that no protected area could hold permanent populations of *A. xanthops* since groups are semi-nomadic and range over enormous territories (Collar *et al.* 1994).

Many of the species added to the threatened list are known to occur sympatrically with other threat-

Table 2. Top Key Areas in Brazil. Those with the area number in bold currently have some form of protected status.

Key Area	No. of threatened spp.	Comments
006 Mamirau	1	The only area where *Crax globulosa* has been recorded regularly in recent years.
008 Borba	1	This area encompasses the entire range of *Todirostrum senex*, rediscovered there more than 150 years after its initial discovery.
012 Serra do Cachimbo	2	The only known site for *Pipra vilasboasi.*
014 Cachoeira Nazaré	1	The only known site for *Clytoctantes atrogularis.*
016 Serra das Araras	2	The only site where *Columbina cyanopis* has been seen recently, and even here its status is uncertain.
022 Rio Gurupi	1	Probably the most important area for *Guaruba guarouba.*
035 Pedra Branca and Fazenda Bananeira	13	One of the most important forests in South America, with 13 threatened species; the type-locality for four species and the only known site for two of these.
036 Pedra Talhada	11	A very important forest for several species endemic to north-east Brazil.
039 Curaçá	1	The only known site for *Cyanopsitta spixii.*
040 Canudos	1	This area embraces virtually the entire population of *Anodorhynchus leari.*
043 Santo Amaro	1	*Pyriglena atra* is restricted to this Key Area.
050 Boa Nova	2	Probably the major area for *Formicivora iheringi* and *Rhopornis ardesiaca.*
051 Serra do Ouricana	9	The area marks the northernmost limit of several threatened species confined to the Atlantic coastal forests of south-east Brazil and also harbours several undescribed taxa (Gonzaga *et al.* 1995, Gonzaga and Pacheco in press),which will undoubtedly qualify as threatened and further elevate the area's importance.
055 Porto Seguro	7	The area supports good populations of seven threatened species, and is a critically important site for *Glaucis dohrnii.*
056 Monte Pascoal	4	At least 11 threatened species are known from this reserve (several not recorded in the last 10 years).
061 Emas	7	A very important national park, especially for grassland species, and the only site where *Caprimulgus candicans* has been observed regularly (but rarely).
062 Brasília	8	Holds many cerrado and grassland species; especially important for the threatened tinamous, *Laterallus xenopterus* and *Scytalopus nova capitalis.*
064 Brejo do Amparo	3	The general area is still the only known site for *Phylloscartes roquettei*, although the habitat is now highly fragmented and it is not clear whether the species survives there.
066 Serra do Cipó	4	The only known site for the recently described *Asthenes luizae* which, although not uncommon in suitable habitat, has a tiny range.
068 Rio Doce	11	This state park has a large number of threatened species and is probably the key site for *Curaeus forbesi.*
072 Serra da Canastra	3	One of the last remaining areas for *Mergus octosetaceus.*
076 Sooretama	13	Supports a high number of threatened species and is probably the key site for several, most notably *Crax blumenbachii.*
077 Linhares	9	A forest adjacent to Sooretama (BR 076) and important for several threatened species.
078 Augusto Ruschi	10	The reserve harbours many threatened species and is the key site at least for *Lipaugus lanioides.*
079 Jatibocas	1	The only site with a twentieth-century record of *Nemosia rourei*, but there is only a slim chance that the species survives there.
084 Desengano	18	More threatened species recorded than any other area in the Neotropics.
086 Itatiaia	13	An important national park, especially for bamboo specialists.
088 Serra dos Órgãos	11	A national park with a high number of threatened species recorded.
094 Angra dos Reis	5	Still the only known site for *Formicivora erythronotus*, although the habitat is now highly fragmented and the species' status is critical.
103 Bairro de Corcovado	10	Large numbers of threatened species sighted in the area.
104 Itapetininga	3	The only area in south-east Brazil where both *Nothura minor* and *Taoniscus nanus* currently occur.
109 Fazenda Intervales	12	This reserve supports a high number of threatened species.
115 Ilhas Comprida and Cananéia	7	The largest breeding population of *Amazona brasiliensis*, at least in São Paulo state.
116 Ilha do Cardoso	15	The second-highest number of threatened species in any Neotropical Key Area has been recorded from this island.
128 Salto do Piraí	5	One of only two sites known for *Hemitriccus kaempferi*, and the only one where it has been recently recorded.

ened birds and in Key Areas already identified, e.g. *Picumnus fulvescens* (BR 028, 033, 035, 036), *P. limae* (BR 026), *Gyalophylax hellmayri* (BR 033, 040, 041), *Myrmotherula minor* (BR 078, 090, 091, 095, 102, 103, 106), *M. unicolor* (BR 088, 090, 091, 095, 102, 103, 108, 111, 113, 116, 128) *M. urosticta* (BR 053, 054, 055, 056, 076, 077, 084) *Onychorhynchus swainsoni* (BR 055, 086, and those in the Serra do Mar such as 102) and *Phylloscartes kronei* (BR 111, 113, 115, 116, 123).

OLD RECORDS AND LITTLE-KNOWN BIRDS

Knowledge of threatened birds in Brazil is constantly improving through fieldwork and museum-based research, but still suffers from the general paucity of recent records for a number of species. The number of new species continuing to be described from Brazil, several of which were discovered in forests close to São Paulo and Rio de Janeiro, demonstrates unequivocally the great importance of increasing fieldwork throughout the country.

For most species listed as threatened we are able to make informed conservation recommendations from the data collected, but the exceptions are those species included within Key Areas but known only from old specimen records or chance observations (from which there can be no guarantees of a current viable population); these include the following.

Mitu mitu, although recorded from around São Miguel dos Campos (BR 037) as recently as the 1980s, is now classified as Extinct in the Wild by Collar *et al.* (1994) owing to chronic habitat loss and hunting in its tiny range within the lowland forests of coastal Alagoas.

Columbina cyanopis has been seen only once recently at Serra das Araras Ecological Station (BR 016), in February 1986, but subsequent searches have not found it there.

Merulaxis stresemanni is known from two specimens taken in coastal Bahia, the last near Ilhéus (BR 052) in 1945.

Scytalopus psychopompus is known from three specimens obtained at two lowland forest sites, in 1944 near Ilhéus (BR 052) and in 1983 at Valença (BR 045), but has not been seen subsequently in spite of searches.

Pipra vilasboasi is still known only from specimens collected in 1957 from the northern edge of the Serra do Cachimbo (BR 012).

Phylloscartes roquettei is known only from the vicinity of Brejo do Amparo (BR 064), the most recent second sighting being in 1977, but subsequent fieldworkers have failed to find it there.

Nemosia rourei is known only from the nineteenth-century type-specimen collected in south-east Minas Gerais and from a 1941 sighting in the vicinity of Jatibocas (BR 079), Espírito Santo, a region which is apparently now deforested (Sick and Teixeira 1979).

OUTLOOK

This analysis has identified 142 Key Areas of which 35 stand out as top priorities, each hosting populations of 10 or more threatened species, or being effectively the only area for a species confined to Brazil (Table 2). These 35 Key Areas are the most important for a number of threatened species and together represent populations of 72 (75%) of the Brazilian threatened birds. All Key Areas (which are individually extremely important for many threatened Brazilian endemics: see 'Key Areas' and 'Key Area protection', above) need some degree of conservation action if the populations of their threatened species are to remain viable. A clear example is the lack of protected areas in the dry forests of Bahia and Minas Gerais, with several species such as *Cyanopsitta spixii*, *Formicivora iheringi*, *Rhopornis ardesiaca* and *Phylloscartes roquettei* not found in any protected Key Areas. For some Key Areas, such as the Serra do Cachimbo (BR 012), there is a requirement to confirm the continued existence or assess the population of a threatened species, and for others, especially those based on old records such as Jatibocas (BR 079), the need is for locating (and defining the extent of) suitable habitat in the vicinity of the Key Area and searching for the 'lost' species.

DATA SOURCES

The above introductory text and the Site Inventory were compiled from information supplied by A. Balmford, C. Balchin, G. Bencke, R. Boçom, C. Bushell, D. Dacol, C. S. Fontana, B. C. Forrester, M. Galetti, J. Goerck, L. P. Gonzaga, N. Guedes, G. Kirwan, P. Martuscelli, F. Olmos, Y. Oniki, J. F. Pacheco, M. Pearman, C. Seger, J. M. C. da Silva, G. Speight, J. Tobias, J. Wall, E. O. Willis, B. M. Whitney, A. Whittaker and N. Varty, as well as from the following references.

Balchin, C. (1994) Summary of species seen during a visit to coastal Brazil. Unpublished (birdwatching) report.

BELTON, W. (1984–1985) Birds of Rio Grande do Sul, Brazil. *Bull. Amer. Mus. Nat. Hist.* 178(4), 180(1).

BENCKE, G. A. (1994) The ecology and conservation of the Blue-bellied Parrot *Triclaria malachitacea* in the forest fragments in Rio Grande do Sul, Brazil. Project proposal to American Bird Conservancy.

BORNSCHEIN, M. R. (1995) Estratégia para o estudo e conservação do 'curiango-do-banhado' *Eleothreptus anomalus* na região metropolitana de Curitiba, Paraná, Brasil. Project proposal to American Bird Conservancy.

BUSHELL, C. (1994) South-eastern Brazil. Unpublished (birdwatching) report.

CLEMENTS, J. F. (1993) Report on a birding trip to Brazil. Unpublished report.

COLLAR, N. J., GONZAGA, L. P., KRABBE, N., MADROÑO NIETO, A., NARANJO, L. G., PARKER, T. A. AND WEGE, D. C. (1992) *Threatened birds of the Americas: the ICBP/IUCN Red Data Book.* Cambridge, U.K.: International Council for Bird Preservation.

COLLAR, N. J., CROSBY, M. J. AND STATTERSFIELD, A. J. (1994) *Birds to watch 2: the world list of threatened birds.* Cambridge, U.K.: BirdLife International (BirdLife Conservation Series no. 4).

COLLAR, N. J., WEGE, D. C. AND LONG, A. J. (in press) Patterns and causes of endangerment in the New World avifauna. In J. V. Remsen, ed. *Natural history and conservation of Neotropical birds.* American Ornithologists' Union (Orn. Monogr.).

DACOL, D. (1994) Birding trip to São Paulo state, Brazil. Unpublished report.

FONTANA, C. S. (1994) Ecologia comportamental de *Heteroxolmis dominicanus* (Vieillot, 1823) (Aves, Tyrrannidae), com ênfase na relação ecológica com *Xanthopsar flavus* (Gmelin, 1788) (Aves, Icteridae), na nordeste do rio Grande do Sul, Brasil. Technical report to BirdLife International, U.S. Fish and Wildlife Service, and Museu de Ciências e Tecnologia da PUC-RS. Unpublished.

FORRESTER, B. C. (1990) Brazil IV, July/August 1990. Unpublished (birdwatching) report.

FORRESTER, B. C. (1992) Brazil VI July/August 1992. Unpublished (birdwatching) report.

FORRESTER, B. C. (1993) *Birding Brazil: a check-list and site guide.* Irvine, Scotland: Forrester.

FORRESTER, B. C. (1994) Brazil VII July/August 1994. Unpublished (birdwatching) report.

FORRESTER, B. C. (1995) Brazil's northern frontier sites: in search of two Rio Branco endemics. *Cotinga* 3: 51–54.

GALETTI, M., MARTUSCELLI, P., OLMOS, F., AND ALEIXO, A. (in prep.) Ecology and conservation of the Jacutinga *Pipile jacutinga* (Cracidae) in southeastern Brazil. *Wilson Bull.*

GONZAGA, L. P. AND PACHECO, J. F. (1995) A new species of *Phylloscartes* (Tyrannidae) from the mountains of southern Bahia, Brazil. *Bull. Brit. Orn. Club.* 115: 88–97.

GONZAGA, L. P., PACHECO, J. F., BAUER, C. AND CASTIGLIONI, G. D. A. (1995) An avifaunal survey of the vanishing montane Atlantic forest of southern Bahia, Brazil. *Bird Conserv. Internatn.* 5(2–3) in press.

IUCN (1992) *Protected areas of the world: a review of national systems,* 4: *Nearctic and Neotropical.* Gland, Switzerland and Cambridge, U.K.: International Union for Conservation of Nature and Natural Resources.

MARTUSCELLI, P. (1995) Ecology and conservation of the Red-tailed Amazon *Amazona brasiliensis* in south-eastern Brazil. *Bird Conserv. Internatn.* 5(2–3) in press.

OLIVER, W. L. R. AND SANTOS, I. B. (1991) *Threatened endemic mammals of the Atlantic Forest region of south-east Brazil.* Jersey, Channel Islands: Jersey Wildlife Preservation Trust (Spec. Sci. Rep. 4).

OLMOS, F. (1993) Birds of Serra da Capivara National Park in the 'caatinga' of north-eastern Brazil. *Bird Conserv. Internatn.* 3: 21–36.

OREN, D. C. AND NOVAES, F. C. (1986) Observations on the Golden Parakeet *Aratinga guarouba* in northern Brazil. *Biol. Conserv.* 36: 329–337.

OREN, D. C. AND DA SILVA, J. M. C. (1987) Cherrie's Spinetail (*Synallaxis cherriei* Gyldenstolpe) (Aves: Furnariidae) in Carajás and Gorotire, Pará, Brazil. *Bol. Mus. Para. Emílio Goeldi,* Ser. Zool. 3: 1–9.

PAYNTER, R. A. AND TRAYLOR, M. A. (1991) *Ornithological gazetteer of Brazil.* Cambridge, Mass.: Museum of Comparative Zoology.

REDFORD, K. H. (1985) Emas National Park and the plight of the Brazilian cerrados. *Oryx* 19: 210–214.

REDFORD, K. H. (1989) Monte Pascoal–indigenous rights and conservation in conflict. *Oryx* 23: 33–36.

SCHERER-NETO, P. AND STRAUBE, F. C. (1995) *Aves do Paraná: história, lista anotada e bibliografia.* Curitiba: privately published.

SICK, H. (1993) *Birds in Brazil: a natural history.* Princeton, New Jersey: Princeton University Press.

SICK, H. AND TEIXEIRA, D. M. (1979) Notas sobre aves brasileiras raras ou ameaçadas de extinção. *Publ. Avuls. Mus. Nac.* 62.

DA SILVA, J. M. C. AND OREN, D. C. (1992) Notes on *Knipolegus franciscanus,* an endemism of cental Brazilian forests. *Goeld. Zool.* 16: 1–9.

DA SILVA, J. M. C., OREN, D. C., ROMA, J. C. AND HENRIQUES, L. M. (in prep.) Composition and avifauna of an Amazonian terra firme savanna, Amapá, Brazil. *Wilson Bull.*

STATTERSFIELD, A. J., CROSBY, M. J., LONG, A. J. AND WEGE, D. C. (in prep.) *Global directory of Endemic Bird Areas.* Cambridge, U.K.: BirdLife International (BirdLife Conservation Series).

TOBIAS, J. A., CATSIS, M. C. AND WILLIAMS, R. S. R. (1993) Notes on scarce birds observed in southern and eastern Brazil: 24 July - 7 September 1993. Unpublished report.

VANZOLINI, P. E. (1992) *A supplement to the ornithological gazetteer of Brazil.* Sao Paulo: Universidade de São Paulo, Museu de Zoologia.

VARTY, N., BENCKE, G. A., BERNARDINI, L. DE M., DA CUNHA, A. S., DIAS, E. V., FONTANA, C. S., GUADAGNIN, D. L., KINDEL, A., KINDEL, E., RAYMUNDO, M. M., RICHTER, M., ROSA, A. AND TOSTES, C. S. (1994a) The ecology and conservation of the Red-spectacled Parrot *Amazona pretrei* in southern Brazil. Final report to BirdLife International. Unpublished.

VARTY, N., BENCKE, G. A., BERNARDINI, L. DE M., DA CUNHA, A. S., FONTANA, C. S., GUADAGNIN, D. L., KINDEL, A., KINDEL, E., RAYMUNDO, M. M., RICHTER, M., ROSA, O. A. AND TOSTES, C. S. (1994b) *Um plano de ação preliminar para o papagaio charão* Amazona pretrei *no sul do Brasil.* Porto Alegre, Brazil: Museu de Ciências e Tecnologia (Special Publ.).

WHITNEY, B. M. AND PACHECO, J. F. (1994) Behavior and

vocalizations of *Gyalophylax* and *Megaxenops* (Furnariidae), two little-known genera endemic to northeastern Brazil. *Condor* 96: 559–565.

WHITNEY, B. M. AND PACHECO, J. F. (1995) Distribution and conservation status of four *Myrmotherula* antwrens (Formicariidae) in the Atlantic forest of Brazil. *Bird Conserv. Internatn.* 5(2–3) in press.

WILLIS, E. O. AND ONIKI, Y. (1990) Levantamento preliminar das aves de inverno em dez áreas do sudoeste de Mato Grosso, Brasil. *Ararajuba* 1: 19–38.

WILLIS, E. O. AND ONIKI, Y. (1992) A new *Phylloscartes* (Tyrannidae) from southeastern Brazil. *Bull. Brit. Orn. Club* 112: 158–165.

YAMASHITA, C. AND FRANÇA, J. T. (1991) A range extension of the Golden Conure *Guaruba guarouba* to Rondônia state, western Amazonia. *Ararajuba* 2: 91–92.

SITE INVENTORY

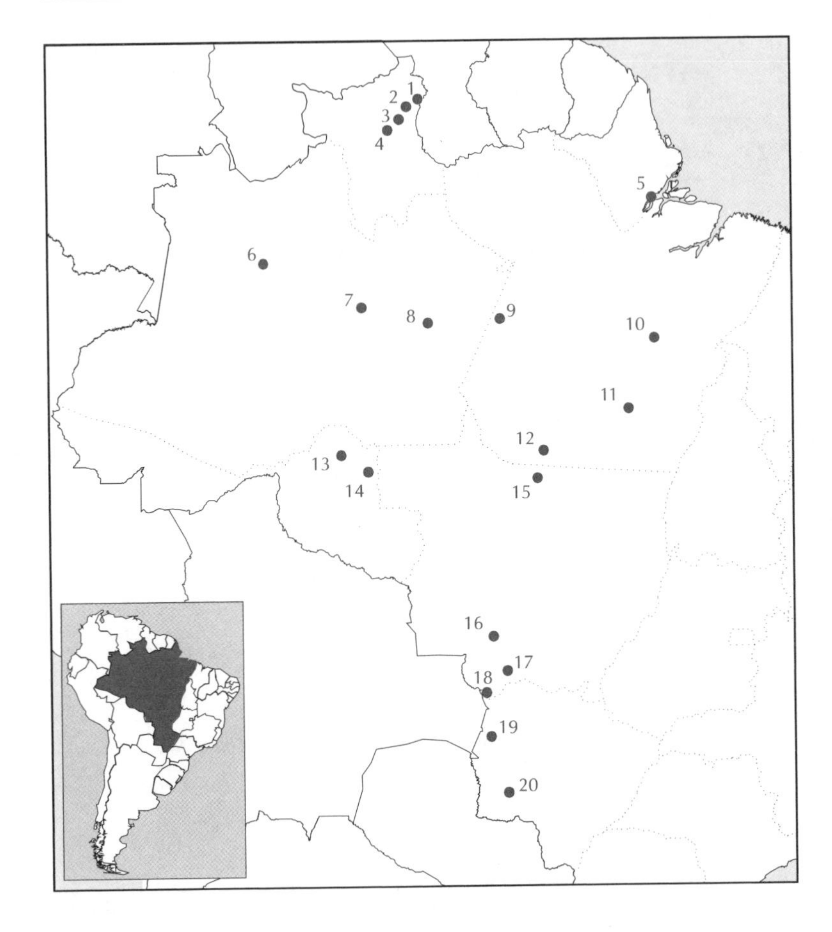

Figure 2. Key Areas in western Brazil.

001 Conceição do Maú
002 Rio Surumu
003 Boa Vista
004 Rio Mucajai
005 Amapá savannas
006 Mamirau
007 Ilhas Codajas
008 Borba
009 Rio Tapajós
010 Serra do Carajás
011 Kayapó
012 Serra do Cachimbo
013 Jamari
014 Cachoeira Nazaré
015 Alta Floresta
016 Serra das Araras
017 Transpantaneira
018 Pantanal Matogrossense
019 Corumbá
020 Serra da Bodoquena

Conceição do Maú (Roraima)

BR 001

3°38′N 59°53′W
EBA B56

Raposa/Serra do Sol (IUCN category VII), 1,401,320 ha

This area is located near Conceição do Maú on the south side of Rio Tacutu, and only a few kilometres from the border with Guyana. The habitat comprises seasonally flooded gallery forest with dense patches of thickets and numerous vines (Forrester 1995).

Synallaxis kollari	1992	A pair observed in July 1992 represented the first sighting of the species since 1956 (Forrester 1995).

Table 3. Matrix of threatened species by Key Area in western Brazil (the area covered by Figure 2). Additional threatened species known from this region but for which no Key Areas have been identified are *Caprimulgus candicans*, *Synallaxis infuscata* and *Conothraupis mesoleuca*.

	001	002	003	004	005	006	007	008	009	010	011	012	013	014	015	016	017	018	019	020	No. of areas
Penelope ochrogaster	–	–	–	–	–	–	–	–	–	–	–	–	–	–	–	–	●	●	–	–	2
Crax globulosa	–	–	–	–	–	●	●	–	–	–	–	–	–	–	–	–	–	–	–	–	2
Columbina cyanopis	–	–	–	–	–	–	–	–	–	–	–	–	–	–	–	●	–	–	–	–	1
Anodorhynchus hyacinthinus	–	–	–	–	–	–	–	–	–	●	●	–	–	–	●	–	●	–	●	●	6
Guaruba guarouba	–	–	–	–	–	–	–	–	●	–	–	–	●	–	–	–	–	–	–	–	2
Synallaxis cherriei	–	–	–	–	–	–	–	–	–	●	●	–	–	–	●	–	–	–	–	–	3
Synallaxis kollari	●	●	–	–	–	–	–	–	–	–	–	–	–	–	–	–	–	–	–	–	2
Clytoctantes atrogularis	–	–	–	–	–	–	–	–	–	–	–	–	–	●	–	–	–	–	–	–	1
Cercomacra carbonaria	–	–	●	●	–	–	–	–	–	–	–	–	–	–	–	–	–	–	–	–	2
Pipra vilasboasi	–	–	–	–	–	–	–	–	–	–	–	●	–	–	–	–	–	–	–	–	1
Euscarthmus rufomarginatus	–	–	–	–	●	–	–	–	–	–	–	●	–	–	–	●	–	–	–	–	3
Todirostrum senex	–	–	–	–	–	–	–	●	–	–	–	–	–	–	–	–	–	–	–	–	1
Sporophila hypochroma	–	–	–	–	–	–	–	–	–	–	–	–	–	–	–	–	–	–	●	–	1
Sporophila nigrorufa	–	–	–	–	–	–	–	–	–	–	–	–	–	–	–	–	–	–	●	–	1
Sporophila palustris	–	–	–	–	–	–	–	–	–	–	–	–	–	–	–	–	●	–	●	–	2
No. of species	1	1	1	1	1	1	1	1	1	2	2	2	1	1	2	2	3	1	4	1	

Rio Surumu (Roraima)

BR 002

Unprotected

3°22′N 60°19′W
EBA B56

Rio Surumu is a small right-bank tributary of the lower Rio Tacutu in northern Roraima, where the main habitat is a mosaic of woodland and savanna with gallery forest along the river.

Synallaxis kollari	1956	One specimen. The species is known from only a handful of records.

Boa Vista (Roraima)

BR 003

Unprotected

2°55′N 60°36′W
EBA B56

Boa Vista town is located along the upper Rio Branco in north-east Roraima. The main habitat is savanna with seasonally flooded gallery forest along the river.

Cercomacra carbonaria	1993	Relatively common in gallery forest on Ilha São José in 1993 (Forrester 1995); previously recorded from river banks in the area.

Rio Mucajai (Roraima)

BR 004

Unprotected

2°32′N 61°02′W
EBA B56

This small tributary of the upper Rio Branco is located c.45 km south of Boa Vista in north-east Roraima. The species below was collected somewhere between the mouth (2°25′N 60°52′W) and the river's confluence with Rio Apiaú (2°39′N 61°12′W), where the principal habitats are terra firme forest and gallery forest (J. M. C. da Silva *in litt.* 1995).

Cercomacra carbonaria	1963	A number of specimens collected in 1962 and 1963.

Amapá savannas (Amapá)

BR 005
0°02′S 51°03′W

Unprotected

This coastal savanna enclave covers c.17,000 km^2 of south-east Amapá. The vegetation includes savannas with gallery forest along the rivers (J. M. C. da Silva *in litt.* 1995).

Euscarthmus rufimarginatus	1990	Two in October 1990 (da Silva *et al.* in prep.).

Mamirau (Amazonas)

BR 006
2°13′S 65°49′W

Mamirau Ecological Station (IUCN category IV), 1,124,000 ha

This enormous reserve is located between Rios Solimões and Japurá, c.150 km north-west of Tefé in central Amazonas. The area embraces several large lakes including Lagos Guedes, Angaiara and Panauá and presumably holds the typical range of Amazonian forest types, especially várzea forest.

Crax globulosa	c.1995	Recorded regularly in recent years (J. F. Pacheco *in litt.* 1994), these being the only recent observations from Brazil.

Ilhas Codajas (Amazonas)

BR 007
3°50′S 62°05′W

Unprotected

The area comprises a cluster of large islands in the Rio Amazonas, located c.250 km south-east of Manaus in western Amazonas.

Crax globulosa	1936	The area is the source of at least 16 specimens, and thus merits investigation to determine the species' current status.

Borba (Amazonas)

BR 008
4°24′S 59°35′W
EBA B43

Unprotected

Borba is on the right bank of the lower Rio Madeira in eastern Amazonas. Land on both sides of this important river needs formal protection as it supports a representative set of Amazonian habitats (white-water and black-water forests, campina) and is the type-locality for a number of birds (B. M. Whitney *in litt.* 1995). The region merits further research, as endemic species (*Todirostrum senex* and the primate *Callithrix chrysoleuca*) have already been found there, even though it has been little studied.

Todirostrum senex	1993	Previously known only from the type-specimen collected in 1830, the species was rediscovered by B. M. Whitney and M. Cohn-Haft and found to be fairly common.

Rio Tapajós (Pará/Amazonas)

BR 009
4°16′S 56°52′W

Amazonia National Park (IUCN category II), 994,000 ha

The national park comprises a vast area of Amazonian forest lying adjacent to the west bank of the Rio Tapajós, 53 km south-west of Itaituba (Forrester 1993). Apparently there is illegal gold-mining taking place within the national park (Forrester 1993).

Guaruba guarouba	1992	Nine recently seen in Amazonia National Park (Forrester 1992, 1993); there were a number of records in the area in the 1980s.

Serra do Carajás (Pará)

BR 010

5°00'S 51°00'W

Tapirape Biological Reserve (IUCN category I), 103,000 ha
Paracana Indigenous Area (IUCN category VII), 351,687 ha

These highlands stretch across c.250 km, lying immediately north of Rio Itacaiuna and east of the middle Rio Xingu in east-central Pará. The area supports extensive stands of unbroken rainforest (Oren and Novaes 1986, Oren and da Silva 1987).

Anodorhynchus hyacinthinus	1990s	Good populations still exist (J. M. C. da Silva *in litt.* 1995).
Synallaxis cherriei	1980s	Found in and adjacent to the Serra dos Carajás at 'Manganês' in the Serra Norte during the mid-1980s.

Kayapó (Pará)

BR 011

7°30'S 52°00'W

Kayapó Indigenous Area (IUCN category VII), 3,204,000 ha

This enormous reserve lies on the east bank of the upper Rio Xingu and stretches east to the Serra dos Gradaús in southern Pará. The area holds rainforest broken by large patches of cerrado (Oren and Novaes 1986).

Anodorhynchus hyacinthinus	1990s	Good populations still exist (B. M. Whitney *in litt.* 1995).
Synallaxis cherriei	1980s	Found within the indigenous reserve at Gorotire (7°40'S 51°15'W), near São Felix do Xingu in the mid-1980s.

Serra do Cachimbo (Pará)

BR 012

9°00'S 55°15'W
EBA B43

Unprotected

This long highland area reaches 500 m and is located between the Rio Tapajós and Rio Xingu in south-central Pará. There is no recent information on the birds of the area, suggesting an urgent need for surveys.

Pipra vilasboasi	1967	The only known site for this species which has not been seen since its discovery.
Euscarthmus rufomarginatus	1955	A female collected in November.

Jamari (Rondônia)

BR 013

9°07'S 62°54'W

Jamari National Forest (IUCN category VIII), 215,000 ha

The habitat of this area, located on the right bank of the Rio Madeira between the Rios Preto and Jamari, comprises submontane open tropical forest on a shield with hilly relief (Yamashita and França 1991).

Guaruba guarouba	1990	Several observations during fieldwork in April and September–November. This locality represents an extension of more than 500 km from the main range in eastern Amazonia.

Cachoeira Nazaré (Rondônia)

BR 014
9°44′S 61°53′W

Unprotected

Lying near the west bank of the Rio Ji-Paraná in north-east Rondônia, this area comprises a large tract of mature terra firme forest dominated by dense vine tangles.

Clytoctantes atrogularis	1986	The type-locality and the only area currently known for this species which must, however, be rare, as only two records were made during many field hours and net-days of study.

Alta Floresta (Mato Grosso)

BR 015
10°00′S 55°30′W

Unprotected

This area is located west of the Rio Teles Pires in north-central Mato Grosso. The habitats are mainly humid tropical forest and savanna.

Anodorhynchus hyacinthinus	1989	A low-density population discovered near Alta Floresta in 1989.
Synallaxis cherriei	1989	At least four pairs held territories in October 1989 along c.400 m of trail through tall bamboo *Guadua* inside 30-m-tall evergreen forest.

Serra das Araras (Mato Grosso)

BR 016
15°39′S 57°13′W

Serra das Araras Federal Ecological Station (IUCN category IV), 115,000 ha

The station is located at the northern end of Serra das Araras in the fork of the Rio Salobra near Palmeiras in south-west Mato Grosso. The main habitat is campo cerrado and dry forests with *Orbignya mertensiana* palms found especially along the rivers (Silva and Oniki 1988, Willis and Oniki 1990).

Columbina cyanopis	1986	One of the only known sites for the species, though it was not found during a survey in 1987.
Euscarthmus rufomarginatus	1988	Recorded twice in December 1988, in open cerrado and campo sujo habitat.

Transpantaneira (Mato Grosso)

BR 017
16°53′S 56°42′W

Unprotected

The Transpantaneira is a road that crosses, for 120 km, the northern Pantanal of Mato Grosso between the towns of Poconé (16°16′S 56°38′W) and Porto Jofré (17°19′S 56°46′W). The habitat comprises seasonally inundated savanna with palms, patches of humid deciduous forest and gallery forest. More surveys are needed during the winter months to gather better information on the status of the wintering *Sporophila* species.

Penelope ochrogaster	1992	This locality is the source of most recent records of the species which appears to be moderately common, especially near Poconé.
Anodorhynchus hyacinthinus	1992	Seen regularly at several places along the road.
Sporophila palustris	1980s	Apparently recorded near Poconé. This area falls within its wintering range.
Sporophila nigrorufa	1980s	A probable sighting between Poconé and Porto Jofré falls within the species' wintering range.

Pantanal Matogrossense (Mato Grosso do Sul)

BR 018

17°40′S 57°30′W

Pantanal Matogrossense National Park (IUCN category II), 135,000 ha

Located in the area between the Rios Paraguaí and São Lourenço (Cuiabá) and their confluence in south-west Mato Grosso, the national park's habitat comprises seasonally inundated savanna with palms, patches of humid deciduous forest and gallery forest. A few individuals of *Anodorhynchus hyacinthinus* may occur in the extreme north of the park, the nearest area with a reasonable population of the species being in the Transpantaneira (BR 017).

Penelope ochrogaster	1992	Recent sightings have all been on the edge of the park (Fazenda Sara, Fazenda Belice, Rita Velha, Morro Campo).

Corumbá (Mato Grosso do Sul)

BR 019

19°15′S 57°20′W

Unprotected

This general area lies east of Corumbá in north-west Mato Grosso do Sul. The habitat comprises seasonally inundated savanna with palms, patches of humid deciduous forest and gallery forest. More surveys are needed during the winter months to gather better information on the status of the wintering *Sporophila* species.

Anodorhynchus hyacinthinus	1990s	Recorded from this part of the pantanal, where the species is the focus of projects on several private fazendas.
Sporophila hypochroma	1979	A sighting east of Corumbá in October 1979 falls within the wintering range.
Sporophila nigrorufa	1979	The only certain record is a male seen east of Corumbá in October 1979; this falls within the species' wintering range.
Sporophila palustris	1980s	Recorded from Corumbá, within the wintering range.

Serra da Bodoquena (Mato Grosso do Sul)

BR 020

21°15′S 56°42′W

Unprotected

Located in south-west Mato Grosso do Sul near the border with Paraguay, the area is dominated by tropical dry forest, but also supports cerrado and gallery forest. During the last 20 years the forest has been converted rapidly to pasture and other forms of agriculture.

Anodorhynchus hyacinthinus	1994	A small population found around the town of Bonito (J. M. C. da Silva and N. Guedes *in litt.* 1995), although more study is needed to evaluate the true status here.

Axixá (Maranhão)

BR 021

2°51′S 44°04′W

Unprotected

Gallery forest is presumed to comprise the main habitat in this area on the west bank of the Rio Munim, c.10 km before the river drains into Baía de São José in northern Maranhão.

Herpsilochmus pectoralis	1980s	Recorded regularly in the 1980s.

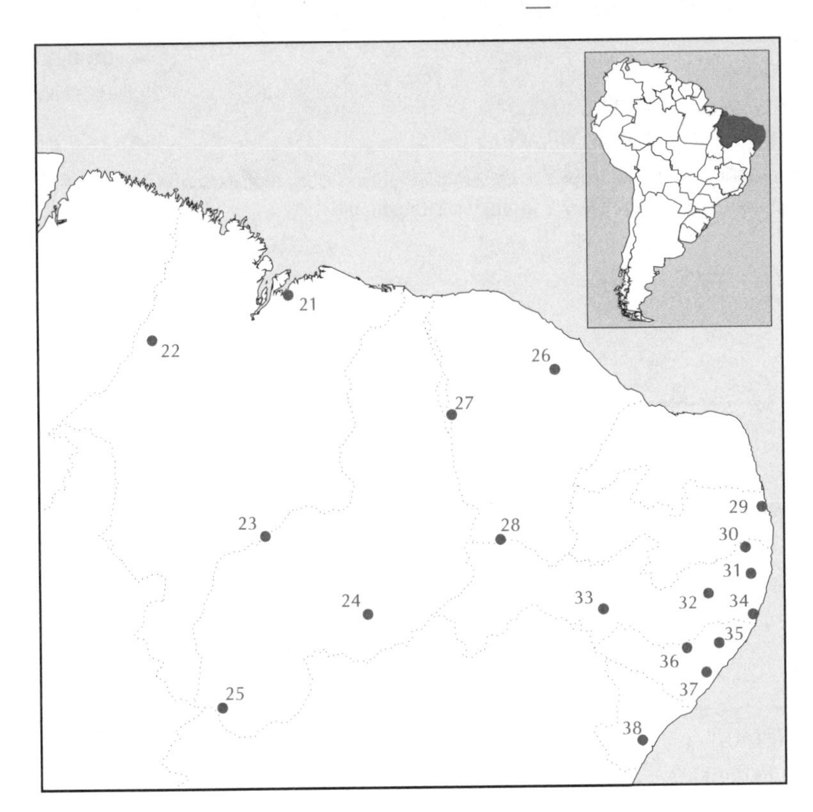

Figure 3. Key Areas in north-east Brazil.

021 Axixá
022 Rio Gurupi
023 Uruçuí-Una
024 Serra da Capivara
025 Parnaguá and Corrente
026 Guaramirangá
027 Tiangúa
028 Chapada de Araripe
029 Mamanguape
030 Água Azul
031 Tapacurá
032 Serra dos Cavalos
033 Serra Negra
034 Saltinho
035 Pedra
036 Pedra Talhada
037 São Miguel dos Campos
038 Serra de Itabaiana

Table 4. Matrix of threatened species by Key Area in north-east Brazil (the area covered by Figure 3). All threatened species known from this region are listed.

	021	022	023	024	025	026	027	028	029	030	031	032	033	034	035	036	037	038	No. of areas
Leucopternis lacernulata	–	–	–	–	–	–	–	–	–	–	–	–	–	–	–	●	–	–	1
Mitu mitu	–	–	–	–	–	–	–	–	–	–	–	–	–	–	–	–	●	–	1
Anodorhynchus hyacinthinus	–	–	●	–	●	–	–	–	–	–	–	–	–	–	–	–	–	–	2
Guaruba guarouba	–	●	–	–	–	–	–	–	–	–	–	–	–	–	–	–	–	–	1
Touit surda	–	–	–	–	–	–	–	–	●	–	–	–	–	–	●	●	●	–	4
Amazona rhodocorytha	–	–	–	–	–	–	–	–	–	–	–	–	–	–	–	–	●	–	1
Xiphocolaptes falcirostris	–	–	–	–	●	●	–	–	–	–	–	–	–	–	–	–	–	–	2
Synallaxis infuscata	–	–	–	–	–	–	–	–	–	–	–	●	–	●	●	●	–	–	4
Philydor novaesi	–	–	–	–	–	–	–	–	–	–	–	–	–	–	●	–	–	–	1
Megaxenops parnaguae	–	–	–	●	●	–	–	●	–	–	–	–	●	–	–	–	–	–	4
Myrmotherula snowi	–	–	–	–	–	–	–	–	–	–	–	–	–	–	●	–	–	–	1
Herpsilochmus pectoralis	●	–	–	–	–	–	–	–	–	–	–	–	–	–	–	–	–	●	2
Terenura sicki	–	–	–	–	–	–	–	–	–	●	–	–	–	–	●	●	–	–	3
Myrmeciza ruficauda	–	–	–	–	–	–	–	–	–	–	●	●	–	●	●	●	–	–	5
Carpornis melanocephalus	–	–	–	–	–	–	–	–	–	–	–	–	–	–	●	–	–	–	1
Iodopleura pipra	–	–	–	–	–	–	–	–	●	–	–	–	–	–	●	–	–	–	2
Xipholena atropurpurea	–	–	–	–	–	–	–	–	●	–	●	–	–	–	●	●	●	–	5
Phylloscartes ceciliae	–	–	–	–	–	–	–	–	–	–	–	–	–	–	●	●	–	–	2
Hemitriccus mirandae	–	–	–	–	–	●	●	–	–	–	–	–	–	–	–	●	–	–	3
Tangara fastuosa	–	–	–	–	–	–	–	–	–	–	●	●	–	●	●	●	●	–	6
Curaeus forbesi	–	–	–	–	–	–	–	–	–	–	–	–	–	–	●	●	–	–	2
Carduelis yarrellii	–	–	–	●	–	–	–	–	●	–	●	–	●	●	●	●	–	–	7
No. of species	1	1	1	2	3	2	1	1	4	1	4	3	2	4	13	11	5	1	

Rio Gurupi (Maranhão)

BR 022
3°39′S 46°41′W

Gurupi Forest Reserve (IUCN category VI), 1,600,000 ha
Alto Turiaçu Indigenous Area (IUCN category VII), 530,524 ha

These reserves are located east of the Rio Gurupi which flows along the Maranhão–Pará border. They hold extensive humid forest with trees up to 40 m tall and embrace upland areas such as Serras da Desordem and do Tiracambu.

Guaruba guarouba	1980s	This is an important area for the species.

Uruçuí-Una (Piauí)

BR 023
7°14′S 44°33′W

Uruçuí-Una Federal Ecological Station (IUCN category IV), 135,000 ha

A large reserve south of Uruçuí in western Piauí, close to the border with Maranhão. The main habitats are cerrado and grasslands with smaller amounts of gallery forest.

Anodorhynchus hyacinthinus	1994	Recorded recently within the park (F. Olmos and P. Martuscelli *in litt.* 1995).

Serra da Capivara (Piauí)

BR 024
8°40′S 42°35′W

Serra da Capivara National Park (IUCN category II), 97,933 ha

This reserve encompasses a variety of terrain including the Serra and Chapada da Capivara, whose steep-walled escarpment marks the southern boundary, and the Serra da Congo which forms the central spine of the reserve. The main vegetation comprises different kinds of caatinga, much of it very dense, with some patches of semi-deciduous forest in the canyons (Olmos 1993).

Megaxenops parnaguae	c.1995	Fairly common in the park in all kinds of caatinga (Olmos 1993), also just outside the park in July 1994 (Forrester 1994).
Carduelis yarrellii	1980s	Rarely recorded from Fazenda Veneza and from the extreme north-west and south-east of the park, in both caatinga and cultivated areas (Olmos 1993).

Parnaguá and Corrente (Piauí)

BR 025
10°20′S 45°24′W

Unprotected

Located in southernmost Piauí, this area was originally covered by tropical dry forest and arboreal caatinga. Most of this important tropical dry forest region has been modified during the last 30 years and only a few patches persist (J. M. C. da Silva *in litt.* 1995). Surveys are needed to identify the best areas of remaining habitat.

Anodorhynchus hyacinthinus	1958	One specimen.
Xiphocolaptes falcirostris	1927	A number of specimens.
Megaxenops parnaguae	1927	A number of specimens.

Guaramirangá (Ceará)

BR 026

4°15′S 38°56′W

Serra de Baturité State Environment Protection Area (IUCN category V), 32,690 ha

The town is located at 800 m in the Serra de Baturité, a range of hills c.100 km long in north-central Ceará. The protected area lies close to the town, and the habitats include semi-deciduous and dry forests.

Hemitriccus mirandae	1991	There are old specimens, but the only recent record is of a bird seen in September.
Xiphocolaptes falcirostris	1987	There are old specimens, but the only recent record is of a bird seen in March.

Tiangúa (Ceará/Piauí)

BR 027

5°03′S 40°55′W

Unprotected

This area is located in the Serra de Ibiapaba, a small outlying range of the Brazilian highlands. The mountains run north to south for c.200 km in the western part of Ceará and reach 900 m.

Hemitriccus mirandae	1991	A specimen collected in 1910 and a recent sighting from Fazenda Gameleira.

Chapada de Araripe (Ceará)

BR 028

7°20′N 40°00′W
EBA B46

Araripe National Forest (IUCN category VIII), 38,262 ha

An upland area reaching 800 m in southernmost Ceará. On the chapada the vegetation is dense woodland up to 12 m high with abundant woody vines, but around its base the habitat is dry caatinga brush (Whitney and Pacheco 1994). In addition to *Megaxenops parnaguae* the area is important for Tawny Piculet *Picumnus fulvescens*. (See 'Recent changes to the threatened list', above.)

Megaxenops parnaguae	1990s	Up to 10 seen in the national forest (Whitney and Pacheco 1994).

Mamanguape (Paraíba)

BR 029

6°47′S 35°00′W

Unprotected

This area, lying within a heavily deforested region, is located on the right bank of the Rio Mamanguape, c.42 km north-west of João Pessoa in east-central Paraíba.

Touit surda	1957	One specimen.
Iodopleura pipra	1989	Two collected in the vicinity of the town.
Xipholena atropurpurea	1957	A number of specimens collected in July.
Carduelis yarrellii	1957	One specimen.

Água Azul (Pernambuco)

BR 030

7°31′S 35°19′W
EBA B47

Unprotected

Located just south of Timbaúba in north-east Pernambuco, the area held one of the last remaining tracts of upland Atlantic forest in the state in the 1980s, but the current condition of the forests is unknown.

Terenura sicki	1989	Eight seen in upland forest.

Tapacurá (Pernambuco)

BR 031

Tapacurá Ecological Station (private), 350 ha

8°00'S 35°13'W
EBA B47

This small private reserve is located just 35 km west of Recife in eastern Pernambuco, and comprises humid forest. Birds which were probably *Hemitriccus mirandae* have been observed in November.

Myrmeciza ruficauda	1979	A specimen from Fazenda São Bento Tapera is close to the present-day ecological station.
Xipholena atropurpurea	1903	A record from São Lourenço is close to the present-day station.
Tangara fastuosa	1980s	Apparently uncommon at the ecological station in the 1980s.
Carduelis yarrellii	1977	Recorded from the station.

Serra dos Cavalos (Pernambuco)

BR 032

Serra dos Cavalos UFPE Ecological Station (private), 450 ha

8°21'S 36°02'W
EBA B47

This small reserve is located near Caruaru in eastern Pernambuco, and is run by the Universidade Federal de Pernambuco (UFPE). Surveys are needed to confirm the current status of the threatened birds.

Synallaxis infuscata	1979	Recorded between 1974 and 1979.
Myrmeciza ruficauda	1986	Apparently one sighting.
Tangara fastuosa	1980s	The population increased in the late 1980s owing to policing of local bird markets. Birds confiscated at the market have in the past been released at this site.

Serra Negra (Pernambuco)

BR 033

Serra Negra Federal Biological Reserve (IUCN category I), 1,000 ha

8°37'S 38°03'W
EBA B46

The Serra Negra is a small plateau at c.1,050 m near Inajá in south-central Pernambuco. The reserve embraces the entire plateau around which the main habitat is caatinga, with more humid forest persisting on the plateau itself.

Megaxenops parnaguae	1980	Collected on the border of the reserve at 1,100 m.
Carduelis yarrellii	1980s	Recorded from the reserve.

Saltinho (Pernambuco)

BR 034

Saltinho Biological Reserve (IUCN category I), 500 ha

8°44'S 35°11'W
EBA B47

This small biological reserve, located in south-west Pernambuco in Município Rio Formoso, represents one of the last areas of Atlantic forest in the state, although much of it is in fact regenerating second growth. The area is nevertheless important for several threatened species endemic to north-east Brazil.

Synallaxis infuscata	1980	A number of specimens.
Myrmeciza ruficauda	1986	One sighting.
Tangara fastuosa	1992	Apparently uncommon. Birds confiscated from markets were released here in 1992.
Carduelis yarrellii	1980s	There were records during the early 1980s.

Pedra (Serra) Branca and Fazenda Bananeira (Alagoas)

BR 035
9°15′S 35°50′W
EBA B47

Murici Biological Reserve (IUCN category I), 3,000 ha

Lying in the Chapada de Borborena on the coastal escarpment in central north-east Alagoas, this area is one of the main remnants (c.3,000 ha) of upland Atlantic forest left in the state, though it is highly fragmented. Fazenda Pedra Branca has lost much of its forest, and the best areas now lie within Fazenda Bananeira where c.1,200 ha is being managed as a reserve with the support of the owners. The forest is still threatened, especially by fires spreading from adjacent sugarcane plantations. This small forest fragment is nevertheless one of the most important sites for threatened birds in the Neotropics.

Touit surda	1986	Several records of small flocks in the mid-1980s.
Synallaxis infuscata	1990s	Probably the largest remaining population—which is in the largest continuous tract of forest in its range.
Philydor novaesi	1990s	The only known site where, however, numbers must be very small.
Myrmotherula snowi	1990s	The only known site; apparently restricted to upland forest at c.550 m.
Terenura sicki	1990s	The type-locality; apparently restricted to the remaining good forest in Fazenda Bananeira.
Myrmeciza ruficauda	1990s	Apparently confined to Fazenda Bananeira, where it is uncommon.
Carpornis melanocephalus	1993	Up to six seen together.
Iodopleura pipra	1991	Seen on a number of occasions through the early 1990s.
Xipholena atropurpurea	1994	A male and two females seen in July 1994 (Forrester 1994).
Phylloscartes ceciliae	1990s	The type-locality; the species is now confined to Fazenda Bananeira and Pedra Talhada Federal Biological Reserve (BR 036).
Tangara fastuosa	1994	A number of records in the 1980s, and not uncommon during 1994 (Forrester 1994).
Curaeus forbesi	1989	Several sightings in the 1980s.
Carduelis yarrellii	1994	Seen on the edge of the remaining forest.

Pedra Talhada (Alagoas)

BR 036
9°20′S 36°28′W
EBA B47

Pedra Talhada Federal Biological Reserve (IUCN category I), 4,469 ha

This area in the Serra das Guaribas at Quebrangulo is one of the last remaining upland Atlantic forest areas within Alagoas and north-east Brazil. Pedra Talhada became a protected area only in 1989 following efforts to preserve the forests in the Serra das Guaribas.

Leucopternis lacernulata	1992	One of the only records for Alagoas.
Touit surda	1995	At least two seen in February 1995 (G. Kirwan *in litt.* 1995).
Synallaxis infuscata	1995	Not uncommon, with observations annually since the 1980s.
Terenura sicki	1990	A small but healthy population existed in the late 1980s, but the species was not seen during searches in July 1994 (Forrester 1994) and February 1995 (G. Kirwan *in litt.* 1995).
Myrmeciza ruficauda	1995	A number of records in the late 1980s, but now uncommon, with visiting birdwatchers having seen only one in July 1994 and a pair in February 1995 (Forrester 1994, G. Kirwan *in litt.* 1995).
Phylloscartes ceciliae	1995	Found here in 1989 and seen annually since. One of only two known localities.
Hemitriccus mirandae	1995	Recorded regularly since the late 1980s, apparently confined to the best forests on the reserve; up to four in mature forest in July 1994 (Forrester 1994) and three seen in February 1995 (G. Kirwan *in litt.* 1995).
Xipholena atropurpurea	1994	One seen in July 1994 (Forrester 1994).
Tangara fastuosa	1995	Not uncommon, but numbers must be quite small; seen in February 1995 (G. Kirwan *in litt.* 1995).
Curaeus forbesi	1995	A stronghold; most sightings come from the open pasture with scattered trees near the park headquarters (e.g. Forrester 1994, G. Kirwan *in litt.* 1995).
Carduelis yarrellii	1995	A maximum of eight, usually in small flocks, seen on two days in February 1995 (G. Kirwan *in litt.* 1995). A single male seen on an open hillside near the park in July 1994 (Forrester 1994).

São Miguel dos Campos (Alagoas)

BR 037

9°47′S 36°05′W
EBA B47

Unprotected

This area is located along the Rio São Miguel, c.20 km from the coast and 55 km south-west of Maceió in the lowlands of south-east Alagoas. Some of the last remnants of Atlantic forest in Alagoas existed in this area in the 1980s, but it is likely that even these have been cleared for sugarcane. Any forest remaining in the area needs immediate protection.

Mitu mitu	c.1988	Probably extinct in the wild, but the last sighting came from this area, with an unconfirmed report in 1988.
Touit surda	1951	One collected in this general area.
Amazona rhodocorytha	1984	One collected from a flock of six.
Xipholena atropurpurea	1983	One record.
Tangara fastuosa	1979	One record.

Serra de Itabaiana (Sergipe)

BR 038

11°00′S 37°20′W

Serra de Itabaiana Federal Ecological Station (IUCN category IV), 1,752 ha

Located between Salgada and Itaporanga d'Ajuda in southern Sergipe, the main habitat in this small reserve is dry deciduous forest. Further surveys may find other threatened species.

Herpsilochmus pectoralis	1991	An adult male singing in September.

Curaçá (Bahia)

BR 039

9°00′S 39°54′W
EBA B46

Unprotected

Comprising the caraiba gallery woodland adjoining the Rio São Francisco of northern Bahia, the area supports the last known wild individual of *Cyanopsittaca spixii* and, although not formally part of the protected-area network, is safeguarded through the conservation measures being made towards saving the macaw. Access is strictly controlled.

Cyanopsittaca spixii	1995	A group of three was discovered in the 1980s but all were believed captured in 1987; however, a single male was found in 1990 (still present in May 1995) and has been monitored closely since. A captive female was released at the site in March 1995.

Canudos (Bahia)

BR 040

9°55′S 38°40′W
EBA B46

Raso da Catarina Federal Ecological Reserve (IUCN category I), 200,000 ha

This region within Canudos in northern Bahia has undergone considerable modification for agriculture, including cattle- and goat-raising. Some extensive caatinga woodland, rich with terrestrial bromeliads, remains on a number of the large, private fazendas.

Anodorhynchus leari	1995	The area encompasses most of the range of the largest remaining population of the species, including nesting and feeding sites. Conservation measures are being undertaken but the species is still vulnerable to trappers.

Figure 4. Key Areas in eastern Brazil.

039 Curaçá
040 Canudos
041 Jeremoabo
042 Morro do Chapéu
043 Santo Amaro
044 Chapada Diamantina
045 Valença
046 Jacaré
047 Coribe
048 Jequié
049 Palmas de Monte Alto
050 Boa Nova
051 Serra do Ouricana
052 Ilhéus
053 Una
054 Barrolândia
055 Porto Seguro
056 Monte Pascoal
057 Araguaia
058 Chapada dos Veadeiros
059 Rio Paranã
060 Cristalina
061 Emas
062 Brasília
063 Itacarambi and Mocambinho
064 Brejo do Amparo
065 Almenara
066 Serra do Cipó
067 Caratinga
068 Rio Doce
069 Mangabeiras
070 Peti
071 Serra do Caraça
072 Serra da Canastra
073 Ibitipoca
074 Córrego Grande
075 Córrego do Veado
076 Sooretama
077 Linhares
078 Augusto Ruschi
079 Jatibocas
080 Duas Bocas

Jeremoabo (Bahia)

BR 041

10°04′S 38°30′W

Unprotected

The habitat in this area, located c.21 km west of Jeremoabo in northern Bahia (Forrester 1994), is tall riparian dry forest, better developed than other caatinga forests in the region.

Herpsilochmus pectoralis	1995	Regularly seen in the 1990s (e.g. Forrester 1994, G. Kirwan *in litt.* 1995, J. Wall *in litt.* 1995).

Table 5. Matrix of threatened species by Key Area in eastern Brazil (the area covered by Figure 4).

	039	040	041	042	043	044	045	046	047	048	049	050	051	052	053
Nothura minor	–	–	–	–	–	–	–	–	–	–	–	–	–	–	–
Taoniscus nanus	–	–	–	–	–	–	–	–	–	–	–	–	–	–	–
Mergus octosetaceus	–	–	–	–	–	–	–	–	–	–	–	–	–	–	–
Leucopternis lacernulata	–	–	–	–	–	–	–	–	–	–	–	–	–	–	–
Harpyhaliaetus coronatus	–	–	–	–	–	–	–	–	–	–	–	–	–	–	–
Penelope ochrogaster	–	–	–	–	–	–	–	–	–	–	–	–	–	–	–
Pipile jacutinga	–	–	–	–	–	–	–	–	–	–	–	–	–	–	–
Crax blumenbachii	–	–	–	–	–	–	–	–	–	–	–	–	–	–	●
Laterallus xenopterus	–	–	–	–	–	–	–	–	–	–	–	–	–	–	–
Claravis godefrida	–	–	–	–	–	–	–	–	–	–	–	–	–	–	–
Anodorhynchus hyacinthinus	–	–	–	–	–	–	–	–	–	–	–	–	–	–	–
Anodorhynchus leari	–	●	–	–	–	–	–	–	–	–	–	–	–	–	–
Cyanopsitta spixii	●	–	–	–	–	–	–	–	–	–	–	–	–	–	–
Aratinga auricapilla	–	–	–	●	–	●	–	–	–	–	–	–	●	–	–
Pyrrhura cruentata	–	–	–	–	–	●	–	–	–	–	–	–	●	–	–
Touit surda	–	–	–	–	–	–	–	●	–	–	–	–	●	–	–
Amazona rhodocorytha	–	–	–	–	–	–	–	–	–	–	–	–	–	–	–
Amazona vinacea	–	–	–	–	–	–	–	–	–	–	–	–	–	–	–
Triclaria malachitacea	–	–	–	–	–	–	–	–	–	–	–	–	–	–	–
Caprimulgus candicans	–	–	–	–	–	–	–	–	–	–	–	–	–	–	–
Eleothreptus anomalus	–	–	–	–	–	–	–	–	–	–	–	–	–	–	–
Glaucis dohrnii	–	–	–	–	–	–	–	–	–	–	–	–	–	–	–
Jacamaralcyon tridactyla	–	–	–	–	–	–	–	–	–	–	–	–	–	–	–
Xiphocolaptes falcirostris	–	–	–	–	–	–	–	–	●	–	–	–	–	–	–
Asthenes luizae	–	–	–	–	–	–	–	–	–	–	–	–	–	–	–
Thripophaga macroura	–	–	–	–	–	–	–	●	–	–	–	–	●	–	–
Megaxenops parnaguae	–	–	–	●	–	●	–	–	●	–	●	–	–	–	–
Dysithamnus plumbeus	–	–	–	–	–	–	–	–	–	–	–	–	–	–	–
Herpsilochmus pectoralis	–	–	●	–	–	●	–	–	–	–	–	–	–	–	–
Formicivora iheringi	–	–	–	–	–	–	–	–	–	●	–	●	–	–	–
Pyriglena atra	–	–	–	–	●	–	–	–	–	–	–	–	–	–	–
Rhopornis ardesiaca	–	–	–	–	–	–	–	–	–	●	–	●	–	–	–
Myrmeciza ruficauda	–	–	–	–	–	–	–	–	–	–	–	–	–	–	–
Merulaxis stresemanni	–	–	–	–	–	–	–	–	–	–	–	–	–	●	–
Scytalopus novacapitalis	–	–	–	–	–	–	–	–	–	–	–	–	–	–	–
Scytalopus psychopompus	–	–	–	–	–	–	●	–	–	–	–	–	–	●	–
Laniisoma elegans	–	–	–	–	–	–	–	–	–	–	–	–	●	–	–
Carpornis melanocephalus	–	–	–	–	–	–	–	–	–	–	–	–	●	–	–
Iodopleura pipra	–	–	–	–	–	–	–	–	–	–	–	–	●	–	–
Lipaugus lanioides	–	–	–	–	–	–	–	–	–	–	–	–	●	–	–
Cotinga maculata	–	–	–	–	–	–	–	–	–	–	–	–	–	–	–
Xipholena atropurpurea	–	–	–	–	–	–	–	–	–	–	–	–	–	–	●
Euscarthmus rufomarginatus	–	–	–	–	–	–	–	–	–	–	–	–	–	–	–
Phylloscartes roquettei	–	–	–	–	–	–	–	–	–	–	–	–	–	–	–
Hemitriccus furcatus	–	–	–	–	–	–	–	–	–	–	–	–	●	–	–
Platyrinchus leucoryphus	–	–	–	–	–	–	–	–	–	–	–	–	–	–	–
Poospiza cinerea	–	–	–	–	–	–	–	–	–	–	–	–	–	–	–
Sporophila falcirostris	–	–	–	–	–	–	–	–	–	–	–	–	–	–	●
Sporophila frontalis	–	–	–	–	–	–	–	–	–	–	–	–	–	–	–
Sporophila hypochroma	–	–	–	–	–	–	–	–	–	–	–	–	–	–	–
Sporophila palustris	–	–	–	–	–	–	–	–	–	–	–	–	–	–	–
Nemosia rourei	–	–	–	–	–	–	–	–	–	–	–	–	–	–	–
Curaeus forbesi	–	–	–	–	–	–	–	–	–	–	–	–	–	–	–
No. of species	1	1	1	2	1	4	1	2	2	2	1	2	9	2	3

Morro do Chapéu (Bahia)

BR 042

Unprotected

11°33′S 41°09′W

A small mountain in the Serra do Tombador of the Chapada Diamantina plateau, east of Morro do Chapéu city in central Bahia. The habitat is dry forest, some of it only 1 m tall.

Aratinga auricapilla	1995	An apparent pair in dense caatinga below the summit, February 1995 (G. Kirwan *in litt.* 1995).
Megaxenops parnaguae	1995	A pair seen in a mixed flock of furnariids and antbirds in dense caatinga along the road to the summit of Morro do Chapéu in February 1995; one was also seen in October 1994 (G. Kirwan *in litt.* 1995, J. Wall *in litt.* 1995).

054	055	056	057	058	059	060	061	062	063	064	065	066	067	068	069	070	071	072	073	074	075	076	077	078	079	080	No. of areas
–	–	–	–	–	–	–	•	•	–	–	–	–	–	–	–	–	–	–	–	–	–	–	–	–	–	–	2
–	–	–	–	–	–	•	–	•	–	–	–	–	–	–	–	–	–	–	–	–	–	–	–	–	–	–	2
–	–	–	–	•	–	–	•	–	–	–	–	–	–	–	–	–	–	•	–	–	–	–	–	–	–	–	3
–	–	•	–	–	–	–	–	–	–	–	–	•	–	•	–	–	–	–	–	•	–	•	•	•	–	–	7
–	–	–	–	•	–	–	•	•	–	–	–	–	–	–	–	–	–	–	–	–	–	–	–	–	–	–	3
–	–	–	•	–	–	–	–	–	–	–	–	–	–	–	–	–	–	–	–	–	–	–	–	–	–	–	1
–	–	•	–	–	–	–	–	–	–	–	–	–	–	•	–	–	–	–	–	•	–	–	–	–	–	–	3
–	–	•	–	–	–	–	–	–	–	–	–	–	–	•	–	–	–	–	–	•	–	•	•	–	–	–	6
–	–	–	–	–	–	–	–	•	–	–	–	–	–	–	–	–	–	–	–	–	–	–	–	–	–	–	1
–	–	–	–	–	–	–	–	–	–	–	–	–	–	–	–	–	–	–	–	–	–	–	–	•	–	–	1
–	–	–	•	–	•	–	–	–	–	–	–	–	–	–	–	–	–	–	–	–	–	–	–	–	–	–	2
–	–	–	–	–	–	–	–	–	–	–	–	–	–	–	–	–	–	–	–	–	–	–	–	–	–	–	1
–	–	–	–	–	–	–	–	–	–	–	–	–	–	–	–	–	–	–	–	–	–	–	–	–	–	–	1
–	–	•	–	–	–	–	–	–	•	–	–	–	•	•	–	–	–	•	–	–	–	–	–	–	–	–	8
•	•	•	–	–	–	–	–	–	–	–	–	–	•	•	–	–	–	–	–	•	•	•	•	–	–	–	11
•	•	•	–	–	–	–	–	–	–	–	–	–	–	–	–	–	–	–	–	•	–	•	–	•	–	–	8
–	•	•	–	–	–	–	–	–	–	–	–	–	–	•	–	–	–	–	•	•	•	•	•	•	–	–	9
–	–	–	–	–	–	–	–	–	•	•	–	–	•	–	–	–	–	–	•	–	–	–	–	–	–	–	4
–	–	–	–	–	–	–	–	–	–	–	–	–	–	–	–	–	–	–	–	–	–	•	–	•	–	•	3
–	–	–	–	–	–	–	•	–	–	–	–	–	–	–	–	–	–	–	–	–	–	–	–	–	–	–	1
–	–	–	–	–	–	–	–	•	–	–	–	–	–	–	–	–	–	–	–	–	–	–	–	–	–	–	1
–	•	•	–	–	–	–	–	–	–	–	–	–	–	–	–	–	–	–	–	•	–	–	•	–	–	–	4
–	–	–	–	–	–	–	–	–	–	–	–	–	•	•	–	–	–	–	–	–	–	–	–	–	–	–	2
–	–	–	–	–	–	–	–	–	•	•	–	–	–	–	–	–	–	–	–	–	–	–	–	–	–	–	3
–	–	–	–	–	–	–	–	–	–	–	–	•	–	–	–	–	–	–	–	–	–	–	–	–	–	–	1
–	–	–	–	–	–	–	–	–	–	–	–	–	–	–	–	–	–	–	–	–	–	•	–	–	–	–	3
–	–	–	–	–	•	–	–	–	•	–	–	–	–	–	–	–	–	–	–	–	–	–	–	–	–	–	6
–	–	–	–	–	–	–	–	–	–	–	–	–	•	•	–	–	–	–	–	–	–	•	•	•	–	–	5
–	–	–	–	–	–	–	–	–	–	–	–	–	–	–	–	–	–	–	–	–	–	–	–	–	–	–	2
–	–	–	–	–	–	–	–	–	–	–	•	–	–	–	–	–	–	–	–	–	–	–	–	–	–	–	3
–	–	–	–	–	–	–	–	–	–	–	–	–	–	–	–	–	–	–	–	–	–	–	–	–	–	–	1
–	–	–	–	–	–	–	–	–	–	–	–	–	–	–	–	–	–	–	–	–	–	–	–	–	–	–	2
–	–	–	–	–	–	–	–	–	–	–	–	–	–	–	–	–	–	–	–	–	•	•	–	–	–	–	2
–	–	–	–	–	–	–	–	–	–	–	–	–	–	–	–	–	–	–	–	–	–	–	–	–	–	–	1
–	–	–	–	–	–	–	–	•	–	–	–	•	–	–	–	–	•	•	–	–	–	–	–	–	–	–	4
–	–	–	–	–	–	–	–	–	–	–	–	–	–	–	–	–	–	–	–	–	–	–	–	–	–	–	2
–	–	–	–	–	–	–	–	–	–	–	–	–	•	–	–	–	•	–	–	–	–	–	–	•	–	–	4
–	•	•	–	–	–	–	–	–	–	–	–	–	–	–	–	–	–	–	–	•	–	•	•	–	–	•	7
–	–	–	–	–	–	–	–	–	–	–	–	–	–	–	–	–	–	–	–	–	–	–	–	–	–	–	1
–	–	–	–	–	–	–	–	–	–	–	–	–	•	•	–	–	•	–	–	–	–	•	–	•	–	•	7
•	•	•	–	–	–	–	–	–	–	–	–	–	–	•	–	–	–	–	–	•	–	•	•	–	–	–	7
•	•	•	–	–	–	–	–	–	–	–	–	–	–	–	–	–	–	–	–	•	–	•	•	–	–	–	7
–	–	–	–	–	–	–	–	•	–	–	–	–	–	–	•	–	–	–	–	–	–	–	–	–	–	–	2
–	–	–	–	–	–	–	–	–	–	•	–	–	–	–	–	–	–	–	–	–	–	–	–	–	–	–	1
–	–	–	–	–	–	–	–	–	–	–	–	–	–	–	–	–	–	–	–	–	–	–	–	–	–	–	1
–	–	–	–	–	–	–	–	–	–	–	–	–	–	–	–	–	–	–	–	–	–	–	–	•	–	–	1
–	–	–	–	–	–	–	•	•	–	–	–	•	–	–	•	•	–	–	–	–	–	–	–	–	–	–	5
–	–	–	–	–	–	–	–	–	–	–	–	–	–	–	–	–	–	–	–	–	–	–	–	–	–	–	1
–	–	–	–	–	–	–	–	–	–	–	–	–	–	–	–	–	–	–	–	–	–	–	–	•	–	–	1
–	–	–	–	–	–	–	•	–	–	–	–	–	–	–	–	–	–	–	–	–	–	–	–	–	–	–	1
–	–	–	–	–	–	–	•	–	–	–	–	–	–	–	–	–	–	–	–	–	–	–	–	–	–	–	1
–	–	–	–	–	–	–	–	–	–	–	–	–	–	–	–	–	–	–	–	–	–	–	–	–	•	–	1
–	–	–	–	–	–	–	–	–	–	–	–	–	–	•	–	–	–	–	–	–	–	–	–	–	–	–	1
4	7	11	2	2	2	1	7	8	4	3	1	4	7	11	2	1	3	3	2	10	3	13	9	10	1	3	

Santo Amaro (Bahia)

BR 043

Unprotected

12°32′S 38°43′W
EBA B52

The town of Santo Amaro is located on the coastal plain near Baía de Todos os Santos c.55 km north-west of Salvador. There is virtually no Atlantic forest left in the vicinity though there are tiny but very important patches along the road west of town.

Pyriglena atra	1994	The only known site for the species; 1994 records were from second-growth forest with a dense tangled understorey and a high density of palms located on steep slopes by a stream (Forrester 1994, J. Wall *in litt.* 1995).

Chapada Diamantina (Bahia)

BR 044

Chapada Diamantina National Park (IUCN category II), 152,000 ha

12°34′S 41°15′W

The national park, covering an extensive part of the southern end of the Chapada Diamantina and the Serra do Sincora, lies in the northern Serra do Espinhaço range of south-central Bahia. The terrain consists of plateaus with steep cliffs, and the main habitat is campos rupestres, with humid forest along the slopes and caatinga and dry forest in the valleys (J. M. C. da Silva *in litt.* 1995).

Aratinga auricapilla	1989	Recorded from Lençóis.
Pyrrhura cruentata	1980s	Recorded from Andara (J. M. C. da Silva *in litt.* 1995).
Megaxenops parnaguae	1980s	Recorded from Lençóis and Andara (J. M. C. da Silva *in litt.* 1995).
Herpsilochmus pectoralis	1980s	Recorded from Ibiquera (J. M. C. da Silva *in litt.* 1995).

Valença (Bahia)

BR 045

Unprotected

13°22′S 39°05′W
EBA B52

The town of Valença is located near the coast, c.80 km south-west of Salvador. There are few tracts of forest remaining in this area and the priority should be to search, in what persists, for the species below.

Scytalopus psychopompus	1983	One of only two known sites for the species which, however, has been recorded just once.

Jacaré (Bahia)

BR 046

Unprotected

13°35′S 39°12′W
EBA B52

The area of concern lies south-west of the small village of Jacaré, c.31 km south of Valença in east-central Bahia. The area holds a good and quite extensive tract of humid forest though logging is now occurring (Forrester 1994, B. C. Forrester *in litt.* 1995).

Touit surda	1994	More than 30 seen in July 1994 (Forrester 1994).
Thripophaga macroura	1994	Up to six birds and a nest found in July 1994 (Forrester 1994).

Coribe (Bahia)

BR 047

Unprotected

13°45′S 44°04′W

The vegetation around Coribe, on the left bank of the Rio São Francisco in south-west Bahia, is mainly dry forest which is being rapidly destroyed. *Phylloscartes roquettei* should be looked for here, as the habitat is similar to Brejo do Amparo (BR 064), the species' only known site (J. M. C. da Silva *in litt.* 1995).

Xiphocolaptes falcirostris	1987	Common in the understorey of dry forest (J. M. C. da Silva *in litt.* 1995).
Megaxenops parnaguae	1990	Common in the second-growth dry forest (J. M. C. da Silva *in litt.* 1995).

Jequié (Bahia)

BR 048

13°50′S 40°05′W
EBA B48

Unprotected

The town of Jequié is located on the Rio das Contas, 200 km south-west of Salvador in east-central Bahia. The area is important for the dry forest in the vicinity of town; *Acacia*-scrub-covered hillside with patches of large terrestrial bromeliads lies to the south-east.

Formicivora iheringi	1980s	Seen in dense tangles in vine-rich forest c.15 m tall, with closely spaced small and medium trees and a dense cover of large terrestrial bromeliads.
Rhopornis ardesiaca	1988	Recorded both north and south-east of Jequié; the last record was 4 km north of town in January.

Palmas de Monte Alto (Bahia)

BR 049

14°17′S 43°20′W

Unprotected

The dominant vegetation at this site, located on the right bank of Rio São Francisco in southern Bahia, is tropical dry forest and caatinga, though it is being rapidly destroyed (J. M. C. da Silva *in litt.* 1995).

Megaxenops parnaguae	1991	Common in the borders of dry forest and tall caatinga (J. M. C. da Silva *in litt.* 1995).

Boa Nova (Bahia)

BR 050

14°22′S 40°11′W
EBA B48

Unprotected

Centred on the small town of Boa Nova in south-central Bahia, the vegetation is dry deciduous forest (mata-de-cipó) characterized by a fairly open understorey with patches of huge terrestrial bromeliads and a high density of vines, but these forests now are highly fragmented and surrounded by agricultural land. The area is adjacent to Serra do Ouricana (BR 051).

Formicivora iheringi	1995	Recorded annually over the last 10 years; favours mata-de-cipó interiors with vine tangles and patches of terrestrial bromeliads.
Rhopornis ardesiaca	1995	Recorded annually over the last 10 years; favours the floor of mata-de-cipó where terrestrial bromeliads abound.

Serra do Ouricana (Bahia)

BR 051

14°24′S 40°10′W
EBA B52

Unprotected

Located 7–13 km south-east of Boa Nova, the Serra do Ouricana is a small range of hills in south-central Bahia reaching c.1,100 m. Humid forest covers the seaward slope and marks the western and highest elevational limits of the Atlantic forest in the region (Gonzaga *et al.* 1995). The great importance of this area has been made apparent only through surveys carried out during the 1990s. Additional to the long list of threatened species found there are a number of undescribed taxa, including a new *Phylloscartes* and a new *Synallaxis* (Gonzaga *et al.* 1995, Gonzaga and Pacheco 1995).

Aratinga auricapilla	1993	Recorded recently; no information on status (Gonzaga *et al.* 1995).
Pyrrhura cruentata	1993	Recorded recently; no information on status (Gonzaga *et al.* 1995).
Touit surda	1995	Small numbers seen in recent years (Gonzaga *et al.* 1995, G. Kirwan *in litt.* 1995).
Glaucis dohrni	1994	At least two birds have been identified as this species (J. Wall *in litt.* 1995), but the records are best considered unconfirmed due to the lack of other sightings despite the many observers visiting the area (L. P. Gonzaga *in litt.* 1995).

cont.

Thripophaga macroura	1995	Seen regularly in humid habitat.
Carpornis melanocephalus	1993	Recorded recently; no information on status (Gonzaga *et al.* 1995).
Laniisoma elegans	1993	Recently recorded (Gonzaga *et al.* 1995). The first precise locality for Bahia and the northernmost site for the species.
Iodopleura pipra	1993	Recorded recently; no information on status (Gonzaga *et al.* 1995).
Lipaugus lanioides	1993	Recorded recently; no information on status (Gonzaga *et al.* 1995).
Hemitriccus furcatus	1995	Found in January 1993 and seen several times since (e.g. Tobias *et al.* 1993, Balchin 1994, Gonzaga *et al.* 1995); a northward range extension of c.1,000 km.

Ilhéus (Bahia)

BR 052

14°49'S 39°02'W
EBA B52

Unprotected

Ilhéus is an Atlantic port city located in the southern part of Bahia. The very poorly known tapaculo species listed below have been collected in the vicinity, so any remaining forest need to be surveyed for them.

Merulaxis stresemanni	1945	A female collected in May.
Scytalopus psychopompus	1944	A male collected in December. There is only one other site for the species.

Una (Bahia)

BR 053

15°12'S 39°07'W
EBA B52

Una Federal Biological Reserve (IUCN category I), 11,400 ha

The reserve lies on the coast of south-central Bahia, 10 km north of Una town. The tropical forest is notably different in form from that further south, being much less tall.

Crax blumenbachii	1986	Last record was of seven individuals.
Carpornis melanocephalus	1994	A bird heard calling from low, second-growth swamp forest in October 1994 was attributed to this species (J. Wall *in litt.* 1995).
Xipholena atropurpurea	1986	The last sighting was of a group of at least five.
Sporophila falcirostris	1987	One record, possibly of a vagrant. The only record for the state.

Barrolândia (Bahia)

BR 054

16°04'S 39°11'W
EBA B52

Barrolândia 'Gregório Bondar' CEPLAC Experimental Station (private), 710 ha

Of the two tracts of Atlantic coastal forest comprising this area, one is a proposed reserve owned by the Companhia Vale do Rio Doce (CVRD), the other is an experimental station. The CVRD-owned forest is situated near Rio Jequitinhonha c.7 km east of Barrolândia and 42 km south-west of Belmonte in central Bahia, with the Gregório Bondar experimental station immediately to the south. Both areas received some protection in the 1980s but hunting is noted to be prevalent at Gregório Bondar (Oliver and Santos 1991).

Pyrrhura cruentata	1995	A group of four seen (G. Kirwan *in litt.* 1995).
Touit surda	1995	A maximum of three seen in selectively logged forest (G. Kirwan *in litt.* 1995).
Cotinga maculata	1995	At least two recorded in January 1995 (G. Kirwan *in litt.* 1995).
Xipholena atropurpurea	1995	A maximum of 19 seen (G. Kirwan *in litt.* 1995).

Porto Seguro (Bahia)

BR 055

16°27′S 39°15′W
EBA B52

Florestas Rio Doce SA Forestry Reserve (private), 6,000 ha
CEPLAC Pau-Brasil Ecological Station (IUCN category I), area unknown

These two forest reserves are located at sea-level c.10 km inland from the seaside resort of Porto Seguro. They are now largely isolated from other forest patches, and together form one of the most important forest tracts in Bahia. A possible sighting of *Lipaugus lanioides* in 1989 needs confirming.

Pyrrhura cruentata	1994	Apparently relatively common; several sightings in 1994 (Forrester 1994, J. Wall *in litt.* 1994).
Touit surda	1994	A flock of eight seen in July 1994 (Forrester 1994).
Amazona rhodocorytha	1994	Regularly recorded in 1994, e.g. at least 40 birds in July (Forrester 1994).
Glaucis dohrnii	1993	A study in February 1993 noted the species relatively frequently, especially visiting flowering *Heliconia* (Y. Oniki *in litt.* 1994).
Carpornis melanocephalus	1994	Recorded a number of times in the last 10 years (Bushell 1994).
Cotinga maculata	1994	A number of recent sightings (e.g. Tobias *et al.* 1993, Bushell 1994).
Xipholena atropurpurea	1994	Numerous records, and described as fairly common to common (Bushell 1994); a stronghold in Bahia.

Monte Pascoal (Bahia)

BR 056

16°50′S 39°19′W
EBA B52

Monte Pascoal National Park (IUCN category II), 22,500 ha

The park is located in southern Bahia, extending from the coast deep inland to around Monte Pascoal (586 m). Habitats include mangroves, dune vegetation and tall humid forest. Conflicts have arisen between the conservation of the habitats within the park and the rights of local people; habitat destruction and hunting were still occurring in the late 1980s (Redford 1989, Forrester 1994). Two records of *Lipaugus lanioides* from the park require confirmation.

Leucopternis lacernulata	1990	A pair displaying in June.
Pipile jacutinga	1977	Recorded from the park.
Crax blumenbachii	1986	The last record was of a pair.
Aratinga auricapilla	1994	A flock of c.12 seen from the viewpoint near the visitor centre in September 1994 (Bushell 1994).
Pyrrhura cruentata	1993	An important site for the species.
Touit surda	1977	A pair seen.
Amazona rhodocorytha	1987	Apparently present during the 1980s. Birds have been poached from this area.
Glaucis dohrnii	1977	Recorded from the park.
Carpornis melanocephalus	1994	Recorded in July and September 1994 (Bushell 1994, Forrester 1994).
Cotinga maculata	1990	Two records.
Xipholena atropurpurea	1993	A handful of records.

Araguaia (Tocantins)

BR 057

10°45′S 50°10′W

Araguaia National Park (IUCN category II), 562,312 ha

The Araguaia National Park is in the northern half of Ilha Bananal, the world's largest river island, formed between the Rio Araguaia and the Braço Menor do Rio Araguaia, in west-central Tocantins. The habitat is a transition vegetation between humid Amazon forest and dry forest of the Brazilian central plateau.

Penelope ochrogaster	1980s	Several sight records of small numbers in the park during the late 1970s and early 1980s.
Anodorhynchus hyacinthinus	1980s	A small, low-density population was present in the 1980s.

Chapada dos Veadeiros (Goiás)

BR 058
14°01′S 47°40′W

Chapada dos Veadeiros National Park (IUCN category II), 60,000 ha

This national park is located in the Alto Paraíso de Goiás, south-west of Cavalcanti in east-central Goiás. The area is centred on the upper Rio Preto and Serra Santana, covers hilly ground ranging from 600 to 1,650 m and has extensive areas of open cerrado.

Mergus octosetaceus	1987	Pairs observed on five occasions between October 1986 and January 1987.
Harpyhaliaetus coronatus	1991	One collected in 1883 and a pair seen in grassy cerrado in 1991.

Rio Paranã (Goiás)

BR 059
14°05′S 46°22′W

Unprotected

Located along the Rio Paranã c.200 km north-west of Brasília in eastern Goiás, this region is covered covered by dry forest, extensive gallery forest and cerrado, but the forest is being destroyed rapidly (J. M. C. da Silva *in litt.* 1993).

Anodorhynchus hyacinthinus	1993	Old specimens were collected in this region and one bird was seen at São Domingos in 1993 (J. M. C. da Silva *in litt.* 1993).
Megaxenops parnaguae	1993	Rare in second-growth dry forest (J. M. C. da Silva *in litt.* 1993).

Cristalina (Goiás)

BR 060
16°45′S 47°36′W

Unprotected

An area located c.110 km south-east of Brasília in south-east Goiás, the main vegetation being gallery forest, savanna and cerrado. Surveys are needed to assess the current status of the habitat and of the species below.

Taoniscus nanus	1966	Two records, the only ones for the state.

Emas (Goiás)

BR 061
18°10′S 53°00′W

Emas National Park (IUCN category II), 131,868 ha

This reserve is situated on the border of Goiás and Mato Grosso do Sul, c.130 km south-west of Jataí. About 60% of the park is grassland, most of the rest being cerrado, plus areas of gallery forest, dry forest and grass swamps with *Mauritia* palms (Redford 1985).

Nothura minor	1987	Known to occur, but greatly outnumbered by Spotted Nothura *N. maculata*.
Harpyhaliaetus coronatus	1991	Recorded several times in recent years.
Mergus octosetaceus	1990	A pair observed in August.
Caprimulgus candicans	1990	Found in open cerrado burnt in the previous year (or two), often favouring areas with a scattering of very low shrubs and prostrate palms.
Poospiza cinerea	1988	Recorded from the park.
Sporophila hypochroma	1990	A wintering male seen by a bird-tour group.
Sporophila palustris	1990	A handful of sightings of wintering birds in the late 1970s and early 1980s, and six males seen in 1990.

Brasília (Distrito Federal)

BR 062
15°47'S 47°55'W

Brasília National Park (IUCN category II), 28,000 ha
Norte Brasília State Ecological Park (IUCN category II), 50,046 ha

The national park is located immediately north-west of Brasília on the Central Plateau in central Distrito Federal, and it encompasses the upper Rio Torto and Rio Bananal, where the principal habitat is cerrado.

Nothura minor	1990	Up to three singing in an area of c.20 ha but none was found in a larger area of similar and seemingly suitable habitat.
Taoniscus nanus	1980s	Presumed resident but very few records.
Harpyhaliaetus coronatus	1991	A handful of records during the last 20 years.
Laterallus xenopterus	1980s	One collected in 1978, and birds seen subsequently at Santa Maria barrage.
Eleothreptus anomalus	1978	Apparently only one record (a specimen).
Scytalopus novacapitalis	1990s	Regularly recorded in the national park.
Euscarthmus rufomarginatus	1989	Recorded from this area.
Poospiza cinerea	1989	Recorded from this area.

Itacarambi and Mocambinho (Minas Gerais)

BR 063
15°05'N 44°07'W

Unprotected

Itacarambi is on the left bank and Mocambinho on the right bank of the Rio São Francisco in northern Minas Gerais. The main habitat in the area is caatinga and dry deciduous forest.

Amazona vinacea	1986	Seen on near Itacarambi; a significant northward range extension.
Aratinga auricapilla	1987	Common at Fazenda Olhos d'Água (J. M. C. da Silva *in litt.* 1995).
Xiphocolaptes falcirostris	1987	Several records from Fazenda Olhos d'Água, Itacarambi.
Megaxenops parnaguae	1990	Seen several times in dense caatinga near Mocambinho in September–October 1990.

Brejo do Amparo (Minas Gerais)

BR 064
15°29'S 44°22'W
EBA B49

Unprotected

An area on the left bank of the Rio São Francisco in northern Minas Gerais, the main habitat being caatinga and dry deciduous forest. It is one of the only sites for Caatinga Black-tyrant *Knipolegus aterrimus franciscanus* which da Silva and Oren (1992) suggest should be given full species status and is listed by Collar *et al.* (1994) as a potential addition to the IUCN threatened list.

Amazona vinacea	1986	Seen on the left bank of the Rio São Francisco near Januária in November 1986.
Phylloscartes roquettei	1977	This is the only record of the species anywhere in recent times, but subsequent searches have failed to find it (J. M. C. da Silva *in litt.* 1993).
Xiphocolaptes falcirostris	1986	Recorded from the area.

Almenara (Minas Gerais)

BR 065
16°11'S 40°42'W
EBA B48

Unprotected

Almenara is on the left bank of the Rio Jequitinhonha, 46 km north-east of Jequitinhonha, in north-east Minas Gerais. Dry forest with dense vine tangles is the important habitat for the species below.

Formicivora iheringi	1986	Found in December 1973.

Serra do Cipó (Minas Gerais)

BR 066

19°14'S 43°53'W
EBA B50

Serra do Cipó National Park (IUCN category II), 33,800 ha

The national park lies at the southern end of the Serra do Espinhaço, c.100 km north-east of Belo Horizonte in south-central Minas Gerais. The higher parts of the park support campo grassland while at lower elevations the main habitat is cerrado and gallery forest. An extension of the park boundary by c.20 km^2 would enclose the entire known range of *Asthenes luizae*.

Leucopternis lacernulata	1980	Recorded from the park.
Asthenes luizae	1995	The only known site for this species which was only recently described to science. Considered not uncommon in a surveyed 10 km^2 of suitable habitat which, however, lies just outside the national park boundaries.
Scytalopus novacapitalis	1978	One sighting near Alto do Palácio.
Poospiza cinerea	1995	Regularly recorded from open scrub in the upland.

Caratinga (Minas Gerais)

BR 067

19°47'S 41°45'W
EBA B52

Caratinga Reserve (Fazenda Montes Claros) (private), 880 ha

Fazenda Montes Claros is a farm situated between Caratinga and Ipanema in eastern Minas Gerais. This isolated area of forest stands on steep hills, ridges and valleys from 400 to 680 m, but perhaps as little as 20% is primary. The forest has no official protection but the owner has banned hunting and has agreed informally to maintain the area for conservation.

Amazona vinacea	1990	Found surprisingly numerous by a visiting bird-tour group.
Aratinga auricapilla	1990	Recorded a number of times during the 1980s, some observers finding it fairly common.
Pyrrhura cruentata	1990	Fairly common.
Jacamaralcyon tridactyla	1986	The only record from a protected area, although Forrester (1993) suggests that it may now be extinct here.
Dysithamnus plumbeus	1990	A number of sightings in the late 1980s.
Laniisoma elegans	1980s	Several sightings in the 1980s.
Lipaugus lanioides	1980s	Apparently recorded at this site.

Rio Doce (Minas Gerais)

BR 068

19°50'S 42°31'W
EBA B52

Rio Doce State Forest Park (IUCN category II), 35,973 ha

Located at the confluence of the Rio Piracicaba and the Rio Doce, immediately south of Ipatinga and 200 km east of Belo Horizonte in eastern Minas Gerais, this reserve supports one of the largest tracts of humid Atlantic forest in south-east Brazil. Much is unsurveyed owing to the low number of trails and multitude of small lakes which prevent easy access.

Leucopternis lacernulata	1980s	Recorded from the reserve.
Pipile jacutinga	1980s	Five seen at two sites near Dionísio are apparently the only records.
Crax blumenbachii	1980s	Seen near Dionísio and below São Caetano on the Rio Pomba.
Aratinga auricapilla	1981	Recorded from the reserve.
Pyrrhura cruentata	1980s	Recorded in 1940 on the lower Rio Piracicaba at the confluence with Rio Doce; apparently still there in the 1980s.
Amazona rhodocorytha	1980s	Seen during the 1980s.
Jacamaralcyon tridactyla	1940	Collected on the lower Rio Piracicaba, presumably within or close to the state park.
Dysithamnus plumbeus	1986	Seen near the Piracicaba confluence.
Lipaugus lanioides	1988	Recorded twice in August 1988.

cont.

Cotinga maculata	1940	One collected on the Lower Rio Suaçuí, close to the present-day state park.
Curaeus forbesi	1990s	Recorded from the park, and known from historical records. A very important site for the species, and the only area in the south of its range where it is noted regularly.

Mangabeiras (Minas Gerais)

BR 069

19°55'S 43°53'W

Mangabeiras Park (status and area unknown)

Located in south-central Minas Gerais, c.10 km east of the state capital Belo Horizonte, the area comprises tall woodland on the edge of a plateau with rocky outcrops and cerrado.

Euscarthmus rufomarginatus	1989	Recorded only once during eight months of regular visits (A. Whittaker *in litt.* 1993).
Poospiza cinerea	1989	Seen regularly during eight months of visits to the park, feeding young and carrying nest-material (A. Whittaker *in litt.* 1993).

Peti (Minas Gerais)

BR 070

20°00'S 43°03'W

Peti (Santa Bárbara) State Ecological Station (IUCN category IV), 2,712 ha

This reserve is located 15 km south-east of Santa Bárbara and 120 km east of Belo Horizonte in east-central Minas Gerais. The habitat is mainly second-growth scrub, but also includes marsh and cerrado.

Poospiza cinerea	1990	Recently recorded from the reserve.

Serra do Caraça (Minas Gerais)

BR 071

20°08'S 43°50'W

Caraça Natural Park (status and area unknown)

Serra do Caraça lies 25 km south of Santa Bárbara and east of Caraça at the southern end of the Serra do Espinhaço in east-central Minas Gerais. The mountains comprise forested slopes interspersed with grassland, the lower slopes tending to support drier forest.

Scytalopus novacapitalis	1980s	A number of sightings in the 1980s have been attributed to this species.
Laniisoma elegans	1980s	Several records in the 1980s.
Lipaugus lanioides	1990	Occasionally recorded; one sighting was at 1,450 m.

Serra da Canastra (Minas Gerais)

BR 072

20°15'S 46°37'W
EBAs B52, B53

Serra da Canastra National Park (IUCN category II), 71,525 ha

Located 330 km west of Belo Horizonte in south-west Minas Gerais, the park comprises a grassy highland plateau with steep escarpments supporting patches of gallery forest.

Mergus octosetaceus	1995	Seen at two sites in the park in July 1995 (A. Balmford *in litt.* 1995). One of the only sites where the species is often recorded, but the population is in serious danger. Between 1981 and 1985 only two pairs were found in the park, with a third pair c.50 km downstream. Breeding was still occurring in the reserve in 1990 when two ducklings were observed.
Aratinga auricapilla	1995	Irregularly recorded, the last sightings being of 18 seen along the road to Casca d'Anta in February 1995 (G. Kirwan *in litt.* 1995).
Scytalopus novacapitalis	1995	Recorded annually during the last five years (e.g. Forrester 1990, Tobias *et al.* 1993, G. Speight *in litt.* 1994, G. Kirwan *in litt.* 1995).

Ibitipoca (Minas Gerais)

BR 073
21°33′S 43°55′W

Ibitipoca State Park (IUCN category II), 1,488 ha

This protected area is located in Lima Duarte município in south-central Minas Gerais. The park's main habitat is forest, but most of it is secondary.

Amazona rhodocorytha	1980s	Seen regularly in the 1980s.
Amazona vinacea	1980s	One or two seen regularly in the 1980s.

Córrego Grande (Espírito Santo)

BR 074
18°16′S 39°48′W
EBA B52

Córrego Grande Federal Biological Reserve (IUCN category I), 1,504 ha

This small reserve lies on the flat coastal plain c.15 km east of the Vitória–Salvador main road in the extreme northern part of Espírito Santo (the northern boundary of the reserve lies along the border with Bahia). The reserve principally supports humid Atlantic forest and although much of it appears to be fairly disturbed a sizeable area of good forest (c.500 ha) lies adjacent to the south-west boundary.

Leucopternis lacernulata	1986	One sighting probably attributable to this species.
Pipile jacutinga	1973	Strong evidence that the species is extinct in this area.
Crax blumenbachii	1980s	Not found in October 1986, and almost certainly extinct.
Pyrrhura cruentata	1987	Recorded from the reserve.
Touit surda	1979	Known from a sighting of three birds.
Amazona rhodocorytha	1986	Seen in October 1986.
Glaucis dohrnii	1973	Formerly thought to be a stronghold but now almost certainly extinct.
Carpornis melanocephalus	1986	Recorded from the reserve.
Cotinga maculata	1973	Recorded from the reserve.
Xipholena atropurpurea	1970	Recorded from the reserve.

Córrego do Veado (Espírito Santo)

BR 075
18°16′S 40°12′W
EBA B52

Córrego do Veado Federal Biological Reserve (IUCN category I), 2,392 ha

This reserve is located on flat terrain north of the lower Rio São Mateus, c.10 km north-east of Pinheiros in northern Espírito Santo, and 25 km south of the state border with Bahia. The semi-deciduous forest is entirely surrounded by pasture, and there is a permanent threat of fire spreading into the forest from this farmland. Indeed a fire in the early 1970s caused considerable damage to much of the reserve.

Amazona rhodocorytha	1986	Present in small numbers in 1986.
Pyrrhura cruentata	1986	A flock of 12 seen in 1986.
Myrmeciza ruficauda	1986	Recorded from the reserve.

Sooretama (Espírito Santo)

BR 076

19°03'S 40°00'W
EBA B52

Sooretama Federal Biological Reserve (IUCN category I), 24,000 ha

This reserve lies at sea-level on the Rio Cupido, c.45 km north of Linhares (BR 077) in east-central Espírito Santo. Together with Linhares it comprises one of the largest remaining tracts of forest in Espírito Santo. There are extensive marshlands in and around the reserve, and some of the forest is secondary.

Species	Year	Notes
Leucopternis lacernulata	1995	Regularly recorded, e.g. two seen in a six-day visit in 1995 (G. Kirwan *in litt.* 1995).
Crax blumenbachii	1995	The stronghold for the species (e.g. G. Kirwan *in litt.* 1995).
Pyrrhura cruentata	1995	A stronghold; flocks of more than 30 seen regularly (e.g. G. Kirwan *in litt.* 1995).
Touit surda	1994	A flock of at least eight present throughout the late 1980s; two seen in August 1994 (Forrester 1994).
Amazona rhodocorytha	1995	Still present in the reserve, and apparently fairly common (e.g. G. Kirwan *in litt.* 1995).
Triclaria malachitacea	1939	Within the present-day reserve, known only from Córrego Cupido, Rio Barra Seca.
Thripophaga macroura	1995	A stronghold for the species, but uncommon (e.g. G. Kirwan *in litt.* 1995).
Dysithamnus plumbeus	1995	Arguably the stronghold for the species, at least in Espírito Santo (e.g. G. Kirwan *in litt.* 1995).
Myrmeciza ruficauda	1995	Recorded a number of times in the last 10 years (e.g. G. Kirwan *in litt.* 1995), but apparently not as common as the subspecies *M. ruficauda soror* in Alagoas and Pernambuco states.
Carpornis melanocephalus	1995	A key site for the species in Espírito Santo (e.g. G. Kirwan *in litt.* 1995).
Lipaugus lanioides	1987	Apparently recorded only irregularly from the reserve.
Cotinga maculata	1995	A small population persists in the reserve (e.g. G. Kirwan *in litt.* 1995).
Xipholena atropurpurea	1995	Seen regularly in small numbers (e.g. G. Kirwan *in litt.* 1995).

Linhares (Espírito Santo)

BR 077

19°10'S 39°55'W
EBA B52

CVRD Forestry Reserve of Linhares (private), 22,000 ha

This humid forest reserve is located at sea-level, north of Linhares in east-central Espírito Santo. It lies adjacent to Sooretama (BR 076), with which it forms the most important lowland forest block in the state.

Species	Year	Notes
Leucopternis lacernulata	1980s	A number of sightings, mainly in the 1980s.
Crax blumenbachii	c.1995	A stronghold for the species; perhaps as many as 100 birds.
Pyrrhura cruentata	1993	A stronghold for the species.
Amazona rhodocorytha	1980s	Recorded from the reserve.
Glaucis dohrnii	1988	Recorded from the reserve, although a search in August 1991 did not find it.
Dysithamnus plumbeus	1991	A key site for the species along with adjacent Sooretama (BR 076).
Carpornis melanocephalus	c.1995	Recorded annually since 1986.
Cotinga maculata	c.1995	Recorded annually since 1984.
Xipholena atropurpurea	c.1995	Regularly recorded; together with Sooretama, this area is a stronghold.

Augusto Ruschi (Nova Lombardia) (Espírito Santo)

BR 078
19°50′S 40°32′W
EBA B52

Augusto Ruschi (Nova Lombardia) Federal Biological Reserve (IUCN category I), 4,492 ha

Situated 8 km north-east of Santa Teresa in east-central Espírito Santo, this reserve ranges from 500 to 1,200 m above sea-level. Humid montane forest rich in epiphytes and *Euterpe* palms covers most of the terrain, and there appear to be no problems with poaching or encroachment of agriculture.

Leucopternis lacernulata	1994	Recorded annually, up to 900 m.
Claravis godefrida	1986	One winter record.
Triclaria malachitacea	1994	Seasonally present in palm groves in the forest (E. O. Willis *per* B. C. Forrester *in litt.* 1995).
Touit surda	1995	Regular sightings involve flocks of up to 14 birds (e.g. G. Kirwan *in litt.* 1995).
Amazona rhodocorytha	1991	Recorded 1986 and 1991.
Dysithamnus plumbeus	1990	Recorded.
Laniisoma elegans	1987	Known from a small number of records during the 1970s and 1980s.
Lipaugus lanioides	1995	Probably one of the species' main strongholds; found nesting in January 1995 (G. Kirwan *in litt.* 1995).
Platyrinchus leucoryphus	1995	Recorded annually 1984–1995; probably a key site for the species (e.g. G. Kirwan *in litt.* 1995).
Sporophila frontalis	1979	One seen in September, probably a vagrant; apparently the only one for the state.

Jatibocas (Espírito Santo)

BR 079
20°05′N 40°55′W
EBA B52

Unprotected

This area is located 66 km west-north-west of Vitória and 87 km north-north-east of Cachoeiro de Itapemirim in south-central Espírito Santo. Apparently this region of Espírito Santo is now deforested (Sick and Teixeira 1979), but a survey to locate any remaining forest patches would seem worthwhile as the area is one of only two old localities for the species below.

Nemosia rourei	1941	A flock of eight seen in the canopy of montane forest at 900–1,000 m was not associated with other birds (Sick 1993).

Duas Bocas (Espírito Santo)

BR 080
20°17′S 40°30′W
EBA B52

Duas Bocas State Biological Reserve (IUCN category IV), 2,910 ha

The vegetation of this small reserve, located near Cariacica in south-east Espírito Santo, consists of humid forest. The area merits further surveys to assess the status of the species listed below.

Triclaria malachitacea	1980s	Regularly recorded from near Cariacica during the 1980s.
Carpornis melanocephalus	1980s	Recorded from this area.
Lipaugus lanioides	1987	Seen at 700 m in 1987; apparently the only confirmed record for the reserve.

Bate-pau (Rio de Janeiro)

BR 081
20°57′S 42°03′W

Unprotected

This area is located c.30 km north-west of Itaperuna in north-west Rio de Janeiro state. The habitat comprises dry forest woodlots in an otherwise mainly cultivated region.

Jacamaralcyon tridactyla	1988	One record.

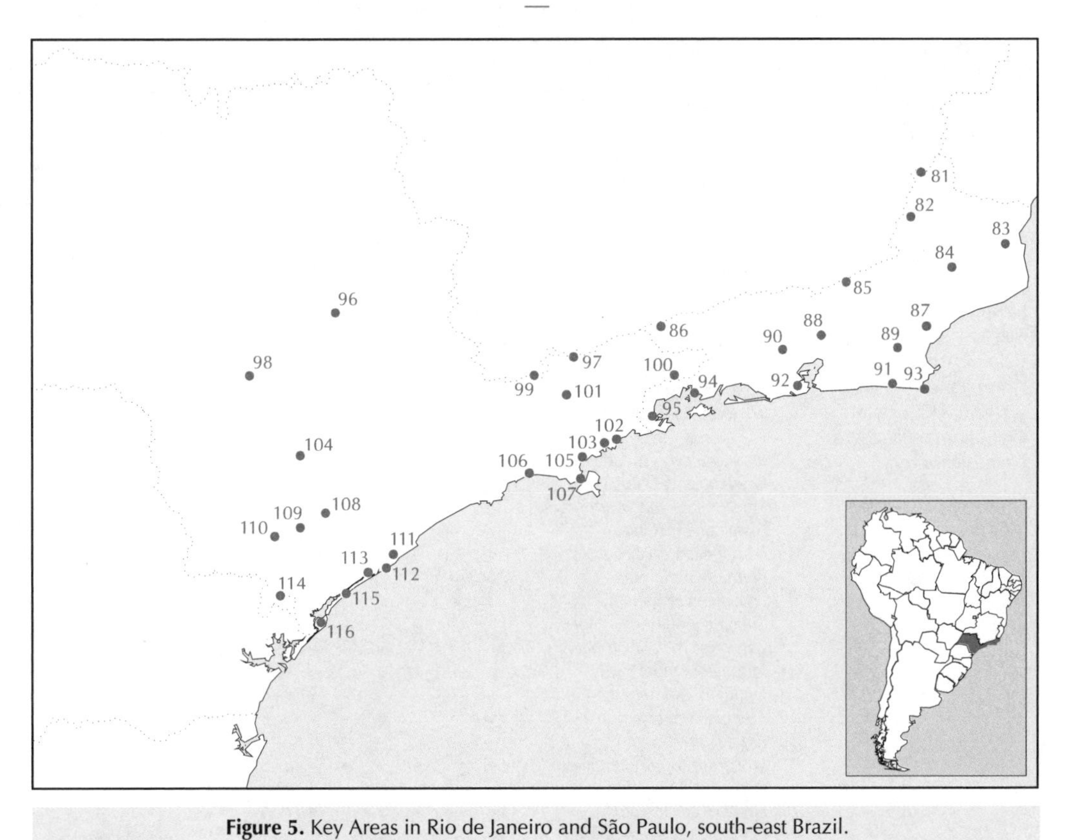

Figure 5. Key Areas in Rio de Janeiro and São Paulo, south-east Brazil.

081 Bate-pau
082 Miracema
083 Campelo
084 Desengano
085 Somidorou
086 Itatiaia
087 Fazenda União
088 Serra dos Órgãos
089 Poço das Antas
090 Serra do Tinguá
091 Maçambaba
092 Tijuca
093 Cabo Frio
094 Angra dos Reis
095 Pedra Branca
096 Itirapina
097 Campos do Jordão
098 Fazenda Rio Pardo
099 Poncianos
100 Serra da Bocaína
101 Taubaté
102 Ubatuba
103 Bairro de Corcovado
104 Itapetininga
105 Caraguatatuba
106 Boracéia and Bertioga
107 Ilha São Sebastião
108 Carlos Botelho
109 Fazenda Intervales
110 Alto Ribeira
111 Foz do Rio Ribeira
112 Juréia
113 Iguape
114 Jacupiranga
115 Ilhas Comprida and Cananéia
116 Ilha do Cardoso
117 Telémaco Borba
118 Adrianópolis

Miracema (Rio de Janeiro)

BR 082

21°22'S 42°09'W

Unprotected

This area is situated along the Rio Pomba valley (a tributary of the Rio Paraíba do Sul) near Miracema, c.40 km south-west of Itaperuna in northern Rio de Janeiro. The habitat comprises dry forest woodlots in an otherwise mainly cultivated region.

Jacamaralcyon tridactyla	1988	One record.

Table 6. Matrix of threatened species by Key Area in Rio de Janeiro and São Paulo, south-east Brazil (the area covered by Figure 5). Additional threatened species known from this region but for which no Key Areas have been identified are *Mergus octosetaceus*, *Crax blumenbachii*, *Aratinga auricapilla*, *Calyptura cristata*, *Cotinga maculata* and *Nemosia rourei*.

	081	082	083	084	085	086	087	088	089	090	091	092	093	094	095
Nothura minor	–	–	–	–	–	–	–	–	–	–	–	–	–	–	–
Taoniscus nanus	–	–	–	–	–	–	–	–	–	–	–	–	–	–	–
Leucopternis lacernulata	–	–	–	•	–	–	–	•	•	–	–	•	–	–	–
Harpyhaliaetus coronatus	–	–	–	–	–	–	–	–	–	•	–	–	–	–	–
Pipile jacutinga	–	–	–	–	–	•	–	•	–	–	–	–	–	–	–
Coturnicops notatus	–	–	–	–	–	–	–	–	–	–	–	–	–	–	–
Claravis godefrida	–	–	–	–	–	•	–	•	–	–	–	–	–	–	–
Pyrrhura cruentata	–	–	–	•	–	–	•	–	–	–	–	–	–	–	–
Touit melanonota	–	–	–	•	–	•	–	•	–	–	–	•	–	–	•
Touit surda	–	–	–	•	–	•	•	•	–	–	–	–	–	–	–
Amazona brasiliensis	–	–	–	–	–	–	–	–	–	–	–	–	–	–	–
Amazona rhodocorytha	–	–	–	•	–	–	•	–	–	–	–	–	–	–	–
Amazona vinacea	–	–	–	–	–	–	–	–	–	–	–	–	–	–	–
Triclaria malachitacea	–	–	–	•	–	–	–	–	–	•	–	–	–	–	–
Eleothreptus anomalus	–	–	–	–	–	–	–	–	–	–	–	–	–	–	–
Jacamaralcyon tridactyla	•	•	•	–	•	–	–	–	–	–	–	–	–	–	–
Dryocopus galeatus	–	–	–	–	–	–	–	–	–	–	–	–	–	–	–
Thripophaga macroura	–	–	–	•	–	–	–	–	–	–	–	–	–	–	–
Biatas nigropectus	–	–	–	•	–	•	–	•	–	–	–	–	–	–	–
Formicivora erythronotos	–	–	–	–	–	–	–	–	–	–	–	–	–	•	–
Formicivora littoralis	–	–	–	–	–	–	–	–	–	–	•	–	•	–	–
Laniisoma elegans	–	–	–	•	–	•	•	–	–	•	–	•	–	–	–
Tijuca condita	–	–	–	–	–	–	–	•	–	•	–	–	–	–	–
Carpornis melanocephalus	–	–	–	–	–	–	–	–	–	–	–	–	–	–	–
Iodopleura pipra	–	–	–	•	–	•	–	•	–	–	–	–	–	•	–
Lipaugus lanioides	–	–	–	–	–	•	–	•	–	–	–	–	–	–	–
Xipholena atropurpurea	–	–	–	•	–	–	–	–	–	–	–	–	–	–	–
Piprites pileatus	–	–	–	–	–	•	–	–	–	–	–	–	–	–	–
Euscarthmus rufomarginatus	–	–	–	–	–	–	–	–	–	–	–	–	–	–	–
Phylloscartes paulistus	–	–	–	•	–	–	–	–	•	–	–	–	–	–	–
Hemitriccus furcatus	–	–	–	•	–	•	–	–	–	–	–	–	–	–	–
Platyrinchus leucoryphus	–	–	–	•	–	–	–	–	–	–	–	•	–	•	–
Anthus nattereri	–	–	–	–	–	–	–	–	–	–	–	–	–	–	–
Sporophila falcirostris	–	–	–	•	–	•	–	–	–	•	–	–	–	–	–
Sporophila frontalis	–	–	–	•	–	•	–	•	•	–	–	•	–	–	–
Tangara peruviana	–	–	–	•	–	–	–	–	–	–	•	•	•	•	–
Dacnis nigripes	–	–	–	•	–	•	•	•	–	•	–	•	–	•	–
No. of species	1	1	1	18	1	13	5	11	3	6	2	7	2	5	1

Campelo (Rio de Janeiro)

BR 083

21°37′S 41°13′W

Unprotected

This area is found near the Rio Muriaé valley (a tributary of the Rio Paraíba do Sul), c.30 km south-east of Itaperuna in northern Rio de Janeiro, where the habitat comprises dry forest woodlots in an otherwise mainly cultivated region.

Jacamaralcyon tridactyla	1988	One record.

096	097	098	099	100	101	102	103	104	105	106	107	108	109	110	111	112	113	114	115	116	No. of areas
●	–	●	–	–	–	–	–	●	–	–	–	–	–	–	–	–	–	–	–	–	3
–	–	–	–	–	–	–	–	●	–	–	–	–	–	–	–	–	–	–	–	–	1
–	–	–	–	–	–	●	●	–	–	●	–	–	–	–	–	–	●	●	●	●	11
–	–	–	–	●	–	–	–	–	–	–	–	–	–	–	–	–	–	–	–	–	2
–	–	–	–	●	–	–	–	–	●	●	●	●	●	●	–	●	–	●	–	●	12
–	–	–	–	–	●	–	–	–	–	–	–	–	–	–	–	–	–	–	–	–	1
–	–	–	–	–	–	–	●	–	–	●	–	–	–	–	–	–	–	–	–	–	4
–	–	–	–	–	–	–	–	–	–	–	–	–	–	–	–	–	–	–	–	–	2
–	–	–	–	–	–	–	●	–	–	–	–	–	–	–	●	–	●	–	●	●	10
–	–	–	●	–	–	–	–	–	–	●	●	–	●	–	–	–	–	●	–	●	10
–	–	–	–	–	–	–	–	–	–	–	–	–	–	–	●	●	●	●	●	●	6
–	–	–	–	●	–	–	–	–	–	–	–	–	–	–	–	–	–	–	–	–	3
–	●	–	●	–	–	–	–	–	–	–	–	–	●	●	–	–	–	●	–	–	5
–	–	–	–	●	–	–	–	–	–	●	–	●	●	●	–	●	–	●	–	●	10
–	–	●	–	–	–	–	–	●	–	–	–	–	–	–	–	–	–	–	–	–	2
–	–	–	–	–	–	–	–	–	–	–	–	–	–	–	–	–	–	–	–	–	4
–	–	–	–	–	–	–	–	–	–	–	–	●	●	–	–	–	–	–	–	●	3
–	–	–	–	–	–	–	–	–	–	–	–	–	–	–	–	–	–	–	–	–	1
–	–	–	–	–	–	–	–	–	–	–	–	–	●	–	–	–	–	–	–	–	4
–	–	–	–	–	–	–	–	–	–	–	–	–	–	–	–	–	–	–	–	–	1
–	–	–	–	–	–	–	–	–	–	–	–	–	–	–	–	–	–	–	–	–	2
–	–	–	–	●	–	–	●	–	–	●	●	–	–	–	–	–	–	–	–	●	10
–	–	–	–	–	–	–	–	–	–	–	–	–	–	–	–	–	–	–	–	–	2
–	–	–	–	–	–	–	–	–	–	●	–	●	●	–	●	●	●	●	–	●	8
–	–	–	–	–	–	●	●	–	–	–	–	–	–	–	–	–	●	–	–	–	7
–	–	–	–	–	–	●	–	–	–	–	–	–	●	●	–	–	–	●	–	–	6
–	–	–	–	–	–	–	–	–	–	–	–	–	–	–	–	–	–	–	–	–	1
–	●	–	–	●	–	–	–	–	–	–	–	–	–	–	–	–	–	–	–	–	4
●	–	–	–	–	–	–	–	–	–	–	–	–	–	–	–	–	–	–	–	–	1
–	–	–	–	–	–	●	●	–	–	●	–	●	●	–	–	–	–	●	–	●	9
–	–	–	–	–	–	●	●	–	●	●	–	–	–	–	–	–	–	–	–	–	6
–	–	–	–	–	–	●	●	–	–	–	–	●	●	●	–	●	–	–	–	●	10
–	–	●	–	–	–	–	–	–	–	–	–	–	–	–	–	–	–	–	–	–	1
–	–	–	–	–	–	●	●	–	–	–	–	–	–	–	●	●	●	–	●	●	10
–	–	–	–	–	–	●	●	–	●	●	●	–	●	–	●	●	●	●	●	●	17
–	–	–	–	–	–	–	–	–	●	●	●	–	–	–	–	●	●	–	●	●	12
–	–	–	–	–	–	●	–	–	–	–	–	–	●	●	–	–	–	–	●	●	12
2	2	3	2	6	1	9	10	3	4	11	5	6	12	6	5	8	8	10	7	15	

BR 084

Desengano (Rio de Janeiro)

21°50′S 41°45′W
EBA B52

Desengano State Park (IUCN category V), 22,500 ha

This area is located c.30 km west of Campos in eastern Rio de Janeiro, and consists mainly humid forest. This general area probably has the highest number of threatened species recorded in the New World. There have been unsubstantiated reports from local people of *Crax blumenbachii* in the reserve. A good number of the records come from outside the park boundaries in lowlands that are being converted, so the situation is actually serious.

Leucopternis lacernulata	c.1987	Apparently found in the park and adjacent areas in the 1980s.
Pyrrhura cruentata	1988	Small flocks seen near the park, but occurrence within it has not been confirmed.
Touit melanonota	1988	A number of records in the late 1980s are all from near the park boundaries.
Touit surda	1988	Three birds seen within the park in 1987; sightings in 1988 from São Julião in the surrounding lowlands.
Amazona rhodocorytha	1990s	Pairs recorded in all months since 1986 from lowlands near the park.
Triclaria malachitacea	c.1987	Apparently recorded within the park after 1986.
Thripophaga macroura	c.1987	Birds have been found at up to 800 m, apparently north-east of Santa Maria Madalena.
Biatas nigropectus	1987	Recorded at 840 m at both Rifa and Vermelho.

cont.

Laniisoma elegans	1987	One record of a bird singing.
Iodopleura pipra	1989	A handful of sightings in the late 1980s.
Xipholena atropurpurea	1986	One seen at Morro do Rifa (700 m); one of the only observations in Rio de Janeiro state.
Phylloscartes paulistus	1987	Recorded from Morumbeca do Imbé.
Hemitriccus furcatus	1980s	Apparently 3–4 pairs have been found regularly since 1986, especially near Santa Maria Madalena.
Platyrinchus leucoryphus	1989	Recorded at three sites in the park in the late 1980s: Morumbeca do Imbé, Marreiros and Ribeirão Macapá.
Sporophila falcirostris	1986	Recorded from Ribeirão Vermelho and Santa Maria Madalena in 1986.
Sporophila frontalis	1987	Seen at Ribeirão Vermelho at 800 m.
Tangara peruviana	1987	One sighting at Ribeirão Vermelho.
Dacnis nigripes	1988	At least four seen at Santa Maria Madalena.

Somidorou (Rio de Janeiro)

BR 085

Unprotected

21°58′S 42°48′W

Somidorou is located c.40 km north-east of Três Rios in the Rio Paraíba do Sul valley in north-central Rio de Janeiro. The habitat in this area consists of small woodlots of secondary dry deciduous forest surrounded by agricultural pasturelands. The woodland is highly threatened by continuing encroachment of cultivation and timber extraction (Tobias *et al.* 1993).

Jacamaralcyon tridactyla	1993	Seen in a woodlot of c.5 ha in July and December 1993 (Tobias *et al.* 1993, Balchin 1994).

Itatiaia (Rio de Janeiro/Minas Gerais)

BR 086

Itatiaia National Park (IUCN category II), 30,000 ha

22°23′S 44°38′W
EBA B53

This national park is located c.130 km north-west of Rio de Janeiro city at the northern end of Serra da Mantiqueira, western Rio de Janeiro. The park includes Pico das Agulhas Negras (2,787 m), Brazil's second-highest mountain, and extensive areas of montane forest.

Pipile jacutinga	1978	A handful of sightings.
Claravis godefrida	1991	One bird seen in September 1991 is apparently the most recent record.
Touit melanonota	1991	A number of old and several recent records from 1,200 m.
Touit surda	1988	Recorded from Serrinha, close to the national park.
Biatas nigropectus	1995	Recorded from a handful of sites within and near the park; most recently a pair seen along the Três Picas trail at 1,200 m in January (G. Kirwan *in litt.* 1995).
Laniisoma elegans	1995	Occasionally recorded, most recently a single male at c.1,300 m in January (G. Kirwan *in litt.* 1995).
Iodopleura pipra	1988	Recorded close to the park in Itatiaia town.
Lipaugus lanioides	1979	A number of old specimens, and a sighting at 1,100 m in 1979.
Piprites pileatus	1995	The national park is a stronghold for the species, which is not uncommon at mid- to upper elevations.
Hemitriccus furcatus	1990	A number collected in 1950 and a record in 1990.
Sporophila falcirostris	1989	A handful of records in the 1980s.
Sporophila frontalis	1987	The last record from this area is from 1987, although when there was a bamboo flowering in June–July 1952 and September–October 1985 'thousands' of birds were present (flocks of 30 or more).
Dacnis nigripes	1987	One record in 1953 and a pair seen in 1987.

Fazenda União (Rio de Janeiro)

BR 087
22°23′N 42°00′W
EBA B52

Unprotected

This important lowland (10–350 m) humid forest area covers c.3,000 ha and is located near Rocha Leão in northern Rio de Janeiro (Whitney and Pacheco 1995). The forest is owned by the Rede Ferroviária Federal SA (RFFSA) and is maintained primarily as a source of timber for railway sleepers. Although RFFSA restricts access to Fazenda União, it is not officially protected in spite of calls for such measures, such as at the Atlantic Forest Workshop in 1990 (Whitney and Pacheco 1995). As well as supporting the threatened species listed below it is the only site where *Myrmotherula minor*, *M. unicolor* and *M. urosticta* survive together (see 'Recent changes to the threatened list', above).

Amazona rhodocorytha	1992	Recorded from the area.
Pyrrhura cruentata	1990	Recorded in August 1989 and September and November 1990.
Touit surda	1990s	Seen and tape-recorded recently (B. M. Whitney *in litt.* 1995).
Laniisoma elegans	1990	Seen in forest below 100 m in July 1990.
Dacnis nigripes	1990	Concentrations, possibly of hundreds, have been seen in winter (e.g. July 1990).

Serra dos Órgãos (Rio de Janeiro)

BR 088
22°28′S 43°03′W
EBAs B52, B53

Serra dos Órgãos National Park (IUCN category II), 11,000 ha

This national park, immediately east of Teresópolis in the Serra dos Órgãos (2,318 m) in central Rio de Janeiro, has extensive montane forest (including cloud-forest), with montane grassland on the highest peaks. It has been suggested that *Myrmotherula fluminensis* could occur; the species is known only from the type-specimen collected near Santo Aleixo in 1982, close to the slopes of Serra dos Órgãos.

Pipile jacutinga	1980s	Historically known from the area and in the 1980s still believed to be present locally but in very reduced numbers.
Leucopternis lacernulata	1980s	Recorded from the foothills of this area.
Claravis godefrida	1980s	One seen at 1,400 m in December 1980, and others reported more recently. Additional records are from Teresópolis, Socavão, Fazenda Comari and Boa Fé, and Ingá (1984), all adjacent to the park boundary.
Touit melanonota	1991	A pair seen at 1,200 m in July.
Touit surda	unknown	Mentioned for this general area (e.g. Teresópolis).
Biatas nigropectus	1991	A number of records from the park and surrounding area from the late 1980s.
Iodopleura pipra	1991	Recorded from the park.
Lipaugus lanioides	1942	An old record near the park at Teresópolis.
Tijuca condita	1993	Recorded almost annually in the last few years from the elfin forest confined to the highest parts. One of only two sites known for the species.
Sporophila frontalis	1986	A number of old sightings around Teresópolis and a recent record from the park itself.
Dacnis nigripes	1984	Sightings in the park and surrounding foothills.

Poço das Antas (Rio de Janeiro)

BR 089
22°35′N 42°17′W
EBA B52

Poço das Antas Federal Biological Reserve (IUCN category I), 5,000 ha

A small reserve located in the inner edge of the coastal plain of Rio de Janeiro state. It comprises mainly secondary forest with smaller areas of scrub and marsh.

Leucopternis lacernulata	1981	Uncommon.
Phylloscartes paulistus	1981	Uncommon.
Sporophila frontalis	1988	Recorded from Fazenda Bandeirantes, Silva Jardim (adjacent to the reserve), August 1988.

Serra do Tinguá (Rio de Janeiro)

BR 090

Tinguá Federal Biological Reserve (IUCN category I), 26,000 ha

22°36'S 43°26'W
EBAs B52, B53

This area embraces a small mountain range rising to 1,500 m above the coastal plain, immediately west of Petrópolis and only 40 km north of Rio de Janeiro. The reserve comprises mature secondary forest and primary montane forest, with cloud-forest on the highest peaks.

Harpyhaliaetus coronatus	1980	An adult and an immature, apparently on migration over the area.
Triclaria malachitacea	1986	Recorded at 700 m in 1980 and 1981 and at Xerén (120 m) in 1986.
Laniisoma elegans	1981	Rare in 1981.
Tijuca condita	1981	Suitable habitat remains unchanged since it was found here in 1980. One of only two sites known for the species.
Sporophila falcirostris	1955	One collected.
Dacnis nigripes	1981	One record.

Maçambaba (Rio de Janeiro)

BR 091

Maçambaba State Environment Protection Area (IUCN category V), 70 ha
Maçambaba State Reserve (status and area unknown)

22°55'S 42°20'W
EBA B52

The coastal dunes and associated vegetation around Maçambaba village are located 100 km east of Rio de Janeiro and c.15 km west of Cabo Frio. Among the sand-dunes are clumps of low-lying restinga scrub, with some woodland patches.

Formicivora littoralis	1995	Seen regularly in dense restinga (e.g. Tobias *et al.* 1993, Forrester 1994, G. Kirwan *in litt.* 1995).
Tangara peruviana	1994	Found in small woodland patches in August 1993 and 1994 (Tobias *et al.* 1993, Forrester 1994).

Tijuca (Rio de Janeiro)

BR 092

Tijuca National Park (IUCN category II), 3,200 ha

22°56'S 43°17'W
EBA B52

This small national park, containing mainly secondary forest, is centred on Pico de Tijuca and located immediately south-west of Rio de Janeiro city.

Leucopternis lacernulata	1980	Recorded from the park.
Touit melanonota	1987	Several records in the 1980s of just a handful of birds.
Laniisoma elegans	1988	A number of records through the 1980s.
Platyrinchus leucoryphus	1988	Recorded adjacent to the national park at Horto Florestal.
Sporophila frontalis	1991	A few records.
Tangara peruviana	1979	Seen above Alto da Boa Vista in 1979.
Dacnis nigripes	1987	Single birds seen in 1985 and 1987.

BR 093

Cabo Frio (Rio de Janeiro)

22°58′S 42°01′W
EBA B52

Guapi-Mirim Federal Environment Protection Area (IUCN category V), 14,340 ha
Jacarepia State Reserve (status and area unknown)

Located 130 km east of Rio de Janeiro, this area includes the headland of Cabo Frio and the small island of the same name immediately offshore. The important habitat is the restinga forest on the coastal sand-dunes.

Formicivora littoralis	1990	Apparently abundant, especially on Cabo Frio island.
Tangara peruviana	1990	Small numbers.

BR 094

Angra dos Reis (Rio de Janeiro)

23°00′S 44°18′W
EBA B52

Unprotected

This area is located on the north-east coast of Baía da Ilha Grande, c.115 km west of Rio de Janeiro. The remaining forest patches are small and very fragmented, and urgent conservation measures are needed if the total extinction of *Formicivora erythronotus* is to be avoided.

Formicivora erythronotus	1995	This is the only site currently known for the species and the population is critically low; the last sighting was of a pair in March 1995 (G. Kirwan *in litt.* 1995).
Iodopleura pipra	1993	One seen at Frade (Clements 1993).
Platyrinchus leucoryphus	1990	Seen at Rio Florestão, near Angra dos Reis, in December 1990.
Dacnis nigripes	1993	A number of recent records.
Tangara peruviana	1993	A pair seen at Frade (Clements 1993).

BR 095

Pedra Branca (Rio de Janeiro)

23°13′S 44°43′W
EBA B52

Pedra Branca State Park (IUCN category II), 12,500 ha

This area is located c.10 km north of Paratí on the western side of Baía da Ilha Grande in south-west Rio de Janeiro, very close to the border with São Paulo.

Touit melanonota	1988	A handful of records in the late 1980s involving only a few individuals.

BR 096

Itirapina (São Paulo)

22°15′S 47°49′W

Itirapina State Ecological Station (IUCN category IV), 2,300 ha

Lying between the towns of Rio Claro and São Carlos in central São Paulo state, this area comprises cerrado and campo grassland with some mixed dry forest. It is one of the few remaining grasslands in São Paulo.

Nothura minor	1993	The only modern-day locality in São Paulo, and source of most historical records. The last records were of a bird singing in September 1993 (F. Olmos and P. Martuscelli *in litt.* 1995) and one in December 1993 (Dacol 1994)
Euscarthmus rufomarginatus	1993	Recorded during a short visit in December (Dacol 1994). This is the only modern-day locality in São Paulo.

Campos do Jordão (São Paulo)

BR 097

Campos do Jordão State Park (IUCN category II), 8,286 ha

22°40′S 45°29′W
EBA B54

This state park is located at 1,500–2,000 m on the north-east slope of the Serra da Mantiqueira in south-east São Paulo. The main habitat is forest dominated by *Araucaria angustifolia* and *Podocarpus lambertii*.

Amazona vinacea	1992	A count of 120 individuals in 1992 (F. Olmos and P. Martuscelli *in litt.* 1995) and similar numbers recorded during surveys in the late 1970s.
Piprites pileatus	1992	Three seen in mixed *Araucaria* forest in July 1992 (F. Olmos and P. Martuscelli *in litt.* 1995).

Fazenda Rio Pardo (São Paulo)

BR 098

Unprotected

22°50′S 48°40′W

Fazenda Rio Pardo is located 40 km west of Botugatu in central São Paulo. The main habitats are cerrado, campo cerrado and grassland (F. Olmos and P. Martuscelli *in litt.* 1995).

Nothura minor	c.1995	Recorded recently (F. Olmos and P. Martuscelli *in litt.* 1995); no information on status.
Eleothreptus anomalus	1956	Apparently collected there (W. Bokerman *per* F. Olmos and P. Martuscelli *in litt.* 1995).
Anthus nattereri	1992	Two seen in inundated grassland in July (F. Olmos and P. Martuscelli *in litt.* 1995).

Poncianos (São Paulo)

BR 099

Unprotected

22°50′S 45°52′W
EBAs B52, B54

This area is sited in the mountains 35 km north-west of Taubaté in eastern São Paulo close to the border with Minas Gerais. The main habitat is montane forest with cloud-forest and *Araucaria* forest (F. Olmos and P. Martuscelli *in litt.* 1995).

Touit surda	c.1995	Recorded recently (F. Olmos and P. Martuscelli *in litt.* 1995); no information on status.
Amazona vinacea	c.1995	Recorded recently (F. Olmos and P. Martuscelli *in litt.* 1995); no information on status.

Serra da Bocaína (São Paulo)

BR 100

Serra da Bocaína National Park (IUCN category II), 100,000 ha

22°50′S 44°30′W
EBAs B52, B53

This national park is centred on the Serra de Bocaina plateau which lies on the São Paulo–Rio de Janeiro state border. The park has rocky beaches, humid forest up to 1,800 m and alpine grasslands and *Araucaria* forest on the peaks to 2,080 m. In spite of the park's proximity to Rio de Janeiro, few biological surveys have been carried out.

Pipile jacutinga	1977	Recorded from the park.
Harpyhaliaetus coronatus	1988	One shot by a farmer in 1987 and a sighting in 1988.
Amazona rhodocorytha	1994	Four seen in the lowlands near the sea in June 1994.
Triclaria malachitacea	1989	Recorded at 1,200 m.
Laniisoma elegans	1980s	Recorded from the park.
Piprites pileatus	1989	One collected in 1951 and sightings at 940 m and 1,200 m in 1989.

BR 101

Taubaté (São Paulo)

23°01′S 45°33′W

Unprotected

Taubaté is a largely agricultural region located in the Rio Paraíba drainage in east-central São Paulo. The principal habitat is marshland and flooded rice-fields.

Coturnicops notatus	1984	Apparently resident; specimens collected in 1976, 1982 and 1984.

BR 102

Ubatuba (São Paulo)

23°26′S 45°04′W
EBA B52

Ubatuba Experimental Station (private reserve), area unknown

This area comprises the Ubatuba Experimental Station, and Fazendas Capricórnio and Angelina, at sea-level north of Ubatuba in eastern São Paulo. The region holds an important area of humid forest with patches of bamboo.

Leucopternis lacernulata	1994	A handful of records, the last a bird calling from a treetop in August 1994 (J. Goerck *in litt.* 1995).
Touit melanonota	1970s	A single sighting of parrotlets was believed to be of this species; no subsequent records but it has been seen at nearby Bairro de Corcovado (BR 103).
Iodopleura pipra	1995	Recorded regularly (e.g. G. Kirwan *in litt.* 1995); nests found November–December 1993 (Balchin 1994, A. Whittaker *in litt.* 1994).
Lipaugus lanioides	1943	An old specimen and an unconfirmed sighting from São Sebastião in May 1990, apparently close to the station.
Phylloscartes paulistus	1993	Recorded a number of times in the late 1970s and early 1990s.
Hemitriccus furcatus	1995	Found in forest around Fazendas Capricórnio and Angelina (e.g. G. Kirwan *in litt.* 1995).
Platyrinchus leucoryphus	1945	Known from the area, and been seen at nearby Bairro de Corcovado (BR 103) in 1991.
Sporophila falcirostris	1991	Up to 30 seen and heard; recent presence of singing birds suggests sporadic breeding.
Sporophila frontalis	1991	Seen a number of times in recent years; more than 100 (possibly many more) present in October 1991.
Dacnis nigripes	1993	A number of records in the general area and specifically at Fazenda Capricórnio; a flock of 12, one of four and a pair building a nest were found in November 1993 (A. Whittaker *in litt.* 1995).

BR 103

Bairro de Corcovado (São Paulo)

28°11′S 45°11′W
EBAs B52, B53

Serra do Mar State Park (IUCN category II), 314,800 ha

This area of tall humid forest, with montane forest at higher elevations, is centred on the Pico do Corcovado, a prominent peak on the coastal escarpment in Ubatuba municipality in eastern São Paulo. The area is in the eastern portion of the enormous Serra do Mar State Park which protects forest above 100 m elevation, but hunting, poaching, orchid-harvesting, fires and sand-mining were all observed during recent fieldwork (J. Goerck *in litt.* 1995).

Leucopternis lacernulata	1994	One flying over the peak at 1,000 m in September 1994 (J. Goerck *in litt.* 1995).
Claravis godefrida	1991	Seen in a bamboo thicket in September 1991 (J. Goerck *in litt.* 1995).
Touit melanonota	1991	A pair seen in September 1991 (J. Goerck *in litt.* 1995).
Iodopleura pipra	1994	One seen in July 1994 (J. Goerck *in litt.* 1995).
Laniisoma elegans	1994	Seen several times in 1994, all above 500 m (J. Goerck *in litt.* 1995).
Phylloscartes paulistus	1991	Seen several times during a few days of fieldwork in September 1991 (J. Goerck *in litt.* 1995).
Hemitriccus furcatus	1991	Seen several times in *Merostachys* bamboo thickets during a few days' fieldwork in September 1991 (J. Goerck *in litt.* 1995).

cont.

Platyrinchus leucoryphus	1991	Seen twice in September 1991 (J. Goerck *in litt.* 1995).
Sporophila falcirostris	1994	Seen in areas of seeding bamboo in November and December, but less common than *S. frontalis* (J. Goerck *in litt.* 1995).
Sporophila frontalis	1994	Seen in areas of seeding bamboo in November and December (J. Goerck *in litt.* 1995).

Itapetininga (São Paulo)

BR 104

Itapetininga State Park (IUCN category II), 3,500 ha

23°35'S 48°10'W
EBA B52

This state park, containing mainly grassland and cerrado, is located immediately west of Itapetininga along the Rio Itapetininga in south-central São Paulo. It is one of the best remaining grasslands in São Paulo state.

Nothura minor	1994	Recorded recently (F. Olmos and P. Martuscelli *in litt.* 1995); no information on status.
Taoniscus nanus	1994	Recorded recently (F. Olmos and P. Martuscelli *in litt.* 1995); no information on status.
Eleothreptus anomalus	1964	One collected in October 1964.

Caraguatatuba (São Paulo)

BR 105

Serra do Mar State Park nucleus (IUCN category II), 13,700 ha

23°40'S 45°24'W
EBA B52

Caraguatatuba is a nucleus zone (core area) of Serra do Mar State Park (314,800 ha) in eastern São Paulo. It comprises montane forest between 100 and 800 m (F. Olmos and P. Martuscelli *in litt.* 1995).

Pipile jacutinga	1994	Recorded only twice in recent years (F. Olmos and P. Martuscelli *in litt.* 1995).
Hemitriccus furcatus	1994	Recorded recently only in the lowland forest (F. Olmos and P. Martuscelli *in litt.* 1995); no information on status.
Sporophila frontalis	1994	Recorded recently (F. Olmos and P. Martuscelli *in litt.* 1995); no information on status.
Tangara peruviana	1994	Recorded recently (F. Olmos and P. Martuscelli *in litt.* 1995); no information on status.

Boracéia and Bertioga (São Paulo)

BR 106

Boracéia Biological Station (private reserve), area unknown

23°45'S 45°55'W
EBA B52

The biological station lies at c.800 m near Biritiba Mirim, c.80 km east of São Paulo city. Bertioga is adjacent to Boracéia on the sandy, swampy ground of the coastal lowlands and covers 6,000 ha. The area has lowland and montane forest (much of it secondary, however) and restinga forest along the coast.

Leucopternis lacernulata	c.1995	Restricted to the lowland forest at Bertioga (F. Olmos and P. Martuscelli *in litt.* 1995).
Pipile jacutinga	1988	Recorded from the area.
Claravis godefrida	1987	A pair seen at Boracéia.
Touit surda	1987	Recorded recently in lowland forest at Bertioga (F. Olmos and P. Martuscelli *in litt.* 1995).
Triclaria malachitacea	1988	Uncommon in the 1980s.
Laniisoma elegans	1945	Not encountered in over 100 days of fieldwork during the late 1980s.
Carpornis melanocephalus	c.1995	Found in lowland forest to 200 m at Bertioga (F. Olmos and P. Martuscelli *in litt.* 1995).
Phylloscartes paulistus	1987	One observation.
Hemitriccus furcatus	c.1995	Found only in the lowland forest at Bertioga (F. Olmos and P. Martuscelli *in litt.* 1995).
Sporophila frontalis	c.1995	Recorded as a migrant at Bertioga (F. Olmos and P. Martuscelli *in litt.* 1995).
Tangara peruviana	c.1995	Recorded at Bertioga (F. Olmos and P. Martuscelli *in litt.* 1995).

BR 107

Ilha São Sebastião (São Paulo)

23°48′S 45°25′W
EBA B52

Ilha Bela State Park (IUCN category II), 27,025 ha

This large, mountainous island (highest point 1,379 m) lies a short distance from the coast, 110 km east of São Paulo. It still holds some areas of humid forest.

Pipile jacutinga	1994	Displaying birds observed (Galetti *et al.* in prep.).
Touit surda	c.1994	Found only in summer (F. Olmos and P. Martuscelli *in litt.* 1995).
Laniisoma elegans	1901	Recorded from the island.
Sporophila frontalis	c.1994	A migrant (F. Olmos and P. Martuscelli *in litt.* 1995).
Tangara peruviana	c.1994	A rare migrant (F. Olmos and P. Martuscelli *in litt.* 1995).

BR 108

Carlos Botelho (São Paulo)

24°07′S 47°55′W
EBAs B52, B53

Carlos Botelho State Park (IUCN category II), 37,797 ha
Sete Barras State Reserve (status unknown), 15,646 ha

This area is located 10–30 km north of Sete Barras in the Serra Paranapicaba of south-central São Paulo, and comprises humid forest below 1,000 m. The two reserves are adjacent and, together with Fazenda Intervales (BR 109) and Alto Ribeira (BR 110) to the west and Jacupiranga (BR114) to the south-west, form one of the largest remaining tracts of coastal forest in south-east Brazil.

Pipile jacutinga	1995	Found at both Carlos Botelho and Sete Barras; breeding confirmed (Galetti *et al.* in prep.).
Triclaria malachitacea	1992	Recorded from Carlos Botelho.
Dryocopus galeatus	1988	Recorded at 900 m.
Carpornis melanocephalus	1995	Recorded at Sete Barras (F. Olmos and P. Martuscelli *in litt.* 1995).
Phylloscartes paulistus	1990	Uncommon at Carlos Botelho and Sete Barras.
Platyrinchus leucoryphus	1993	Recorded recently at 750 m in montane forest (F. Olmos and P. Martuscelli *in litt.* 1995); no information on status.

BR 109

Fazenda Intervales (São Paulo)

24°15′S 48°10′W
EBAs B52, B53

Fazenda Intervales State Reserve (status unknown), 38,000 ha

This area is located 30 km north-west of Sete Barras and 30 km south-east of Capão Bonito in the Serra Paranapicaba of south-central São Paulo. The reserve holds Atlantic humid forest from 50 to 1,100 m, and, together with Carlos Botelho (BR 108) to the east, Alto Ribeira (BR 110) to the west and Jacupiranga (BR 114) to the south-west, forms one of the largest remaining tracts of coastal forest in south-east Brazil.

Pipile jacutinga	1995	Density estimates based on 290 km strip transect was 2.24 individuals/km^2, giving a maximum possible density of 851 individuals within the reserve, but the total population may be lower as densities could be lower in parts of the reserve which are less well protected (Galetti *et al.* in prep.).
Touit surda	1995	One record from Saibadela (A. Aleixo *in litt.* 1995).
Amazona vinacea	1995	Only one sighting during fieldwork in 1994–1995 (A. Aleixo *in litt.* 1995), but small flocks seen during the species' breeding season in 1993 (F. Olmos and P. Martuscelli *in litt.* 1995).
Triclaria malachitacea	1994	Most regularly recorded at Saibadela and Carmo at 700–800 m (F. Olmos and P. Martuscelli *in litt.* 1995).
Dryocopus galeatus	1990	Recorded from the area.
Biatas nigropectus	1994	Resident at two sites in the reserve but rarely recorded (F. Olmos and P. Martuscelli *in litt.* 1995).

cont.

Carpornis melanocephalus	1995	Common species in primary forest at Saibadela (A. Aleixo *in litt.* 1995).
Lipaugus lanioides	1994	Common in the Carmo valley (F. Olmos and P. Martuscelli *in litt.* 1995).
Phylloscartes paulistus	1995	Uncommon at Saibadela, normally seen at forest edge, but also in the canopy of primary forest (A. Aleixo *in litt.* 1995).
Platyrinchus leucoryphus	1995	Uncommon at Saibadela (A. Aleixo *in litt.* 1995) and seen recently in the Carmo valley (F. Olmos and P. Martuscelli *in litt.* 1995).
Sporophila frontalis	1994	Two records, from bamboo stands (F. Olmos and P. Martuscelli *in litt.* 1995).
Dacnis nigripes	1990s	Recorded from the area.

Alto Ribeira (São Paulo)

BR 110

Alto Ribeira State Park (IUCN category II), 37,712 ha

24°20′S 48°25′W
EBA B52

Lying at 200–900 m, 35 km south of Capão Bonito in the Serra Paranapicaba of south-central São Paulo, the park holds Atlantic humid forest (F. Olmos and P. Martuscelli *in litt.* 1995). Together with Fazenda Intervales (BR 109) and Carlos Botelho (BR 108) to the east, and Jacupiranga (BR 114) to the south-west, it forms one of the largest remaining tracts of coastal forest in São Paulo and, indeed, south-east Brazil.

Pipile jacutinga	1994	Recorded recently (Galetti *et al.* in prep.); no information on status.
Amazona vinacea	1994	Breeds in the reserve (F. Olmos and P. Martuscelli *in litt.* 1995).
Triclaria malachitacea	1994	Recorded recently (F. Olmos and P. Martuscelli *in litt.* 1995); no information on status.
Lipaugus lanioides	1994	Recorded recently (F. Olmos and P. Martuscelli *in litt.* 1995); no information on status.
Platyrinchus leucoryphus	1994	Recorded recently (F. Olmos and P. Martuscelli *in litt.* 1995); no information on status.
Dacnis nigripes	1992	Recorded recently (F. Olmos and P. Martuscelli *in litt.* 1995); no information on status.

Foz do Rio Ribeira (São Paulo)

BR 111

Cananéia–Iguape–Peruíbe Environmental Protection Area (IUCN category V), 202,832 ha

24°30′S 47°15′W

The mouth of the Rio Ribeira is c.20 km north-east of Iguape, south-east São Paulo. The area includes c.1,000 ha of flooded forest, swamp and lagoons (F. Olmos and P. Martuscelli *in litt.* 1995).

Touit melanonota	1994	Recorded recently (F. Olmos and P. Martuscelli *in litt.* 1995); no information on status.
Amazona brasiliensis	1994	Recorded recently (F. Olmos and P. Martuscelli *in litt.* 1995); no information on status.
Carpornis melanocephalus	1994	Recorded recently (F. Olmos and P. Martuscelli *in litt.* 1995); no information on status.
Sporophila falcirostris	1994	Recorded recently (F. Olmos and P. Martuscelli *in litt.* 1995); no information on status.
Sporophila frontalis	1994	Recorded recently (F. Olmos and P. Martuscelli *in litt.* 1995); no information on status.

Juréia (São Paulo)

BR 112

Juréia-Itatins Ecological Station (IUCN category IV), 80,000 ha

24°40′S 47°10′W
EBA B52

This ecological station is located 30 km south-west of Peruíbe along the coast of south-east São Paulo. The reserve extends from sea-level to 1,350 m and holds sand-plain and montane forests (F. Olmos and P. Martuscelli *in litt.* 1995).

Pipile jacutinga	c.1995	Breeds with young birds found in December (Galetti *et al.* in prep.).
Amazona brasiliensis	c.1995	About 40 birds are resident in the mangrove–lowland forest ectone. One of the only areas where breeding birds do not suffer from poaching (Martuscelli 1995).

cont.

Triclaria malachitacea	c.1995	Recorded recently (F. Olmos and P. Martuscelli *in litt.* 1995); no information on status.
Carpornis melanocephalus	c.1995	Recorded recently (F. Olmos and P. Martuscelli *in litt.* 1995); no information on status.
Platyrinchus leucoryphus	c.1995	Recorded recently (F. Olmos and P. Martuscelli *in litt.* 1995); no information on status.
Sporophila falcirostris	c.1995	Recorded recently (F. Olmos and P. Martuscelli *in litt.* 1995); no information on status.
Sporophila frontalis	c.1995	Recorded recently (F. Olmos and P. Martuscelli *in litt.* 1995); no information on status.
Tangara peruviana	c.1994	Recorded recently (F. Olmos and P. Martuscelli *in litt.* 1995); no information on status.

BR 113

Iguape (São Paulo)

24°40'S 47°30'W
EBA B52

Cananéia–Iguape–Peruíbe Environmental Protection Area (IUCN category V), 202,832 ha

This area is located close to Iguape and opposite Ilha Comprida along the coast of south-east São Paulo. It covers c.1,500 ha of various habitats including sand-plain forest, flooded forest and montane forest.

Leucopternis lacernulata	c.1995	Recorded recently (F. Olmos and P. Martuscelli *in litt.* 1995).
Touit melanonota	1994	Recorded recently (F. Olmos and P. Martuscelli *in litt.* 1995); no information on status.
Amazona brasiliensis	1994	Recorded recently (F. Olmos and P. Martuscelli *in litt.* 1995); no information on status.
Carpornis melanocephalus	1994	Recorded recently (F. Olmos and P. Martuscelli *in litt.* 1995); no information on status.
Iodopleura pipra	c.1960	One specimen.
Sporophila falcirostris	1994	Recorded recently (F. Olmos and P. Martuscelli *in litt.* 1995); no information on status.
Sporophila frontalis	1994	Recorded recently (F. Olmos and P. Martuscelli *in litt.* 1995); no information on status.
Tangara peruviana	1994	Recorded recently (F. Olmos and P. Martuscelli *in litt.* 1995); no information on status.

BR 114

Jacupiranga (São Paulo)

24°53'S 48°22'W
EBA B52

Jacupiranga State Park (IUCN category II), 110,000 ha

This state park is located 35 km south-west of Jacupiranga in the coastal foothills of extreme south-west São Paulo, close to the state border with Paraná. The park ranges from sea-level to 1,350 m and is covered by sand-plain forest (at lower elevations) and humid Atlantic forests (F. Olmos and P. Martuscelli *in litt.* 1995). The northern boundary of the reserve lies close to Alto Ribeira (BR110).

Pipile jacutinga	c.1995	Recorded recently (F. Olmos and P. Martuscelli *in litt.* 1995); no information on status.
Leucopternis lacernulata	c.1995	Recorded only twice in recent years (F. Olmos and P. Martuscelli *in litt.* 1995).
Touit surda	c.1995	Recorded recently (F. Olmos and P. Martuscelli *in litt.* 1995); no information on status.
Amazona brasiliensis	c.1995	Winters in the lowlands of the south-east part of the reserve (Martuscelli 1995, F. Olmos and P. Martuscelli *in litt.* 1995).
Amazona vinacea	1994	Resident; population estimated at 160 in 1994 (F. Olmos and P. Martuscelli *in litt.* 1995).
Triclaria malachitacea	c.1995	Recorded recently (F. Olmos and P. Martuscelli *in litt.* 1995); no information on status.
Carpornis melanocephalus	c.1995	Recorded recently from lowland forest at 300 m (F. Olmos and P. Martuscelli *in litt.* 1995).
Lipaugus lanioides	c.1995	Recorded recently (F. Olmos and P. Martuscelli *in litt.* 1995); no information on status.
Phylloscartes paulistus	c.1995	Recorded recently (F. Olmos and P. Martuscelli *in litt.* 1995); no information on status.
Sporophila frontalis	c.1995	Recorded recently (F. Olmos and P. Martuscelli *in litt.* 1995); no information on status.

Ilhas Comprida and Cananéia (São Paulo)

BR 115

Cananéia–Iguape–Peruíbe Environmental Protection Area (IUCN category V), 202,832 ha

24°54′S 47°38′W
EBA B52

Ilha Comprida is a deltaic, long, thin low-lying island (60 × 5 km) running immediately offshore and parallel to the coastline just south of Iguape. The smaller Ilha Cananéia lies seaward of Comprida. The main habitat on Comprida is sand-plain forest (F. Olmos and P. Martuscelli *in litt.* 1995), with mangrove, caixeta *Tabebuia* and gerivá palms *Arecastrum romanzoffianum* growing on swampy terrain.

Leucopternis lacernulata	c.1995	Four recent records on Ilha Comprida (F. Olmos and P. Martuscelli *in litt.* 1995).
Touit melanonota	c.1995	Recorded from Ilha Comprida only in summer (F. Olmos and P. Martuscelli *in litt.* 1995).
Amazona brasiliensis	c.1995	The largest breeding population in São Paulo is on Ilha Comprida (Martuscelli 1995); it breeds on Ilha Cananéia but poaching is so intense that 87 birds raised only two fledglings in 1990/91 and four the following season (Martuscelli 1995).
Sporophila falcirostris	c.1995	Restricted to swamp forests and lagoons on Ilha Comprida in summer (F. Olmos and P. Martuscelli *in litt.* 1995).
Sporophila frontalis	c.1995	Recorded recently from Ilha Comprida (F. Olmos and P. Martuscelli *in litt.* 1995); no information on status.
Tangara peruviana	c.1995	Resident on both islands (F. Olmos and P. Martuscelli *in litt.* 1995).
Dacnis nigripes	c.1995	One recorded in winter on Ilha Comprida (F. Olmos and P. Martuscelli *in litt.* 1995).

Ilha do Cardoso (São Paulo)

BR 116

Ilha do Cardoso State Park (IUCN category II), 22,500 ha

25°08′S 47°58′W
EBA B52

This protected area is a large island located close to the coast in southernmost São Paulo. The island rises to 800 m, with habitat comprising sand-plain forest through to humid montane forest.

Leucopternis lacernulata	1992	Restricted to the sand-plain forest, making occasional movements inland.
Pipile jacutinga	1990s	Breeds with young observed in December (Galetti *et al.* in prep.). Migrates altitudinally from 100 to 900 m.
Amazona brasiliensis	1993	Population estimated at 94–98 birds (Martuscelli 1995). A stronghold for the species and one of its only well protected sites.
Touit melanonota	1991	Occupies the sand-plain forest, although habitat destruction is affecting the population.
Touit surda	1992	Flocks of up to 12 seen recently.
Triclaria malachitacea	1992	Apparently still resident.
Dryocopus galeatus	1992	Numbers appear very small but stable.
Laniisoma elegans	1992	The population is evidently small but apparently stable, occurring in the lowland forest below 100 m.
Carpornis melanocephalus	1994	Recorded regularly (F. Olmos and P. Martuscelli *in litt.* 1995); no information on status.
Phylloscartes paulistus	1993	A small population (F. Olmos and P. Martuscelli *in litt.* 1995); no information on status.
Platyrinchus leucoryphus	1994	Recorded recently (F. Olmos and P. Martuscelli *in litt.* 1995); no information on status.
Sporophila falcirostris	1994	Recorded recently (F. Olmos and P. Martuscelli *in litt.* 1995); no information on status.
Sporophila frontalis	1992	Periodically fairly common.
Tangara peruviana	1993	A small population is likely to persist, although suitable habitat is rather limited (F. Olmos and P. Martuscelli *in litt.* 1995).
Dacnis nigripes	1991	Recorded from the island.

Telémaco Borba (Paraná)

BR 117
24°12'S 50°30'W

Unprotected

This area comprises humid forests and plantations of *Pinus* and *Eucalyptus* in east-central Paraná. The forests are owned by Klabin Paper and Cellulose Industries.

Amazona vinacea	1995	Flocks of 15–20 seen in January 1995 (C. Seger and R. Boçom *in litt.* 1995).

Adrianópolis (Paraná)

BR 118
24°45'S 48°40'W

Lauráceas State Park (status unknown), 23,000 ha

This protected area lies near Adrianópolis in the mountains of eastern Paraná, close to the border with São Paulo. The reserve holds extensive tracts of montane forest, some of which is degraded but much remains in an excellent state (C. Seger and R. Boçom *in litt.* 1995).

Amazona vinacea	1995	Flocks of more than 100 seen in March 1995 (C. Seger and R. Boçom *in litt.* 1995).

Campina Grande do Sul (Paraná)

BR 119
25°11'S 48°50'W

Unprotected

This general area lies in the foothills of the Serra do Mar north-east of Curitiba in extreme eastern Paraná, adjoining the border with São Paulo. The area supports dense and mixed broadleaf forests.

Amazona vinacea	1995	Flocks of up to 50 seen in February 1995 (C. Seger and R. Boçom *in litt.* 1995).

Guaragueçaba (Paraná)

BR 120
25°20'S 48°25'W
EBA B52

Guaragueçaba Special Protection Area (IUCN category V), 291,500 ha

This reserve is located c.100 km east of Curitiba and immediately north of Guaragueçaba in south-east Paraná, close to the border with São Paulo. The area ranges from sea-level to 1,000 m and the main habitats are sand-plain forest and montane forest (F. Olmos and P. Martuscelli *in litt.* 1995). The reserve's northern boundary lies close to Jacupiranga State Park (BR 114), thus joining it with the other Key Areas identified in the Serra Paranapicaba of south-central São Paulo.

Pipile jacutinga	c.1995	Recorded recently (F. Olmos and P. Martuscelli *in litt.* 1995); no information on status.
Leucopternis lacernulata	c.1995	Recorded recently (F. Olmos and P. Martuscelli *in litt.* 1995); no information on status.
Amazona brasiliensis	c.1995	Recorded recently (F. Olmos and P. Martuscelli *in litt.* 1995); no information on status.
Carpornis melanocephalus	c.1995	Recorded recently (F. Olmos and P. Martuscelli *in litt.* 1995); no information on status.
Lipaugus lanioides	c.1995	Recorded recently (F. Olmos and P. Martuscelli *in litt.* 1995); no information on status.
Platyrinchus leucoryphus	c.1995	Recorded recently (F. Olmos and P. Martuscelli *in litt.* 1995); no information on status.

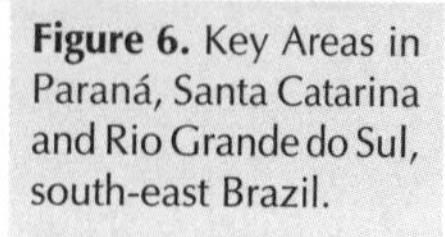

Figure 6. Key Areas in Paraná, Santa Catarina and Rio Grande do Sul, south-east Brazil.

119 Campina Grande do Sul
120 Guaragueçaba
121 Palmeira
122 Curitiba
123 Ilha do Superagüi and Ilha das Peças
124 Ilha do Mel
125 Santa Cruz
126 Iguaçu
127 General Carneiro
128 Salto do Piraí
129 Itajaí
130 Brusque
131 São Joaquim
132 Espigão Alto
133 Aracuri
134 Carazinho–Passo Fundo–Soledade–Salto do Jacui
135 Vacaria–Bom Jesus
136 Serra Geral
137 Aparados da Serra
138 São Francisco do Paula
139 Monte Alverne
140 Encruzilhada do Sul
141 Caçapava do Sul
142 Santana da Boa Vista

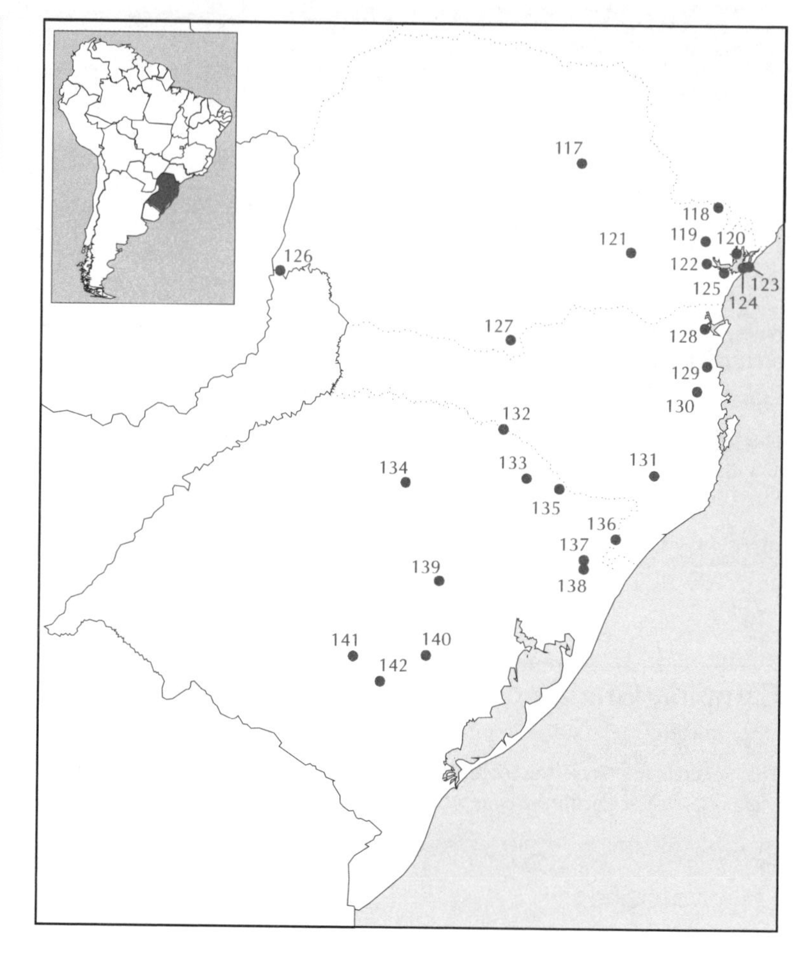

Table 7. Matrix of threatened species by Key Area in Paraná, Santa Catarina and Rio Grande do Sul, south-east Brazil (the area covered by Figure 6). Additional threatened species known from this region but for which no Key Areas have been identified are *Larus atlanticus, Anodorhynchus glaucus, Yetapa risora* and *Gubernatrix cristata.*

	117	118	119	120	121	122	123	124	125	126	127	128	129	130	131	132	133	134	135	136	137	138	139	140	141	142	No. of areas
Leucopternis lacernulata	–	–	–	●	–	–	–	–	●	–	–	●	–	–	–	–	–	–	–	–	–	–	–	–	–	–	3
Pipile jacutinga	–	–	–	●	–	–	–	–	–	●	–	–	–	–	–	–	–	–	–	–	–	–	–	–	–	–	2
Claravis godefrida	–	–	–	–	–	–	–	–	–	●	–	–	–	–	–	–	–	–	–	–	–	–	–	–	–	–	1
Amazona brasiliensis	–	–	–	●	–	–	●	●	–	–	–	–	–	–	–	–	–	–	–	–	–	–	–	–	–	–	3
Amazona pretrei	–	–	–	–	–	–	–	–	–	–	–	–	–	–	–	●	●	●	●	●	●	●	–	●	●	●	10
Amazona vinacea	●	●	●	–	●	–	–	–	–	–	●	–	●	–	●	●	●	●	●	●	●	●	–	–	–	–	14
Triclaria malachitacea	–	–	–	–	–	–	–	–	–	–	–	–	–	–	–	–	–	–	–	–	●	–	●	–	–	–	2
Eleothreptus anomalus	–	–	–	–	–	●	–	–	–	–	–	–	–	–	–	–	–	–	–	–	–	–	–	–	–	–	1
Dryocopus galeatus	–	–	–	–	–	–	–	–	–	●	–	–	–	–	–	–	–	–	–	–	–	–	–	–	–	–	1
Biatas nigropectus	–	–	–	–	–	–	–	–	–	●	–	–	–	–	–	–	–	–	–	–	–	–	–	–	–	–	1
Carpornis melanocephalus	–	–	–	●	–	–	●	–	–	–	–	–	–	–	–	–	–	–	–	–	–	–	–	–	–	–	2
Lipaugus lanioides	–	–	–	●	–	–	–	–	–	–	–	–	–	–	–	–	–	–	–	–	–	–	–	–	–	–	1
Piprites pileatus	–	–	–	–	–	–	–	–	–	–	–	–	–	–	–	–	–	–	–	–	●	–	–	–	–	–	1
Phylloscartes paulistus	–	–	–	–	–	–	–	–	–	●	–	●	–	–	–	–	–	–	–	–	–	–	–	–	–	–	2
Hemitriccus kaempferi	–	–	–	–	–	–	–	–	–	–	–	●	–	●	–	–	–	–	–	–	–	–	–	–	–	–	2
Platyrinchus leucoryphus	–	–	–	●	–	–	–	–	●	●	–	●	–	–	–	–	–	–	–	–	–	–	–	–	–	–	4
Anthus nattereri	–	–	–	–	–	–	–	–	–	–	–	–	–	–	–	–	–	●	–	–	–	●	–	–	–	–	2
Sporophila frontalis	–	–	–	–	–	–	●	–	–	–	–	–	●	–	–	–	–	–	–	–	–	–	–	–	–	–	2
Dacnis nigripes	–	–	–	–	–	–	–	–	–	–	–	–	●	–	–	–	–	–	–	–	–	–	–	–	–	–	1
Xanthopsar flavus	–	–	–	–	–	–	–	–	–	–	–	–	–	–	–	–	–	–	–	–	●	●	–	–	–	–	2
Sturnella militaris	–	–	–	–	–	–	–	–	–	–	–	●	–	–	–	–	–	–	–	–	–	–	–	–	–	–	1
No. of species	1	1	1	6	1	1	3	1	2	6	1	5	3	1	1	2	2	3	2	2	5	4	1	1	1	1	

Palmeira (Paraná)

BR 121
25°20′S 49°50′W

Unprotected

The habitat in this general area, lying immediately west of Palmeira and 70 km west of Curitiba in eastern Paraná, comprises *Araucaria* groves of 20–50 ha among open areas of agricultural fields (C. Seger and R. Boçom *in litt.* 1995).

Amazona vinacea	1994	Present April–July in flocks of up to 30; a few pairs nest in the forest (C. Seger and R. Boçom *in litt.* 1995).

Curitiba (Paraná)

BR 122
25°28′S 48°49′W
EBA B52

Cambuí Biological Reserve (status and area unknown)

The large city of Curitiba is c.100 km inland from the coast in eastern Paraná. Cambui Biological Reserve is near São José dos Pinhais on the outskirts of Curitiba.

Eleothreptus anomalus	c.1994	A sighting at Cambui Biological Reserve in 1986, and recently recorded near Curitiba (Bornschein 1995, Scherer-Neto and Straube 1995).

Ilha do Superagüi and Ilha das Peças (Paraná)

BR 123
25°30′S 48°15′W
EBA B52

Ilha do Superagüi National Park (IUCN category), 21,400 ha

The national park includes Ilha do Superagüi and the adjacent Ilha das Peças, located in coastal south-east Paraná close to the border with São Paulo. The principal habitat is sand-plain forest but there is montane forest at 300 m on Ilha do Superagüi (F. Olmos and P. Martuscelli *in litt.* 1995). An extension of the park boundary to include the tiny but important Ilha dos Pinheiros was suggested by Collar *et al.* (1992).

Amazona brasiliensis	c.1995	Recorded recently (F. Olmos and P. Martuscelli *in litt.* 1995); in 1984–1985 the population was estimated at up to 250 on Ilha do Superagüi and 179–343 on Ilha das Peças, both populations roosting on the nearby unprotected Ilha dos Pinheiros.
Carpornis melanocephalus	c.1995	Recorded recently (F. Olmos and P. Martuscelli *in litt.* 1995); no information on status.
Sporophila frontalis	c.1995	Recorded recently (F. Olmos and P. Martuscelli *in litt.* 1995); no information on status.

Ilha do Mel (Paraná)

BR 124
25°31′N 48°20′W
EBA B52

Ilha do Mel Ecological Station (IUCN category IV), 2,240 ha

The island is immediately offshore from Ponta do Sul and south of Ilha das Peças (BR 123) at the entrance to Baía de Paranaguá, in coastal south-east Paraná. The main habitat is sand-plain forest.

Amazona brasiliensis	1980s	An estimated 69–241 birds in 1984–1985.

Santa Cruz (Paraná)

BR 125

Santa Cruz Forest Reserve (status and area unknown)

25°35′S 48°35′W
EBA B52

This reserve is located 15 km south-west of Paranaguá in south-east Paraná. Presumably the reserve holds lowland humid forest, which should be surveyed to assess the current status of the threatned species.

Leucopternis lacernulata	1985	One record.
Platyrinchus leucoryphus	1946	One specimen.

Iguaçu (Paraná)

BR 126

Iguaçu National Park (IUCN category II), 170,000 ha

25°43′S 54°25′W
EBA B52

This large national park is located on the northern side of the Rio Iguaçu near the famous Iguaçu Falls in south-west Paraná. The park comprises humid forest which, together with Iguazú National Park (AR 15) and Urugua-í Provincial Park (AR 16) across the border in Argentina, forms one of the most extensive tracts within the southern part of the Atlantic forest domain.

Pipile jacutinga	1980s	Recorded throughout the 1980s.
Claravis godefrida	1981	One record, apparently the only one from the state, although seen in the adjacent Iguazú National Park (AR 15) in 1991.
Dryocopus galeatus	1993	All recent records are from the Poço Preto trail (e.g. Tobias *et al.* 1993). A stronghold for the species, especially when considered together with the contiguous Iguazú National Park (AR 15) in Argentina.
Biatas nigropectus	1990	Seen along the Poço Preto trail, but only a handful of sightings within the reserve and the state.
Phylloscartes paulistus	1993	A stronghold with a moderately large population, possibly in the low thousands. Regularly recorded (e.g. Forrester 1993); up to 12 can be noted in a day along the Poço Preto trail.
Platyrinchus leucoryphus	1990	Small numbers recorded from 1985.

General Carneiro (Paraná)

BR 127

Unprotected

26°28′S 51°25′W

This large area lies west of Caçador and Porto União in southern Paraná, and adjacent to the border with Santa Catarina. The principal habitats are evergreen broadleaf and *Araucaria angustifolia* forest. Much of the original forest has been destroyed or heavily degraded, but still represents an important example of mixed broadleaf forest within the state (C. Seger and R. Boçom *in litt.* 1995).

Amazona vinacea	1995	Currently the most important area in Santa Catarina, with a maximum population estimated at 400 birds (C. Seger and R. Boçom *in litt.* 1995).

Salto do Piraí (Santa Catarina)

BR 128
26°18′S 48°50′W
EBA B52

Protected

This area of humid forest is located 8 km north-north-west of Vila Nova, near Joinville in north-east Santa Catarina. It is protected by Centrais Elétricas de Santa Catarina as a catchment area for the Salto do Piraí hydroelectric power station. More surveys should be carried out in the neighbouring hillsides to establish which are the best remaining tracts of forest.

Leucopternis lacernulata	1991	A single sighting in July 1991.
Phylloscartes paulistus	1992	Seen in July 1991 and March 1992 (M. Pearman *in litt.* 1995).
Hemitriccus kaempferi	1991	The type-specimen collected in 1929 was the only record of the species until it was rediscovered in July 1991. This remains the only area where it is known.
Platyrinchus leucoryphus	1987	Apparently only one record.
Sturnella militaris	1991	A wintering flock of more than 35 seen in a stubble field between Salto do Piraí and Vila Nova in July 1991.

Itajaí (Santa Catarina)

BR 129
26°47′S 48°48′W
EBA B52

Unprotected

This area comprises c.5,000 ha of humid lowland and montane forest from sea-level to 1,500 m located on the coastal escarpment of eastern Santa Catarina, north-east of Itajaí.

Amazona vinacea	1992	Recorded recently (F. Olmos and P. Martuscelli *in litt.* 1995); no information on status.
Sporophila frontalis	1992	Recorded recently (F. Olmos and P. Martuscelli *in litt.* 1995); no information on status.
Dacnis nigripes	1992	Recorded recently (F. Olmos and P. Martuscelli *in litt.* 1995); no information on status.

Brusque (Santa Catarina)

BR 130
27°06′S 48°56′W
EBA B52

Unprotected

The town of Brusque is on the left bank of the Rio Itajaí-Mirim, 23 km south-east of Blumenau in east-central Santa Catarina. The area is near sea-level on the coastal plain, and the dominant habitat would be humid Atlantic forest, although this has mostly been cleared. Surveys should be carried out in any remaining forest to determine whether *Hemitriccus kaempferi* survives.

Hemitriccus kaempferi	1950	One specimen. This is one of only two sites known for the species.

São Joaquim (Santa Catarina)

BR 131
28°11′S 49°30′W
EBA B54

São Joaquim National Park (IUCN category II), 49,300 ha

This national park is located east of São Joaquim city in the Serra Geral of south-east Santa Catarina. The uplands comprise campo grassland and *Araucaria* groves, with some areas of montane forests.

Amazona vinacea	c.1995	Breeds in the area and feeds on the *Araucaria* seeds during winter (N. Varty *in litt.* 1995).

Espigão Alto (Rio Grande do Sul)

BR 132
27°36′S 51°30′W
EBA B54

Espigão Alto State Park (IUCN category II), 1,319 ha

This reserve is located c.10 km north-west of Baracão, adjacent to the southern bank of Rio Pelotas on the border between Rio Grande do Sul and Santa Catarina.

Amazona pretrei	1993	An important feeding area in winter (N. Varty *in litt.* 1995).
Amazona vinacea	1993	An important feeding area in winter, with birds possibly breeding (N. Varty *in litt.* 1995).

Aracuri (Rio Grande do Sul)

BR 133
28°14′S 51°11′W
EBA B54

Aracuri Federal Ecological Station (IUCN category IV), 272 ha

This small reserve of grassland with *Araucaria* groves is located 15 km south of Esmeralda in northern Rio Grande do Sul. The reserve was established to protect a roost of *Amazona pretrei* which, however, now use other sites. Nevertheless, the reserve is still an important feeding site for *A. pretrei* in winter.

Amazona pretrei	1993	Although roosting elsewhere, feeds regularly on the reserve in winter and is dispersed throughout the region in the breeding season (N. Varty *in litt.* 1995).
Amazona vinacea	1993	An occasional winter visitor (N. Varty *in litt.* 1995).

Carazinho–Passo Fundo–Soledade–Salto do Jacui (Rio Grande do Sul)

BR 134
28°18′S 52°48′W
EBA B54

Carazinho Municipal Park (status unknown), 206 ha
Passo Fundo National Forest (IUCN category VIII), 1,260 ha

This area is located on the planalto of north-central Rio Grande do Sul in the municipalities of Carazinho, Passo Fundo, Soledade and Salto do Jacui. The key habitat for the threatened parrots are the *Araucaria* groves, most of which are unprotected, but also of importance are the roosting areas for *Amazona pretrei* such as Carazinho Municipal Park. Passo Fundo National Forest holds *Araucaria* forests in which the parrots feed but which receives no conservation measures on the ground (Varty *et al.* 1994b).

Amazona pretrei	1994	Carazinho Municipal Park is a summer roost, with a maximum in 1992/93 of c.600, though only 140 in 1993/94 (Varty *et al.* 1994a). Another important roost is near Capão Bonito, south-east of Salto do Jacui. During the day in the breeding season birds are widely dispersed throughout forest patches in the region.
Amazona vinacea	1994	Breeds at low density in forest throughout the area.
Anthus nattereri	1978	One of the only modern records for Rio Grande do Sul.

Vacaria–Bom Jesus (Rio Grande do Sul)

BR 135
28°22′S 50°45′W
EBA B54

Unprotected

This area is on the planalto of north-east Rio Grande do Sul in the Vacaria and Bom Jesus municipalities. The most important habitat is *Araucaria angustifolia* forest—feeding habitat for both the species below. Muitos Capoês, the major winter roost for *Amazona pretrei*, is on private land (Varty *et al.* 1994b).

Amazona pretrei	1994	The Muitos Capoês roost held 7,290 birds in July 1993 (winter) and 4,600 in January 1994, with birds from the roost travelling up to 70 km north to the forested areas along the Rio Pelotas (Varty *et al.* 1994a). Other smaller roosts exist in the area.
Amazona vinacea	1994	A winter visitor in reasonable numbers, especially around and east of Bom Jesus where extensive *Araucaria* forest remains (N. Varty *in litt.* 1995).

Serra Geral (Rio Grande do Sul)

BR 136
29°00′S 50°00′W
EBA B54

Serra Geral National Park (IUCN category II), 17,300 ha

This national park is located mainly in Cambará do Sul municipality in the southern Serra Geral of north-east Rio Grande do Sul. The park holds extensive *Araucaria* forest but conservation measures are yet to be established on the ground (Varty *et al.* 1994b). Together with Aparados da Serra National Park (BR 137) it comprises a large and important tract of upland forest.

Amazona pretrei	1994	Breeds in the woods on the plateau, feeding also in the escarpment forests in winter (N. Varty *in litt.* 1995).
Amazona vinacea	1994	More commonly seen in forests on the escarpment than on the plateau (N. Varty *in litt.* 1995).

Aparados da Serra (Rio Grande do Sul/Santa Catarina)

BR 137
29°16′S 50°25′W
EBA B54

Aparados da Serra National Park (IUCN category II), 12,250 ha

This protected area is located in Cambará do Sul municipality of north-east Rio Grande do Sul and the adjacent Praia Grande municipality of extreme southern Santa Catarina. The park is in the southern Serra Geral where the principal geographical features are flat upland plateaus deeply incised by steep-sided canyons, notably the Itaimbezinho canyon in this national park. The area contains important stands of *Araucaria* forest on the plateaus with semi-deciduous forest on the slopes. Together with Serra Geral National Park (BR 136) and São Francisco de Paula (BR 138) this comprises a large and important tract of upland forest.

Amazona pretrei	1994	Restricted mainly to the forests on the plateau, where it is commoner in winter (N. Varty *in litt.* 1995).
Amazona vinacea	1994	Prefers forests on the slopes rather than on the plateau (N. Varty *in litt.* 1995); seen more frequently than *A. pretrei.*
Triclaria malachitacea	c.1987	Recorded from just north of the park at Cambará do Sul.
Piprites pileatus	1991	At least three sightings in the 1980s.
Xanthopsar flavus	1993	A group of eight seen just outside the southern entrance to the park in August 1993 (Tobias *et al.* 1993). Seen regularly over several years on the periphery of the park.

São Francisco de Paula (Rio Grande do Sul)

BR 138
29°23′S 50°20′W
EBA B54

São Francisco de Paula National Forest (IUCN category VIII), 1,138 ha

This area, c.20 km east of São Francisco de Paula in north-east Rio Grande do Sul, lies at 900 m in the uplands of the southern Serra Geral where the habitat consists of wet grassland marshes and *Araucaria* forest. The area includes unprotected marshes such as Banhados das Capivaras and do Fundo in Morrinho, and forests within the national forest reserve.

Amazona pretrei	1994	Breeds in small numbers, especially in woods on the plateau (N. Varty *in litt.* 1995).
Amazona vinacea	1994	Prefers forests on the escarpment to those on the plateau (N. Varty *in litt.* 1995).
Anthus nattereri	1979	One of the only modern records for Rio Grande do Sul.
Xanthopsar flavus	1994	Observed monthly between August 1992 and July 1993 at Banhado das Capivaras, with a maximum of 234 birds in April (Fontana 1994).

Monte Alverne (Rio Grande do Sul)

BR 139
29°33'S 52°20'W

Unprotected

The area is located at 500 m, c.20 km north-west of Santa Cruz do Sul in the central escarpment of north-central Rio Grande do Sul. The habitat comprises plateau forest with an understorey rich in palmito palms *Euterpe edulis*.

Triclaria malachitacea	1994	Recorded for a number of years. This is the main site for a study of the species' ecology (Bencke 1994).

Encruzilhada do Sul (Rio Grande do Sul)

BR 140
30°30'S 52°30'W
EBA B54

Podocarpus State Park (IUCN category II), 3,645 ha

This area is in the south-east hills of Rio Grande do Sul state at c.400 m. There are still some extensive areas of *Podocarpus lambertii*, notably in the Podocarpus State Park. Although designated as a protected area, conservation measures are yet to be established on the ground (Varty *et al.* 1994a). Formal protection should be secured for the *Amazona pretrei* roost site at Serra dos Pedrosas (see below). This area, together with Caçapava do Sul (BR 141) and Santana da Boa Vista (BR 142), forms the main breeding area for the species in Rio Grande do Sul.

Amazona pretrei	1994	One of the main breeding areas. In January 1994 3,200 roosted at Serra dos Pedrosas, a site used for many years (Varty *et al.* 1994a).

Caçapava do Sul (Rio Grande do Sul)

BR 141
30°31'S 53°29'W
EBA B54

Unprotected

There are still some extensive areas of *Podocarpus lambertii* forest in this area, located at c.400 m in the Serra do Sudeste of south-east Rio Grande do Sul. Together with Encruzilhada do Sul (BR 140) and Santana da Boa Vista (BR142), it forms the main breeding area for *Amazona pretrei* in Rio Grande do Sul.

Amazona pretrei	1993	An important breeding and feeding area with a small roost north of the city (Varty *et al.* 1994a).

Santana da Boa Vista (Rio Grande do Sul)

BR 142
30°53'S 53°07'W
EBA B54

Unprotected

This area is located in the Serra do Sudeste of south-east Rio Grande do Sul at c.450 m. Much of the land is used for mixed agriculture and cattle-raising, but patches of *Podocarpus* forest still remain. A priority is to secure protection for the *Amazona pretrei* roost site in the Serra dos Vargas and the remaining woodlands where it breeds. This area, together with Encruzilhada do Sul (BR 140) and Caçapava do Sul (BR 141), forms the main breeding area for the species in Rio Grande do Sul.

Amazona pretrei	1994	One of the species' main breeding areas. In January 1993 5,310 birds roosted at Serra dos Vargas, but only a few were using the site in January 1994 (Varty *et al.* 1994a).

■

■ CHILE

Horned Coot *Fulica cornuta*

CHILE supports an avifauna of 450 species (Pearman 1995), of which 12 (3%) are endemic to the country, 29 (6%) have restricted ranges (Stattersfield *et al.* in prep.) and 13 (3%) are threatened (Collar *et al.* 1992). The analysis has identified 18 Key Areas for the threatened birds in Chile (see 'Introduction', p. 11, for criteria).

THREATENED BIRDS

Of the 13 Chilean species which Collar *et al.* (1992) considered at risk of extinction, four are confined to the country (see Table 1). Habitat loss is a major reason for the threatened state of six (46%) of the species and is the sole reason for five (38%), with introduced predators or competitors a threat to six (46%). There is no predominant habitat or altitudinal zone that supports an unusually high number of threatened species but the high proportion of seabirds and waterbirds (six, 46%) is noteworthy. The distributions of threatened birds and their relationship to Endemic Bird Areas (EBAs) are shown in Figure 1.

KEY AREAS

The 18 Key Areas that have been identified would, if adequately protected, help ensure the conservation of 12 (92%) of Chile's threatened species, including all those that are confined to and resident in the country (always accepting that important new populations and areas may yet be found). Of these areas five each support three threatened species, while the rest hold just one. Four of the multiple-species Key Areas (CL 02–05) support the same threatened (and restricted-range) species of the south Peruvian and north Chilean Pacific slope EBA (B32), while the fifth is Islas Robinson Crusoe and Santa Clara (CL 18). However, as important as these areas are for the conservation of Chilean threatened species, they should not detract from the significance of the remaining single-species Key Areas, as these are of major importance for the remaining six species in the country.

Tables 1 and 2 show that four species occur in just one Chilean Key Area, two of these species—*Sephanoides fernandensis* in CL 17 and *Aphrastura masafuerae* in CL 18—being endemic to the Juan Fernández archipelago and thus totally reliant for their survival on the integrity of habitat on their respective islands (see 'Outlook', below).

Not represented within the Key Area analysis is *Numenius borealis*, which formerly ranged from Arica to Chiloé in the austral summer, but there are no twentieth-century records.

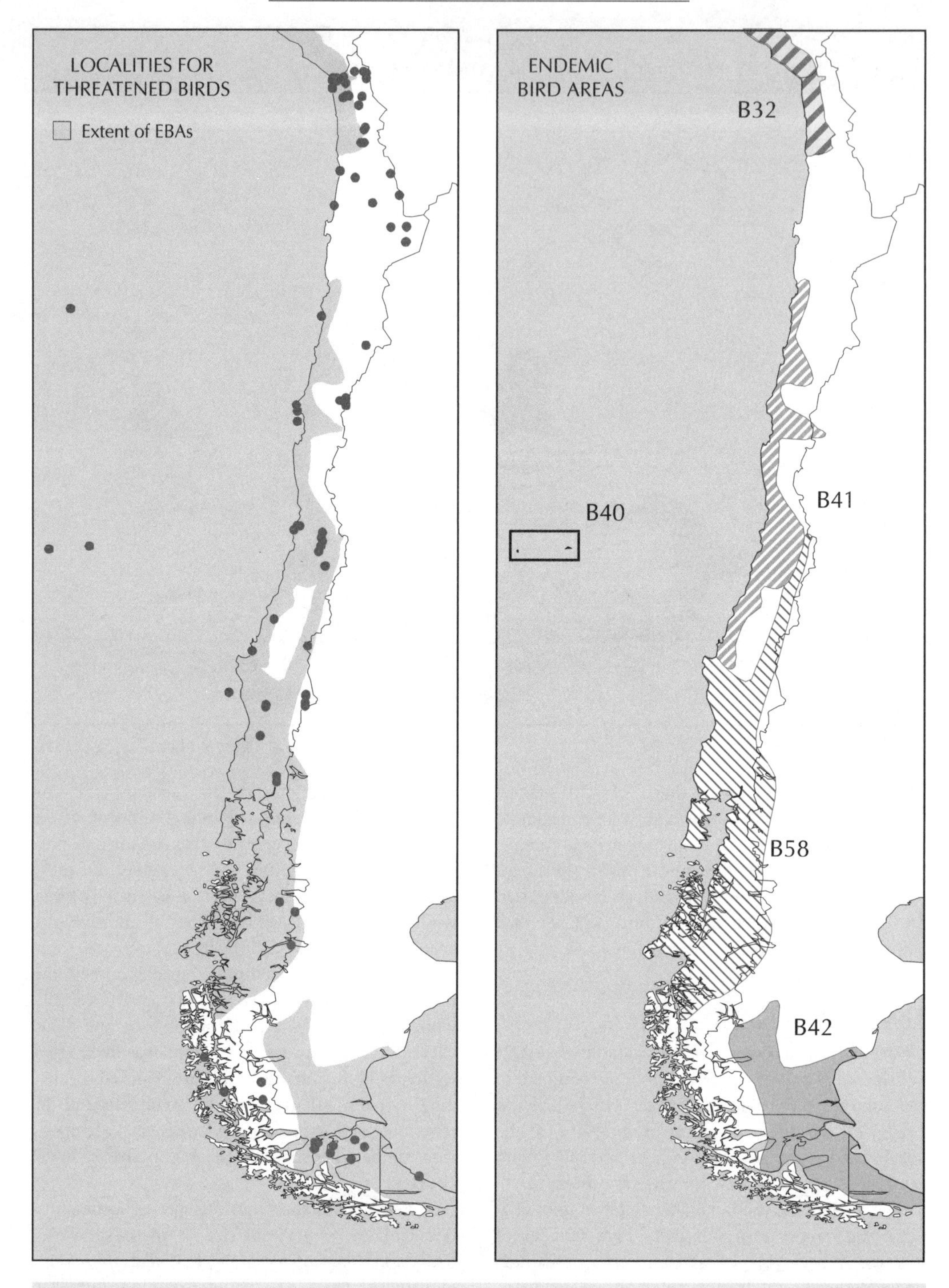

Figure 1. The localities where threatened birds have been recorded in Chile and their relationship to Endemic Bird Areas (EBAs).

- There are five threatened birds in Chile which have restricted ranges and they are the principal members of two of the five EBAs, which fall within the country, as listed below (figures are numbers of these species in each EBA).

B32 South Peruvian and north Chilean Pacific slope (3)
B40 Islas Juan Fernández (2)
B41 Central Chile (0)
B42 Tierra del Fuego and the Falklands (0)
B58 Valdivian forests of central Chile and Argentina (0)

Table 1. Coverage of threatened species by Key Area. Areas in bold presently have some form of protection.

	Key Areas occupied	No. of Key Areas protected	Total nos. of Key Areas	
			Chile	Neotropics
Pink-footed Shearwater *Puffinus creatopus* [E]	**13**,**17**	2	2	2
Defilippe's Petrel *Pterodroma defilippiana* [E]	16,**17**	1	2	2
Peruvian Diving-petrel *Pelecanoides garnotii*	**09**,**11**	2	2	3
Austral Rail *Rallus antarcticus*	**14**	1	1	2
Horned Coot *Fulica cornuta*	08,10	0	2	6
Eskimo Curlew *Numenius borealis*	—	0	0	0
Juan Fernández Firecrown *Sephanoides fernandensis* [E]	**17**	1	1	1
Chilean Woodstar *Eulidia yarrellii*	02,03,04,05	0	4	5
Masafuera Rayadito *Aphrastura masafuerae* [E]	**18**	1	1	1
Austral Canastero *Asthenes anthoides*	**12**,**15**	2	2	5
White-tailed Shrike-tyrant *Agriornis andicola*	**01**	1	1	11
Slender-billed Finch *Xenospingus concolor*	02,03,04,05	0	4	7
Tamarugo Conebill *Conirostrum tamarugense*	02,03,04,05,**06**,**07**	2	6	8

[E] Endemic to Chile

KEY AREA PROTECTION

Currently, 11 (61%) of Chile's Key Areas are under some form of protection, six (33%) as national parks or Biosphere Reserves (IUCN categories II and IX). The seven unprotected Key Areas (39% of the total) require appropriate conservation measures if the populations of their threatened species are to survive. However, even the formally protected areas remain under threat and, in many, habitat degradation continues unchecked; effective management of activities undertaken within them is thus essential.

With 7 (39%) of the Key Areas currently lacking any form of protection, the long-term survival of a number of threatened species must be in question. Indeed, three Chilean threatened species (*Fulica cornuta*, *Eulidia yarrellii* and *Xenospingus concolor*) are not currently known from any protected Key Area (Table 1). This illustrates that there are no protected Key Areas (in fact no protected areas at all) within the arid valleys of northern Chile at the centre of the south Peruvian and north Chilean Pacific slope EBA (B32).

RECENT CHANGES TO THE THREATENED LIST

With the publication of Collar *et al.* (1994) two seabirds (White-vented Storm-petrel *Oceanites gracilis* and Ringed Storm-petrel *Oceanodroma hornbyi*) and two waterbirds (Andean Flamingo *Phoenicopterus andinus* and Puna Flamingo *P. jamesi*) were added to the Chilean threatened list, but data for these have not been included in the Site Inventory (see 'Introduction', p. 12). Additionally, Westland Petrel *Procellaria westlandica*, an endemic breeder to New Zealand and listed as threatened by Collar *et al.* (1994), has been recorded recently on at least two occasions in Chilean waters (Howell and Webb 1995, Pearman 1995).

Table 2. Matrix of threatened species by Key Area.

	01	02	03	04	05	06	07	08	09	10	11	12	13	14	15	16	17	18	No. of areas
Pterodroma defilippiana	–	–	–	–	–	–	–	–	–	–	–	–	–	–	–	●	●	–	2
Puffinus creatopus	–	–	–	–	–	–	–	–	–	–	–	–	●	–	–	–	●	–	2
Pelecanoides garnotii	–	–	–	–	–	–	–	–	●	–	●	–	–	–	–	–	–	–	2
Fulica cornuta	–	–	–	–	–	–	–	●	–	●	–	–	–	–	–	–	–	–	2
Rallus antarcticus	–	–	–	–	–	–	–	–	–	–	–	–	–	●	–	–	–	–	1
Numenius borealis	–	–	–	–	–	–	–	–	–	–	–	–	–	–	–	–	–	–	0
Eulidia yarrellii	–	●	●	●	●	–	–	–	–	–	–	–	–	–	–	–	–	–	4
Sephanoides fernandensis	–	–	–	–	–	–	–	–	–	–	–	–	–	–	–	–	●	–	1
Aphrastura masafuerae	–	–	–	–	–	–	–	–	–	–	–	–	–	–	–	–	–	●	1
Asthenes anthoides	–	–	–	–	–	–	–	–	–	–	–	●	–	–	●	–	–	–	2
Agriornis andicola	●	–	–	–	–	–	–	–	–	–	–	–	–	–	–	–	–	–	1
Xenospingus concolor	–	●	●	●	●	–	–	–	–	–	–	–	–	–	–	–	–	–	4
Conirostrum tamarugense	–	●	●	●	●	●	●	–	–	–	–	–	–	–	–	–	–	–	6
No. of species	1	3	3	3	3	1	1	1	1	1	1	1	1	1	1	1	3	1	

Although these changes represent a high percentage difference (24%), all the species added to the list were classified as Near Threatened in Collar *et al.* (1992), the new criteria and/or recent data having promoted their upgrading.

The additions to the Chilean threatened species list will have little distributional impact on the Key Area analysis. The Andean lakes already chosen for *Fulica cornuta* (CL 08 and CL 10) are major sites for both the flamingo species, and the lakes within Lauca National Park (CL 01) are also important for them (Scott and Carbonell 1986, Pearman 1995). Both species of storm-petrel are found in Chilean waters though the nesting grounds of *Oceanodroma hornbyi* are unknown, and the only nest ever found of *Oceanites gracilis* was on the Chilean island of Chungungo.

OLD RECORDS AND LITTLE-KNOWN BIRDS

A number of threatened species in Chile suffer from a general paucity of recent records. Most species are little known somewhere within their range, but recent records from other areas mean that we are able to make informed recommendations for their conservation. The exceptions are species known only from old specimen records or chance observations (from which there can be no guarantees of a current viable population), and in Chile this applies to *Rallus antarcticus*, known in Chile mainly from a number of specimens collected in the 1800s; the records from Lago Todos los Santos (CL 14) in the 1960s are among the most recent sightings within its range.

OUTLOOK

This analysis has identified 18 Key Areas, each of which is of major importance for the conservation of threatened birds in Chile. However, certain Key Areas stand out as top priorities: the seven listed in Table 3 are the most important Key Areas for nine (69%) of the Chilean threatened species. These and the remaining Key Areas (which are individually extremely important for many threatened Chilean endemics: see 'Key Areas' and 'Key Area protection', above) all need some degree of conservation action if the populations of their threatened birds are to remain viable. For some areas there is a requirement to confirm the continued existence or assess the population of a threatened species (e.g. *Rallus antarcticus* at Lago Todos los Santos, CL 14). The Site Inventory highlights the potential importance of each Key Area for one or more threatened species, and the fact that each has been identified points to a need for conservation action, often in the form of protection—but in the case of the threatened seabirds specific measures are also required to keep introduced predators away from breeding colonies.

Table 3. Top Key Areas in Chile. Those with the area number in bold currently have some form of protected status.

Key Area		No. of threatened spp.	Comments
02	Lluta valley	3	Probably the major area for *Eulidia yarrellii* and *Xenospingus concolor.*
06	Salar de Pintados	1	The only site where *Conirostrum tamarugense* has been found in good numbers, its population being estimated in the thousands.
11	Isla Choros	1	The principal nesting area for *Pelecanoides garnotii* in Chile.
12	Ñuble	1	This national park holds a significant population of *Asthenes anthoides.*
16	Desventuradas Islands	1	Probably the major breeding site for *Pterodroma defilippiana.*
17	Islas Robinson Crusoe and Santa Clara	3	Robinson Crusoe holds the only extant population of *Sephanoides fernandensis,* and nesting sites for threatened seabirds are found on both islands.
18	Isla Alejandro Selkirk	1	The island supports the endemic *Aphrastura masafuerae.*

DATA SOURCES

The above introductory text and the Site Inventory were compiled from information supplied by C. Estades, S. N. G. Howell and M. Pearman, as well as from the following references.

Collar, N. J., Gonzaga, L. P., Krabbe, N., Madroño Nieto, A., Naranjo, L. G., Parker, T. A. and Wege, D. C. (1992) *Threatened birds of the Americas: the ICBP/IUCN Red Data Book.* Cambridge, U.K.: International Council for Bird Preservation.

Collar, N. J., Crosby, M. J. and Stattersfield, A. J. (1994) *Birds to watch 2: the world list of threatened birds.* Cambridge, U.K.: BirdLife International (BirdLife Conservation Series no. 4).

COLLAR, N. J., WEGE, D. C. AND LONG, A. J. (in press) Patterns and causes of endangerment in the New World avifauna. In J. V. Remsen, ed. *Natural history and conservation of Neotropical birds*. American Ornithologists' Union (Orn. Monogr.).

ESTADES, C., GABELLA, J. P. AND ROTTMANN, J. (1994) Nota sobre el canastero del sur *Asthenes anthoides* en la reserva nacional Ñuble. *Bol. Chil. Orn.* 1: 31–32.

IUCN (1992) *Protected areas of the world: a review of national systems*, 4: *Nearctic and Neotropical*. Gland, Switzerland and Cambridge, U.K.: International Union for Conservation of Nature and Natural Resources.

HOWELL, S. N. G. AND WEBB, S. (1995) Noteworthy bird observations from Chile. *Bull. Brit. Orn. Club* 115: 57–66.

JEHL, J. R. (1973) The distribution of marine birds in Chilean waters in winter. *Auk* 90: 114–135.

MILLIE, W. R. (1963) Brief notes on the birds of San Ambrosio and San Felix Islands, Chile. *Ibis* 105: 563–566.

PAYNTER, R. A. (1988) *Ornithological gazetteer of Chile*. Cambridge, Mass.: Museum of Comparative Zoology.

PEARMAN, M. (1995) *The essential guide to birding in Chile*. Belper, U.K.: Worldwide Publications.

SCOTT, D. A. AND CARBONELL, M. (1986) *A directory of Neotropical wetlands*. Cambridge and Slimbridge, U.K.: International Union for Conservation of Nature and Natural Resources and International Waterfowl Research Bureau.

STATTERSFIELD, A. J., CROSBY, M. J., LONG, A. J. AND WEGE, D. C. (in prep.) *Global directory of Endemic Bird Areas*. Cambridge, U.K.: BirdLife International (BirdLife Conservation Series).

SITE INVENTORY

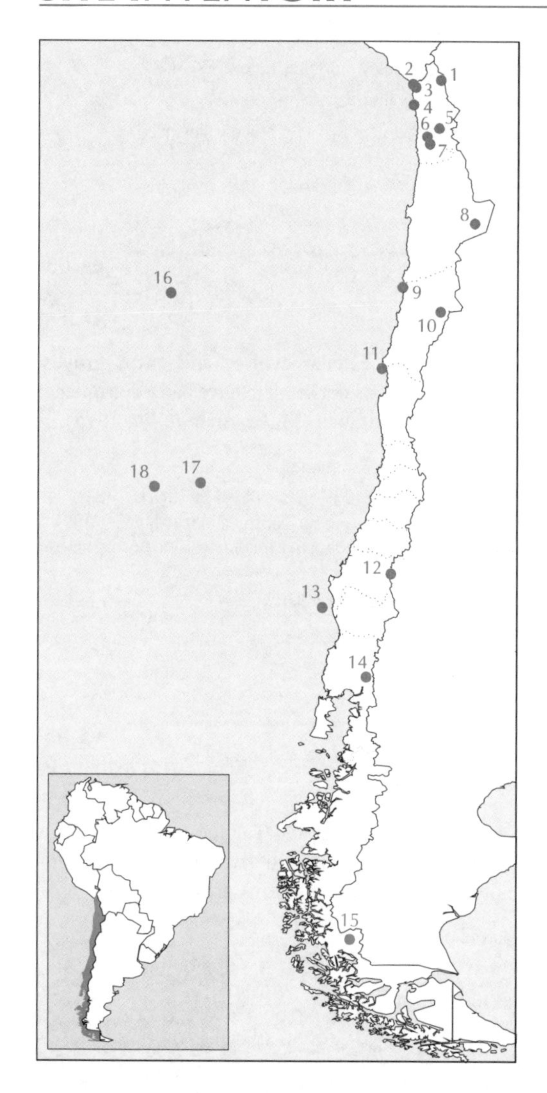

Figure 2. Key Areas in Chile.

01 Lauca
02 Lluta valley
03 Azapa valley
04 Camarones valley
05 Mamiña
06 Salar de Pintados
07 Salar de Bellavista
08 Lagunas Meñique and Miscanti
09 Isla Pan de Azúcar
10 Laguna Santa Rosa
11 Isla Choros
12 Ñuble
13 Isla Mocha
14 Lago Todos los Santos
15 Torres del Paine
16 Islas Desventuradas
17 Islas Robinson Crusoe and Santa Clara
18 Isla Alejandro Selkirk

Lauca (Tarapacá)

CL 01
18°15′S 69°10′W

Lauca National Park (IUCN category II), 137,883 ha

This national park is located high on the Andean altiplano at 3,200–6,342 m in extreme north-east Chile on the border with Peru and Bolivia. The landscape is dominated by two volcanoes, Pomerape (6,282 m) and Parinacota (6,342 m), and the habitat is puna grassland with small areas of cushion-plant bog and several large lakes, notably Parinacota, Cotacotani and Chungara (Pearman 1995).

Agriornis andicola	1992	One seen 2 km south-east of Parinacota village in December 1992 (Howell and Webb 1995), but the species has also been recorded around Laguna Chungara.

Lluta valley (Tarapacá)

CL 02
18°24′S 70°19′W
EBA B32

Unprotected

Like the Azapa valley (CL 03), the Lluta valley forms within the Atacama desert of northern Chile one of the only continuous strips of vegetation connecting the high altiplano to the Pacific coast. The landscape is highly modified with extensive olive and citrus groves, but the threatened species are still well represented in the area (M. Pearman *in litt.* 1995).

Eulidia yarrellii	1991	Locally common in the lower part of the valley, normally below 800 m.
Xenospingus concolor	1991	Locally common in the lower part of the valley.
Conirostrum tamarugense	1990	Recent sightings involve six birds recorded from two sites at the valley head.

Azapa valley (Tarapacá)

CL 03
18°31′S 70°11′W
EBA B32

Unprotected

The Azapa valley, like the Lluta valley (CL 02), forms a major green corridor across the Atacama desert of northern Chile. The valley is now highly modified with extensive olive and citrus groves, although this does not seem to have had a detrimental effect on the avifauna (M. Pearman *in litt.* 1995).

Eulidia yarrellii	1991	A stronghold for the species, which is locally common up to 720 m.
Xenospingus concolor	1991	Found from sea-level inland to Molinos and appears to be at least locally common: e.g. at 200–240 m near San Miguel, in only a few hours observation in January 1991, M. Pearman (*in litt.* 1995) found c.20 birds (including five immatures) and heard at least another 10.
Conirostrum tamarugense	1991	Apparently the only recent record is of one seen near San Miguel.

Camarones valley (Tarapacá)

CL 04
19°11′S 70°17′W
EBA B32

Unprotected

The Camarones valley is c.70 km south of Arica in north-central Tarapacá. The species below were recorded from the mid-elevations of the valley between Conanoxa (at 400 m) and Taltape (at 780 m). This desert river valley supports riverine scrub and agricultural areas including citrus groves.

Eulidia yarrellii	1970	Observations come from near Chupicilca.
Xenospingus concolor	1986	Four collected in 1986.
Conirostrum tamarugense	1975	Two sightings at 400 and 780 m.

Mamiña (Tarapacá)

CL 05

Unprotected

20°05′S 69°14′W
EBA B32

Mamiña is a small settlement at 2,600 m, 100 km east of Iquique in central Tarapacá. The area is in the arid zone of northern Chile, so the vegetation presumably comprises xerophytic scrub.

Eulidia yarrellii	1969	Known only from observations in the late 1960s.
Xenospingus concolor	1985	Specimens collected from Mamiña and Palca in 1985.
Conirostrum tamarugense	1969	Known only from observations in the late 1960s at 2,600–2,950 m.

Salar de Pintados (Tarapacá)

CL 06

Pampa del Tamarugal National Reserve (IUCN category IV), 100,650 ha

20°24′S 69°44′W
EBA B32

This area comprises the disjunct central sector of the Pampa del Tamarugal National Reserve, 70 km south-east of Iquique in central Tarapacá. The environment is very arid, with vegetation comprising only tamarugo *Prosopis tamarugo* plantations, established in 1946 (C. Estades *in litt.* 1993).

Conirostrum tamarugense	1990s	A population, probably in the thousands, and almost certainly the largest known for the species, was recently discovered (C. Estades *in litt.* 1993).

Salar de Bellavista (Tarapacá)

CL 07

Pampa del Tamarugal National Reserve (IUCN category IV), 100,650 ha

20°41′S 69°37′W
EBA B32

This area comprises the disjunct southern sector of the Pampa del Tamarugal National Reserve, 50 km east-south-east of Iquique in central Tarapacá. The environment is very arid, with vegetation comprising only tamarugo *Prosopis tamarugo* plantations, established in 1970 (C. Estades *in litt.* 1993).

Conirostrum tamarugense	1990s	A very low-density population, possibly representing the southern limit for the species (C. Estades *in litt.* 1993).

Lagunas Meñique and Miscanti (Antofagasta)

CL 08

Unprotected

23°43′S 67°47′W

These two hypersaline lakes at c.4,250 m in the Andean altiplano are 7 km (by track) from the Toconao–San Antonio de las Cobres (trans-Andean) road in east-central Antofagasta.

Fulica cornuta	1992	A stronghold for the species in Chile: up to 100 pairs breed at Laguna Meñique and 20 pairs on Laguna Miscanti (Pearman 1995, S. N. G. Howell *in litt.* 1993).

Isla Pan de Azúcar (Atacama)

CL 09

Pan de Azúcar National Park (IUCN category II), 43,754 ha

26°09′S 70°42′W

This small island, c.1.5 km long, is little more than a rock with only a sparse covering of vegetation. It lies 4 km offshore from Pan de Azúcar village in north-west Atacama.

Pelecanoides garnotii	1991	About 200 active burrows have been found on the north-east side of island.

Laguna Santa Rosa (Atacama)

CL 10

27°05′S 69°10′W

Unprotected

Laguna Santa Rosa consists of three small saline lakes totalling 70 ha, high in the Andean altiplano (3,750 m) of north-east Atacama, c.170 km east of Caldera.

Fulica cornuta	1991	159 counted in May 1991.

Isla Choros (Atacama)

CL 11

29°14′S 71°33′W

Pingüino National Reserve (IUCN category IV), area unknown

Isla Choros is a small island (c.4 km long) lying 7 km off the Chilean coast, 80 km north-west of Coquimbo and c.15 km south-west of Punta Carrizal.

Pelecanoides garnotii	1991	About 300 active nests found in the last census in 1991.

Ñuble (Ñuble)

CL 12

37°05′S 71°10′W

Ñuble National Reserve (IUCN category IV), 55,948 ha

The Ñuble National Reserve is located at the northern end of Laguna de la Laja in the Andes of eastern Ñuble near the border with Argentina. The main habitat is *Festuca*-dominated grassland and patches of *Nothofagus antarctica* scrub. The reserve is threatened by fire and overgrazing (Estades *et al.* 1994).

Asthenes anthoides	1994	Found in good numbers in the Río Blanquillo valley at 1,250 m in March 1994; a density of 2,265 birds per km^2 was estimated, and, given that its favoured habitat covers thousands of hectares in the reserve, the population could be enormous—though the species may occur only seasonally as much of the reserve is snow-covered in winter (Estades *et al.* 1994).

Isla Mocha (Arauco)

CL 13

38°22′S 73°56′W

Isla Mocha National Reserve (IUCN category IV), 2,368 ha

This island measuring c.15 × 7 km lies 33 km from the Chilean coast of central Arauco. It is mountainous, and at least half the land area is covered by temperate forest.

Puffinus creatopus	1980s	One of the only known colonies of the species.

Lago Todos los Santos (Llanquihue)

CL 14

41°00′S 72°10′W

Vicente Pérez Rosales National Park (IUCN category II), 226,305 ha
Puyehue National Park (IUCN category II), 107,000 ha

This freshwater lake, covering 12,500 ha, is located c.90 km south-east of Osorno in north-east Llanquihue. Surrounding it are some areas of riverine marsh, seasonally flooded meadows and large areas of Valdivian forest comprising *Araucaria araucana* and southern beech *Nothofagus dombeyi*.

Rallus antarcticus	1960s	Suggested in the 1960s to breed in the reedbeds around the lake at Peulla, Puntiagudo and the mouth of Río Cayutué, but not found during searches in the late 1980s.

Torres del Paine (Magallanes)

CL 15

51°00′S 72°48′W

Torres del Paine National Park (IUCN category II), 181,414 ha

The large national park of Torres del Paine covers much of the eastern part of Magallanes province from 48° to 51°30′S. The area holds extensive tracts of southern temperate forest and Patagonian shrub-steppes with bunch-grass.

Asthenes anthoides	1990	Reported in fair numbers during January 1990.

Islas Desventuradas

CL 16

26°23′S 80°05′W

Unprotected

These volcanic islands, comprising Islas San Ambrosio, San Felix and González, lie 900 km west of continental Chile. Isla San Ambrosio is the largest (2.5 × 0.8 km), with land reaching 1,000 m on its summit plateau. The islands support important breeding colonies of several species of seabird.

Pterodroma defilippiana	1970	Nesting has been recorded on San Ambrosio (and González islet) where a cavern containing 300–400 birds was found in 1962 (Millie 1963), but where in 1970 the cliffs were teeming with 10,000 or more birds (Jehl 1973).

Islas Robinson Crusoe and Santa Clara

CL 17

33°37′S 78°51′W
EBA B40

Juan Fernández Biosphere Reserve (IUCN category IX), 9,290 ha

These volcanic islands, which are part of the Juan Fernández archipelago, lie 680 km west of continental Chile. Isla Robinson Crusoe, at 21 km long, covers 4,711 ha with land reaching 915 m on the dormant El Yunque. Isla Santa Clara is an outlier of Robinson Crusoe, being sited just 2 km south-west of it and covering 233 ha. The islands have a highly endemic flora, and the threatened bird species are all endemic to these islands or the wider Juan Fernández archipelago.

Pterodroma defilippiana	1991	Formerly bred on both islands but now probably only on Santa Clara, where the population in 1991 was estimated at 100–200 birds.
Puffinus creatopus	1991	The population on Santa Clara in 1991 was estimated at 2,000–3,000 pairs, and on Robinson Crusoe in 1986 at a few thousand pairs.
Sephanoides fernandensis	1991	The species is now restricted to Robinson Crusoe, with a population estimated at c.1,000 birds in a census conducted by Corporación Nacional Forestal (CONAF) in 1991.

Isla Alejandro Selkirk

CL 18

33°45′S 80°45′W
EBA B40

Juan Fernández Biosphere Reserve (IUCN category IX), 9,290 ha

The isolated Isla Alejandro Selkirk, with its highly endemic flora, lies 182 km west of Isla Robinson Crusoe (CL 17); it covers 4,464 ha, and land reaches 1,380 m at Los Innocentes. The island has a highly endemic flora, and there are also important seabird colonies including the only ones known for both Juan Fernández Petrel *Pterodroma externa* and Stejneger's Petrel *P. longirostris*. The endemic race of *Sephanoides fernandensis* is extinct, the last sighting of it having been in 1908.

Aphrastura masafuerae	1986	Endemic to the island and inhabits the native *Dicksonia externa* tree-fern forest which survives on the south-west slopes; a brief survey estimated the population at 500 birds, and not exceeding 1,000.

■ COLOMBIA

Black Inca *Coeligena prunellei*

COLOMBIA has the richest avifauna of any country in the world, with more than 1,700 species having been recorded (Hilty and Brown 1986, Downing 1993), of which c.170 (10%) are migrants (primarily from breeding areas in North America) (Hilty and Brown 1986), at least 66 (4%) are endemic to the country, 193 (11%) have restricted ranges (Stattersfield *et al.* in prep.) and 57 (3.5%) are threatened (Collar *et al.* 1992). This analysis has identified 69 Key Areas for the threatened birds in Colombia (see '*Key Areas*: the book', p. 11, for criteria).

THREATENED BIRDS

Of the 57 Colombian species which Collar *et al.* (1992) considered at risk of extinction, 30 are confined to the country (see Table 1), making Colombia third in rank in the Americas for numbers of threatened birds (Collar *et al.* in press). These species rely primarily on wet forest habitats (77% of species) within the subtropical and temperate zones (i.e. above 500 m), and are almost exclusively threatened by habitat loss and, to a lesser but significant extent, by hunting (Collar *et al.* in press). The distributions of these 57 threatened birds and their relationship to Endemic Bird Areas are shown in Figure 1.

KEY AREAS

The 69 Key Areas that have been identified would, if adequately protected, help ensure the conservation of 54 (95%) of Colombian threatened species, including all those confined to the country, although it should be noted that important new populations and areas may yet be found. Of the 69 Key Areas, 39 are important for two or more (up to 10) threatened species, and these are therefore the most efficient areas (currently known) in which to conserve Colombia's threatened birds (see 'Outlook', below): the 10 areas which each harbour five or more threatened species (see Table 3) together represent potential sanctuaries for 33 threatened species, 58% of the total number.

From Tables 1 and 2 it can be seen that 16 species occur in just one Colombian Key Area, five of which (*Eriocnemis mirabilis*, *Crypturellus saltuarius*, *Thryothorus nicefori*, *Sporophila insulata* and *Vireo caribaeus*) are endemic to the country, and are thus totally reliant for their survival on the integrity of

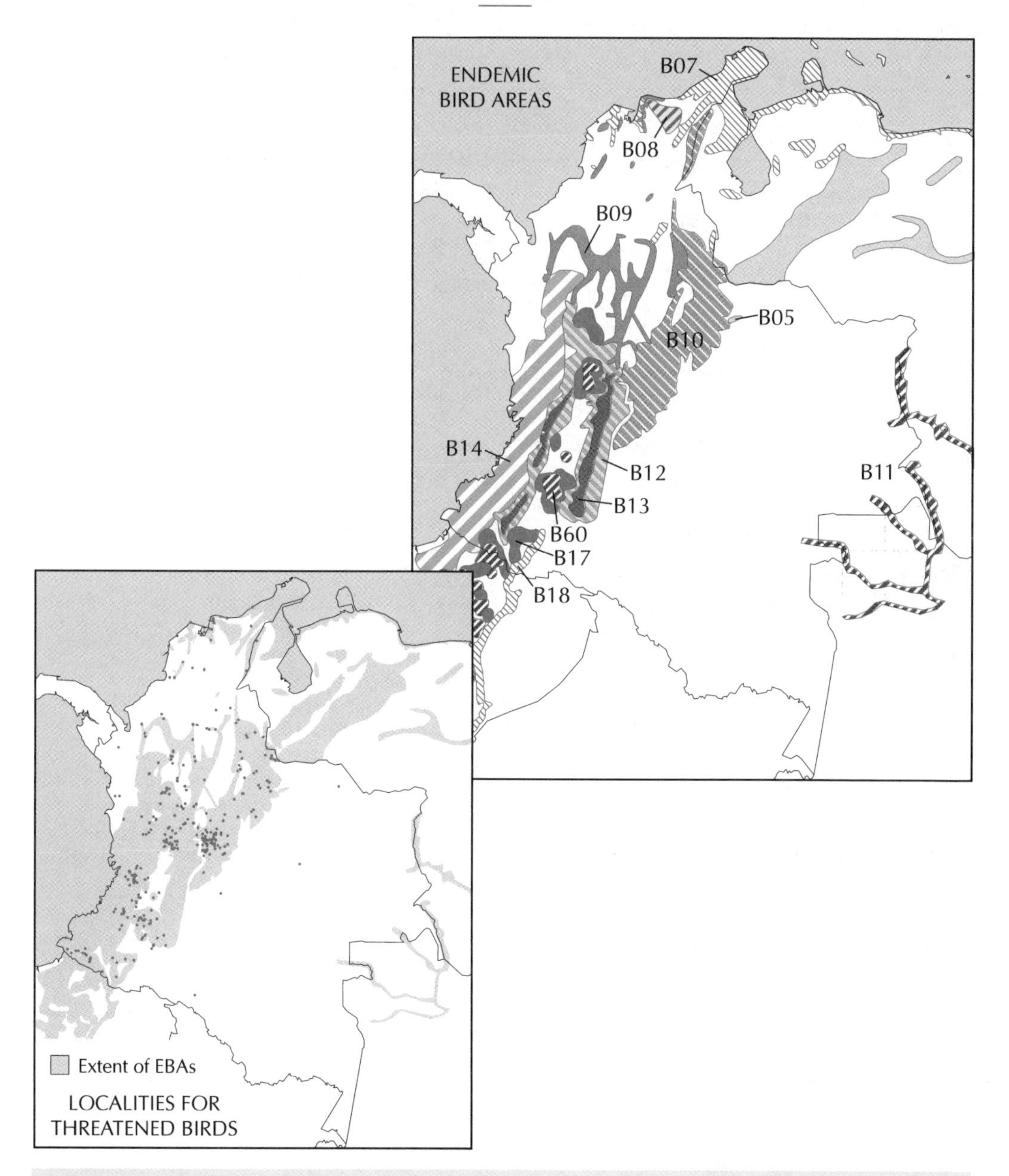

Figure 1. The localities where threatened birds have been recorded in Colombia and their relationship to Endemic Bird Areas (EBAs).

- Most (50) of Colombia's threatened birds have restricted ranges and thus mainly occur together in various combinations within EBAs, which are listed below (figures are numbers of these species in each EBA). The three Andean cordilleras dominate in terms of threatened bird distributions, but of particular importance are the EBAs in the East Andes (B10), the subtropical inter-Andean zone (B12) and the wet forests of the Chocó and Pacific-slope Andes (B14).

A19 Darién and Urabá lowlands (3)
A20 East Panama and Darién highlands (0)
B07 Caribbean dry zone (1)
B08 Santa Marta mountains (0) *
B09 Nechí lowlands (5)
B10 East Andes of Colombia (15)
B11 Upper Río Negro and Orinoco white-sand forest (0)
B12 Subtropical inter-Andean Colombia (10)
B13 Dry inter-Andean valleys (3)
B14 Chocó and Pacific-slope Andes (15)
B17 North Central Andean forests (5)
B18 East Andes of Ecuador and northern Peru (4)
B19 Napo and upper Amazon lowlands (0)
B60 Central Andean páramo (5)

* See 'Recent changes to the threatened list'

Table 1. Coverage of threatened species by Key Area. Areas in bold currently have some form of protected status.

	Key Areas occupied	No. of Key Areas protected	Total nos. of Key Areas	
			Colombia	Neotropics
Black Tinamou *Tinamus osgoodi*	**59**	1	1	4
Chocó Tinamou *Crypturellus kerriae*	**17**	1	1	3
Magdalena Tinamou *Crypturellus saltuarius*[E]	05	0	1	1
Colombian Grebe *Podiceps andinus*[E]	39,40,41,44,**45**	1	5	5
Plumbeous Forest-falcon *Micrastur plumbeus*	**53**,64,**65**	2	3	4
Cauca Guan *Penelope perspicax*[E]	**25**,**51**,**53**,54,55	3	5	5
Northern Helmeted Curassow *Pauxi pauxi*	02,**30**,**38**	2	3	10
Blue-billed Curassow *Crax alberti*[E]	06,**08**,09	1	3	3
Wattled Curassow *Crax globulosa*	**61**	1	1	4
Gorgeted Wood-quail *Odontophorus strophium*[E]	32,**35**,42	1	3	3
Bogotá Rail *Rallus semiplumbeus*[E]	39,40,41,42,44,**45**,47, **49**	2	8	8
Speckled Crake *Coturnicops notatus*	**62**	1	1	8
Tolima Dove *Leptotila conoveri*[E]	27,28,58	0	3	3
Golden-plumed Parakeet *Leptosittaca branickii*	**24**,**25**,**26**,28,**53**,**56**,**59**, **67**,**68**	8	9	22
Yellow-eared Parrot *Ognorhynchus icterotis*	28,**53**,**56**,57,**59**,**66**	4	6	7
Flame-winged Parakeet *Pyrrhura calliptera*[E]	**30**,**35**,**38**,39,**48**,**49**,50	5	7	7
Rufous-fronted Parakeet *Bolborhynchus ferrugineifrons*[E]	**24**,**25**,**26**	3	3	3
Spot-winged Parrotlet *Touit stictoptera*	**62**	1	1	8
Rusty-faced Parrot *Hapalopsittaca amazonina*	**22**,**48**,**49**,**56**,**59**	5	5	8
Fuertes's Parrot *Hapalopsittaca fuertesi*[E]	**24**,**26**	2	2	2
Banded Ground-cuckoo *Neomorphus radiolosus*	**52**,**53**,**65**	3	3	4
White-chested Swift *Cypseloides lemosi*	55	0	1	1
Sapphire-bellied Hummingbird *Lepidopyga lilliae*[E]	03,**04**	1	2	2
Chestnut-bellied Hummingbird *Amazilia castaneiventris*[E]	07,31,33,37	0	4	4
Black Inca *Coeligena prunellei*[E]	**35**,36,42,43	1	4	4
Turquoise-throated Puffleg *Eriocnemis godini*	—	0	0	0
Colourful Puffleg *Eriocnemis mirabilis*[E]	**53**	1	1	1
Hoary Puffleg *Haplophaedia lugens*	**65**,**66**	2	2	4
Coppery-chested Jacamar *Galbula pastazae*	69	0	1	6
White-mantled Barbet *Capito hypoleucus*[E]	07,09,**14**,**15**,**23**	3	5	5
Chestnut-throated Spinetail *Synallaxis cherriei*	—	0	0	4
Recurve-billed Bushbird *Clytoctantes alixii*	07,**08**,09	1	3	4
Speckled Antshrike *Xenornis setifrons*	**17**	1	1	5
Bicoloured Antvireo *Dysithamnus occidentalis*	**52**,**53**	2	2	4
Moustached Antpitta *Grallaria alleni*[E]	**26**,**59**	2	2	2
Giant Antpitta *Grallaria gigantea*	**56**	1	1	4
Brown-banded Antpitta *Grallaria milleri*[E]	**25**,**26**	2	2	2
Bicoloured Antpitta *Grallaria rufocinerea*[E]	**22**,**25**,**26**,28,**56**	4	5	5
Hooded Antpitta *Grallaricula cucullata*	**53**,**59**	2	2	3
Antioquia Bristle-tyrant *Phylloscartes lanyoni*[E]	09,**15**,**23**	2	3	3
Ochraceous Attila *Attila torridus*	—	0	0	11
Apolinar's Wren *Cistothorus apolinari*[E]	**38**,39,40,41,42,44,**45**, 47	2	8	8
Nicéforo's Wren *Thryothorus nicefori*[E]	34	0	1	1
Tumaco Seedeater *Sporophila insulata*[E]	63	0	1	1
Yellow-headed Brush-finch *Atlapetes flaviceps*[E]	28,57	0	2	2
Tanager-finch *Oreothraupis arremonops*	**11**,21,**52**,**53**,**66**	4	5	6
Yellow-green Bush-tanager *Chlorospingus flavovirens*	**52**,**65**	2	2	3
Gold-ringed Tanager *Buthraupis aureocincta*[E]	19,**20**,21	1	3	3
Black-and-gold Tanager *Buthraupis melanochlamys*[E]	10,19,**20**,21	1	4	4
Masked Mountain-tanager *Buthraupis wetmorei*	**56**	1	1	7
Multicoloured Tanager *Chlorochrysa nitidissima*[E]	**20**,**25**,**51**,**52**,**53**	5	5	5
Scarlet-breasted Dacnis *Dacnis berlepschi*	64,**65**	1	2	5
Turquoise Dacnis *Pseudodacnis hartlaubi*[E]	42,43,**46**,**51**	2	4	4
San Andrés Vireo *Vireo caribaeus*[E]	01	0	1	1
Baudó Oropendola *Psarocolius cassini*[E]	16,**17**,18	1	3	3
Red-bellied Grackle *Hypopyrrhus pyrohypogaster*[E]	**08**,09,**11**,12,13,**20**,**24**, **25**,**26**,28,**59**,60	7	12	12
Red Siskin *Carduelis cucullatus*	29	0	1	4

[E] Endemic to Colombia

habitat in their respective Key Areas (see 'Outlook'). Of these species, the latter four occur on their own within their respective Key Areas, emphasizing that, in these cases, targeted single-species site-based conservation is a necessity. At least four threatened Colombian endemics (*Odontophorus strophium* in CO 35, *Bolborhynchus ferrugineifrons* in CO 24, *Hapalopsittaca fuertesi* in CO 26 and *Grallaria milleri* in CO 25), although recorded from a number of Key Areas, appear to be effectively dependent on forests within single Key Areas (i.e. those where sizeable populations exist, or from where the only recent records derive), and three other endemics (*Lepidopyga lilliae*, *Grallaria alleni* and *Atlapetes flaviceps*) appear each to be reliant on just two Key Areas.

Not represented within the Colombian Key Area analysis are three of the country's threatened species, as follows.

Eriocnemis godini is known in Colombia only from two 'Bogotá' trade-skins taken during the nineteenth century. The species has been recorded with certainty only from Ecuador where, however, there have been no confirmed records this century.

Synallaxis cherriei is known in Colombia from two specimens collected during 1967 in Putumayo department. The precise locality has not been traced, and without further evidence of its presence in this area, priorities for the conservation of this bird should be confined to Brazil, Ecuador and Peru.

Attila torridus is essentially confined to humid lowland and foothill forest in western Ecuador, and conservation efforts should be concentrated there and in adjacent north-westernmost Peru. There is just one record from Colombia, a specimen collected in Nariño during 1958.

KEY AREA PROTECTION

Currently, only 27 (39%) of Colombia's Key Areas have some form of protected status (i.e. as national, regional or municipal parks, or private or watershed reserves), 15 of these as national parks (IUCN category II). The 42 Key Areas (61% of the total) that are currently unprotected require attention in the form of appropriate conservation measures if the populations of their threatened species are to survive. However, even the formally protected areas remain under threat and, in many, habitat degradation and hunting continue unchecked; effective management of activities undertaken within them is thus required.

With 42 (61%) of the Key Areas currently not under any form of protection, the long-term survival of a number of threatened species must be in question. Ten Colombian threatened species are not currently recorded from any protected Key Area (see Table 1): *Crypturellus saltuarius*, *Leptotila conoveri*, *Cypseloides lemosi*, *Amazilia castaneiventris*, *Galbula pastazae*, *Thryothorus nicefori*, *Sporophila insulata*, *Atlapetes flaviceps*, *Vireo caribaeus* and *Carduelis cucullatus*. With the exception of *Cypseloides lemosi*, *Galbula pastazae* and *Carduelis cucullatus*, these species are endemic to Colombia (five being known from just one Key Area: see 'Key Areas', above), and Key Areas supporting populations of any of them should be regarded as high conservation priorities. A further eight threatened Colombian endemics are known only from one protected Key Area (see Table 1) which in each case will potentially be of major importance for the continued survival of the species in question (see 'Outlook', below). Species that fall into this category include *Crax alberti*, *Odontophorus strophium*, *Lepidopyga lilliae*, *Coeligena prunellei*, *Eriocnemis mirabilis*, *Buthraupis aureocincta*, *B. melanochlamys* and *Psarocolius cassini*.

RECENT CHANGES TO THE THREATENED LIST

With the publication of Collar *et al.* (1994), two of the 57 threatened species (*Haplophaedia lugens* and *Synallaxis cherriei*) were dropped from the Colombian threatened species list (Collar *et al.* 1992), with 11 added: Baudó Guan *Penelope ortoni*, Brown Wood-rail *Aramides wolfi*, Military Macaw *Ara militaris*, Santa Marta Parakeet *Pyrrhura viridicata* (endemic), Bogotá Sunangel *Heliangelus zusii* (endemic), Five-coloured Barbet *Capito quinticolor* (endemic), Cundinamarca Antpitta *Grallaria kaestneri* (endemic), Chestnut-bellied Cotinga *Doliornis remseni*, Long-wattled Umbrellabird *Cephalopterus penduliger*, Santa Marta Bush-tyrant *Myiotheretes pernix* (endemic) and Chocó Vireo sp. (endemic); these additional species have not, however, been included in the Site Inventory (see '*Key Areas*: the book', p. 12). *Cypseloides lemosi*, previously thought endemic to the southern end of the Cauca valley, has recently been recorded from Napo province, Ecuador (Collar *et al.* 1994).

Although these changes represent a high percentage difference (23%), three of the species added (*Heliangelus zusii*, *Grallaria kaestneri* and *Doliornis remseni*) were described as new to science after the 1992 listing, with the formal description of Chocó Vireo *Vireo* sp. imminent (Salaman and Stiles in press). The other species, with the exception of *Ara militaris*, were listed as Near Threatened in Collar *et al.* (1992), with new criteria and/or new data having led to their upgrading. For example, the plight of the two Santa Marta endemics was highlighted during

Table 2. Matrix of threatened species by Key Area.

	01	02	03	04	05	06	07	08	09	10	11	12	13	14	15	16	17	18	19	20	21	22	23	24	25	26	27	28	29	30	31
Tinamus osgoodi	–	–	–	–	–	–	–	–	–	–	–	–	–	–	–	–	–	–	–	–	–	–	–	–	–	–	–	–	–	–	–
Crypturellus kerriae	–	–	–	–	–	–	–	–	–	–	–	–	–	–	–	–	●	–	–	–	–	–	–	–	–	–	–	–	–	–	–
Crypturellus saltuarius	–	–	–	–	●	–	–	–	–	–	–	–	–	–	–	–	–	–	–	–	–	–	–	–	–	–	–	–	–	–	–
Podiceps andinus	–	–	–	–	–	–	–	–	–	–	–	–	–	–	–	–	–	–	–	–	–	–	–	–	–	–	–	–	–	–	–
Micrastur plumbeus	–	–	–	–	–	–	–	–	–	–	–	–	–	–	–	–	–	–	–	–	–	–	–	–	–	–	–	–	–	–	–
Penelope perspicax	–	–	–	–	–	–	–	–	–	–	–	–	–	–	–	–	–	–	–	–	–	–	–	–	●	–	–	–	–	–	–
Pauxi pauxi	–	●	–	–	–	–	–	–	–	–	–	–	–	–	–	–	–	–	–	–	–	–	–	–	–	–	–	–	–	●	–
Crax alberti	–	–	–	–	–	●	–	●	●	–	–	–	–	–	–	–	–	–	–	–	–	–	–	–	–	–	–	–	–	–	–
Crax globulosa	–	–	–	–	–	–	–	–	–	–	–	–	–	–	–	–	–	–	–	–	–	–	–	–	–	–	–	–	–	–	–
Odontophorus strophium	–	–	–	–	–	–	–	–	–	–	–	–	–	–	–	–	–	–	–	–	–	–	–	–	–	–	–	–	–	–	–
Rallus semiplumbeus	–	–	–	–	–	–	–	–	–	–	–	–	–	–	–	–	–	–	–	–	–	–	–	–	–	–	–	–	–	–	–
Coturnicops notatus	–	–	–	–	–	–	–	–	–	–	–	–	–	–	–	–	–	–	–	–	–	–	–	–	–	–	–	–	–	–	–
Leptotila conoveri	–	–	–	–	–	–	–	–	–	–	–	–	–	–	–	–	–	–	–	–	–	–	–	–	–	–	●	●	–	–	–
Leptosittaca branickii	–	–	–	–	–	–	–	–	–	–	–	–	–	–	–	–	–	–	–	–	–	–	–	●	●	●	–	●	–	–	–
Ognorhynchus icterotis	–	–	–	–	–	–	–	–	–	–	–	–	–	–	–	–	–	–	–	–	–	–	–	–	–	–	–	●	–	–	–
Pyrrhura calliptera	–	–	–	–	–	–	–	–	–	–	–	–	–	–	–	–	–	–	–	–	–	–	–	–	–	–	–	–	–	●	–
Bolborhynchus ferrugineifrons	–	–	–	–	–	–	–	–	–	–	–	–	–	–	–	–	–	–	–	–	–	–	–	●	●	●	–	–	–	–	–
Touit stictoptera	–	–	–	–	–	–	–	–	–	–	–	–	–	–	–	–	–	–	–	–	–	–	–	–	–	–	–	–	–	–	–
Hapalopsittaca amazonina	–	–	–	–	–	–	–	–	–	–	–	–	–	–	–	–	–	–	–	–	–	●	–	–	–	–	–	–	–	–	–
Hapalopsittaca fuertesi	–	–	–	–	–	–	–	–	–	–	–	–	–	–	–	–	–	–	–	–	–	–	–	●	–	●	–	–	–	–	–
Neomorphus radiolosus	–	–	–	–	–	–	–	–	–	–	–	–	–	–	–	–	–	–	–	–	–	–	–	–	–	–	–	–	–	–	–
Cypseloides lemosi	–	–	–	–	–	–	–	–	–	–	–	–	–	–	–	–	–	–	–	–	–	–	–	–	–	–	–	–	–	–	–
Lepidopyga lilliae	–	–	●	●	–	–	–	–	–	–	–	–	–	–	–	–	–	–	–	–	–	–	–	–	–	–	–	–	–	–	–
Amazilia castaneiventris	–	–	–	–	–	–	●	–	–	–	–	–	–	–	–	–	–	–	–	–	–	–	–	–	–	–	–	–	–	–	●
Coeligena prunellei	–	–	–	–	–	–	–	–	–	–	–	–	–	–	–	–	–	–	–	–	–	–	–	–	–	–	–	–	–	–	–
Eriocnemis godini	–	–	–	–	–	–	–	–	–	–	–	–	–	–	–	–	–	–	–	–	–	–	–	–	–	–	–	–	–	–	–
Eriocnemis mirabilis	–	–	–	–	–	–	–	–	–	–	–	–	–	–	–	–	–	–	–	–	–	–	–	–	–	–	–	–	–	–	–
Haplophaedia lugens	–	–	–	–	–	–	–	–	–	–	–	–	–	–	–	–	–	–	–	–	–	–	–	–	–	–	–	–	–	–	–
Galbula pastazae	–	–	–	–	–	–	–	–	–	–	–	–	–	–	–	–	–	–	–	–	–	–	–	–	–	–	–	–	–	–	–
Capito hypoleucus	–	–	–	–	–	–	●	–	●	–	–	–	–	●	●	–	–	–	–	–	–	–	●	–	–	–	–	–	–	–	–
Synallaxis cherriei	–	–	–	–	–	–	–	–	–	–	–	–	–	–	–	–	–	–	–	–	–	–	–	–	–	–	–	–	–	–	–
Clytoctantes alixii	–	–	–	–	–	–	●	●	●	–	–	–	–	–	–	–	–	–	–	–	–	–	–	–	–	–	–	–	–	–	–
Xenornis setifrons	–	–	–	–	–	–	–	–	–	–	–	–	–	–	–	–	●	–	–	–	–	–	–	–	–	–	–	–	–	–	–
Dysithamnus occidentalis	–	–	–	–	–	–	–	–	–	–	–	–	–	–	–	–	–	–	–	–	–	–	–	–	–	–	–	–	–	–	–
Grallaria alleni	–	–	–	–	–	–	–	–	–	–	–	–	–	–	–	–	–	–	–	–	–	–	–	–	–	●	–	–	–	–	–
Grallaria gigantea	–	–	–	–	–	–	–	–	–	–	–	–	–	–	–	–	–	–	–	–	–	–	–	–	–	–	–	–	–	–	–
Grallaria milleri	–	–	–	–	–	–	–	–	–	–	–	–	–	–	–	–	–	–	–	–	–	–	–	–	●	●	–	–	–	–	–
Grallaria rufocinerea	–	–	–	–	–	–	–	–	–	–	–	–	–	–	–	–	–	–	–	–	–	●	–	–	●	●	–	●	–	–	–
Grallaricula cucullata	–	–	–	–	–	–	–	–	–	–	–	–	–	–	–	–	–	–	–	–	–	–	–	–	–	–	–	–	–	–	–
Phylloscartes lanyoni	–	–	–	–	–	–	–	–	●	–	–	–	–	–	●	–	–	–	–	–	–	–	●	–	–	–	–	–	–	–	–
Attila torridus	–	–	–	–	–	–	–	–	–	–	–	–	–	–	–	–	–	–	–	–	–	–	–	–	–	–	–	–	–	–	–
Cistothorus apolinari	–	–	–	–	–	–	–	–	–	–	–	–	–	–	–	–	–	–	–	–	–	–	–	–	–	–	–	–	–	–	–
Thryothorus nicefori	–	–	–	–	–	–	–	–	–	–	–	–	–	–	–	–	–	–	–	–	–	–	–	–	–	–	–	–	–	–	–
Sporophila insulata	–	–	–	–	–	–	–	–	–	–	–	–	–	–	–	–	–	–	–	–	–	–	–	–	–	–	–	–	–	–	–
Atlapetes flaviceps	–	–	–	–	–	–	–	–	–	–	–	–	–	–	–	–	–	–	–	–	–	–	–	–	–	–	–	●	–	–	–
Oreothraupis arremonops	–	–	–	–	–	–	–	–	–	–	●	–	–	–	–	–	–	–	–	–	●	–	–	–	–	–	–	–	–	–	–
Chlorospingus flavovirens	–	–	–	–	–	–	–	–	–	–	–	–	–	–	–	–	–	–	–	–	–	–	–	–	–	–	–	–	–	–	–
Buthraupis aureocincta	–	–	–	–	–	–	–	–	–	–	–	–	–	–	–	–	–	–	●	●	●	–	–	–	–	–	–	–	–	–	–
Buthraupis melanochlamys	–	–	–	–	–	–	–	–	–	●	–	–	–	–	–	–	–	–	●	●	●	–	–	–	–	–	–	–	–	–	–
Buthraupis wetmorei	–	–	–	–	–	–	–	–	–	–	–	–	–	–	–	–	–	–	–	–	–	–	–	–	–	–	–	–	–	–	–
Chlorochrysa nitidissima	–	–	–	–	–	–	–	–	–	–	–	–	–	–	–	–	–	–	–	●	–	–	–	–	●	–	–	–	–	–	–
Dacnis berlepschi	–	–	–	–	–	–	–	–	–	–	–	–	–	–	–	–	–	–	–	–	–	–	–	–	–	–	–	–	–	–	–
Pseudodacnis hartlaubi	–	–	–	–	–	–	–	–	–	–	–	–	–	–	–	–	–	–	–	–	–	–	–	–	–	–	–	–	–	–	–
Vireo caribaeus	●	–	–	–	–	–	–	–	–	–	–	–	–	–	–	–	–	–	–	–	–	–	–	–	–	–	–	–	–	–	–
Psarocolius cassini	–	–	–	–	–	–	–	–	–	–	–	–	–	–	–	●	●	●	–	–	–	–	–	–	–	–	–	–	–	–	–
Hypopyrrhus pyrohypogaster	–	–	–	–	–	–	–	●	●	–	●	●	●	–	–	–	–	–	–	●	–	–	–	●	●	●	–	●	–	–	–
Carduelis cucullatus	–	–	–	–	–	–	–	–	–	–	–	–	–	–	–	–	–	–	–	–	–	–	–	–	–	–	–	–	●	–	–
No. of species	1	1	1	1	1	1	3	3	5	1	2	1	1	1	2	1	3	1	2	4	3	2	2	4	7	7	1	6	1	2	1

work on the *Global directory of Endemic Bird Areas* (Stattersfield *et al.* in prep.), when it was demonstrated that the Sierra Nevada de Santa Marta had been severely deforested. In spite of its national park designation, only 15% of the sierra's original vegetation remains unaltered, with just the northern slope still relatively intact (L. G. Olarte *in litt.* 1993, L. M. Renjifo *in litt.* 1993), and thus two of the rarest

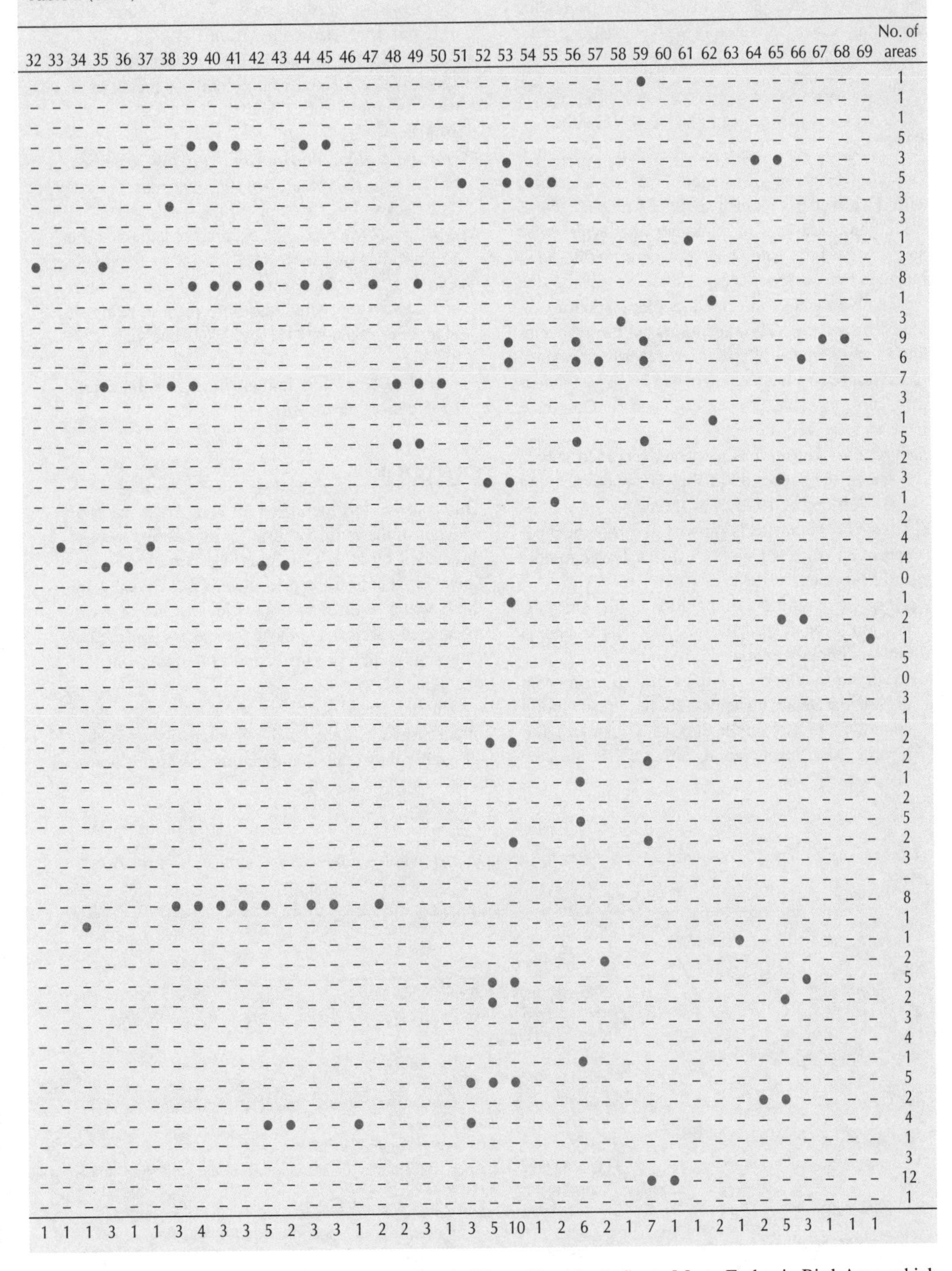

Table 2 (cont.)

32	33	34	35	36	37	38	39	40	41	42	43	44	45	46	47	48	49	50	51	52	53	54	55	56	57	58	59	60	61	62	63	64	65	66	67	68	69	No. of areas
–	–	–	–	–	–	–	–	–	–	–	–	–	–	–	–	–	–	–	–	–	–	–	–	–	–	–	●	–	–	–	–	–	–	–	–	–	–	1
–	–	–	–	–	–	–	–	–	–	–	–	–	–	–	–	–	–	–	–	–	–	–	–	–	–	–	–	–	–	–	–	–	–	–	–	–	–	1
–	–	–	–	–	–	–	–	–	–	–	–	–	–	–	–	–	–	–	–	–	–	–	–	–	–	–	–	–	–	–	–	–	–	–	–	–	–	1
–	–	–	–	–	–	–	●	●	●	–	–	●	●	–	–	–	–	–	–	–	–	–	–	–	–	–	–	–	–	–	–	–	–	–	–	–	–	5
–	–	–	–	–	–	–	–	–	–	–	–	–	–	–	–	–	–	–	–	–	●	–	–	–	–	–	–	–	–	–	–	●	●	–	–	–	–	3
–	–	–	–	–	–	–	–	–	–	–	–	–	–	–	–	–	–	–	●	–	●	●	●	–	–	–	–	–	–	–	–	–	–	–	–	–	–	5
–	–	–	–	–	–	●	–	–	–	–	–	–	–	–	–	–	–	–	–	–	–	–	–	–	–	–	–	–	–	–	–	–	–	–	–	–	–	3
–	–	–	–	–	–	–	–	–	–	–	–	–	–	–	–	–	–	–	–	–	–	–	–	–	–	–	–	–	–	–	–	–	–	–	–	–	–	3
–	–	–	–	–	–	–	–	–	–	–	–	–	–	–	–	–	–	–	–	–	–	–	–	–	–	–	–	–	●	–	–	–	–	–	–	–	–	1
●	–	–	●	–	–	–	–	–	–	●	–	–	–	–	–	–	–	–	–	–	–	–	–	–	–	–	–	–	–	–	–	–	–	–	–	–	–	3
–	–	–	–	–	–	–	●	●	●	●	–	●	●	–	●	–	●	–	–	–	–	–	–	–	–	–	–	–	–	–	–	–	–	–	–	–	–	8
–	–	–	–	–	–	–	–	–	–	–	–	–	–	–	–	–	–	–	–	–	–	–	–	–	–	–	–	–	–	●	–	–	–	–	–	–	–	1
–	–	–	–	–	–	–	–	–	–	–	–	–	–	–	–	–	–	–	–	–	–	–	–	–	–	●	–	–	–	–	–	–	–	–	–	–	–	3
–	–	–	–	–	–	–	–	–	–	–	–	–	–	–	–	–	–	–	–	–	●	–	–	●	–	–	●	–	–	–	–	–	–	–	●	●	–	9
–	–	–	–	–	–	–	–	–	–	–	–	–	–	–	–	–	–	–	–	–	●	–	–	●	●	–	●	–	–	–	–	–	–	●	–	–	–	6
–	–	–	●	–	–	●	●	–	–	–	–	–	–	–	–	●	●	●	–	–	–	–	–	–	–	–	–	–	–	–	–	–	–	–	–	–	–	7
–	–	–	–	–	–	–	–	–	–	–	–	–	–	–	–	–	–	–	–	–	–	–	–	–	–	–	–	–	–	–	–	–	–	–	–	–	–	3
–	–	–	–	–	–	–	–	–	–	–	–	–	–	–	–	–	–	–	–	–	–	–	–	–	–	–	–	–	–	●	–	–	–	–	–	–	–	1
–	–	–	–	–	–	–	–	–	–	–	–	–	–	–	–	●	●	–	–	–	–	–	–	●	–	–	●	–	–	–	–	–	–	–	–	–	–	5
–	–	–	–	–	–	–	–	–	–	–	–	–	–	–	–	–	–	–	–	–	–	–	–	–	–	–	–	–	–	–	–	–	–	–	–	–	–	2
–	–	–	–	–	–	–	–	–	–	–	–	–	–	–	–	–	–	–	–	●	●	–	–	–	–	–	–	–	–	–	–	–	●	–	–	–	–	3
–	–	–	–	–	–	–	–	–	–	–	–	–	–	–	–	–	–	–	–	–	–	–	●	–	–	–	–	–	–	–	–	–	–	–	–	–	–	1
–	–	–	–	–	–	–	–	–	–	–	–	–	–	–	–	–	–	–	–	–	–	–	–	–	–	–	–	–	–	–	–	–	–	–	–	–	–	2
–	●	–	–	–	●	–	–	–	–	–	–	–	–	–	–	–	–	–	–	–	–	–	–	–	–	–	–	–	–	–	–	–	–	–	–	–	–	4
–	–	–	●	●	–	–	–	–	–	●	●	–	–	–	–	–	–	–	–	–	–	–	–	–	–	–	–	–	–	–	–	–	–	–	–	–	–	4
–	–	–	–	–	–	–	–	–	–	–	–	–	–	–	–	–	–	–	–	–	–	–	–	–	–	–	–	–	–	–	–	–	–	–	–	–	–	0
–	–	–	–	–	–	–	–	–	–	–	–	–	–	–	–	–	–	–	–	–	●	–	–	–	–	–	–	–	–	–	–	–	–	–	–	–	–	1
–	–	–	–	–	–	–	–	–	–	–	–	–	–	–	–	–	–	–	–	–	–	–	–	–	–	–	–	–	–	–	–	–	●	●	–	–	–	2
–	–	–	–	–	–	–	–	–	–	–	–	–	–	–	–	–	–	–	–	–	–	–	–	–	–	–	–	–	–	–	–	–	–	–	–	–	●	1
–	–	–	–	–	–	–	–	–	–	–	–	–	–	–	–	–	–	–	–	–	–	–	–	–	–	–	–	–	–	–	–	–	–	–	–	–	–	5
–	–	–	–	–	–	–	–	–	–	–	–	–	–	–	–	–	–	–	–	–	–	–	–	–	–	–	–	–	–	–	–	–	–	–	–	–	–	0
–	–	–	–	–	–	–	–	–	–	–	–	–	–	–	–	–	–	–	–	–	–	–	–	–	–	–	–	–	–	–	–	–	–	–	–	–	–	3
–	–	–	–	–	–	–	–	–	–	–	–	–	–	–	–	–	–	–	–	–	–	–	–	–	–	–	–	–	–	–	–	–	–	–	–	–	–	1
–	–	–	–	–	–	–	–	–	–	–	–	–	–	–	–	–	–	–	–	●	●	–	–	–	–	–	–	–	–	–	–	–	–	–	–	–	–	2
–	–	–	–	–	–	–	–	–	–	–	–	–	–	–	–	–	–	–	–	–	–	–	–	–	–	–	●	–	–	–	–	–	–	–	–	–	–	2
–	–	–	–	–	–	–	–	–	–	–	–	–	–	–	–	–	–	–	–	–	–	–	–	●	–	–	–	–	–	–	–	–	–	–	–	–	–	1
–	–	–	–	–	–	–	–	–	–	–	–	–	–	–	–	–	–	–	–	–	–	–	–	–	–	–	–	–	–	–	–	–	–	–	–	–	–	2
–	–	–	–	–	–	–	–	–	–	–	–	–	–	–	–	–	–	–	–	–	–	–	–	●	–	–	–	–	–	–	–	–	–	–	–	–	–	5
–	–	–	–	–	–	–	–	–	–	–	–	–	–	–	–	–	–	–	–	–	●	–	–	–	–	–	●	–	–	–	–	–	–	–	–	–	–	2
–	–	–	–	–	–	–	–	–	–	–	–	–	–	–	–	–	–	–	–	–	–	–	–	–	–	–	–	–	–	–	–	–	–	–	–	–	–	3
–	–	–	–	–	–	–	–	–	–	–	–	–	–	–	–	–	–	–	–	–	–	–	–	–	–	–	–	–	–	–	–	–	–	–	–	–	–	
–	–	–	–	–	–	●	●	●	●	●	–	●	●	–	●	–	–	–	–	–	–	–	–	–	–	–	–	–	–	–	–	–	–	–	–	–	–	8
–	–	●	–	–	–	–	–	–	–	–	–	–	–	–	–	–	–	–	–	–	–	–	–	–	–	–	–	–	–	–	–	–	–	–	–	–	–	1
–	–	–	–	–	–	–	–	–	–	–	–	–	–	–	–	–	–	–	–	–	–	–	–	–	–	–	–	–	–	–	●	–	–	–	–	–	–	1
–	–	–	–	–	–	–	–	–	–	–	–	–	–	–	–	–	–	–	–	–	–	–	–	–	●	–	–	–	–	–	–	–	–	–	–	–	–	2
–	–	–	–	–	–	–	–	–	–	–	–	–	–	–	–	–	–	–	–	●	●	–	–	–	–	–	–	–	–	–	–	–	–	●	–	–	–	5
–	–	–	–	–	–	–	–	–	–	–	–	–	–	–	–	–	–	–	–	●	–	–	–	–	–	–	–	–	–	–	–	–	●	–	–	–	–	2
–	–	–	–	–	–	–	–	–	–	–	–	–	–	–	–	–	–	–	–	–	–	–	–	–	–	–	–	–	–	–	–	–	–	–	–	–	–	3
–	–	–	–	–	–	–	–	–	–	–	–	–	–	–	–	–	–	–	–	–	–	–	–	–	–	–	–	–	–	–	–	–	–	–	–	–	–	4
–	–	–	–	–	–	–	–	–	–	–	–	–	–	–	–	–	–	–	–	–	–	–	–	●	–	–	–	–	–	–	–	–	–	–	–	–	–	1
–	–	–	–	–	–	–	–	–	–	–	–	–	–	–	–	–	–	–	●	●	●	–	–	–	–	–	–	–	–	–	–	–	–	–	–	–	–	5
–	–	–	–	–	–	–	–	–	–	–	–	–	–	–	–	–	–	–	–	–	–	–	–	–	–	–	–	–	–	–	–	●	●	–	–	–	–	2
–	–	–	–	–	–	–	–	–	–	●	●	–	–	●	–	–	–	–	●	–	–	–	–	–	–	–	–	–	–	–	–	–	–	–	–	–	–	4
–	–	–	–	–	–	–	–	–	–	–	–	–	–	–	–	–	–	–	–	–	–	–	–	–	–	–	–	–	–	–	–	–	–	–	–	–	–	1
–	–	–	–	–	–	–	–	–	–	–	–	–	–	–	–	–	–	–	–	–	–	–	–	–	–	–	–	–	–	–	–	–	–	–	–	–	–	3
–	–	–	–	–	–	–	–	–	–	–	–	–	–	–	–	–	–	–	–	–	–	–	–	–	–	–	●	●	–	–	–	–	–	–	–	–	–	12
–	–	–	–	–	–	–	–	–	–	–	–	–	–	–	–	–	–	–	–	–	–	–	–	–	–	–	–	–	–	–	–	–	–	–	–	–	–	1
1	1	1	3	1	1	3	4	3	3	5	2	3	3	1	2	2	3	1	3	5	10	1	2	6	2	1	7	1	1	2	1	2	5	3	1	1	1	

species confined to this area are now listed (21 restricted-range species occur in these mountains).

The addition of the two threatened Santa Marta species emphasizes the importance of the Sierra Nevada de Santa Marta Endemic Bird Area, which should in future be considered a priority for bird conservation. Apart from these two species, the recent additions to the Colombian threatened species list will not have any major distributional impact on the Key Area analysis, as most of the species are known to occur sympatrically with at least one other threatened bird.

OLD RECORDS AND LITTLE-KNOWN BIRDS

Our knowledge of threatened birds in Colombia suffers from the general paucity of recent records from many areas and of a number of species. Most species are little known somewhere within their range, but recent records from other areas mean that we are able to make informed recommendations for their conservation. The exceptions are those species known only from old specimen records or chance observations (from which there can be no guarantees of a current viable population), and in Colombia these include the following.

Crypturellus kerriae has not been recorded subsequent to the collection of the type-specimen and one other during 1912 from Baudó.

Crypturellus saltuarius is known only from the type-specimen collected in 1943 in the lower middle Río Magdalena valley.

Podiceps andinus, though included in this analysis, has not been recorded since the 1970s and is almost certainly extinct.

Crax globulosa chiefly occupies the Amazon basin in Peru and Brazil, but appears to be almost wholly unknown; in Colombia there are just two old records, and recent reports around Chiribiquete and Araracuate (Collar *et al.* 1994).

Coturnicops notatus is known in Colombia from only one specimen collected in 1959.

Amazilia castaneiventris is known primarily from specimens collected at a number of localities in the Serranía de San Lucas during the 1950s and 1960s.

Clytoctantes alixii, known from very few localities, has not been recorded since 1965, and most records date back to the 1940s or earlier.

Xenornis setifrons has not been recorded since the collection of two specimens in 1940 at Baudó.

Grallaria alleni is known only from the type-localities of its two subspecies, the only records being of specimens taken in 1911 and 1971 (but see p. 173, EC 13).

Psarocolius cassini is known from just three pre-1945 collecting records.

OUTLOOK

This analysis has identified 69 Key Areas, each of which is of major importance for the conservation of threatened birds in Colombia. Certain Key Areas, however, stand out as priorities (Table 3); they are those which each host populations of six or more threatened species, and those for species confined to Colombia which are (effectively) known from just one area.

The 10 areas named in Table 3 are the most important Key Areas for 13 threatened species, but altogether they contain populations of 27 threatened

Table 3. Top Key Areas in Colombia. Those with the area number in bold currently have some form of protected status.

Key Area		No. of threatened spp.	Comments
01	Isla de San Andrés	1	*Vireo caribaeus* is endemic to this island.
25	Ucumarí	7	Supports the only known population of *Grallaria milleri*, along with six other threatened birds.
26	Alto Quindío–Laguneta	7	Holds the only confirmed population of *Hapalopsittaca fuertesi*, along with six other threatened birds.
28	Río Toche	6	Especially important for two unprotected Colombian endemics, *Leptotila conoveri* and *Atlapetes flaviceps*.
34	San Gil	1	*Thryothorus nicefori* is confined entirely to this Key Area.
35	Guanentá–Alto Río Fonce	3	Holds the only confirmed population of *Odontophorus strophium*, along with two other threatened birds.
53	Munchique	10	The only known area for *Eriocnemis mirabilis* and possibly the most important Colombian Key Area for *Neomorphus radiolosus* and *Oreothraupis arremonops*.
56	Puracé	6	All Colombian records of *Buthraupis wetmorei* are from this Key Area.
59	Cueva de los Guácharos	7	The only Colombian record of *Tinamus osgoodi* is from this Key Area.
63	Isla de Tumaco and Isla Boca Grande	1	The entire range of *Sporophila insulata* lies within this Key Area.

species (47% of the Colombian total). These, and the remaining Key Areas (which are individually extremely important for many threatened Colombian endemics: see 'Key Areas' and 'Key Area protection', above), all need some degree of conservation action if the populations of their threatened birds are to remain viable. For many areas there is a requirement to confirm the continued existence or assess the population of a threatened species, and for others, especially those areas whose inclusion is based on old records (see 'Old records and little-known birds', above), the need is to locate and define the extent of suitable habitat in the vicinity of the Key Area.

DATA SOURCES

The above introductory text and the Site Inventory (below) were compiled from information supplied by G. I. Andrade, C. Downing, J. Farthing, M. G. Kelsey, E. Londoño, A. J. Negret, M. Pearman, L. M. Renjifo, P. Salaman, F. G. Stiles and F. Uribe, as well as from the following references.

ANDRADE, G. I. AND REPIZZO, A. (1994) Guanentá–Alto Río Fonce Fauna and Flora Sanctuary: a new protected area in the Colombian East Andes. *Cotinga* 2: 42–43.

COLLAR, N. J., GONZAGA, L. P., KRABBE, N., MADROÑO NIETO, A., NARANJO, L. G., PARKER, T. A. AND WEGE, D. C. (1992) *Threatened birds of the Americas: the ICBP/IUCN Red Data Book* (Third edition, part 2). Cambridge, U.K.: International Council for Bird Preservation.

COLLAR N. J., CROSBY, M. J. AND STATTERSFIELD, A. J. (1994) *Birds to watch 2: the world list of threatened birds*. Cambridge, U.K.: BirdLife International (BirdLife Conservation Series no.4).

COLLAR, N. J., WEGE, D. C. AND LONG, A. J. (in press) Patterns and causes of endangerment in the New World avifauna. In J. V. Remsen, ed. *Natural history and conservation of Neotropical birds*. American Ornithologists' Union (Orn. Monogr.).

DOWNING, C. (1993) *A field checklist of the birds of Colombia*. Cambridge, U.K: privately published.

HILTY, S. L. AND BROWN, W. L. (1986) *A guide to the birds of Colombia*. Princeton, New Jersey: Princeton University Press.

IUCN (1992) *Protected areas of the world: a review of national systems. Volume 4: Nearctic and Neotropical.* Gland, Switzerland and Cambridge, U.K.: International Union for Conservation of Nature and Natural Resources.

KRABBE, N., DESMET, G., GREENFIELD, P., JÁCOME, M., MATHEUS, J. C. AND SORNOZA M., F. (1994) Giant Antpitta *Grallaria gigantea. Cotinga* 2: 32–34.

LOZANO, E. (1993) Observaciones sobre la ecología y el comportamiento de *Rallus semiplumbeus* en el humedal de la Florida sabana de Bogota. Unpublished report.

NEGRET, A. J. (1991) Reportes recientes en el Parque Nacional Munchique de aves consideradas raras o amenazadas de extinción. *Nov. Colombianas* 3: 39–46.

PAYNTER, R. A. AND TRAYLOR, M. A. (1981) *Ornithological gazetteer of Colombia.* Cambridge, Mass.: Museum of Comparative Zoology.

RENJIFO, L. M. (1994) First records of the Bay-vented Cotinga *Doliornis sclateri. Bull. Brit. Orn. Club* 114: 101–103.

SALAMAN, P. AND STILES, F. G. (in press) A distinctive new species of vireo (Vireonidae) from the Chocó region of western Colombia. *Ibis*.

SMITH, J., ED. (1994) University of Bristol Utría Sound expedition 1991/1992. Unpublished report.

STATTERSFIELD, A. J., CROSBY, M. J., LONG, A. J. AND WEGE, D. C. (in prep.) *A global directory of Endemic Bird Areas*. Cambridge, U.K.: BirdLife International (BirdLife Conservation Series).

SITE INVENTORY

Isla de San Andrés (Intendencia de San Andrés y Providencia)

CO 01

12°32′N 81°42′W

Unprotected

This Caribbean island, just 52 km² in extent, lies c.200 km off the coast of Nicaragua. Brushy pastures and inland mangrove swamp are the most important vegetation and are now restricted to the southern third of the island. The northern two-thirds contains most of the human population, the area around the capital and major tourist resort of San Andrés being totally urbanized.

Vireo caribaeus	1989	Seemingly common in suitable habitat, now primarily restricted to the southern third of the island.

Cerro Pintado (Guajira)

CO 02

Unprotected

10°29′N 72°55′W
EBA B07

The peak of Cerro Pintado rises to 3,000 m in the northern Sierra de Perijá on the Guajira/Cesar and Colombia/Venezuela borders (it is contiguous with the Sierra de Perijá National Park, VE 01). Montane forest on the upper slopes remains intact, though the species mentioned below is under severe pressure from local hunters, and from an unfavourable security situation (M. Pearman *in litt.* 1995).

Pauxi pauxi	1987	Reliable reports by local hunters who had recently shot the species (M. Pearman *in litt.* 1995).

Mouth of the Río Ranchería (Guajira)

CO 03

Unprotected

11°34′N 72°54′W
EBA B07

Mangroves at this site on the Caribbean coast of northernmost Colombia, just east of Ríohacha, form a small but potentially important area for the species noted here. Other rivers supporting mangroves along this stretch of coast would also be worth investigating, especially those to the north in Portete and Honda bays (G. I. Andrade *in litt.* 1994).

Lepidopyga lilliae	1974	A mangrove specialist; almost nothing is known of its status at this site.

Ciénaga Grande de Santa Marta and Isla de Salamanca (Magdalena)

CO 04

10°50′N 74°25′W
EBA B07

Isla de Salamanca National Park (IUCN category II), 21,000 ha
Ciénaga Grande de Santa Marta Fauna and Flora Sanctuary (IUCN category IV), 23,000 ha

This large area of interconnected saline lakes east of the Río Magdalena supports large but dwindling stands of mangroves, a large proportion of which died between 1986 and 1992 (F. G. Stiles *in litt.* 1994). Development projects have caused obstructions to both tidal and freshwater flow in this area

Figure 2 (opposite). Key Areas in Colombia.

01 Isla de San Andrés (not shown)
02 Cerro Pintado
03 Mouth of the Río Ranchería
04 Ciénaga Grande de Santa Marta and Isla de Salamanca
05 Ayacucho
06 Serranía de San Jacinto
07 Serranía de San Lucas
08 Paramillo
09 Puerto Valdivia
10 Yarumal
11 Las Orquideas
12 Barbosa–Santo Domingo road
13 Fizebad, upper Río Negro
14 Punchina dam
15 Río Claro
16 Río Truandó
17 Serranía del Baudó
18 Río Dubasa
19 Alto de Pisones–Mistrató
20 Cerro Tatamá
21 Alto de los Galápagos
22 Río Blanco watershed
23 La Victoria
24 Los Nevados
25 Ucumarí
26 Alto Quindío–Laguneta
27 Juntas
28 Río Toche
29 Cúcuta
30 Paramó de Tamá
31 Lebrija
32 Cuchilla del Ramo
33 Portugal
34 San Gil
35 Guanentá–Alto Río Fonce
36 Cerro Carare
37 Soatá
38 El Cocuy
39 Laguna de Tota
40 Laguna de Fúquene
41 Laguna de Cucunubá
42 Laguna de Pedropalo
43 Bojacá
44 Laguna de la Herrera
45 Laguna de la Florida
46 Salto del Tequendama
47 Laguna Chisacá
48 Valle de Jesús forest
49 Páramo de Chingaza
50 Monterredondo–El Calvario
51 Bosque de Yotoco
52 Los Farallones de Cali
53 Munchique
54 Valle del Patía
55 Santa Helena–Cerro Munchique
56 Puracé
57 Río de la Plata
58 San Agustín
59 Cueva de los Guácharos
60 Recinas–Florencia
61 Sierra de Chiribiquete
62 Serranía de la Macarena
63 Isla de Tumaco and Isla Boca Grande
64 La Guayacana
65 Río Ñambi
66 La Planada
67 Volcán Galeras and Río San Francisco
68 Laguna de la Cocha
69 El Carmen

(leading to salinity imbalances), and this, combined with pollution and mangrove clearance, has left the area in urgent need of attention, the area on the west side of the ciénaga being especially damaged (F. G. Stiles *in litt.* 1994). Isla de Salamanca National Park is currently in a critically run-down condition, offering little if any protection to its fauna and flora, although current PRO-CIENAGA projects are aimed at alleviating some of the problems (G. I. Andrade *in litt.* 1994, P. Salaman *in litt.* 1995).

Lepidopyga lilliae	1994	The centre of abundance for this species, which has been recorded from a number of localities including the national park and sanctuary (most recently a female nest-building in 1994: P. Salaman *in litt.* 1995).

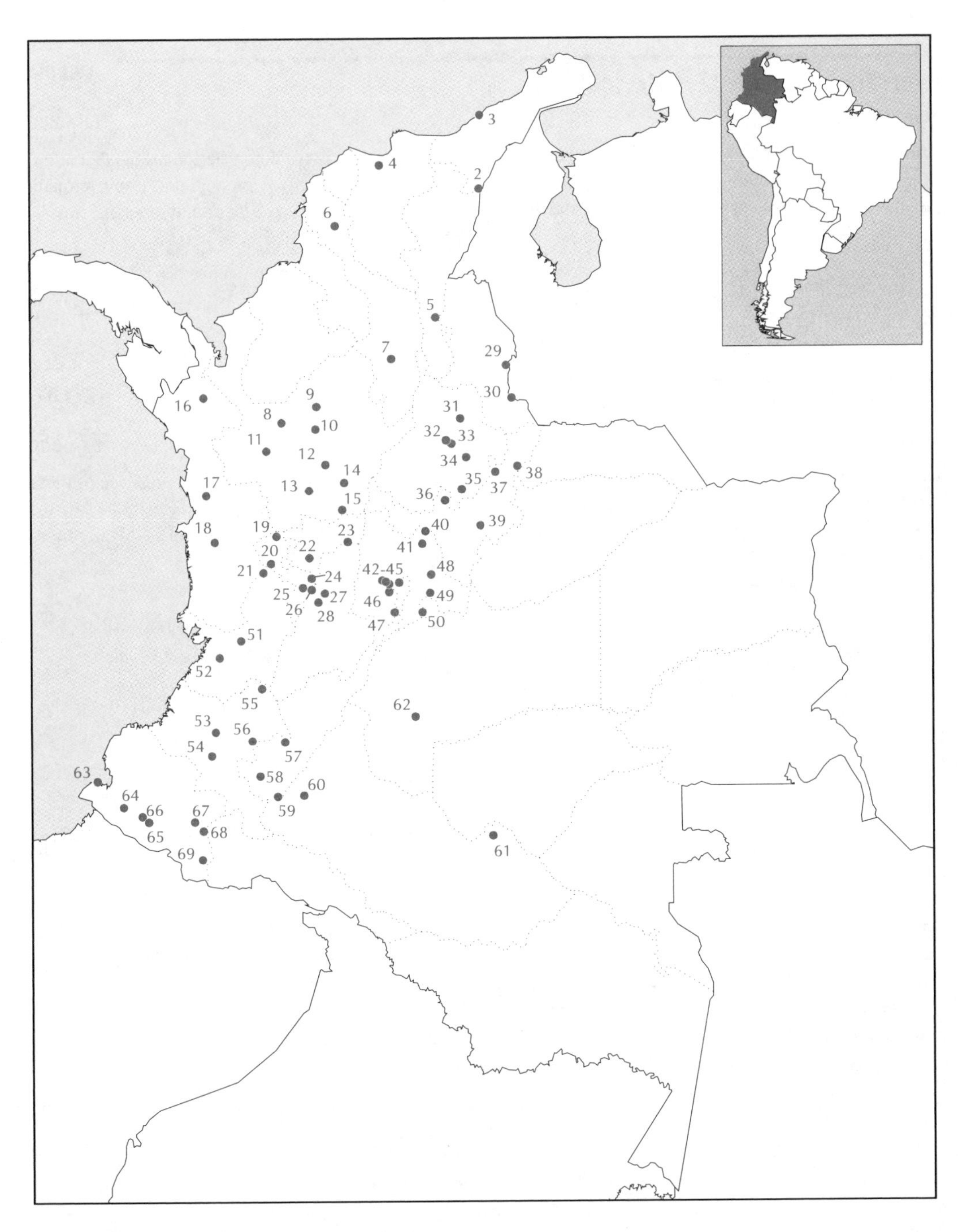

Ayacucho (Cesar)

CO 05

Unprotected

8°36′N 73°35′W
EBA B09

An area traditionally covered by low, dry deciduous forest and savannas, on the lower middle Río Magdalena valley. Extensive clearance of the dry forests for cattle-ranching and agriculture is a serious threat to this ecosystem which is primarily under private ownership.

Crypturellus saltuarius	1943	The type- and only known locality.

Serranía de San Jacinto (Bolívar)

CO 06

Unprotected

9°57′N 75°05′W
EBA B09

An isolated, low range of hills that supported moist forest on the Bolívar/Sucre border; the northern end of the hills (in Bolívar) was apparently the most important. Extensive deforestation means that attention should be focused on surveying any remnant forest patches for *Crax alberti* and other species.

Crax alberti	c.1975	Known from very few localities after the 1950s, and had become very rare by the 1970s (last recorded from San Juan Nepomuceno). Searches are urgently needed.

Serranía de San Lucas (Bolívar)

CO 07

Unprotected

8°00′N 74°15′W
EBA B09

The eastern slope of the northern half of the Serranía de San Lucas, between c.100 and 800 m above sea-level. Records of threatened species from this serranía all date from the late 1940s. The entire northern half of the serranía needs to be surveyed for remnant forest patches and checked for the presence of these and other threatened (and endemic) birds, but the unfavourable security situation makes this impractical (F. G. Stiles *in litt.* 1994).

Amazilia castaneiventris	1947	The bird collected at Norosí in March 1947 is the only record from the serranía.
Capito hypoleucus	1947	The five birds collected from c.800 m at Volador represent the only records from this area.
Clytoctantes alixii	1947	The record of 10 birds from c.600 m at Santa Rosa in April–May 1947 suggests that this species was at least locally common.

Paramillo (Córdoba/Antioquia)

CO 08

Paramillo National Park (IUCN category II), 460,000 ha

7°04′N 75°55′W
EBAs B09, B14

At the northernmost end of the West Andes, this national park embraced lowland tropical forest up to the páramo zone, the lowlands now being especially important as they are the most threatened (G. I. Andrade *in litt.* 1994). None of the threatened species has been recorded from this park since the late 1940s, and fieldwork is urgently needed to confirm their continued presence. The numerous settlements, extensive deforestation and unfavourable security situation present problems for both conservation and fieldwork (G. I. Andrade *in litt.* 1994, F. G. Stiles *in litt.* 1994).

Crax alberti	1949	Recorded just once on the northern edge of the park within the lowlands of the Río Verde Indigenous Reserve.
Clytoctantes alixii	1949	Four collected at 185 m on the Río Salvajín.
Hypopyrrhus pyrohypogaster	1915	One bird recorded at 1,500 m near Peque.

CO 09

Puerto Valdivia (Antioquia)

7°18′N 75°23′W
EBAs B09, B12

Unprotected

Comprising a short stretch of the Cauca valley close to the point where the Río Cauca emerges from between the West and Central Andes, this area also includes the lowlands between Puerto Valdivia and Valdivia to the south. Extensive forest destruction for pasture has confined any remaining forest to ridge-crests and isolated patches, in which the presence of the following species needs to be ascertained.

Crax alberti	pre-1942	Mentioned for this locality.
Capito hypoleucus	1948	Primarily collected around Valdivia; four taken there in May 1948.
Clytoctantes alixii	1914	Two collected at 180 m.
Phylloscartes lanyoni	1948	Two collected 12 km below Puerto Valdivia at c.500 m.
Hypopyrrhus pyrohypogaster	1962	Birds were seen at 800 m above Puerto Valdivia in 1962, 12 collected at 1,200 m near Valdivia in 1948.

CO 10

Yarumal (Antioquia)

6°58′N 75°24′W
EBA B12

Unprotected

At the northernmost end of the Central Andes, on the saddle between the Cauca and Nechí drainages, this area embraces localities to the north between Valdivia and Yarumal. There has been almost total destruction of forest around this locality.

Buthraupis melanochlamys	pre-1978	Fifteen collected between 1,500 and 2,290 m at Valdivia in June 1948, although Yarumal has been mentioned for the species.

CO 11

Las Orquideas (Antioquia)

6°39′N 76°09′W
EBA B14

Las Orquideas National Park (IUCN category II), 32,000 ha

Lying near the northern end of the West Andes, Las Orquideas includes the area south-west of Frontino (Hacienda Potreros) on the north-east edge of the park. There have been no records of either of the two following species from in or around the park since 1950, suggesting that searches need to be made to confirm their continued presence.

Oreothraupis arremonops	1950	Known only from Hacienda Potreros, where one was collected at c.2,000 m.
Hypopyrrhus pyrohypogaster	1950	Known only from Hacienda Potreros, where five were collected at c.2,000 m.

CO 12

Barbosa–Santo Domingo road (Antioquia)

6°27′N 75°15′W
EBA B12

Unprotected

At the northern end of the Central Andes, north-east of Medellín, where at c.1,800 m there is still some humid montane forest. There has been extensive clearance of forest around Medellín, so remaining tracts are of utmost importance—but are also in great demand.

Hypopyrrhus pyrohypogaster	1990	Eight birds seen at 1,800 m.

Fizebad, upper Río Negro (Antioquia)

CO 13

6°04′N 75°30′W
EBA B12

Unprotected

This area on the upper Río Negro (south of Medellín and east of Caldas) in the northern Central Andes comprises a small patch of old secondary growth (humid forest) surrounded by a plantation of *Pinus patulla*. Deforestation in the area around Medellín has been extensive, and remnant patches of primary and old secondary growth need to be protected.

Hypopyrrhus pyrohypogaster	1990	A flock, and a nest, recorded at 2,100–2,400 m.

Punchina dam (Antioquia)

CO 14

6°11′N 74°58′W
EBA B09

Punchina watershed reserve (ISA reserve), area unknown

Lying close to San Carlos on the eastern slope of the northern Central Andes, the lower montane forest surrounding this area (i.e. the watershed) is protected by the local electricity company (ISA). This area is undergoing heavy logging which emphasizes the importance of the protected forested watersheds. Jaguas dam near San Rafael (north of Punchina) also holds the species noted below and is similarly protected. The suggestion that *Crax alberti* may be present at Punchina should be investigated (L. M. Renjifo *in litt*. 1994).

Capito hypoleucus	1991	A total of eight birds recorded at 1,200 m up to April 1991.

Río Claro (Antioquia)

CO 15

5°47′N 75°00′W
EBA B09

Río Claro Natural Reserve (private), >100 ha

This area near El Doradal is on the eastern slope of the northern Central Andes, close to the Caldas border. The private Río Claro reserve supports a small patch of humid lowland and foothill forest and is situated in an area undergoing widespread deforestation, highlighting the site's importance.

Capito hypoleucus	1992	Several recent records, with 3–5 birds in 1992 (F. G. Stiles *in litt.* 1994).
Phylloscartes lanyoni	1993	Four seen together in June 1990; one seen August 1993 (M. Pearman *in litt.* 1995).

Río Truandó (Chocó)

CO 16

7°26′N 77°07′W
EBA A19

Unprotected

This area comprises the middle Río Truandó, east of the isolated Serranía de los Saltos, in an area of humid lowland forest, some of which is on sandy riverine deposits. There is an urgent need for ornithological exploration along the rivers here.

Psarocolius cassini	1858	The type-specimen was collected in this area at 'Camp Abert'.

Serranía del Baudó (Chocó)

CO 17
6°0′N 77°05′W
EBA A19

Ensenada Utría National Park (IUCN category II), 54,300 ha

The Serranía del Baudó is primarily a low mountain range (i.e below 1,000 m, but reaching 1,800 m at Alto del Buey), on which there remain large expanses of forest, some of which are protected within the Ensenada Utría National Park. There is no exact information on where in the serranía any of the three threatened species were recorded, and thus it is not known whether they occur in the national park (which is under threat from construction projects, such as road and dam building), although they were not recorded there during a recent survey (Smith 1994). There is an urgent need for further fieldwork in the serranía.

Crypturellus kerriae	1912	Two collected at 450 m, apparently on the Río Baudó.
Xenornis setifrons	1940	Two collected at 550 m beside the Río Baudó.
Psarocolius cassini	1940	Two collected at 275–365 m on the upper Río Baudó.

Río Dubasa (Chocó)

CO 18
5°19′N 76°57′W
EBA A19

Unprotected

The Río Dubasa is a tributary of the Río Baudó, at the southern end of the Serranía de Baudó, and supports humid lowland and gallery forest.

Psarocolius cassini	1945	One female collected and nine others seen at 100 m.

Alto de Pisones–Mistrató (Risaralda)

CO 19
5°24′N 76°00′W
EBA B14

Unprotected

This area is on the western slope of the West Andes, and along the ridge at Alto de Pisones there remains a virgin tract of subtropical cloud-forest, which seemingly extends in a more or less intact state, 24 km south-east to Mistrató. This extensive tract of montane forest is in urgent need of protection, and, although the current security situation in the area could hamper such activity (P. Salaman *in litt.* 1995), CARDER is seeking to conserve Alto de Pisones as a 'Special Management Areas' (E. Londoño and F. Uribe *in litt.* 1995). The Chocó Vireo *Vireo* sp. which is soon to be formally described (Salaman and Stiles in press) also occurs in this area, and two other threatened species, *Chlorochrysa nitidissima* and *Hypopyrrhus pyrohypogaster*, should be searched for.

Buthraupis aureocincta	1993	Fairly common (breeding has been confirmed) at 1,600–1,800 m (e.g. M. Pearman *in litt.* 1995).
Buthraupis melanochlamys	1993	Less common than *B. aureocincta* (breeding has been confirmed), but primarily at 1,400–1,600 m (e.g. M. Pearman *in litt.* 1995).

Cerro Tatamá (Chocó/Valle/Risaralda)

CO 20

Tatamá National Park (IUCN category II), 54,300 ha

5°00′N 76°05′W
EBA B14

Cerro Tatamá rises to 4,000 m in the central West Andes, with the national park covering areas primarily above 2,000 m. The vegetation comprises mainly humid montane forest. The altitudinal distribution of species in this area is centred on the lower reaches of the park, which should ideally be extended to ensure their adequate protection, as forest at these altitudes has largely been removed (J. Farthing *in litt.* 1994).

Buthraupis aureocincta	1946	Known from just four specimens taken in or before 1946; recent visits have failed to locate it (F. G. Stiles *in litt.* 1994).
Buthraupis melanochlamys	1987	Just one recent record, with previous ones dating back to before 1945.
Chlorochrysa nitidissima	1946	Twelve collected at 1,500–2,150 m between December 1945 and February 1946.
Hypopyrrhus pyrohypogaster	1994	Ten collected at 1,500–1,800 m in January 1946; rediscovered here in 1994 (J. Farthing *in litt.* 1994).

Alto de los Galápagos (Chocó/Valle)

CO 21

Unprotected

4°52′N 76°12′W
EBA B14

Alto de los Galápagos is at the highest point (c.2,100 m) on the Cartago–San José del Palmar road where it passes over a ridge on the Chocó/Valle border on the Pacific slope of the central West Andes (J. Farthing *in litt.* 1994). Humid montane forest persists in the vicinity of Alto de los Galápagos (J. Farthing *in litt.* 1994). This area is just to the south of Tatamá National Park (CO 20) and c.10 km east of the Nóvita trail which is in the vicinity of San José del Palmar and should perhaps be considered in any conservation initiatives owing to the (at least historical) presence of both *Buthraupis* species.

Oreothraupis arremonops	1994	Recorded from Alto de los Galápagos (J. Farthing *in litt.* 1994).
Buthraupis aureocincta	1994	Nine recorded at 1,800–2,100 m in the vicinity of Alto de los Galápagos in 1994 (J. Farthing *in litt.* 1994). Two collected at 2,200 m along the Nóvita trail in December 1911.
Buthraupis melanochlamys	1987	Apparently locally common near San José del Palmar during October 1987.

Río Blanco watershed (Caldas)

CO 22

Río Blanco watershed reserve (Manizales water company), area unknown

5°05′N 75°30′W
EBA B17

The Río Blanco watershed lies near Manizales on the north-west side of the Nevado del Ruiz, on the western slope of the Central Andes. At c.2,400 m the vegetation is humid montane forest although there are large plantations of native alder *Alnus acuminata* within which, however, the undergrowth is cleared, preventing natural regeneration (L. M. Renjifo *in litt.* 1994).

Hapalopsittaca amazonina	1991	Two seen at 2,400 m in December 1991 are the most recent records; others refer to flocks above Manizales.
Grallaria rufocinerea	1990	Commonly seen and heard in 1989–1990.

La Victoria (Caldas)

CO 23

La Victoria watershed reserve (private), c.20 ha

5°19′N 74°55′W
EBA B09

Lying on the eastern slope of the Central Andes, La Victoria is midway down the Magdalena valley. Some protection of the humid lowland and foothill forest in the vicinity (west) of La Victoria is afforded by the municipality watershed reserve, although this patch is only c.20 ha in extent and rather dry (M. Pearman *in litt.* 1995). More formal protection of this area, with extensions to any forest patches in the immediate area, would be desirable.

Capito hypoleucus	1993	A number were recorded at 925 m in May 1990, in 1992 (F. G. Stiles *in litt.* 1994), and in August 1993 (M. Pearman *in litt.* 1995).
Phylloscartes lanyoni	1990	One seen at 750 m.

Los Nevados (Caldas/Risaralda/Tolima)

CO 24

Los Nevados National Park (IUCN category II), 58,300 ha

4°47′N 75°28′W
EBAs B12, B17, B60

This national park in the Central Andes embraces the Nevado del Ruiz, Nevado de Santa Isabel and Nevado del Tolima, at altitudes from 3,000 to 5,000 m. The vegetation is primarily temperate forest and páramo. Burning and overgrazing of the páramo and adjacent forest is a serious problem in the southern end of this area. Lower elevations are in urgent need of protection (for many species), and both *Grallaria milleri* and *G. rufocinerea* should be looked for in the park (L. M. Renjifo *in litt.* 1994).

Leptosittaca branickii	1994	Several flocks of 4–16 seen at 3,800 m in the Río Otún valley in September 1993 and November 1994 (P. Salaman *in litt.* 1995).
Bolborhynchus ferrugineifrons	1994	Good numbers are recorded regularly on páramo and in montane forest in the south of the park around Laguna del Otún near the boundary with Ucumarí (CO 25) (L. M. Renjifo *in litt.* 1994, P. Salaman *in litt.* 1995).
Hapalopsittaca fuertesi	1911	Specimens were collected before the designation of the national park.
Hypopyrrhus pyrohypogaster	1889	Known only from an old specimen record.

Ucumarí (Risaralda)

CO 25

Ucumarí Regional Park, 4,240 ha

4°39′N 75°36′W
EBAs B12, B17

Ucumarí Regional Park is on the western slope of the Central Andes, and between 1,800 and 3,900 m it offers excellent protection to an important fragment of upper subtropical forest which extends up to páramo in the buffer zone of Los Nevados National Park (CO 24), close to the Alto Quindío area (CO 26) but in a different watershed.

Penelope perspicax	1994	A small population is currently present (16 birds counted in November 1990); common in plantations 1993–1994 (P. Salaman *in litt.* 1995).
Leptosittaca branickii	1994	Not uncommon in the park, with one flock of 30 seen in November–December 1994 (P. Salaman *in litt.* 1995, M. G. Kelsey verbally 1995).
Bolborhynchus ferrugineifrons	1994	Several large flocks (up to 80 birds) in the montane forest–páramo ecotone at 3,000–3,900 m in 1993–1994 (P. Salaman *in litt.* 1995).
Grallaria milleri	1994	Two birds (the first records since 1942) trapped and released at La Pastora (2,400 m) during May and November 1994 in dense mossy cloud-forest (G. Kattan *in litt.* 1994, P. Salaman *in litt.* 1995).
Grallaria rufocinerea	1994	Seen at La Pastora (2,400 m) in October 1993 and May 1994 (P. Salaman *in litt.* 1995).
Chlorochrysa nitidissima	1994	Records here are the only ones from the Central Andes since 1951. Found in 1993–1994 to be uncommon at 1,900–2,100 m (P. Salaman *in litt.* 1995).
Hypopyrrhus pyrohypogaster	1990	Apparently the only recent record.

Alto Quindío–Laguneta (Quindío)

CO 26

4°37'N 75°28'W
EBAs B12, B17

Alto Quindío Acaime Natural Reserve (private), 200 ha
Cañon del Quindío Natural Reserve (departmental), 4,850 ha

The Laguneta and Alto Quindío areas lie c.10 km apart at the head of the Río Quindío watershed, on the western slope near the crest of the Central Andes, in Salento municipality. The area comprises temperate cloud-forest with wax palms *Ceroxylon* extending up to the páramo zone (M. Pearman *in litt.* 1995), although further down the valley and close to the town of Salento (1,900 m) are the subtropical and mainly deforested Boquía valley and El Roble. Most historical records come from Laguneta and the Boquía valley, with recent records (including the recently described Chestnut-bellied Cotinga *Doliornis remseni*: Renjifo 1994) almost exclusively from the two reserves. A specimen of *Coeligena prunellei* from Salento is now suspected to be in error (A. J. Negret *in litt.* 1994).

Leptosittaca branickii	1993	Present all year in the two reserves with a mean flock size of 19 birds recorded (M. G. Kelsey verbally 1994, M. Pearman *in litt.* 1995).
Bolborhynchus ferrugineifrons	1990	Occurs at low densities.
Hapalopsittaca fuertesi	1991	The only confirmed population surviving, all contained within the two reserves where the average flock size is c.8 birds (the largest flock being 25).
Grallaria alleni	1911	The type-specimen, collected at 2,135 m in the Boquía valley, is the only Central Andean record.
Grallaria milleri	1942	The type-series of six birds was collected in 1911, with two more taken in 1942, all at 2,750–3,150 m.
Grallaria rufocinerea	1990	Found commonly at 2,500–3,150 m during 1989–1991.
Hypopyrrhus pyrohypogaster	1917	All records are old and from the Boquía valley area, giving no indication of abundance.

Juntas (Tolima)

CO 27

4°34'N 75°16'W
EBA B12

Unprotected

Juntas is c.30 km north-north-west of Ibaqué, on the eastern slope of the Central Andes. The species below was found beside a road at the bottom of a steep valley which was covered in secondary forest and coffe plantations.

Leptotila conoveri	1990	One record of birds flushed from the road.

Río Toche (Tolima)

CO 28

4°26'N 75°22'W
EBAs B12, B17

Unprotected

On the eastern slope of the Central Andes, the Río Toche valley includes all localities for threatened species between (and just north of) Toche and west of Ibaqué. Much of this watershed comprises tall secondary growth or cultivated areas, with forest remnants primarily confined to ravines and the steepest slopes. Fundación Herencia Verde is promoting the designation of a private reserve in the valley, and the finca at 'La Carbonera' is already established as a small reserve (D. Uribe *in litt.* 1995).

Leptotila conoveri	1990	Birds were seen on two of six occasions near Tapias.
Leptosittaca branickii	1994	Regularly ranges into this area from the adjacent Alto Quindío area (most recently in December 1994), and almost certainly breeds (M. G. Kelsey verbally 1995).
Ognorhynchus icterotis	1991	A flock of c.10 seen flying over wax palms in May 1991.
Grallaria rufocinerea	1990	Recorded in May 1990.
Atlapetes flaviceps	1993	Found commonly (in small flocks) in secondary growth, most recently in August 1993 (M. Pearman *in litt.* 1995).
Hypopyrrhus pyrohypogaster	1911	Mentioned for this locality.

CO 29

Cúcuta (Norte de Santander)

7°54′N 72°31′W
EBA B10

Unprotected

This area includes localities up to 20 km south of Cúcuta (on the Pamplona highway), between 400 and 1,700 m on the eastern side of the northern East Andes. The excessive and relentless trapping for the cage-bird trade must be stopped if the species below is to survive in Colombia.

Carduelis cucullatus	1986	Very rare in this area and apparently declining owing to trapping.

CO 30

Páramo de Tamá (Norte de Santander)

7°25′N 72°26′W
EBA B10

Tamá National Park (IUCN category II), 48,000 ha

This area is in southernmost Norte de Santander, where the East Andes abut the Venezuela border, and the national park is contiguous with El Tamá National Park on the Venezuelan side. El Tamá National Park in Venezuela may hold a population of *Hapalopsittaca amazonina* which should be searched for on the Colombian side.

Pauxi pauxi	pre-1959	Apparently present on the eastern slope of the park.
Pyrrhura calliptera	pre-1959	Historically recorded from this locality.

CO 31

Lebrija (Santander)

7°07′N 73°13′W
EBA B10

Unprotected

Lebrija is on the western slope of the northern East Andes in an area of lower montane humid forest at 800–1,100 m. There has been progressive deforestation on the lower slopes of the East Andes, although it seems likely that the species below can survive in the forest-edge and secondary-growth habitats that now predominate.

Amazilia castaneiventris	1963	One collected.

CO 32

Cuchilla del Ramo (Santander)

6°48′N 73°26′W
EBA B10

Unprotected

This area is on the western slope of the East Andes, in an area of humid subtropical and temperate oak–Lauraceae forest (at 1,500–2,500 m). The forest in this zone has been largely cleared, although important fragments remain.

Odontophorus strophium	1970	The presence of a male and chicks in May 1970 confirmed breeding.

Portugal (Santander)

CO 33

Unprotected

6°45′N 73°21′W
EBA B10

Portugal is on the western slope of the East Andes, on the west side of the Río Suáraez valley, in an area of lower montane humid forest at 800–1,000 m. However, there has been extensive deforestation, and the continued presence of the species below urgently needs to be confirmed.

Amazilia castaneiventris	1963	Two collected in May 1962, with six there in May 1963, all at 850–950 m.

San Gil (Santander)

CO 34

Unprotected

6°33′N 73°08′W
EBA B10

San Gil is in the upper tropical zone on the western slope of the East Andes, where thick, xeric acacia scrub predominates along the Río Fonce. Widespread cultivation (including coffee plantations, pasture and arable) around this locality has caused the loss of much natural vegetation, although the extent to which this has affected the acacia scrub is unknown. The taxonomic status of the species below has been questioned, and studies to clarify this are urgently required.

Thryothorus nicefori	1989	The only known site for this species. A pair found c.1 km east of San Gil in 1989 is the most recent record, although a number of specimens (in the Museo de la Salle) were collected during the 1950s and 1960s (F. G. Stiles *in litt.* 1994).

Guanentá–Alto Río Fonce (Santander)

CO 35

Guanentá–Alto Río Fonce Fauna and Flora Sanctuary (IUCN category IV), 10,000 ha

6°05′N 73°12′W
EBA B10

At the headwaters of the Río Fonce and north-east of Virolín, on the western slope of the East Andes of Santander department is a band (25 × 50 km) of humid subtropical and temperate oak forest which extends from c.2,200 to 3,900 m and is protected as a wildlife sanctuary (Andrade and Repizzo 1994). This is the only sizeable forest tract remaining in this area. The continued presence of the threatened birds listed here needs to be determined (as does the suggested occurrence of *Pyrrhura calliptera*, *Hapalopsittaca amazonina* and *Cistothorus apolinari*), and suitable forest areas outside the wildlife sanctuary should be identified. The poorly known Mountain Grackle *Macroagelaius subalaris*, classified as Near Threatened, is common in this area (D. Uribe *per* L. M. Renjifo *in litt.* 1994).

Odontophorus strophium	1988	Seven groups of birds heard in the vicinity of Virolín during March 1988.
Pyrrhura calliptera	c.1980	Apparently known to occur.
Coeligena prunellei	c.1990	Found to be not uncommon within 5 km radius of Virolín during March 1988; subsequently recorded as common (G. I. Andrade *in litt.* 1994).

Cerro Carare (Boyacá)

CO 36

Unprotected

5°55′N 73°27′W
EBA B10

Cerro Carare is on the western slope of the East Andes, and embraces humid montane forest (primarily oak), especially at 2,300–2,500 m. The area is now surrounded by intensive crop cultivation and pastureland. Surveys are urgently required, as is protection for the remnant forest.

Coeligena prunellei	1987	Found to be quite common during 1978, with three birds seen in February 1987.

CO 37

Soatá (Boyacá)

6°20′N 72°41′W
EBA B10

Unprotected

This area is on the western slope of the northern East Andes, between 1,600 and 2,000 m, above the left bank of the Río Chicamocha. The critical habitats are apparently the bushy canyons and forest borders, which appear to be quite widespread here.

Amazilia castaneiventris	1953	Ten collected during December 1952 and January 1953.

CO 38

El Cocuy (Boyacá/Arauca/Casanare)

6°25′N 72°21′W
EBA B10

El Cocuy National Park (IUCN category II), 306,000 ha

A large national park on the eastern side of the East Andes, embracing most of the Sierra Nevada del Cocuy, with numerous high-altitude lakes on the western slope. The area is threatened and, despite being a national park, has little real protection (G. I. Andrade *in litt.* 1994).

Pauxi pauxi	1965	Apparently still present on the eastern slope of the park.
Pyrrhura calliptera	1917	Specimens collected near Chinivaque.
Cistothorus apolinari	1971	Known from several localities, though not in great numbers.

CO 39

Laguna de Tota (Boyacá)

5°33′N 72°55′W
EBA B10

Unprotected

This large lake at 3,015 m on the eastern side of the East Andes is in urgent need of environmental management. The surrounding area is now almost totally deforested, and fringing wetland habitat is reduced to less than 175 ha. Introduced fish, intensive agriculture, pollution and sediment run-off have caused serious problems to the lake community.

Podiceps andinus	1977	The last-known locality for a species which is now extinct: 300 were reported in 1968, since when there were records in 1972 and 1977.
Rallus semiplumbeus	1991	The largest-known population, estimated at up to 400 birds in 1991.
Pyrrhura calliptera	1963	Apparently just one record of this species, which has now almost certainly gone from this area.
Cistothorus apolinari	1982	A population of 30–50 pairs was estimated in 1982.

CO 40

Laguna de Fúquene (Cundinamarca)

5°28′N 73°45′W
EBA B10

Unprotected

A large lake (currently 15 km^2: G. I. Andrade *in litt.* 1994) at 2,600 m within the Ubaté valley in the central East Andes, close to the Boyacá border. The lake has a complex marginal vegetation zone, with a wide fringe of tall reeds. Soil erosion from the surrounding deforested hills has reduced water transparency such that there has been an almost total elimination of submerged vegetation. Agricultural activities, excessive hunting, pollution and introduced carp *Cyprinus* exacerbate the problems.

Podiceps andinus	c.1955	Now extinct.
Rallus semiplumbeus	1991	Presence of a low-density population (possibly owing to excessive hunting pressure: G. I. Andrade *in litt.* 1994) was confirmed in 1991.
Cistothorus apolinari	1991	Known to occur, with dense populations suspected in some places.

Laguna de Cucunubá (Cundinamarca)

CO 41
5°17′N 73°48′W
EBA B10

Unprotected

Laguna de Cucunubá is at 2,500 m in the Ubaté valley, covers c.3.5 km^2 and has a complex marginal vegetation zone, although the water quality is extremely poor (the adjacent Laguna Palacio no longer exists: G. I. Andrade *in litt.* 1994). Soil erosion from the surrounding hills has reduced the water transparency such that the submerged vegetation has been totally eliminated, although this may have only indirect consequences for the two extant threatened species. There is significant potential for conservation and management of this lake (G. I. Andrade *in litt.* 1994), which would certainly benefit the threatened bird populations.

Podiceps andinus	c.1970	There are unconfirmed reports of flocks during the early 1970s, although the species is now extinct.
Rallus semiplumbeus	1991	Just one bird was seen in October 1991, although a significant population must be present (G. I. Andrade *in litt.* 1994).
Cistothorus apolinari	1992	Several pairs recorded.

Laguna de Pedropalo (Cundinamarca)

CO 42
4°45′N 74°24′W
EBA B10

Unprotected

Lying at c.2,000 m on the western side of the East Andes, Laguna de Pedropalo is in an area of pasture, oak woodland and humid lower montane forest. The laguna and surrounding forest are in urgent need of some form of protection.

Odontophorus strophium	1954	Apparently the only record from Cundinamarca since 1923.
Rallus semiplumbeus	1991	The first record at this site, presumably of a small population.
Coeligena prunellei	1993	Regularly recorded in small numbers (e.g. M. Pearman *in litt.* 1995). The lake appears to be the local centre of abundance.
Cistothorus apolinari	1981	Only one bird has been recorded (singing) around the lake.
Pseudodacnis hartlaubi	1993	One to five birds recorded on all recent visits (e.g. M. Pearman *in litt.* 1995).

Bojacá (Cundinamarca)

CO 43
4°44′N 74°21′W
EBA B10

Unprotected

Bojacá is at 2,845 m on the western slope of the East Andes, in an area which presumably holds humid montane forest and is certainly deserving of more fieldwork.

Coeligena prunellei	1974	Only one record.
Pseudodacnis hartlaubi	1992	Only one, albeit recent, record.

Laguna de la Herrera (Cundinamarca)

CO 44
4°42′N 74°18′W
EBA B10

Unprotected

This lake, lying north-west of Bogotá in the East Andes, is one of the largest in the Bogotá savanna, although it has decreased greatly in size during recent years. There remains just c.250–350 ha of marsh surrounding the lake, which recent reports suggest no longer has any open water (L. M. Renjifo *in litt.* 1994). The lake and marsh are threatened by hunting, cattle trampling the reedbeds, and irrigation schemes cutting off the supply of water, all suggesting that an integrated conservation plan is urgently needed (G. I. Andrade *in litt.* 1994).

Podiceps andinus	1945	Now extinct.
Rallus semiplumbeus	1992	An estimated 50 territories make this one of the largest-known populations.
Cistothorus apolinari	1989	Recorded singing at several places around the lake.

Laguna de la Florida (Cundinamarca)

CO 45
4°43′N 74°09′W
EBA B10

Laguna Florida Park (public park), 35 ha

Laguna de la Florida and the adjacent Jaboque marsh are at 2,600 m on the outskirts of Bogotá in the East Andes. The lake's closeness to Bogotá, combined with its small size, make it extremely vulnerable to pollution and disturbance. Development projects and illegal settlement pose an extreme threat to the lake and surrounding marshes, which provide excellent opportunities for education, watching and studying birds (G. I. Andrade *in litt.* 1994, P. Salaman *in litt.* 1995).

Podiceps andinus	c.1940	Now extinct.
Rallus semiplumbeus	1994	Population recently estimated at c.55 pairs (Lozano 1993); regularly recorded (e.g. P. Salaman *in litt.* 1995).
Cistothorus apolinari	1994	Regularly recorded (e.g. P. Salaman *in litt.* 1995); local population estimated at 10–15 pairs.

Salto del Tequendama (Cundinamarca)

CO 46
4°35′N 74°18′W
EBA B10

San Antonio de Tequendama reserve (private), area unknown

This area on the western slope of the central East Andes (the Bogotá savanna) includes the Tequendama falls, Finca Rancho Grande and San Antonio de Tequendama, all 15–25 km west of Bogotá. There are remnant patches of humid and lower montane forest close to Laguna de Pedropalo. Extensive clearance of forest near Bogotá means that any patches remaining are extremely important for conservation.

Pseudodacnis hartlaubi	1992	Records (at the finca) come from 1,700–2,200 m where, however, the species was scarce in 1972–1973. Most recently recorded at San Antonio de Tena.

Laguna Chisacá (Cundinamarca)

CO 47
4°17′N 74°13′W
EBA B10

Unprotected

Lying north of Sumapaz National Park, Laguna Chisacá is at 4,000 m at the southern end of the East Andes. Searches should perhaps be extended to Sumapaz National Park for the species listed below.

Rallus semiplumbeus	1960	One specimen is the only record.
Cistothorus apolinari	1973	Numerous specimens collected during the 1950s and 1960s; later records suggest continued presence of a population.

Valle de Jesús forest (Cundinamarca)

CO 48

Valle de Jesús communal reserve, area unknown

4°50′N 73°40′W
EBA B10

This forest, lying above 2,000 m in the humid upper subtropical zone, is just north of Chingaza National Park (east of Bogotá) in the East Andes. At least some of the forest has been selectively felled (G. I. Andrade *in litt.* 1994). Combined with the forest in Chingaza National Park and adjacent reserves, Valle de Jesús deserves formal protection.

Pyrrhura calliptera	c.1992	A population is apparently present.
Hapalopsittaca amazonina	1991	Six to eight seen feeding in secondary forest in July 1991.

Páramo de Chingaza (Cundinamarca)

CO 49

Chingaza National Park (IUCN category II), 50,374 ha
Río Blanco–Olivares Forest Reserve (IUCN category V), 4,900 ha
Carpanta Biological Reserve (private), 1,200 ha

4°34′N 73°41′W
EBA B10

Páramo de Chingaza (with areas to the south and east) is on the eastern side of the East Andes, and is contained within the national park and adjacent reserves. With altitudes from 1,800 to 4,000 m, predominant vegetation ranges from subtropical forest up to páramo. The Río Blanco forest reserve is adjacent to and on the west side of the national park, and protects almost 4,000 ha of forest above 2,600 m (F. G. Stiles *in litt.* 1994). Each of these contiguous protected areas is critically important for the birds mentioned below, and if extended to embrace forest at lower altitudes (towards Key Area CO 50) might include areas suitable for *Grallaria kaestneri* (L. M. Renjifo *in litt.* 1994; see 'Recent changes to the threatened list', above, and CO 50). A proposal for the enlargement of the national park has been made by Fundación Natura (G. I. Andrade *in litt.* 1994).

Rallus semiplumbeus	1991	Small populations exist in the Carpanta and Chingaza reserve.
Pyrrhura calliptera	1994	Groups of 30–60 recorded on the west side of Chingaza and in Río Blanco reserve (F. G. Stiles *in litt.* 1994), and in the upper reaches of the Carpanta reserve; these represent perhaps the most important population currently known (M. Pearman *in litt.* 1995).
Hapalopsittaca amazonina	1992	A small population exists, with a group of five recorded from Río Blanco reserve (F. G. Stiles *in litt.* 1994).

Monterredondo–El Calvario (Cundinamarca)

CO 50

Unprotected

4°17′N 73°48′W
EBA B10

The recently opened road between Monterredondo and El Calvario (in Meta department) is on the eastern slope of the East Andes, 50 km south-east of Bogotá. Near Monterredondo, from c.1,700 to above 2,300 m, is found humid, upper subtropical forest continuing up to páramo. There has been quite extensive disturbance of forest on this slope of the East Andes, although large areas still remain in adjacent watersheds and surrounding slopes, mostly difficult of access (F. G. Stiles *in litt.* 1994). This site is the only known area for the recently described *Grallaria kaestneri* (see 'Recent changes to the threatened list', above).

Pyrrhura calliptera	1991	Flocks of up to 25 birds recorded recently.

Bosque de Yotoco (Valle)

CO 51

Bosque de Yotoco reserve, 560 ha

3°52′N 76°33′W
EBA B12

A forest area at 1,400–1,600 m on the eastern slope of the West Andes. This small reserve is administered by the Corporación Autónoma Regional de Cauca (CVC) and the Universidad Nacional de Colombia. Almost total destruction of humid forest in the middle Cauca valley means that any remaining area has extremely high conservation value.

Penelope perspicax	1989	A small but apparently stable population remains.
Chlorochrysa nitidissima	1979	Recorded from the reserve, though numbers involved are unknown.
Pseudodacnis hartlaubi	1979	One male seen is the only record.

Los Farallones de Cali (Valle)

CO 52

Los Farallones National Park (IUCN category II), 150,000 ha
Alto Anchicayá–Verde CVC watershed reserve, area unknown.

3°37′N 76°53′W
EBA B14

Lying west of Cali, Los Farallones de Cali National Park covers a large area at the southern end of the West Andes (including the watershed reserve, Reserva Hato Viejo and Reserva La Teresita). From 500 to 4,000 m it comprises humid subtropical and temperate forest and páramo (A. J. Negret *in litt.* 1994). Most of the birds listed here are recorded only from the old Cali to Buenaventura road which forms the northern edge of the park. Surveys are urgently needed in the large area of almost pristine habitat that forms the major part of this park, and which also supports a population of *Cephalopterus penduliger* (M. G. Kelsey *in litt.* 1995: see 'Recent changes to the threatened list', above).

Neomorphus radiolosus	1989	One at Alto Anchicayá (c.600 m) remains the only record.
Dysithamnus occidentalis	1990	One collected at 2,200 m.
Oreothraupis arremonops	1994	Recorded from the extreme south of the park (A. J. Negret *in litt.* 1994); previously just one 1970s record.
Chlorospingus flavovirens	1989	Groups of up to five seen at Alto Anchicayá, 1972–1975; one seen in 1989.
Chlorochrysa nitidissima	1992	Numerous recent records of 2–5 birds.

Munchique (Cauca)

CO 53

Munchique National Park (IUCN category II), 44,400 ha
Tambito Nature Reserve (private), c.3,000 ha

2°32′N 76°57′W
EBAs B12, B14

Munchique National Park, including Cerro Munchique at the southern tip of the park, covers an area along the main ridge of the West Andes at the southern end of the Cauca valley. The park mainly comprises humid cloud-forest. The adjacent Tambito (private) reserve (south-west of Cerro Munchique, at 1,300–2,600 m) may be important for many of the birds listed below, and is included within this Key Area (A. J. Negret *in litt.* 1994). Birds recorded at or near 'El Tambo' are also included, as they were probably recorded inside the national park. This area is threatened by colonization, by construction of a hydroelectric plant (MICAY) and by proposed highway development (G. I. Andrade *in litt.* 1994). A specimen of *Grallaria gigantea* from Munchique has been reidentified as *G. squamigera* (Krabbe *et al.* 1994), and records of *Cypseloides lemosi* assumed to be from this area in Collar *et al.* (1992) are now known to be from Santa Helena (CO 55).

Micrastur plumbeus	1938	Although historically recorded within this area, the species is at the upper end of its altitudinal range.
Penelope perspicax	1987	Considered rare, even though extensive forest cover remains; not seen in subsequent searches.

cont.

Leptosittaca branickii	pre-1957	One record of this nomadic species; its status in the park is unknown.
Ognorhynchus icterotis	1978	A flock of 25; no subsequent records.
Neomorphus radiolosus	1988	This area embraces the majority of Colombian records, and is of key importance; the guaranteed integrity of forest on the lower slopes is essential.
Eriocnemis mirabilis	1994	Munchique is the type- and only locality for this uncommon species; two sites known within the park (M. Pearman *in litt.* 1995, P. Salaman *in litt.* 1995).
Dysithamnus occidentalis	1939	Three of the four Colombian records are from this area.
Grallaricula cucullata	1994	One seen at 2,000 m in Tambito reserve in February (P. Salaman *in litt.* 1995).
Oreothraupis arremonops	1994	Most recent Colombian records are from this area, which seems to have a healthy population. In Munchique, it occurs at 2,300–2,600 m in groups of 4–5 (Negret 1991, P. Salaman *in litt.* 1995); regularly recorded in Tambito (A. J. Negret *in litt.* 1994).
Chlorochrysa nitidissima	1993	Recorded in groups of up to six at 1,700 m near El Roblé (A. J. Negret *in litt.* 1994); previously known from records up to 1938.

CO 54

Valle del Patía (Cauca)

Unprotected

2°11′N 77°01′W
EBA B12

Centred around El Hoyo, the Patía valley separates the Central and West Andes south of Popayán. The status of the forest and of its threatened species urgently needs to be assessed, although it appears that there is only 150 ha of dry forest remaining in the area, and this is threatened by burning, cutting and hunting (A. J. Negret *in litt.* 1994).

Penelope perspicax	1990	One record at 600 m near El Hoyo.

CO 55

Santa Helena–Cerro Munchique (Cauca)

Unprotected

3°06′N 76°17′W
EBA B12

Santa Helena and Cerro Munchique lie some kilometres apart towards the southern end of the Cauca valley, on the western slope of the Central Andes, where there remain some small areas of humid forest (e.g. c.250 ha on Cerro Munchique) (A. J. Negret *in litt.* 1994). Almost total destruction of humid forest in the middle Cauca valley means that any remaining has extremely high conservation value.

Penelope perspicax	1989	The recent discovery of a young bird at Santa Helena confirms the presence of a breeding population in this area.
Cypseloides lemosi	c.1993	Birds were recorded within sight of Cerro Munchique in 1962, and again more recently (A. J. Negret *in litt.* 1994), suggesting that a population may persist.

CO 56

Puracé (Cauca/Huila)

Puracé National Park (IUCN category II), 83,000 ha

2°24′N 76°23′W
EBAs B17, B60

The national park sits along the border of Cauca and Huila, in the high Central Andes. The vegetation is primarily humid subtropical, temperate and elfin forest, and páramo, and at present is well protected (A. J. Negret *in litt.* 1994). Some of the birds in this park (e.g. *Hapalopsittaca amazonina*) also occur in the adjacent Finca Merenberg private nature reserve.

Leptosittaca branickii	1989	Flocks of up to 60 recorded in the park in the late 1980s, suggesting a sizeable population.
Ognorhynchus icterotis	1976	A pair seen in 1976 is the only record for the park despite substantial fieldwork; it seems unlikely that there is a viable population.

cont.

Hapalopsittaca amazonina	1973	One record of what was thought to be this species (but possibly *H. fuertesi*).
Grallaria gigantea	1941	One specimen, though suitable habitat still remains. This is one of two old records which are the only ones from Colombia (Krabbe *et al.* 1994).
Grallaria rufocinerea	1970	Despite subsequent fieldwork, and the continued presence of suitable habitat, there is still just one record (a pair).
Buthraupis wetmorei	1980	All Colombian records (two specimens and three sightings) come from the park, suggesting an important population.

Río de la Plata (Huila)

CO 57

Unprotected

2°23′N 75°53′W
EBA B12

This area is at the head of the Magdalena valley, on the eastern slope of the Central Andes, and includes the Río de la Plata valley upriver from La Plata towards Plata Vieja. Deforestation has been almost total, suggesting that the area is no longer likely to support the parrot (A. J. Negret *in litt.* 1994).

Ognorhynchus icterotis	1939	One record, in March.
Atlapetes flaviceps	1967	One record, at c.1,300 m.

San Agustín (Huila)

CO 58

Unprotected

1°53′N 76°16′W
EBA B12

San Agustín is near the headwaters of the Río Magdalena, on the eastern slope of the Central Andes, in the subtropical zone. On the outskirts of town, all forest has been cleared for coffee, bananas and sugarcane, though a small area of forest exists within an archaeological reserve (A. J. Negret *in litt.* 1994).

Leptotila conoveri	1990	Two seen in March 1990.

Cueva de los Guácharos (Huila)

CO 59

Cueva de los Guácharos National Park (IUCN category II), 9,000 ha

1°35′N 76°00′W
EBA B12

The national park is at the headwaters of the Río Suaza in southernmost Huila, on the western slope of the East Andes. The forest in this park is primarily subtropical and cloud-forest, and, although well protected, is increasingly threatened by human encroachment and opium production (P. Salaman *in litt.* 1995).

Tinamus osgoodi	1976	One seen at 2,100 m is one of only two Colombian records.
Leptosittaca branickii	1994	A flock of eight seen at 1,700 m in the Río Suaza valley, July (P. Salaman *in litt.* 1995).
Ognorhynchus icterotis	1975	One sight record of a small group.
Hapalopsittaca amazonina	1976	Records possibly refer to *H. fuertesi*.
Grallaria alleni	1971	One collected at 2,000–2,100 m.
Grallaricula cucullata	1978	Six collected and one seen during 1975–1978.
Hypopyrrhus pyrohypogaster	1994	Found to be common around the park headquarters (P. Salaman *in litt.* 1995).

Recinas–Florencia (Caquetá)

CO 60

Unprotected

1°36′N 75°36′W
EBA B10

Recinas, at 2,250 m (Km 30), is the highest point in the road which drops down the eastern slope of the East Andes to Florencia. The area of importance is at Km 50–53 between the two towns and along the Río Hacha, where much of the forest is intact, although a recently felled area and mining activity suggest that this may change (M. Pearman *in litt.* 1995).

Hypopyrrhus pyrohypogaster	1990	The area is well known for the species, the latest record being of a group of eight.

Sierra de Chiribiquete (Caquetá/Guaviare)

CO 61

Chiribiquete National Park (IUCN category II), 1,280,000 ha

1°00′N 72°45′W

The Sierra de Chiribiquete (c.120 km by 30–40 km) is a mountainous area with plateaus up to 800 m, sitting astride the Caquetá–Guaviare border (F. G. Stiles *in litt.* 1994). The Río Apaporis forms the departmental border which divides the sierra, and supports the forest which is so important for the species below.

Crax globulosa	c.1990	Reported as fairly common, with live birds being collected on the Río Apaporis (Collar *et al.* 1994).

Serranía de la Macarena (Meta)

CO 62

Serranía de la Macarena National Park (IUCN category II), 630,000 ha

2°45′N 73°55′W

This large isolated serranía to the east of the East Andes is surrounded by lowland tropical forest. The serranía rises to over 2,500 m, with rivers flowing all around it, and a large area of lowland forest is incorporated within the park to the east. Almost half the forest is already cut, and invading colonists are putting the remainder, even at higher altitudes, under increasing pressure (F. G. Stiles *in litt.* 1994).

Coturnicops notatus	1959	One collected, March; the only Colombian record.
Touit stictoptera	1950	One collected.

Isla de Tumaco and Isla Boca Grande (Nariño)

CO 63

Unprotected

1°49′N 78°46′W

Two narrow sandy islands off the exposed north-west-facing coast of the Mira peninsula. The natural vegetation is grassland and scrub, but none remains on Isla de Tumaco as human settlement has spread over the entire island. Isla Boca Grande retains c.1 ha of natural habitat, though even this is threatened by tourist developments (P. Salaman *in litt.* 1995).

Sporophila insulata	1994	Formerly known from just four specimens collected in 1912 on Isla de Tumaco; rediscovered (five birds) in December 1994 on Isla Boca Grande (P. Salaman *in litt.* 1995).

La Guayacana (Nariño)

CO 64

1°26′N 78°22′W
EBA B14

Unprotected

La Guayacana is on the lower Pacific slope of the West Andes in the Río Güiza valley, where lowland and foothill wet forest (typical of the Chocó) predominate. There has been widespread destruction of the coastal plain and foothill forest in this region, primarily for palm plantations and cattle grazing (P. Salaman *in litt.* 1995).

Micrastur plumbeus	1959	Two collected at 250 m in 1944, with another at 225 m in 1959.
Dacnis berlepschi	1959	Six collected along this river at 220–250 m; one bird (probably this species) seen in September 1991.

Río Ñambi (Nariño)

CO 65

1°18′N 78°05′W
EBA B14

Río Ñambi Community Nature Reserve (private), c.2,000 ha

The Río Ñambi watershed on the Pacific slope of the West Andes includes tropical to subtropical pluvial forest beside the Pasto–Tumaco highway (near Junín) at 500–1,600 m. There has been widespread destruction of the lowland and foothill forest in this region. The area is known to hold populations of *Penelope ortoni*, *Cephalopterus penduliger* and the Chocó Vireo *Vireo* sp. (P. Salaman *in litt.* 1995: see 'Recent changes to the threatened list', above).

Micrastur plumbeus	1992	Three caught and released in August (P. Salaman *in litt.* 1995).
Neomorphus radiolosus	1993	One seen at 900 m in 1988; reported by local hunters to be not uncommon at 500 m (P. Salaman *in litt.* 1995).
Haplophaedia lugens	c.1986	No details available.
Chlorospingus flavovirens	1993	Two seen regularly at 500 m in July (P. Salaman *in litt.* 1995).
Dacnis berlepschi	c.1980	Recorded above Junín at 1,200 m. A bird probably of this species was found below the Río Ñambi at 700 m in September 1991 (P. Salaman *in litt.* 1995).

La Planada (Nariño)

CO 66

1°13′N 77°59′W
EBA B14

La Planada Nature Reserve (private), 1,650 ha

This area is on the Pacific slope of the West Andes, and includes upper tropical and subtropical wet forest at c.1,800–2,300 m. The La Planada Nature Reserve is a small but extremely important park within this area, which has otherwise undergone widespread deforestation.

Ognorhynchus icterotis	1989	Up to 21 returned to the reserve yearly during the 1980s; no subsequent records.
Haplophaedia lugens	1994	Regularly found to be fairly common in young secondary growth in the reserve (P. Salaman *in litt.* 1995).
Oreothraupis arremonops	1985	One record from within the reserve.

Volcán Galeras and Río San Francisco (Nariño)

CO 67
1°13′N 77°17′W
EBA B17

Galeras Fauna and Flora Sanctuary (IUCN category IV), 17,600 ha

Volcán Galeras (including the Río San Francisco canyon and Yacuanquer valley) is west of Pasto on the western side of the Andes. The intermontane plateaus around Pasto have been extensively cleared for agriculture and *Pinus* plantations (P. Salaman *in litt.* 1995). Only on the most inaccessible slopes of Volcán Galeras do some small fragments of temperate forest remain (P. Salaman *in litt.* 1995).

Leptosittaca branickii	1994	Several flocks of 6–15 seen on three days in September 1994 (P. Salaman *in litt.* 1995); first recorded 1989.

Laguna de la Cocha (Nariño)

CO 68
1°05′N 77°09′W
EBA B17

Isla La Corota National Sanctuary, 20 ha
Many small private reserves (Asociación para el Desarrollo Campesino)

Lying at 2,800 m, south-east of Pasto in eastern Nariño on the eastern slope of the Andes, the lake is the source of the Río Guamués. Mountains surround it on all sides, and there is Andean forest between 2,700 and 3,000 m. About 30 small private reserves protect some of the forest around the lake, with the state reserve protecting forest on an island (L. M. Renjifo *in litt.* 1994).

Leptosittaca branickii	1992	A flock heard on the east shore, in Tungurahua, during April 1992.

El Carmen (Nariño)

CO 69
0°40′N 77°10′W
EBA B18

Unprotected

El Carmen is at c.1,500 m on the eastern slope of the Andes, in the Río Guamués drainage. A large area of pristine forest encompassing the El Carmen area is threatened by recent plans to construct a highway beside the Tumaco–Orito oil pipeline (P. Salaman *in litt.* 1995).

Galbula pastazae	1970	Four collected; the only Colombian record.

■ COSTA RICA

Yellow-billed Cotinga *Carpodectes antoniae*

COSTA RICA, like neighbouring Panama, is part of the land-bridge between the very different avifaunas of North and South America, and in consequence a disproportionately large number of bird species, c.850, have been recorded from this small country (50,900 km^2) and its territorial waters including Cocos Island (Stiles and Skutch 1989). The species total includes c.600 permanent residents and more than 200 regular migrants (primarily from breeding areas in North America) (Stiles and Skutch 1989). Six species are endemic to the country, 78 have restricted ranges (Stattersfield *et al.* in prep.) and four are threatened (Collar *et al.* 1992). This analysis has identified 14 Key Areas for the threatened birds in Costa Rica (see '*Key Areas*: the book', p. 11, for criteria).

THREATENED BIRDS

Four Costa Rican species were considered at risk of extinction by Collar *et al.* (1992), one of which, *Amazilia boucardi*, is confined to the country (see Table 1). Both *A. boucardi* and *Carpodectes antoniae* are dependent on mangroves, the other two threatened birds (and *C. antoniae*, at least seasonally) relying on wet forest (Collar *et al.* in press). All four are found primarily in the lowland tropical zone (0–500 m), with *Cephalopterus glabricollis* breeding in the subtropical zone (up to 2,000 m), and all four are threatened by loss of habitat (Collar *et al.* in press). The distributions of these four threatened birds and their relationship to Endemic Bird Areas are shown in Figure 1.

KEY AREAS

The 14 Costa Rican Key Areas would, if adequately protected, help ensure the conservation of all four of the country's threatened species—always accepting that important new populations and areas may yet be found. Eight of these areas are important for two threatened birds (Tables 1–2), although each Key Area is vitally important for the conservation of the threatened species and habitats that it supports. Just one threatened bird, *Amazilia boucardi*, is endemic to Costa Rica, and is thus totally reliant for its survival on the integrity of the mangroves in the seven Key Areas from which it is known (see 'Out-

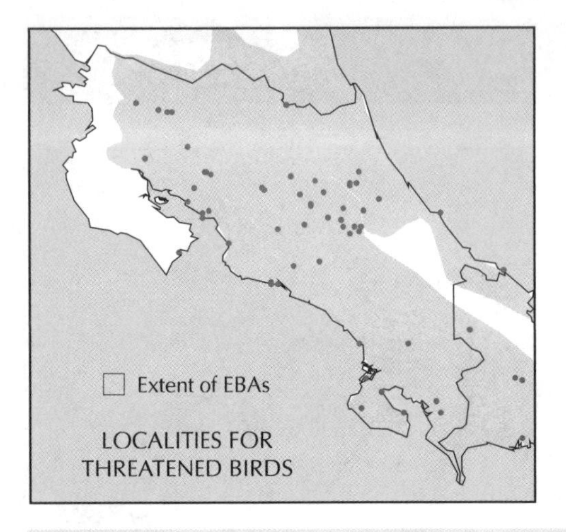

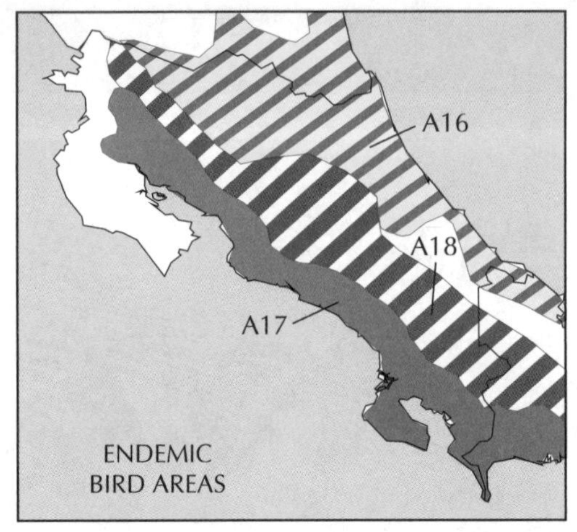

Figure 1. The localities where threatened birds have been recorded in Costa Rica and their relationship to Endemic Bird Areas (EBAs).

- Three Costa Rican threatened species have restricted ranges and thus occur within the various EBAs listed below (figures are numbers of these species in each EBA).

A16 Central American Caribbean slope (1)
A17 South Central American Pacific slope (2)
A18 Costa Rica and Panama highlands (1)
A21 Cocos Island (0: see 'Recent changes to the threatened list', below)

look', below). Although *Cephalopterus glabricollis* appears to be well represented in Costa Rican Key Areas, these are primarily within its breeding grounds, and the species remains relatively exposed when in its winter quarters (see 'Old records and little-known birds', below). The large number of Key Areas selected for *Amazilia boucardi* and *Carpodectes antoniae* reflects the importance that each of these areas potentially has for the continued survival of the two species.

KEY AREA PROTECTION

Costa Rica has placed more than 8% of its territory in national parks and equivalent reserves (Stiles and Skutch 1989), and indeed eight (57%) of the Key Areas currently have some form of protected status, four as national parks or biological reserves (IUCN categories I and II). Outside Costa Rica's protected areas, however, the natural habitats and birds are increasingly threatened, and it is even questionable whether the parks and reserves will survive as pressure on the land becomes more intense (Stiles and Skutch 1989). Thus, effective management is required of activities undertaken within protected Key Areas, but for the six Key Areas (43% of the total) that are currently unprotected attention in the form of appropriate conservation measures is perhaps more urgent if the populations of their threatened species are to survive. All four threatened species are present within at least two protected Key Areas.

RECENT CHANGES TO THE THREATENED LIST

With the publication of Collar *et al.* (1994), seven species (Military Macaw *Ara militaris*, Cocos Cuckoo *Coccyzus ferrugineus*, Turquoise Cotinga *Cotinga ridgwayi*, Three-wattled Bellbird *Procnias tricarunculata*, Cocos Flycatcher *Nesotriccus ridgwayi*,

Table 1. Coverage of threatened species by Key Area. Areas in bold currently have some form of protected status.

	Key Areas occupied	No. of Key Areas protected	Total nos. of Key Areas	
			Costa Rica	Neotropics
Mangrove Hummingbird *Amazilia boucardi* [E]	05,**06**,**07**,08,**10**,12,**13**	4	7	7
Keel-billed Motmot *Electron carinatum*	**01**,02,**03**	2	3	7
Yellow-billed Cotinga *Carpodectes antoniae*	**07**,08,09,**10**,**11**,12,**13**,14	4	8	9
Bare-necked Umbrellabird *Cephalopterus glabricollis*	**01**,02,**03**,**04**	3	4	6

[E] Endemic to Costa Rica

	01	02	03	04	05	06	07	08	09	10	11	12	13	14	No. of areas
Amazilia boucardi	–	–	–	–	●	●	●	●	–	●	–	●	●	–	7
Electron carinatum	●	●	●	–	–	–	–	–	–	–	–	–	–	–	3
Carpodectes antoniae	–	–	–	–	–	–	●	●	●	●	●	●	●	●	8
Cephalopterus glabricollis	●	●	●	●	–	–	–	–	–	–	–	–	–	–	4
No. of species	2	2	2	1	1	1	2	2	1	2	1	2	2	1	

Table 2. Matrix of threatened species by Key Area.

Cocos Finch *Pinaroloxias inornata* and Black-cheeked Ant-tanager *Habia atrimaxillaris*) were added to the Costa Rican threatened species list, with Keel-billed Motmot *Electron carinatum* being relegated to Near Threatened status; the additional species have not, however, been included in the Site Inventory (see '*Key Areas*: the book', p. 12). Three of these recently added species are endemic to Cocos Island and were reclassified (on the basis of new criteria) owing to their ranges being less than 100 km^2 (Collar *et al.* 1994). With three threatened species in such a small area, Cocos Island, which is not currently covered in the Key Area analysis, should in future be considered a high priority for bird conservation. With the exception of *Ara militaris* (which may be sympatric with *Cephalopterus glabricollis* during the non-breeding season), the mainland species added in Collar *et al.* (1994) are each broadly sympatric with the species considered in this analysis, and thus will not have any major impact on the Key Area analysis, although each species should be considered in future conservation strategies or initiatives.

OLD RECORDS AND LITTLE-KNOWN BIRDS

Each of the four threatened species has been relatively regularly and recently (1980s and 1990s) recorded from Costa Rica. However, this disguises the fact that each bird remains poorly known. The status, population and even the distribution of *Amazilia boucardi*, for example, are poorly known, both within the Key Areas and in mangrove areas where its presence is to be expected (e.g. CR 09 Río Sierpe and CR 14 Río Coto). The ecological requirements of *Carpodectes antoniae*, especially those related to seasonal movements and breeding, are essentially unknown, but urgently need elucidation if its conservation is to be assured. Likewise, *Cephalopterus glabricollis* migrates to the Caribbean lowlands outside the breeding season, but very few precise areas are currently known.

OUTLOOK

Each of the 14 Key Areas in Costa Rica would, if adequately protected, help ensure the survival of the country's four threatened species. The guaranteed integrity of the areas currently under some form of protection is essential, but increasing this protection to currently unprotected Key Areas such as those supporting two threatened species would increase the likelihood of long-term survival for each species. Therefore, the protection of at least Volcán Tenorio and Bijagua (CR 02), Parrita–Palo Seco (CR 08) and Puerto Jiménez (CR 12) would be desirable. Surveys are urgently required to determine the status, distribution and ecological requirements of *Amazilia boucardi* and *Carpodectes antoniae*, both within the appropriate Key Areas (see 'Old records and little-known birds', above) and in as-yet-unsurveyed mangroves. Determination of the primary non-breeding areas for *Cephalopterus glabricollis* is also needed if this species is to be adequately protected throughout its life-cycle.

DATA SOURCES

The above introductory text and the Site Inventory (below) were compiled from information supplied by M. Reid, as well as from the following references.

COLLAR, N. J., GONZAGA, L. P., KRABBE, N., MADROÑO NIETO, A., NARANJO, L. G., PARKER, T. A. AND WEGE, D. C. (1992) *Threatened birds of the Americas: the ICBP/IUCN Red Data Book* (Third edition, part 2). Cambridge, U.K.: International Council for Bird Preservation.

COLLAR N. J., CROSBY, M. J. AND STATTERSFIELD, A. J. (1994) *Birds to watch 2: the world list of threatened birds*. Cambridge, U.K.: BirdLife International (BirdLife Conservation Series no.4).

COLLAR, N. J., WEGE, D. C. AND LONG, A. J. (in press) Patterns and causes of endangerment in the New World avifauna. In J. V. Remsen, ed. *Natural history and conservation of Neotropical birds*. American Ornithologists' Union (Orn. Monogr.).

IUCN (1992) *Protected areas of the world: a review of national systems. Volume 4: Nearctic and Neotropical.*

Gland, Switzerland and Cambridge, U.K.: International Union for Conservation of Nature and Natural Resources.
SLUD, P. (1964) The birds of Costa Rica: distribution and ecology. *Bull. Amer. Mus. Nat. Hist.* 128.
STATTERSFIELD, A. J., CROSBY, M. J., LONG, A. J. AND WEGE, D. C. (in prep.) *A global directory of Endemic Bird Areas.* Cambridge, U.K.: BirdLife International (BirdLife Conservation Series).
STILES, F. G. AND SKUTCH, A. F. (1989) *A guide to the birds of Costa Rica.* London: Christopher Helm.

SITE INVENTORY

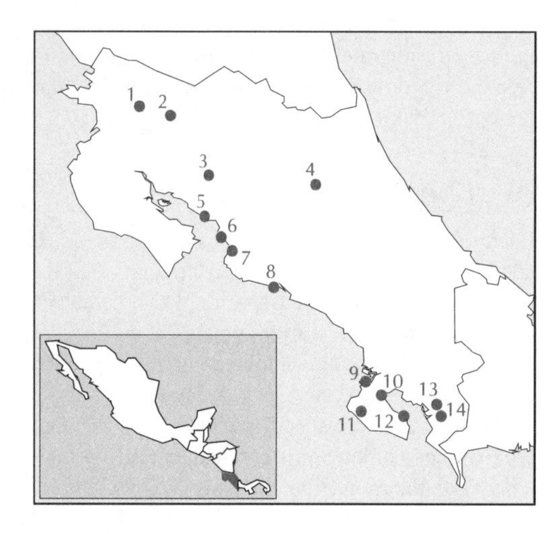

Figure 2. Key Areas in Costa Rica.

01 Rincón de la Vieja
02 Volcán Tenorio and Bijagua
03 Monteverde
04 Braulio Carillo–La Selva–Rara Avis
05 Puntarenas
06 Tivives
07 Río Grande de Tárcoles
08 Parrita–Palo Seco
09 Río Sierpe
10 Rincón
11 Corcovado
12 Puerto Jiménez
13 Golfito
14 Río Coto

Rincón de la Vieja (Guanacaste/Alejuela)

CR 01

Rincón de la Vieja National Park (IUCN category II), 14,083 ha

10°48′N 85°20′W
EBA A18

The area lies at the northern end of the Cordillera de Guanacaste in the north-west corner of Costa Rica, with the national park protecting an area of moist montane and foothill forest (Stiles and Skutch 1989).

Electron carinatum	1988	Known from two areas on the northern slopes at 300–500 m, but densities are low.
Cephalopterus glabricollis	1989	Birds (probably breeding) seen at 1,500 m.

Volcán Tenorio and Bijagua (Guanacaste/Alejuela)

CR 02

Unprotected

10°44′N 85°06′W
EBA A18

Bijagua is at the headwaters of the Río Zapote on the northern (Caribbean) slope of Volcán Tenorio in the Cordillera de Guanacaste. This area embraces tropical moist and subtropical wet forest (Slud 1964), although the remaining extent of either is not known.

Electron carinatum	c.1914	Mentioned for this locality.
Cephalopterus glabricollis	c.1991	Repeated recent records suggest this to be an important area.

Monteverde (Alejuela)

CR 03

10°18′N 84°48′W
EBA A18

Monteverde Biological Reserve (private), c.20,000 ha

The reserve protects an area of cloud- and wet montane forest (with some moist forest lower down) in the Cordillera de Tilarán. It sits on the continental divide, at the head of the Río Peñas Blancas.

Electron carinatum	1986	An individual was noted having pair-bonded with a Broad-billed Motmot *E. platyrhynchum* in the Peñas Blancas valley.
Cephalopterus glabricollis	c.1991	Groups of displaying males regularly recorded in the upper Peñas Blancas valley.

Braulio Carillo–La Selva–Rara Avis (Heredia)

CR 04

10°02–26′N 83°59′W
EBAs A16, A18

Braulio Carillo National Park (IUCN category II), 44,099 ha
La Selva Protection Zone (IUCN category VIII), 2,815 ha

This large, contiguous protected area is situated on the Caribbean slope of the Cordillera Central. It comprises a number of small field or biological stations such as Rara Avis and La Selva, a protected forest, and the large Braulio Carillo National Park which embraces Volcán Barva. The area is relatively intact, and forms Costa Rica's only sizeable protected tract of forest which extends from the Caribbean lowlands in a continuous belt up into montane forest.

Cephalopterus glabricollis	1993	Groups of displaying males regularly recorded in the upper Peñas Blancas valley (e.g. M. Reid *in litt.* 1994).

Puntarenas (Puntarenas)

CR 05

10°00′N 84°50′W
EBA A17

Unprotected

Puntarenas is on a peninsula projecting into the northern side of the Golfo de Nicoya. Areas of mangrove at the base of this peninsula are under severe presure from deforestation and pollution.

Amazilia boucardi	1990	Apparently common, though much less so in recent years owing to pollution and deforestation.

Tivives (Puntarenas)

CR 06

9°51′N 84°42′W
EBA A17

Río Tivives Protection Zone (IUCN category VIII), 2,368 ha

Tivives is at the mouth of the Río Jesús María on the eastern shore of the Golfo de Nicoya. At the base of this peninsula are areas of mangrove which, though protected, are under severe pressure from deforestation and pollution. The extent of *Pelliciera* mangrove, and hence of suitable hummingbird habitat, is relatively small, although still important.

Amazilia boucardi	1990	A small population persists, despite the limited habitat.

Río Grande de Tárcoles (Puntarenas)

CR 07

9°45′N 84°37′W
EBA A17

Carara Biological Reserve (IUCN category I), 4,700 ha

An extensive area of mangrove exists at the mouth of the Río Grande de Tárcoles, which is on the eastern side of the Golfo de Nicoya. This area includes the extensive mangroves in the Pigres–Estero Guacalillos sector, and the Carara Biological Reserve (through which the Río Grande de Tárcoles flows, and which comprises primarily tall humid forest). These localities are all within close proximity, and are important as a contiguous unit for the threatened species, and for the regionally threatened Scarlet Macaw *Ara macao*.

Amazilia boucardi	1990	A sizeable population exists.
Carpodectes antoniae	1990	Recorded, but present status unknown.

Parrita–Palo Seco (Puntarenas)

CR 08

9°29′N 84°18′W
EBA A17

Unprotected

Parrita (at the mouth of the Río Pirrís) and Palo Seco (at the mouth of the Río Palo Seco) are c.5 km apart on a stretch of coast that supports a large expanse of mangrove and adjacent lowland humid forest. Although the large tourist population at Palo Seco would make the protection of particular areas difficult, protection is clearly desirable in the coastal sector and adjacent forest as far inland as Pozo Azul.

Amazilia boucardi	1985	A breeding population is present, although the size is unknown.
Carpodectes antoniae	1989	A large population was previously known to exist, although its status is currently unknown.

Río Sierpe (Puntarenas)

CR 09

8°48′N 83°36′W
EBA A17

Unprotected

The Río Sierpe is just north of the Península de Osa and comprises an important area of mangroves along the coast at its mouth, and adjacent fringing lowland forest along the Río Sierpe inland. Both habitats are essential for *Carpodectes antoniae*, but neither is formally protected at present. There is an urgent need for a survey of the area, primarily for *C. antoniae*, but also for *Amazilia boucardi* which may well be present in the mangroves.

Carpodectes antoniae	1987	Found to be numerous in what is possibly one of its most important breeding areas.

Rincón (Puntarenas)

CR 10

8°42′N 83°29′W
EBA A17

Golfo Dulce Forest Reserve (IUCN category VIII), 67,287 ha

Rincón is at the head of the Golfo Dulce, on the Península de Osa, in an area that appears to have retained a large enough expanse of mangrove to warrant a higher level of protection. The presence of forest tracts adjacent to the mangroves suggest that this area could be important for both species mentioned below, although surveys are urgently needed.

Amazilia boucardi	1971	A lack of recent records highlights the need for a survey of these mangroves, which may harbour a sizeable population of this species.
Carpodectes antoniae	1989	Large groups were once recorded, but the only recent record is of three birds seen in January 1989.

Corcovado (Puntarenas)

CR 11

Corcovado National Park (IUCN category II), 54,568 ha

8°35′N 83°38′W
EBA A17

The national park is situated on the Pacific side of the Península de Osa and comprises large tracts of lowland tropical forest, but as it lacks any large expanses of mangrove it is of only seasonal importance to *Carpodectes antoniae*. However, the park is important for two other forest species, Black-cheeked Ant-tanager *Habia atrimaxillaris* and Turquoise Cotinga *Cotinga ridgwayi* (see 'Recent changes to the threatened list', above).

Carpodectes antoniae	1986	Recorded in the non-breeding season (July–February) of a number of years, most recently in 1986.

Puerto Jiménez (Puntarenas)

CR 12

Unprotected

8°33′N 83°19′W
EBA A17

Puerto Jiménez is on the western side of the Golfo Dulce, at the eastern end of the Península de Osa. The large area of mangrove in this area has been logged for the tall *Rhizophora* mangroves, leaving a predominance of *Pelliciera* species which are favoured by *Amazilia boucardi* (and may explain its local abundance in the immediate area). The lack of recent records of either species suggests that surveys are urgently needed.

Amazilia boucardi	1978	Large breeding population thought to persist, but apparently no recent documented records.
Carpodectes antoniae	1926	Judged locally common but no subsequent records.

Golfito (Puntarenas)

CR 13

Golfito Faunal Refuge (IUCN category IV), 1,350 ha

8°38′N 83°04′W
EBA A17

Golfito is on the northern side of the Golfo Dulce and comprises mangroves (in a poor state) to the north of town, and steeply sloping forests beyond (Stiles and Skutch 1989). Owing to the disturbed and fragmented nature of the habitat, the current status of both threatened species is in urgent need of assessment.

Amazilia boucardi	1986	At least three seen in March 1986, but no estimate of the size of this population.
Carpodectes antoniae	1983	Up to six seen in mangroves.

Río Coto (Puntarenas)

CR 14

Unprotected

8°33′N 83°02′W
EBA A17

The Río Coto flows into the eastern side of the Golfo Dulce, and appears to have retained a large enough expanse of mangroves at its mouth and inland (at least 10 km) to warrant some form of protection. There is an urgent need for a survey, primarily for *Carpodectes antoniae*, but also for *Amazilia boucardi* which may well be present.

Carpodectes antoniae	c.1983	This site is mentioned as one of the species' main nesting areas, although no formal assessment has been made.

■ ECUADOR

Bearded Guan *Penelope barbata*

ECUADOR is home to c.1,530 species of resident and migrant bird (Ortiz and Carrión 1991), of which c.37 (c.2.5%) are endemic to the country, 160 (c.10.5%) have restricted ranges (Stattersfield *et al.* in prep.) and 46 (3%) are threatened (Collar *et al.* 1992). This analysis has identified 50 Key Areas for the threatened birds in Ecuador (see '*Key Areas*: the book', p. 11, for criteria).

THREATENED BIRDS

Of the 46 Ecuadorean species which Collar *et al.* (1992) considered to be at risk of extinction, 11 (24%) are confined to the country (Table 1), making Ecuador fourth in rank in the Americas for numbers of threatened bird species (Collar *et al.* in press). Some 59% of the threatened species rely on wet forest, with dry forest (19.5%) and grasslands (including páramo, 9%) both significant for a number of species (Collar *et al.* in press). The mainland threatened birds are evenly distributed between the tropical, subtropical and temperate zones, where they are primarily threatened by habitat loss and alteration, and to a lesser extent by hunting and the simple fact of their restricted ranges (Collar *et al.* in press). The distributions of these threatened birds and their relationship to Endemic Bird Areas are shown in Figure 1.

KEY AREAS

The 50 Ecuadorean Key Areas would, if adequately protected, help ensure the conservation of 44 (96%) of the threatened species in the country—always accepting that important new populations and areas may yet be found. Of the 50 Key Areas, 42 are important for two or more (up to 10) threatened species, and these are therefore perhaps the most efficient areas currently known in which to conserve Ecuador's threatened birds (see 'Outlook', below). The seven areas which each harbour seven or more threatened species (see Table 3) together represent potential sanctuaries for 23 threatened species, 50% of the total number. However, as vital as these areas are for the conservation of Ecuadorean threatened species, they should not detract from the signifi-

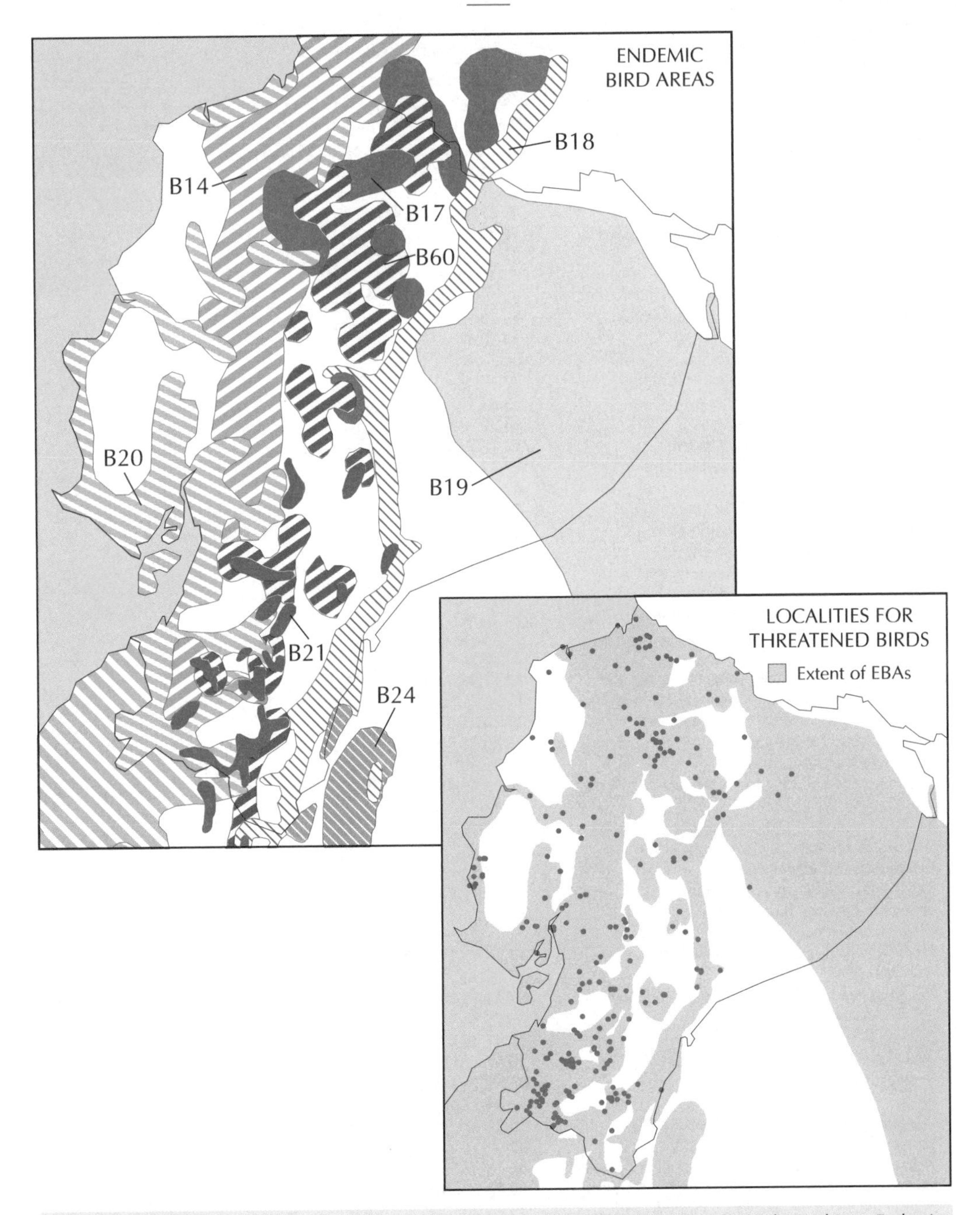

Figure 1. The localities where threatened birds have been recorded in Ecuador and their relationship to Endemic Bird Areas (EBAs).

- Most of the threatened birds in Ecuador have restricted ranges and thus occur together in various combinations within EBAs, which are listed below (figures are numbers of these species in each EBA). Although EBAs throughout the Ecuadorean Andes are extremely important for threatened birds, particularly so are the wet forests of the Chocó and Pacific-slope Andes (B14) and the Tumbesian region of western Ecuador (B20).

B14 Chocó and Pacific-slope Andes (9)
B16 Islas Galápagos (5)
B17 North Central Andean forests (1)
B18 East Andes of Ecuador and northern Peru (3)
B19 Napo and upper Amazon lowlands (0)
B20 Tumbesian western Ecuador and Peru (14)
B21 South Central Andean forests (3)
B24 Sub-Andean ridgetop forests of south-east Ecuador and northern Peru (2)
B60 Central Andean páramo (4)

Table 1. Coverage of threatened species by Key Area. Areas in bold currently have some form of protected status.

	Key Areas occupied	No. of Key Areas protected	Total nos. of Key Areas	
			Ecuador	Neotropics
Grey-backed Hawk *Leucopternis occidentalis*	03,04,**05**,09,**11**,16,**17**,18,**19**,**20**,**21**,25,32,35,36,37,38,40	6	18	20
Lesser Collared Forest-falcon *Micrastur buckleyi*	**12**,**23**	2	2	5
Plumbeous Forest-falcon *Micrastur plumbeus*	**02**	1	1	4
Bearded Guan *Penelope barbata*	32,33,**34**,44,45,**47**	2	6	12
Wattled Curassow *Crax globulosa*	—	0	0	4
Ochre-bellied Dove *Leptotila ochraceiventris*	**19**,**20**,**21**,32,36,37,38,40,41,42,43	3	11	14
Golden-plumed Parakeet *Leptosittaca branickii*	06,24,**27**,**34**,45,**47**	3	6	22
Yellow-eared Parrot *Ognorhynchus icterotis*	**07**	1	1	7
White-necked Parakeet *Pyrrhura albipectus*[E]	**23**,**47**,48	2	3	3
El Oro Parakeet *Pyrrhura orcesi*[E]	25,32	0	2	2
Spot-winged Parrotlet *Touit stictoptera*	**12**,**14**,**22**,**23**,46	4	5	8
Red-faced Parrot *Hapalopsittaca pyrrhops*	**22**,**27**,33,**34**,45,**47**	4	6	7
Banded Ground-cuckoo *Neomorphus radiolosus*	**02**	1	1	4
Turquoise-throated Puffleg *Eriocnemis godini*[E]	—	0	0	0
Black-breasted Puffleg *Eriocnemis nigrivestis*[E]	08,10	0	2	2
Hoary Puffleg *Haplophaedia lugens*	**07**,09	1	2	4
Violet-throated Metaltail *Metallura baroni*[E]	**27**	1	1	1
Neblina Metaltail *Metallura odomae*	44,**47**	1	2	3
Esmeraldas Woodstar *Acestrura berlepschi*[E]	**19**	1	1	1
Little Woodstar *Acestrura bombus*	**07**,**11**,**12**,18,**19**,**22**,29,37,38,41,46,**47**	6	12	15
Coppery-chested Jacamar *Galbula pastazae*	**12**,13,**22**,**23**,**47**	4	5	6
Chestnut-throated Spinetail *Synallaxis cherriei*	15	0	1	4
Blackish-headed Spinetail *Synallaxis tithys*	**19**,**20**,**31**,37,39,43	3	6	7
Henna-hooded Foliage-gleaner *Hylocryptus erythrocephalus*	**19**,**20**,37,38,39,40,41,42,43	2	9	12
Rufous-necked Foliage-gleaner *Syndactyla ruficollis*	26,32,37,38,41,42,43,44	0	8	15
Bicoloured Antvireo *Dysithamnus occidentalis*	**12**,**14**	2	2	4
Grey-headed Antbird *Myrmeciza griseiceps*	36,37,38,43	0	4	7
Giant Antpitta *Grallaria gigantea*	**07**,08,13	1	3	4
Slaty Becard *Pachyramphus spodiurus*	**05**,**11**,16,25,**31**,35	3	6	9
Cinnamon-breasted Tody-tyrant *Hemitriccus cinnamomeipectus*	48	0	1	4
Pacific Royal Flycatcher *Onychorhynchus occidentalis*	**11**,16,**17**,**19**,**20**,**21**,25,26,32	5	9	10
Grey-breasted Flycatcher *Lathrotriccus griseipectus*	**07**,**11**,**17**,**19**,**20**,**21**,25,26,32,35,37,38,39,41,42,43	6	16	20
White-tailed Shrike-tyrant *Agriornis andicola*	30,**47**	1	2	11
Ochraceous Attila *Attila torridus*	04,**11**,16,**17**,**19**,25,32,**34**,37,38	4	10	11
Pale-headed Brush-finch *Atlapetes pallidiceps*[E]	28,29,41	0	3	3
Tanager-finch *Oreothraupis arremonops*	**07**	1	1	6
Yellow-green Bush-tanager *Chlorospingus flavovirens*	**02**	1	1	3
Masked Mountain-tanager *Buthraupis wetmorei*	06,**12**,**22**,24,**47**	3	5	7
Orange-throated Tanager *Wetmorethraupis sterrhopteron*	49	0	1	3
Scarlet-breasted Dacnis *Dacnis berlepschi*	01,**02**,**11**	2	3	5
Saffron Siskin *Carduelis siemiradzkii*	**19**,**20**,39	2	3	4
Islas Galápagos				
Dark-rumped Petrel *Pterodroma phaeopygia*	**50**	1	1	1
Galápagos Cormorant *Nannopterum harrisi*[E]	**50**	1	1	1
Galápagos Hawk *Buteo galapagoensis*[E]	**50**	1	1	1
Floreana Mockingbird *Nesomimus trifasciatus*[E]	**50**	1	1	1
Mangrove Finch *Camarhynchus heliobates*[E]	**50**	1	1	1

[E] Endemic to Ecuador

cance of the remaining Key Areas, many of these being of major importance for the remaining 23 species (eight of which are confined to the country) (see 'Outlook').

From Tables 1 and 2 it can be seen that 15 species each occur in just one Ecuadorean Key Area; six of these species (*Metallura baroni*, *Acestrura berlepschi*, *Nannopterum harrisi*, *Buteo galapagoensis*, *Nesomimus trifasciatus* and *Camarhynchus heliobates*) are endemic to the country, and are thus totally reliant for their survival on the integrity of habitat in their respective single Key Areas (see 'Outlook'). Of almost equal concern are *Pyrrhura orcesi* and *Eriocnemis nigrivestis*, both endemic to Ecuador but known from just two Key Areas. All of the species mentioned above are represented within the Key Areas listed in 'Outlook'.

Not represented within the Ecuadorean Key Area analysis are two of the country's threatened species, as follows (see also p. 243 for a discussion of the status of *Pithys castanea*).

Crax globulosa chiefly occupies the Amazon basin in Peru and Brazil, although the species is known from the Río Negro and Río Napo in Ecuador where however, there have been no reports since 1982 despite much work in the region on cracids, and thus the identification of Key Areas at this time seems premature.

Eriocnemis godini is known with certainty from just one locality in northern Ecuador, in an area that has now been largely cleared of natural vegetation and where it has not been definitely recorded since 1850. It seems likely that it may be extinct, so the efforts currently required are field searches rather than any form of site-based conservation.

KEY AREA PROTECTION

Only 17 (34%) of Ecuador's Key Areas currently have some form of protected status, just six (12% of the total) as national parks or ecological reserves (IUCN categories I and II). The 33 unprotected Key Areas (66% of the total) require attention in the form of appropriate conservation measures if the populations of their threatened species are to survive. However, even the formally protected areas remain under threat and, in many, habitat degradation and hunting continue without check; practical—and effective—management of activities undertaken within them is thus required.

With 33 (66%) of the country's Key Areas currently unprotected, the long-term survival of a number of threatened species must be questioned. Eight Ecuadorean threatened species represented within this national analysis are not currently recorded from any protected Key Area: *Pyrrhura orcesi*, *Eriocnemis nigrivestis*, *Atlapetes pallidiceps*, *Synallaxis cherriei*, *Syndactyla ruficollis*, *Hemitriccus cinnamomeipectus* and *Wetmorethraupis sterrhopteron* (Table 1). The first three of these species are endemic to the country, and Key Areas supporting populations of any of them should be regarded as high conservation priorities (although see 'Old records and little-known birds', below). A further six threatened Ecuadorean endemics (the same six as listed under 'Key Areas', above) are known from just a single protected Key Area (Table 1) which in each case will be of major importance for the continued survival of the species in question (see 'Outlook', below).

RECENT CHANGES TO THE THREATENED LIST

With the publication of Collar *et al.* (1994), six of the 46 threatened species (*Micrastur buckleyi*, *Haplophaedia lugens*, *Metallura odomae*, *Synallaxis cherriei*, *Pachyramphus spodiurus* and *Hemitriccus cinnamomeipectus*) were dropped from the Ecuadorean threatened species list (Collar *et al.* 1992), with the following 10 added: Galápagos Penguin *Spheniscus mendiculus* (endemic), White-vented Storm-petrel *Oceanites gracilis*, Rufous-headed Chachalaca *Ortalis erythroptera*, Baudó Guan *Penelope ortoni*, Brown Wood-rail *Aramides wolfi*, Lava Gull *Larus fuliginosus* (endemic), Military Macaw *Ara militaris*, White-chested Swift *Cypseloides lemosi*, Chestnut-bellied Cotinga *Doliornis remseni* and Long-wattled Umbrellabird *Cephalopterus penduliger*; these additional species have not, however, been included in the Site Inventory (see '*Key Areas*: the book', p. 12).

The changes represent a high percentage difference (35%), although one of the species, *Doliornis remseni*, was described as new to science after the 1992 listing, and *Cypseloides lemosi* was an addition to the Ecuadorean avifauna. The other species, with the exception of *Oceanites gracilis* and *Ara militaris*, were listed as Near Threatened in Collar *et al.* (1992), with new criteria and/or new data having led to their upgrading.

These recent additions will not have any major distributional impact on the Key Area analysis, as each species occurs sympatrically with at least one other Ecuadorean threatened bird. For example, the addition of three Galápagos species simply reinforces the importance of the Galápagos Islands (EC 50), which already plays host to five threatened birds, and similarly Cabeceras de Bilsa (EC 04) supports populations of *Ortalis erythroptera* and *Cephalopterus penduliger* with the nearby Bilsa Biological Reserve (EC 05) known to hold *Aramides wolfi* and *C. penduliger*.

Table 2. Matrix of threatened species by Key Area.

	01	02	03	04	05	06	07	08	09	10	11	12	13	14	15	16	17	18	19	20	21	22	23	24	25	26	27	28	29	30	31	32	33
Leucopternis occidentalis	–	–	•	•	•	–	–	–	•	–	•	–	–	–	–	•	•	•	•	•	•	–	–	–	•	–	–	–	–	–	–	•	–
Micrastur buckleyi	–	–	–	–	–	–	–	–	–	–	–	•	–	–	–	–	–	–	–	–	–	–	•	–	–	–	–	–	–	–	–	–	–
Micrastur plumbeus	–	•	–	–	–	–	–	–	–	–	–	–	–	–	–	–	–	–	–	–	–	–	–	–	–	–	–	–	–	–	–	–	–
Penelope barbata	–	–	–	–	–	–	–	–	–	–	–	–	–	–	–	–	–	–	–	–	–	–	–	–	–	–	–	–	–	–	–	•	•
Crax globulosa	–	–	–	–	–	–	–	–	–	–	–	–	–	–	–	–	–	–	–	–	–	–	–	–	–	–	–	–	–	–	–	–	–
Leptotila ochraceiventris	–	–	–	–	–	–	–	–	–	–	–	–	–	–	–	–	–	–	•	•	•	–	–	–	–	–	–	–	–	–	–	•	–
Leptosittaca branickii	–	–	–	–	–	•	–	–	–	–	–	–	–	–	–	–	–	–	–	–	–	–	–	•	–	–	•	–	–	–	–	–	–
Ognorhynchus icterotis	–	–	–	–	–	–	•	–	–	–	–	–	–	–	–	–	–	–	–	–	–	–	–	–	–	–	–	–	–	–	–	–	–
Pyrrhura albipectus	–	–	–	–	–	–	–	–	–	–	–	–	–	–	–	–	–	–	–	–	–	–	•	–	–	–	–	–	–	–	–	–	–
Pyrrhura orcesi	–	–	–	–	–	–	–	–	–	–	–	–	–	–	–	–	–	–	–	–	–	–	–	–	•	–	–	–	–	–	–	•	–
Touit stictoptera	–	–	–	–	–	–	–	–	–	–	–	•	–	•	–	–	–	–	–	–	–	•	•	–	–	–	–	–	–	–	–	–	–
Hapalopsittaca pyrrhops	–	–	–	–	–	–	–	–	–	–	–	–	–	–	–	–	–	–	–	–	–	•	–	–	–	–	•	–	–	–	–	–	•
Neomorphus radiolosus	–	•	–	–	–	–	–	–	–	–	–	–	–	–	–	–	–	–	–	–	–	–	–	–	–	–	–	–	–	–	–	–	–
Eriocnemis godini	–	–	–	–	–	–	–	–	–	–	–	–	–	–	–	–	–	–	–	–	–	–	–	–	–	–	–	–	–	–	–	–	–
Eriocnemis nigrivestis	–	–	–	–	–	–	–	•	–	•	–	–	–	–	–	–	–	–	–	–	–	–	–	–	–	–	–	–	–	–	–	–	–
Haplophaedia lugens	–	–	–	–	–	–	•	–	•	–	–	–	–	–	–	–	–	–	–	–	–	–	–	–	–	–	–	–	–	–	–	–	–
Metallura baroni	–	–	–	–	–	–	–	–	–	–	–	–	–	–	–	–	–	–	–	–	–	–	–	–	–	–	•	–	–	–	–	–	–
Metallura odomae	–	–	–	–	–	–	–	–	–	–	–	–	–	–	–	–	–	–	–	–	–	–	–	–	–	–	–	–	–	–	–	–	–
Acestrura berlepschi	–	–	–	–	–	–	–	–	–	–	–	–	–	–	–	–	–	–	•	–	–	–	–	–	–	–	–	–	–	–	–	–	–
Acestrura bombus	–	–	–	–	–	–	•	–	–	–	•	•	–	–	–	–	–	•	•	–	–	•	–	–	–	–	–	–	•	–	–	–	–
Galbula pastazae	–	–	–	–	–	–	–	–	–	–	–	•	•	–	–	–	–	–	–	–	–	•	•	–	–	–	–	–	–	–	–	–	–
Synallaxis cherriei	–	–	–	–	–	–	–	–	–	–	–	–	–	–	•	–	–	–	–	–	–	–	–	–	–	–	–	–	–	–	–	–	–
Synallaxis tithys	–	–	–	–	–	–	–	–	–	–	–	–	–	–	–	–	–	–	•	•	–	–	–	–	–	–	–	–	–	–	•	–	–
Hylocryptus erythrocephalus	–	–	–	–	–	–	–	–	–	–	–	–	–	–	–	–	–	–	•	•	–	–	–	–	–	–	–	–	–	–	–	–	–
Syndactyla ruficollis	–	–	–	–	–	–	–	–	–	–	–	–	–	–	–	–	–	–	–	–	–	–	–	–	–	•	–	–	–	–	–	•	–
Dysithamnus occidentalis	–	–	–	–	–	–	–	–	–	–	–	•	–	•	–	–	–	–	–	–	–	–	–	–	–	–	–	–	–	–	–	–	–
Myrmeciza griseiceps	–	–	–	–	–	–	–	–	–	–	–	–	–	–	–	–	–	–	–	–	–	–	–	–	–	–	–	–	–	–	–	–	–
Grallaria gigantea	–	–	–	–	–	–	•	•	–	–	–	–	•	–	–	–	–	–	–	–	–	–	–	–	–	–	–	–	–	–	–	–	–
Pachyramphus spodiurus	–	–	–	–	•	–	–	–	–	–	•	–	–	–	–	•	–	–	–	–	–	–	–	–	•	–	–	–	–	–	•	–	–
Hemitriccus cinnamomeipectus	–	–	–	–	–	–	–	–	–	–	–	–	–	–	–	–	–	–	–	–	–	–	–	–	–	–	–	–	–	–	–	–	–
Onychorhynchus occidentalis	–	–	–	–	–	–	–	–	–	–	•	–	–	–	–	•	•	–	•	•	•	–	–	–	•	•	–	–	–	–	–	•	–
Lathrotriccus griseipectus	–	–	–	–	–	–	•	–	–	–	•	–	–	–	–	–	•	–	•	•	•	–	–	–	•	•	–	–	–	–	–	•	–
Agriornis andicola	–	–	–	–	–	–	–	–	–	–	–	–	–	–	–	–	–	–	–	–	–	–	–	–	–	–	–	–	–	•	–	–	–
Attila torridus	–	–	–	•	–	–	–	–	–	–	•	–	–	–	–	•	•	–	•	–	–	–	–	–	•	–	–	–	–	–	–	•	–
Atlapetes pallidiceps	–	–	–	–	–	–	–	–	–	–	–	–	–	–	–	–	–	–	–	–	–	–	–	–	–	–	–	•	•	–	–	–	–
Oreothraupis arremonops	–	–	–	–	–	–	•	–	–	–	–	–	–	–	–	–	–	–	–	–	–	–	–	–	–	–	–	–	–	–	–	–	–
Chlorospingus flavovirens	–	•	–	–	–	–	–	–	–	–	–	–	–	–	–	–	–	–	–	–	–	–	–	–	–	–	–	–	–	–	–	–	–
Buthraupis wetmorei	–	–	–	–	–	•	–	–	–	–	–	•	–	–	–	–	–	–	–	–	–	•	–	•	–	–	–	–	–	–	–	–	–
Wetmorethraupis sterrhopteron	–	–	–	–	–	–	–	–	–	–	–	–	–	–	–	–	–	–	–	–	–	–	–	–	–	–	–	–	–	–	–	–	–
Dacnis berlepschi	•	•	–	–	–	–	–	–	–	–	•	–	–	–	–	–	–	–	–	–	–	–	–	–	–	–	–	–	–	–	–	–	–
Carduelis siemiradzkii	–	–	–	–	–	–	–	–	–	–	–	–	–	–	–	–	–	–	•	•	–	–	–	–	–	–	–	–	–	–	–	–	–
Islas Galápagos																																	
Pterodroma phaeopygia	–	–	–	–	–	–	–	–	–	–	–	–	–	–	–	–	–	–	–	–	–	–	–	–	–	–	–	–	–	–	–	–	–
Nannopterum harrisi	–	–	–	–	–	–	–	–	–	–	–	–	–	–	–	–	–	–	–	–	–	–	–	–	–	–	–	–	–	–	–	–	–
Buteo galapagoensis	–	–	–	–	–	–	–	–	–	–	–	–	–	–	–	–	–	–	–	–	–	–	–	–	–	–	–	–	–	–	–	–	–
Nesomimus trifasciatus	–	–	–	–	–	–	–	–	–	–	–	–	–	–	–	–	–	–	–	–	–	–	–	–	–	–	–	–	–	–	–	–	–
Camarhynchus heliobates	–	–	–	–	–	–	–	–	–	–	–	–	–	–	–	–	–	–	–	–	–	–	–	–	–	–	–	–	–	–	–	–	–
No. of species	1	4	1	2	2	2	6	2	2	1	7	6	2	2	1	4	4	2	10	7	4	5	4	2	6	3	3	1	2	1	2	8	2

Table 2 (cont.)

34	35	36	37	38	39	40	41	42	43	44	45	46	47	48	49	50	No. of areas
–	●	●	●	●	–	●	–	–	–	–	–	–	–	–	–	–	18
–	–	–	–	–	–	–	–	–	–	–	–	–	–	–	–	–	2
–	–	–	–	–	–	–	–	–	–	–	–	–	–	–	–	–	1
●	–	–	–	–	–	–	–	–	–	●	●	–	●	–	–	–	6
–	–	–	–	–	–	–	–	–	–	–	–	–	–	–	–	–	0
–	–	●	●	●	–	●	●	●	●	–	–	–	–	–	–	–	11
●	–	–	–	–	–	–	–	–	–	●	●	–	●	–	–	–	6
–	–	–	–	–	–	–	–	–	–	–	–	–	–	–	–	–	1
–	–	–	–	–	–	–	–	–	–	–	–	–	●	●	–	–	3
–	–	–	–	–	–	–	–	–	–	–	–	–	–	–	–	–	2
–	–	–	–	–	–	–	–	–	–	–	–	●	–	–	–	–	5
●	–	–	–	–	–	–	–	–	–	–	●	–	●	–	–	–	6
–	–	–	–	–	–	–	–	–	–	–	–	–	–	–	–	–	1
–	–	–	–	–	–	–	–	–	–	–	–	–	–	–	–	–	0
–	–	–	–	–	–	–	–	–	–	–	–	–	–	–	–	–	2
–	–	–	–	–	–	–	–	–	–	–	–	–	–	–	–	–	2
–	–	–	–	–	–	–	–	–	–	–	–	–	–	–	–	–	1
–	–	–	–	–	–	–	–	–	–	●	–	–	●	–	–	–	2
–	–	–	–	–	–	–	–	–	–	–	–	–	–	–	–	–	1
–	–	–	●	●	–	–	●	–	–	–	–	●	●	–	–	–	12
–	–	–	–	–	–	–	–	–	–	–	–	–	●	–	–	–	5
–	–	–	–	–	–	–	–	–	–	–	–	–	–	–	–	–	1
–	–	–	●	–	●	–	–	–	●	–	–	–	–	–	–	–	6
–	–	–	●	●	●	●	●	●	●	–	–	–	–	–	–	–	9
–	–	–	●	●	–	–	●	●	●	●	–	–	–	–	–	–	7
–	–	–	–	–	–	–	–	–	–	–	–	–	–	–	–	–	2
–	–	●	●	●	–	–	–	–	●	–	–	–	–	–	–	–	4
–	–	–	–	–	–	–	–	–	–	–	–	–	–	–	–	–	3
–	●	–	–	–	–	–	–	–	–	–	–	–	–	–	–	–	6
–	–	–	–	–	–	–	–	–	–	–	–	–	–	●	–	–	1
–	–	–	–	–	–	–	–	–	–	–	–	–	–	–	–	–	9
–	●	–	●	●	●	–	●	●	●	–	–	–	–	–	–	–	16
–	–	–	–	–	–	–	–	–	–	–	–	–	●	–	–	–	2
●	–	–	●	●	–	–	–	–	–	–	–	–	–	–	–	–	10
–	–	–	–	–	–	–	●	–	–	–	–	–	–	–	–	–	3
–	–	–	–	–	–	–	–	–	–	–	–	–	–	–	–	–	1
–	–	–	–	–	–	–	–	–	–	–	–	–	–	–	–	–	1
–	–	–	–	–	–	–	–	–	–	–	–	–	●	–	–	–	5
–	–	–	–	–	–	–	–	–	–	–	–	–	–	–	●	–	1
–	–	–	–	–	–	–	–	–	–	–	–	–	–	–	–	–	3
–	–	–	–	–	●	–	–	–	–	–	–	–	–	–	–	–	3
–	–	–	–	–	–	–	–	–	–	–	–	–	–	–	–	●	1
–	–	–	–	–	–	–	–	–	–	–	–	–	–	–	–	●	1
–	–	–	–	–	–	–	–	–	–	–	–	–	–	–	–	●	1
–	–	–	–	–	–	–	–	–	–	–	–	–	–	–	–	●	1
–	–	–	–	–	–	–	–	–	–	–	–	–	–	–	–	●	1
4	3	3	9	8	4	3	6	4	6	4	3	2	9	2	1	5	

OLD RECORDS AND LITTLE-KNOWN BIRDS

Our knowledge of threatened birds in Ecuador is relatively good, there being recent records of most species from somewhere within their ranges, such that we are able to make informed recommendations for their conservation through the identification of appropriate Key Areas. The exceptions are those species known only from old specimen records or chance observations (from which there can be no guarantees of a current viable population), and these include the following.

Micrastur plumbeus, known in Ecuador only from specimens obtained before 1940 at four localities; throughout its range the species is virtually unknown in life.

Crax globulosa (see 'Key Areas', above).

Ognorhynchus icterotis, known with certainty from very few localities and even fewer recent reports, the centre of its distribution being apparently in Colombia.

Eriocnemis godini (see 'Key Areas', above).

Atlapetes pallidiceps, despite specific searches last recorded (at one of its five known localities) in 1969, and feared extinct.

OUTLOOK

This analysis has identified 50 Key Areas, each of which is of major importance for the conservation of threatened birds in Ecuador. However, certain Key Areas stand out as top priorities (Table 3); they comprise those which each host populations of seven or more threatened species, and also those for species which are confined to Ecuador and (effectively) known from just one area.

The 10 areas mentioned in Table 3 are the most important Key Areas for a number of threatened species, and represent populations of 31 (67%) of the Ecuadorean threatened birds. These and the remaining Key Areas (which are individually extremely important for many threatened Ecuadorean endemics: see 'Key Areas' and 'Key Area protection', above) all need some degree of conservation action if the populations of their threatened birds are to remain viable. The seasonal movements of some west Ecuadorean species such as *Leptotila ochraceiventris* and *Lathrotriccus griseipectus* should be considered in any conservation action: only a complementary set of Key Areas will ensure their continued survival.

Although our knowledge of Ecuador's threatened birds is comparatively good, for some areas there is still a requirement to confirm the continued existence or assess the population of a threatened species (see, for example, 'Old records and little-known birds',

Table 3. Top Key Areas in Ecuador. Those with the area number in bold currently have some form of protected status.

Key Area	No. of threatened spp.	Comments
08 Volcán Pichincha	2	The most important area known for *Eriocnemis nigrivestis.*
11 Río Palenque	7	An important protected area for at least five of the threatened species known to occur.
19 Machalilla	10	Arguably one of the two most important Key Areas in Ecuador (the other being Podocarpus), and the only known area for *Acestrura berlepschi.*
20 Cerro Blanco, Guayaquil	7	An important protected area for all seven threatened species known to occur, each of which is, however, found in Machalilla.
27 Cajas and Río Mazan	3	The only known area for *Metallura baroni* and upon which its survival is totally dependent.
32 Buenaventura	8	An important area for at least seven of the threatened species known to occur, but especially for *Pyrrhura orcesi.*
37 Alamor	9	An important area for at least six of the threatened species known to occur.
38 Celica	8	An important area for at least six of the threatened species known to occur.
47 Podocarpus	9	Arguably one of the two most important Key Areas in Ecuador (the other being Machalilla), being especially so for *Pyrrhura albipectus* and *Metallura odomae.*
50 Islas Galápagos	5	Four of the threatened species are endemic to this Key Area, *Pterodroma phaeopygia* also breeding in the Hawaiian islands.

above), and for others, especially the few areas based on old records, the need is for locating and defining the extent of suitable habitat in the vicinity of the Key Area. A more comprehensive and detailed set of site-based recommendations for the conservation of endemic and threatened birds in south-west Ecuador has been documented by Best and Kessler (1995).

DATA SOURCES

The above introductory text and the Site Inventory (below) were compiled from information supplied by K. S. Berg, B. J. Best, P. Boesman, M. Catsis, R. P. Clay, A. Drewitt, S. N. G. Howell, N. Krabbe, R. Naveen, M. Pearman, R. S. Ridgely, M. B. Robbins, G. Speight, J. Tobias, E. P. Toyne, M. Whittingham, R. S. R. Williams, as well as from the following references.

Berg, K. S. (1994) New and interesting records of birds from a dry forest reserve in south-west Ecuador. *Cotinga* 2: 14–19.

Best, B. J. ed. (1992) *The threatened forests of south-western Ecuador.* Leeds, U.K.: Biosphere Publications.

Best, B. J. and Kessler, M. (1995) *Biodiversity and conservation in Tumbesian Ecuador and Peru.* Cambridge, U.K.: BirdLife International.

Chapman, F. M. (1926) The distribution of bird-life in Ecuador. *Bull. Amer. Mus. Nat. Hist.* 55.

Clay, R. P., Jack, S. R. and Vincent, J. P. (1994) A survey of the birds and large mammals of the proposed Jatun Sacha Bilsa Biological Reserve, north-western Ecuador. Unpublished report.

Collar, N. J., Gonzaga, L. P., Krabbe, N., Madroño Nieto, A., Naranjo, L. G., Parker, T. A. and Wege, D. C. (1992) *Threatened birds of the Americas: the ICBP/IUCN Red Data Book* (Third edition, part 2). Cambridge, U.K.: International Council for Bird Preservation.

Collar N. J., Crosby, M. J. and Stattersfield, A. J. (1994) *Birds to watch 2: the world list of threatened birds.* Cambridge, U.K.: BirdLife International (BirdLife Conservation Series no.4).

Collar, N. J., Wege, D. C. and Long, A. J. (in press) Patterns and causes of endangerment in the New World avifauna. In J. V. Remsen, ed. *Natural history and conservation of Neotropical birds.* American Ornithologists' Union (Orn. Monogr.).

Evans, R. J., ed. (1988) An ornithological survey in the province of Esmeraldas in north-west Ecuador—August 1986. Durham University expedition. Unpublished report.

Groombridge, B. (1993) *1994 IUCN Red List of threatened animals.* Gland, Switzerland, and Cambridge, U.K.: International Union for Conservation of Nature and Natural Resources.

IUCN (1992) *Protected areas of the world: a review of national systems. Volume 4: Nearctic and Neotropical.* Gland, Switzerland and Cambridge, U.K.: International Union for Conservation of Nature and Natural Resources.

Kirwan, G., Marlow, T. and Coopmans, P. (in prep.) A review of avifaunal records from Mindo, Pichincha province, north-west Ecuador. *Cotinga.*

Krabbe, N. (1994) The White-tailed Shrike-tyrant, an extinction prone species? *Cotinga* 1: 33–34.

Krabbe, N., Braun, M. J., Jácome, M., Robbins, M. B., Schjørring, S. and Sornoza M., F. (1994a) Black-breasted Puffleg found: extant but seriously threatened. *Cotinga* 1: 8–9.

KRABBE, N., DESMET, G., GREENFIELD, P., JÁCOME, M., MATHEUS, J. C. AND SORNOZA M., F. (1994b) Giant Antpitta *Grallaria gigantea*. *Cotinga* 2: 32–34.

ORTIZ C., F. AND CARRIÓN, J. M. (1991) *Introducción a las aves del Ecuador*. Quito, Ecuador: Fundación Ecuatoriana para la Conservación y el Desarrollo Sostenible.

PARKER, T. A. AND CARR, J. L., EDS. (1992) *Status of forest remnants in the Cordillera de la Costa and adjacent areas of southwestern Ecuador (Rapid Assessment Program)*. Washington, D.C.: Conservation International.

PAYNTER, R. A. (1993) *Ornithological gazetteer of Ecuador*. Second edition. Cambridge, Mass.: Museum of Comparative Zoology.

POULSEN, M. AND WEGE, D. (1994) Coppery-chested Jacamar *Galbula pastazae*. *Cotinga* 2: 60–62.

RASMUSSEN, J. F. AND RAHBEK, C. (1994) *Aves del Parque Nacional Podocarpus: una lista anotada. Birds of Podocarpus National Park: an annotated checklist*. Quito, Ecuador: CECIA.

ROBBINS, M. B., ROSENBERG, G. H. AND SORNOZA M., F. (1994) A new species of cotinga (Cotingidae: *Doliornis*) from the Ecuadorian Andes, with comments on plumage sequences in *Doliornis* and *Ampelion*. *Auk* 111: 1–7.

STATTERSFIELD, A. J., CROSBY, M. J., LONG, A. J. AND WEGE, D. C. (in prep.) *A global directory of Endemic Bird Areas*. Cambridge, U.K.: BirdLife International (BirdLife Conservation Series).

WILLIAMS, R. S. R. AND TOBIAS, J. A. (1994) *The conservation of Ecuador's threatened avifauna: final report of the Amaluza projects, 1990–1991*. Cambridge, U.K.: BirdLife International (Study Report 60).

ZIMMER, J. T. (1953) Studies of Peruvian birds, 63. The hummingbird genera *Oreonympha*, *Schistes*, *Heliothryx*, *Loddigesia*, *Heliomaster*, *Rhodopis*, *Thaumastura*, *Calliphlox*, *Myrtis*, *Myrmia*, and *Acestrura*. *Amer. Mus. Novit.* 1604.

SITE INVENTORY

San Lorenzo (Esmeraldas)

EC 01

1°17′N 78°50′W
EBA B14

Unprotected

The town of San Lorenzo is located at sea-level in extreme northern coastal Esmeraldas, only 20 km from the border with Colombia.

Dacnis berlepschi	1990	One bird recorded south of the town.

Awa (Carchi/Esmeraldas/Imbabura)

EC 02

1°00′N 78°30′W
EBA B14

Awa Forest Reserve Zone (IUCN category VII), 101,000 ha

The reserve zone abuts the Colombian border along the Río Mira drainage. Outside the area, but within a few kilometres of the southern edge, are a number of important localities such as Ventanas, Alto Tambo, El Placer, Lita and Achotal. Most records are from outside the reserve between sea-level and 650 m (little work has been carried out within the reserve), although the sightings west of Maldonado were apparently in (or very near) the reserve.

Micrastur plumbeus	1987	Two collected at 670 m.
Neomorphus radiolosus	1992	One recorded daily near Ventanas.
Chlorospingus flavovirens	1987	Recent sightings, including two birds in 1987, come from the ridges at El Placer (also M. B. Robbins *in litt.* 1994).
Dacnis berlepschi	1986	At least two seen at Ventanas.

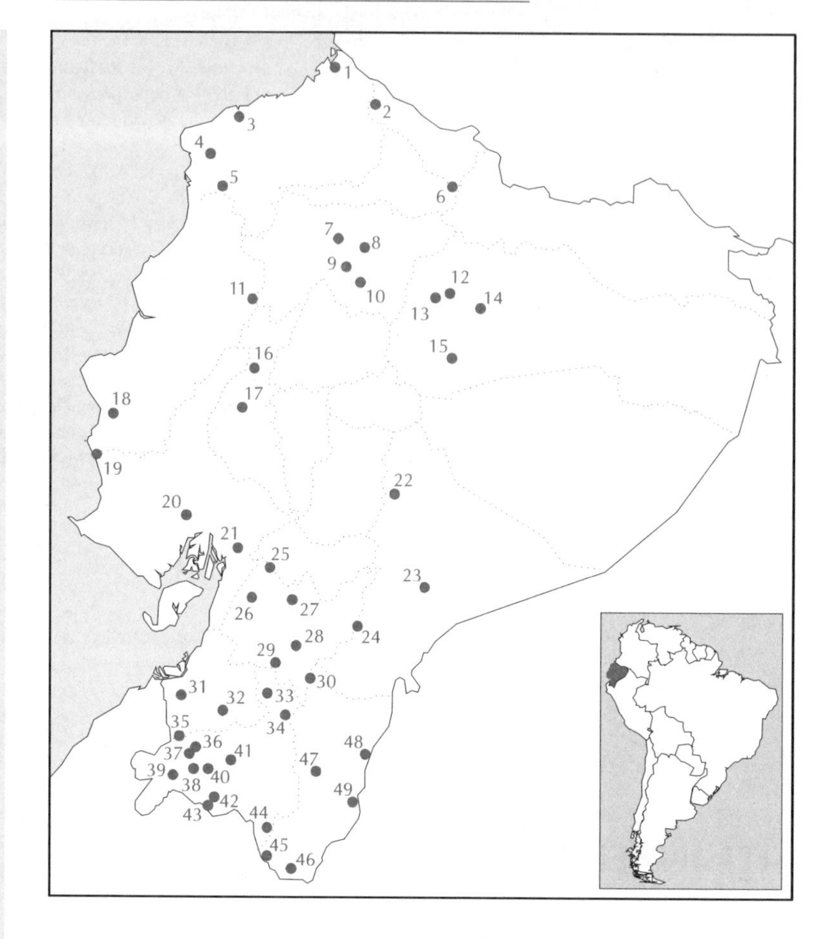

Figure 2. Key Areas in Ecuador.

01 San Lorenzo
02 Awa
03 Cerro Mutiles
04 Cabeceras de Bilsa
05 Bilsa Biological Reserve
06 Cerro Mongus
07 Mindo
08 Volcán Pichincha
09 Chiriboga
10 Atacazo
11 Río Palenque
12 Cayambe–Coca
13 Cordillera de Guacamayo and Hacienda Aragón
14 Volcán Sumaco
15 Archidona
16 Hacienda Pacaritambo
17 Jauneche
18 Cerro Achi
19 Machalilla
20 Cerro Blanco, Guayaquil
21 Manglares-Churute
22 Sangay
23 Cordillera de Cutucú
24 Zapote Najda mountains
25 Manta Real
26 San Miguel de Azuay
27 Cajas and Río Mazan
28 Girón
29 Valle de Yunguilla
30 Bestión
31 Arenillas
32 Buenaventura
33 Selva Alegre
34 Acanama–Saraguro
35 Puyango
36 Vicentino
37 Alamor
38 Celica
39 Sabanillas
40 El Empalme
41 Catacocha
42 Sozoranga
43 Tambo Negro
44 Angashcola
45 Lagunillas
46 Zumba
47 Podocarpus
48 Chinapinza
49 Shaime
50 Islas Galápagos (not shown)

Cerro Mutiles (Esmeraldas)

EC 03

Unprotected

0°54′N 79°37′W
EBA B20

This hill (200–300 m) lies at the northernmost end of the Cordillera de la Costa, a few kilometres south-east of Esmeraldas (Parker and Carr 1992). The southern slope of the hill, mostly covered in fairly typical semi-deciduous moist forest, is maintained by the Jardín Tropical of the Universidad Técnica 'Luis Vargas Torres' (Parker and Carr 1992). Up to eight pairs of Rufous-headed Chachalaca *Ortalis erythroptera* were found in an area of c.2 km² (Parker and Carr 1992; see 'Recent changes to the threatened list', above).

Leucopternis occidentalis	1991	One sight record (Parker and Carr 1992).

Cabeceras de Bilsa (Esmeraldas)

EC 04

Unprotected

0°37′N 79°51′W
EBAs B14, B20

This ridge system (600–800 m) east of the village of San José de Bilsa, and near the headwaters of the Río Bilsa in the Montañas de Muisne (south of Esmeraldas), comprises what was until very recently pristine, extremely wet forest. The bird community is rich and includes populations of *Ortalis erythroptera* and Long-wattled Umbrellabird *Cephalopterus penduliger* (Parker and Carr 1992; see 'Recent changes to the threatened list', above).

Leucopternis occidentalis	1991	Sight records in 1991 only (Parker and Carr 1992).
Attila torridus	1991	Sight records in 1991 only (Parker and Carr 1992).

Bilsa Biological Reserve (Esmeraldas)

EC 05

Jatun Sacha Bilsa Biological Reserve (private), 800 ha

0°22′N 79°45′W
EBAs B14, B20

The reserve lies c.30 km south of the Cabeceras de Bilsa, in the Mache mountains between Muisne on the coast and Quinende inland (Clay *et al.* 1994). It is situated at 400–700 m, and comprises extremely wet pristine forest in two watersheds, those of the Ríos Dogola and Río Cube (Clay *et al.* 1994). The area has a rich bird community, including populations of Brown Wood-rail *Aramides wolfi* and *Cephalopterus penduliger* (Clay *et al.* 1994; see 'Recent changes to the threatened list', above). A campaign is under way to expand the reserve to 5,000 ha (*Cotinga* 1995, 3: 70–71).

Leucopternis occidentalis	1994	One seen over the forest (Clay *et al.* 1994).
Pachyramphus spodiurus	1994	Recorded from the reserve (R. P. Clay *in litt.* 1995).

Cerro Mongus (Carchi)

EC 06

Unprotected

0°22′N 77°52′W
EBA B60

Cerro Mongus, in the East Andes of south-east Carchi, rises to over 3,500 m. An initiative to protect the forested watershed has been instigated by villagers (R. S. R. Williams *in litt.* 1995). Present in this area in 1992 and 1994 was the recently described Chestnut-bellied Cotinga *Doliornis remseni* (Robbins *et al.* 1994, A. Lewis *in litt.* 1995; see 'Recent changes to the threatened list', above).

Leptosittaca branickii	1992	Recorded in March 1992 (M. B. Robbins *in litt.* 1994).
Buthraupis wetmorei	1994	Uncommon on the western slope in 1992 (M. B. Robbins *in litt.* 1994) but in 1994 judged more common than at Podocarpus (EC 47) (R. S. R. Williams *in litt.* 1995).

Mindo (Pichincha)

EC 07

Mindo Nambillo Protection Forest (IUCN category VIII), 19,200 ha

0°02′S 78°48′W
EBAs B14, B20, B21

The Mindo area is c.25 km north-west of Quito on the western slope of Volcán Pichincha at c.1,500 m. The area includes the Ríos Cinto and Nambillo, and Tandayapa. Much forest clearance has occurred around Mindo although most of this has been south and east of the village (Kirwan *et al.* in prep.).

Ognorhynchus icterotis	1970	A caged bird on sale in a local market was supposedly captured near Mindo.
Haplophaedia lugens	1990	Numerous recent sightings.
Acestrura bombus	1987	A specimen collected below Hacienda Santa Rosa.
Grallaria gigantea	1994	Birds seen on a number of occasions at c.1,300 m, south-east of Mindo (S. N. G. Howell *in litt.* 1994, Krabbe *et al.* 1994b).

cont.

Lathrotriccus griseipectus	1914	One collected at 1,830 m.
Oreothraupis arremonops	1987	Collected at 1,200 m in c.1900 and 1939; birds seen above Tandayapa in 1987.

Volcán Pichincha (Pichincha)

EC 08

Unprotected

0°06′S 78°35′W
EBA B60

This mountain in north-west Ecuador has two main summits with numerous ridge-crests, of which the most important in this context are those on the northern slope, such as Cerro Pugsi, Loma Gramalote, Frutillas, Yanococha and Cerro Alaspungo. In the humid temperate zone, especially on the ridge-crests, much of the natural vegetation has been completely cleared.

Eriocnemis nigrivestis	1993	Three seen in February–March 1993 at 3,300 m; other records from the early 1980s (Krabbe *et al.* 1994a).
Grallaria gigantea	1958	Collected at a number of localities (2,000–3,350 m) in 1936–1937, and on Cerro El Castillo in 1958.

Chiriboga (Pichincha)

EC 09

Unprotected

0°15′S 78°44′W
EBAs B14, B20

Chiriboga is at c.1,500 m on the Pacific slope of the Andes along the old Quito–Santo Domingo de los Colorados road. About 800 ha of primary subtropical forest is 'protected' by a private landowner.

Leucopternis occidentalis	1986	One record of a bird apparently of this species (Evans 1988).
Haplophaedia lugens	1986	Recorded in 1980 (at 1,900 m) and 1986.

Atacazo (Pichincha)

EC 10

Unprotected

0°22′S 78°37′W
EBA B60

Atacazo is a mountain rising to c.4,400 m, located c.30 km south-west of Quito, and therefore a similar distance from Volcán Pichincha (EC 08), the only other area known for the species below.

Eriocnemis nigrivestis	1983	Specimens taken in 1898, and a female (apparently this species) seen at 3,500 m in October 1983.

Río Palenque (Pichincha)

EC 11

Río Palenque Scientific Centre (private), 167 ha

0°30′S 79°30′W
EBAs B14, B20

The scientific centre is in the lowlands (c.200 m) of western Ecuador, near the Manabí/Los Ríos border. The private reserve has only 97 ha of pristine forest still standing.

Leucopternis occidentalis	1977	Formerly not rare, but may no longer be present.
Acestrura bombus	c.1991	A population of unknown size was recorded recently.
Onychorhynchus occidentalis	1994	Two found in a mixed-species flock at the edge of forest (M. Catsis *in litt.* 1994).
Lathrotriccus griseipectus	c.1989	Several recent sightings.
Attila torridus	1994	Recent records include a pair tape-recorded (A. Lewis *in litt.* 1995).
Pachyramphus spodiurus	c.1992	Recently seen in the reserve.
Dacnis berlepschi	1990	One sight record.

EC 12

Cayambe–Coca (Napo/Imbabura/Pichincha)

0°27′S 77°53′W
EBA B18

Cayambe–Coca Ecological Reserve (IUCN category I), 403,103 ha

The reserve is north-east of Quito and includes numerous localities along its southern and eastern boundaries, such as Papallacta, Baeza and San Rafael. The elevation ranges from 800 to 5,900 m, thus covering all major vegetation zones.

Micrastur buckleyi	c.1923	One collected below Baeza.
Touit stictoptera	1979	A flock of six seen at San Rafael.
Acestrura bombus	1877	Two collected.
Galbula pastazae	1994	Recorded most recently at San Rafael falls, with three there in 1993 and one in 1994 (G. Speight *in litt.* 1994, A. Drewitt *in litt.* 1995).
Dysithamnus occidentalis	1960	One collected, apparently from Baeza.
Buthraupis wetmorei	1992	Recorded at Papallacta Pass (*Cotinga* 1994, 1: 27).

EC 13

Cordillera de Guacamayo and Hacienda Aragón (Napo)

0°29–43′S 78°00′W
EBA B17

Unprotected

The Cordillera de Guacamayo (which rises to c.2,500 m) is a spur extending south-west from Volcán Sumaco (EC 14) to the main Andean range in western Napo. Hacienda Aragón lies in the upper Río Cosanga drainage on the northern slope of the cordillera, with wet mossy forest at 2,200–2,300 m c.1–3 km south of the hacienda. The forest around the hacienda itself has mostly been cleared, and much that remains is secondary (Krabbe *et al.* 1994b), although other areas of forest persist on the southern slope and along the ridge trail. Birds which are almost certainly *Grallaria alleni* (see p. XXX) have been seen and tape-recorded at 2,200 m along the ridge trail (P. Coopmans *in litt.* 1995, A. Lewis *in litt.* 1995), and confirmation would make this Key Area the only one currently known for the species.

Galbula pastazae	1994	A female seen at 2,000 m on the southern slope (A. Lewis *in litt.* 1995).
Grallaria gigantea	1993	Three collected at 2,200–2,300 m, 1992–1993 (Krabbe *et al.* 1994b).

EC 14

Volcán Sumaco (Napo)

0°34′S 77°38′W
EBA B18

Sumaco Protection Forest (IUCN category VIII), 100,045 ha

This somewhat isolated peak (3,900 m) in western Napo, to the east of the main Andean range, is still mostly pristine, with an intact corridor of habitats from the lowlands to above the treeline. Cultivation of naranjilla by the increasing human population poses a serious threat to the forest at c.1,000 m and above.

Touit stictoptera	1990	Flocks of c.10 seen along the new Hollín–Loreto road from the late 1980s.
Dysithamnus occidentalis	1990	A reasonable density of birds.

EC 15

Archidona (Napo)

0°57′S 77°52′W
EBA B17

Unprotected

The town is at c.700 m in the Amazonian foothill zone of eastern Ecuador. An area of dense humid secondary woodland to the south-east appears to support a small population of the following species.

Synallaxis cherriei	1991	A pair tape-recorded in 1989.

Hacienda Pacaritambo (Los Ríos/Guayas)

EC 16

Unprotected

1°02′S 79°29′W
EBA B20

Hacienda Pacaritambo is in the lowlands of northern Guayas, close to Quevedo (in Los Ríos), and embraces (or embraced) a small area of tropical deciduous forest.

Leucopternis occidentalis	c.1884	One collected before 1884.
Onychorhynchus occidentalis	1963	Recorded in May 1962 and February 1963.
Attila torridus	1962	One record.
Pachyramphus spodiurus	c.1963	Recorded once during 1962–1963.

Jauneche (Los Ríos)

EC 17

Jauneche Biological Research Station (private), 138 ha

1°20′S 79°35′W
EBA B20

The scientific station, in north-west Los Ríos, is one of the only surviving examples of moist forest in lowland south-west Ecuador. The reserve, managed by the University of Guayaquil, is an extremely important but all-too-small island of forest, surrounded by densely settled agricultural land.

Leucopternis occidentalis	1991	Recorded as rare.
Onychorhynchus occidentalis	1991	Recorded as fairly common (Williams and Tobias 1994).
Lathrotriccus griseipectus	1991	Many sightings from July, also September–October.
Attila torridus	1991	One sight record.

Cerro Achi (Manabí)

EC 18

Unprotected

1°23′S 80°38′W
EBA B20

This small, isolated ridge c.15 km long forms part of the Cordillera de la Costa in western Ecuador and lies between the towns of Puerto Cayo and Jipijapa, north of Machalilla National Park. It reaches 600 m and comprises dry to moist forest.

Leucopternis occidentalis	1991	One record.
Acestrura bombus	1991	A population of unknown size recorded at 550–600 m.

Machalilla (Guayas/Manabí)

EC 19

Machalilla National Park (IUCN category II), 55,000 ha

1°42′S 80°46′W
EBA B20

Sitting on the border between north-west Guayas and southern Manabi in the middle portion of the Cordillera de la Costa, the park covers the area from the coast to the highest ridges of the cordillera at 800 m, and, as the only national park in western Ecuador, covers important but small areas of fog forest and dry forest. Cerro San Sebastián is biologically the richest part of the park, which is threatened from continuing habitat destruction (Parker and Carr 1992).

Leucopternis occidentalis	1993	At least two pairs seen in 1991 (Parker and Carr 1992); three birds found in 1993 (P. Boesman *in litt.* 1994).
Leptotila ochraceiventris	1993	Found to be rare in January 1991 (Parker and Carr 1992), with six noted in December 1993 (P. Boesman *in litt.* 1994).
Acestrura berlepschi	1991	One seen in March 1990, with four present in January 1991 on a ridge near Río Ayampe, just outside the national park. The only area in which this species is represented.
Acestrura bombus	1993	Found to be uncommon in January 1991 (Parker and Carr 1992); two seen in December 1993 (P. Boesman *in litt.* 1994).

cont.

Onychorhynchus occidentalis	1991	Rare on Cerro San Sebastián (Parker and Carr 1992).
Lathrotriccus griseipectus	1991	Fairly common (Parker and Carr 1992).
Attila torridus	1991	Apparently fairly common within the national park.
Synallaxis tithys	1991	Found to be uncommon (Parker and Carr 1992).
Hylocryptus erythrocephalus	1991	Rare on Cerro San Sebastián (Parker and Carr 1992).
Carduelis siemiradzkii	1993	Fairly common on Cerro San Sebastián in 1991 (Parker and Carr 1992); several groups in December 1993 (P. Boesman *in litt.* 1994).

EC 20

Cerro Blanco, Guayaquil (Guayas)

2°10'S 80°02'W
EBA B20

Cerro Blanco Protected Forest (private), 2,000 ha

This recently established reserve of Fundación Natura and the company Cemento Nacional, lying 15 km west of Guayaquil in the Cordillera de Chongón at the southern end of the Cordillera de la Costa, has an important area of dry to moist tropical forest. Adjacent remnant forest areas urgently need similar protection.

Leucopternis occidentalis	1994	Regularly recorded; a pair thought to be nesting in 1991 (Berg 1994).
Leptotila ochraceiventris	1994	Present seasonally (December–April), and often recorded during this period (Parker and Carr 1992, Berg 1994, A. Lewis *in litt.* 1995).
Onychorhynchus occidentalis	1995	Found breeding in April 1993 (Berg 1994), and nest-building in early 1995 (A. Lewis *in litt.* 1995).
Lathrotriccus griseipectus	1995	Recorded with some regularity on one particular trail (Berg 1994, A. Lewis *in litt.* 1995).
Synallaxis tithys	1993	Found to be uncommon in 1991.
Hylocryptus erythrocephalus	1993	Recorded a number of times 1992–1993 (Berg 1994, P. Boesman *in litt.* 1994).
Carduelis siemiradzkii	1992	Several groups of 18–25 birds are present in the reserve (Parker and Carr 1992, Berg 1994, G. Speight *in litt.* 1994).

EC 21

Manglares-Churute (Guayas)

2°25'S 79°37'W
EBA B20

Manglares-Churute Ecological Reserve (IUCN category I), 35,042 ha

Lying on the coast of Guayas province, south of Guayaquil, the reserve comprises lowland tropical forest, mangroves and marshes and extends from sea-level (where it is dry and dedicated to cattle) to c.900 m, the forest becoming increasingly humid with altitude (Best 1992, K. S. Berg *in litt.* 1995). The extent of forest is unknown, although most of it has been selectively logged, with cattle-grazing affecting the understorey vegetation (Best 1992, K. S. Berg *in litt.* 1995).

Leucopternis occidentalis	1994	Recorded five times on the partly forested slopes (150–500 m) of Cerro Masvale (K. S. Berg *in litt.* 1995).
Leptotila ochraceiventris	1991	Recorded during surveys in January 1991 (Best and Kessler 1995).
Onychorhynchus occidentalis	1994	Recorded in January 1991 (Best and Kessler 1995), with eight sightings at 100–300 m in August–September 1994 (K. S. Berg *in litt.* 1995).
Lathrotriccus griseipectus	1994	Seen twice in a low, thick secondary tangle at c.100 m (K. S. Berg *in litt.* 1995).

EC 22

Sangay (Morona-Santiago/Chimborazo)

2°00'S 78°20'W
EBAs B18, B21, B60

Sangay National Park (IUCN category II), 517,725 ha

The national park lies east of Riobamba and embraces the eastern Andean mountains of Tungurahua and Sangay. With altitudes ranging from 1,000 to 5,400 m, this area covers most vegetation zones.

Touit stictoptera	1979	Flock of 12 seen at 1,600 m in the upper Río Upano valley.
Hapalopsittaca pyrrhops	1976	Two or three seen on three occasions in October 1976.

cont.

Acestrura bombus	1940	Two collected in September 1937 and October 1940.
Galbula pastazae	1939	Birds have been collected in the vicinity of Macas, at 1,000–1,200 m.
Buthraupis wetmorei	1929	Recorded above 3,350 m in the south-eastern end of the Culebrillas valley.

Cordillera de Cutucú (Morona-Santiago)

EC 23

Cordillera de Cutucú Protection Forest (IUCN category VIII), 311,500 ha

2°43′S 78°05′W
EBA B18

This semi-isolated mountain ridge lies east of the main Andean chain in south-east Ecuador. The area is south and east of Macas between Logroño and Yaupi, and reaches altitudes of more than 2,000 m.

Micrastur buckleyi	1938	Two collected at 1,800 m.
Pyrrhura albipectus	1984	Several flocks of 4–10 at 1,200–1,700 m.
Touit stictoptera	1984	Sighted in flocks of fewer than 10 birds.
Galbula pastazae	1984	One collected at 1,525 m.

Zapote Najda mountains (Morona-Santiago)

EC 24

Unprotected

3°01′S 78°38′W
EBAs B28, B60

This area, including the locality San Vicente, is on the eastern slope of the Andes near the Azuay border. The mountains reach more than 3,200 m (near the Gualaceo–General P. Limón G. road), and a large tract of temperate forest remains along the road where it adjoins an extensive páramo zone (M. Pearman *in litt.* 1995).

Leptosittaca branickii	1987	Sightings in June 1984 and 1987.
Buthraupis wetmorei	1993	Two collected in 1984, subsequent records at 3,200–3,350 m, including four in 1990 (M. Pearman *in litt.* 1995) and an adult and immature in December 1993 (P. Boesman *in litt.* 1994).

Manta Real (Cañar)

EC 25

Unprotected

2°34′S 79°21′W
EBA B20

Manta Real is a small village at the base of the Andes, south-east of Guayaquil. The village is at 250 m, but mountains (such as Cerro San José Chico) rise steeply to c.1,200 m to the east and south. There is a considerable amount of forest remaining on these slopes, which are either just to the north of or within the 28,000 ha Cordillera de Molleturo Protected Forest. Logging was being carried out in September 1994 (K. S. Berg *in litt.* 1995).

Leucopternis occidentalis	1993	Uncommon in this area during 1991, though up to five have been seen together subsequently (K. S. Berg *in litt.* 1995).
Pyrrhura orcesi	1993	Numerous recent sightings—at 300 m (once), but mostly from forest at 500–1,000 m (K. S. Berg *in litt.* 1995).
Onychorhynchus occidentalis	1993	Numerous recent sightings include two nests, one of which fledged young in April 1993 (K. S. Berg *in litt.* 1995).
Lathrotriccus griseipectus	1993	Recently recorded in c.5-ha patches of moist lowland forest as well on the humid forested slopes (K. S. Berg *in litt.* 1995).
Attila torridus	1993	Found to be rare in 1991, with one seen in a cacao plantation in 1993 (K. S. Berg *in litt.* 1995).
Pachyramphus spodiurus	1991	A male collected at 300 m.

EC 26

San Miguel de Azuay (Azuay)

2°48′S 79°30′W
EBA B20

Unprotected

San Miguel de Azuay is on the Pacific slope of the Andes in north-west Azuay, at c.900 m, with (primary) lower montane tropical forest nearby. The Cordillera de Molleturo Protected Forest is within 10 km of San Miguel, and *Hapalopsittaca pyrrhops* has been recorded recently there.

Syndactyla ruficollis	1992	Recorded three times during surveys in January 1992 (M. Whittingham *in litt.* 1992).
Onychorhynchus occidentalis	1992	A pair seen at a part-built nest in January 1992.
Lathrotriccus griseipectus	1992	One seen.

EC 27

Cajas and Río Mazan (Azuay)

2°49′S 79°10′W
EBAs B21,B60

Cajas National Recreation Area (IUCN category V), 28,808 ha
Río Mazan Cloud-forest Reserve (private), 700 ha

Río Mazan reserve is at the edge of the much larger Cajas recreation area. Both are c.15 km north-west of Cuenca, and embrace *Polylepis* woodland, cloud-forest and páramo vegetation at 3,000–4,000 m.

Leptosittaca branickii	1988	Up to 1987, flocks of up to 50 recorded; local population estimated at several hundred birds.
Hapalopsittaca pyrrhops	1995	One to five seen on 10 occasions in August–September 1987; flock of five seen February 1995 (N. Krabbe *in litt.* 1995).
Metallura baroni	1994	Twelve mist-netted at Río Mazan in 1986; small numbers seen recently (e.g. P. Boesman *in litt.* 1994, A. Lewis *in litt.* 1995). The only known area for the species.

EC 28

Girón (Azuay)

3°10′S 79°08′W
EBA B20

Unprotected

Girón is at the head of the Río Girón, c.40 km south-west of Cuenca. At 2,100 m it lies within an area of inter-Andean arid tableland, a region that has been subjected to widespread habitat destruction.

Atlapetes pallidiceps	1939	One collected at 2,100 m, June 1939.

EC 29

Valle de Yunguilla (Azuay)

3°18′S 79°18′W
EBA B20

Unprotected

This area is apparently in the lower part of the arid valley containing the Río Giron in southern Azuay and has suffered from extensive loss of the dry forest habitat.

Acestrura bombus	1940	Four collected 1939–1940.
Atlapetes pallidiceps	1961	Three collected in July 1939, two others in August 1955 and April 1961.

Bestión (Azuay)

EC 30

3°25′S 79°01′W
EBA B60

Unprotected

Bestión is in south-east Azuay on the plain of the Río Shingata, with a semi-humid ridge bordering the arid central valley (Krabbe 1994). This ridge, at c.3,000 m, is covered in bunch grass and chaparral, with *Puya* scattered throughout (Krabbe 1994).

Agriornis andicola	1992	Three collected in 1921, one in 1992 (Krabbe 1994).

Arenillas (El Oro)

EC 31

3°33′S 80°04′W
EBA B20

Arenillas Military Reserve (private), 20,000 ha

This reserve, the largest area of intact dry forest and thorn scrub in south-west Ecuador, lies to the north of Arenillas and west of Machala, abutting the Peruvian border in westernmost El Oro.

Synallaxis tithys	1991	Three seen in July 1991.
Pachyramphus spodiurus	1990	Four recorded just south of Arenillas.

Buenaventura (El Oro)

EC 32

3°40′S 79°44′W
EBA B20

Unprotected

The Hacienda Buenaventura is 9 km west of Piñas on the Machala road, at 900–1,050 m. It covers c.3,000 ha, two-thirds of which is cattle pasture, with very humid cloud-forest occurring in patches across the rest. The largest contiguous forest patch covers less than 1,000 ha, the surrounding area being even more fragmented. Protection of forest in this area has been the subject of much discussion by Fundación Natura (R. S. R. Williams *in litt.* 1995).

Penelope barbata	1920	The type-specimen was collected here.
Leucopternis occidentalis	1994	Several pairs apparently resident (e.g. P. Boesman *in litt.* 1994, A. Lewis *in litt.* 1995).
Leptotila ochraceiventris	1990	Three seen in July 1990.
Pyrrhura orcesi	1994	Usually recorded in flocks of 5–10, though 40–50 birds have been noted at c.1,300 m (e.g. A. Lewis *in litt.* 1995).
Syndactyla ruficollis	1990	Two seen, July 1990.
Onychorhynchus occidentalis	1991	Recorded in the late 1980s; one in 1991 (Williams and Tobias 1994).
Lathrotriccus griseipectus	1991	One collected in 1990, one seen in 1991.
Attila torridus	1991	Recorded from the late 1980s.

Selva Alegre (Loja)

EC 33

3°32′S 79°22′W
EBA B21

Unprotected

Selva Alegre is in the Cordillera de Chilla, 25 km from Saraguro. The area comprises fragmented humid forest on the slopes of a small valley, extending for c.4 km^2 at altitudes from 2,850 to 3,050 m. There is a dense population of indian farmers and much felling and/or disturbance of remnant forest.

Penelope barbata	1989	Five to ten pairs observed in May 1989.
Hapalopsittaca pyrrhops	1995	Up to 18 seen in April 1992, with a nesting pair found in December 1994 (N. Krabbe *in litt.* 1995).

Acanama–Saraguro (Loja)

EC 34

Huashapamba Protection Forest, c.300 ha

3°42′S 79°13′W
EBAs B20, B21

The Acanama area is c.7 km south-east of Saraguro (and c.7 km from San Lucas), on the eastern slope of the Cordillera Cordoncillo. There are two forest patches in a strongly human-influenced landscape at 3,000–3,200 m. The humid montane forest totals just over 1 km². This forest area is currently well protected (by three Saraguro Indian cooperatives) as a watershed reserve (Williams and Tobias 1994). Lower altitude humid forest was presumably present nearer to San Lucas, at least historically. A team from the Academy of Natural Sciences Philadelphia worked an area at 3,175 m, just below a military antenna station, where species classified as Near Threatened, such as Crescent-faced Antpitta *Grallaricula lineifrons* and Orange-banded Flycatcher *Myiophobus lintoni*, were recorded (M. B. Robbins *in litt.* 1994).

Penelope barbata	1992	Birds seen in 1989, with eight (and three juveniles) in September 1991 (Williams and Tobias 1994) and two in March 1992 (M. B. Robbins *in litt.* 1994).
Leptosittaca branickii	1992	Flocks of 10–35 recorded from 1990 (e.g. Williams and Tobias 1994, M. B. Robbins *in litt.* 1994).
Hapalopsittaca pyrrhops	1992	A group of c.20 seen at 3,200 m in 1991, with others in March–May 1992 (M. B. Robbins *in litt.* 1994, N. Krabbe *in litt.* 1995).
Attila torridus	pre-1929	Recorded as having occurred, but is unlikely still to do so at the altitude at which forest persists.

Puyango (Loja)

EC 35

Unprotected

3°52′S 80°05′W
EBA B20

Puyango lies on the Loja–El Oro border, 18 km north-north-west of Alamor, at c.300 m on the Río Puyango. The area comprises patches of dry deciduous and (lower down) semi-deciduous forest. The area is almost contiguous with Tumbes National Forest (PE 01).

Leucopternis occidentalis	1921	One specimen.
Lathrotriccus griseipectus	1919	One specimen.
Pachyramphus spodiurus	1992	A male seen at 300 m.

Vicentino (Loja)

EC 36

Unprotected

3°57′S 79°57′W
EBA B20

Areas east and south-west of Vicentino (which lies to the north-east of Alamor, in western Loja) comprise small patches of humid cloud-forest lying between 900 and 1,450 m, separated by pastures and crop-lands.

Leucopternis occidentalis	1991	At least one pair present.
Leptotila ochraceiventris	1991	Birds heard in February 1991.
Myrmeciza griseiceps	1991	Five singing in February 1991.

Alamor (Loja)

EC 37

Unprotected

4°00′S 80°00′W
EBA B20

Alamor is situated on a ridge between the Río Tumbez and Río Alamor. Between 1,200 and 1,450 m, patches of humid cloud-forest exist (mainly confined to steep ravines) to the north-east of Alamor on the road to Vicentino (including the forest patches along Quebrada Las Vegas) with slightly drier forest patches to the south of town.

Species	Year	Notes
Leucopternis occidentalis	1991	Three seen together at Quebrada Las Vegas.
Leptotila ochraceiventris	1991	Several records from a number of localities in late 1991.
Acestrura bombus	pre-1926	Four collected (Chapman 1926).
Synallaxis tithys	1991	Two birds present.
Hylocryptus erythrocephalus	1991	Up to three seen.
Syndactyla ruficollis	1991	Not uncommon in 1991 (Williams and Tobias 1994).
Myrmeciza griseiceps	1991	Recorded just north of Alamor (N. Krabbe *in litt.* 1991).
Lathrotriccus griseipectus	pre-1926	One collected (Chapman 1926).
Attila torridus	1991	Recorded February and August 1991.

Celica (Loja)

EC 38

Unprotected

4°07′S 79°58′W
EBA B20

This area includes the forest patches at 1,800–2,100 m in the Cordillera de Celica along the Celica–Alamor road, up to 6 km west and north-west of Celica, although the largest patches are close to Guachanamá to the north-east of the town (B. J. Best *in litt.* 1994). The area is densely populated and the humid montane forest patches rarely exceed 10 ha.

Species	Year	Notes
Leucopternis occidentalis	1989	Daily sightings of several birds in August 1989.
Leptotila ochraceiventris	1990	Seen up until at least December 1990.
Acestrura bombus	pre-1926	Three collected (Chapman 1926).
Hylocryptus erythrocephalus	1992	Four territorial birds in April 1992.
Syndactyla ruficollis	1992	Fairly common and encountered almost daily in August 1991 (Williams and Tobias 1994), and noted as present in 1992 (M. B. Robbins *in litt.* 1994).
Myrmeciza griseiceps	1989	One collected and another seen.
Lathrotriccus griseipectus	pre-1926	One collected at Guainche (Chapman 1926).
Attila torridus	1992	Birds recorded at 1,700 m (M. B. Robbins *in litt.* 1994).

Sabanillas (Loja)

EC 39

Unprotected

4°10′S 80°08′W
EBA B20

Sabanillas is in the valley of the Río Alamor, south-west Loja. Areas up to 10 km south of Sabanillas (at 350–750 m) comprise degraded *Acacia* woodland and, higher up, deciduous forest. This area deserves more survey work, as other threatened species (e.g. *Leucopternis occidentalis*) are almost certainly present (M. B. Robbins *in litt.* 1994).

Species	Year	Notes
Synallaxis tithys	1992	Recorded c.10 km south of Sabanillas.
Hylocryptus erythrocephalus	1992	At least eight heard c.10 km south of Sabanillas.
Lathrotriccus griseipectus	1991	Specimens collected at 500 m.
Carduelis siemiradzkii	1992	Fairly common c.10 km south of Sabanillas.

El Empalme (Loja)

EC 40
4°07′S 79°51′W
EBA B20

Unprotected

El Empalme lies within the Río Catamayo valley at 800–900 m. The area comprises *Ceiba*-dominated deciduous forest, mostly with a heavily degraded understorey, but with some more intact areas in ravines.

Leucopternis occidentalis	1989	One pair seen.
Leptotila ochraceiventris	1991	One bird seen, late 1991.
Hylocryptus erythrocephalus	1991	One bird seen.

Catacocha (Loja)

EC 41
4°03′S 79°40′W
EBA B20

Unprotected

Catacocha is in central Loja, in the upper Río Casanga valley, and c.6 km to the north-west is a narrow strip of semi-deciduous lower montane forest covering c.1.5 km² at 1,400–1,750 m. The forest is surrounded by crop-lands and cattle pasture.

Leptotila ochraceiventris	1991	Heard and seen in March 1991.
Acestrura bombus	1990	Two seen at c.1,850 m, west of Catamayo.
Hylocryptus erythrocephalus	1992	At least eight birds, and an active nest, were found in April 1992.
Syndactyla ruficollis	1991	Small numbers (Best 1992).
Lathrotriccus griseipectus	1992	Three or four singing at 1,650 m, early April 1992 (M. B. Robbins *in litt.* 1994).
Atlapetes pallidiceps	1969	Six collected apparently in the Casanga valley during December 1968 and January 1969, although the presence of the species in this area has been questioned.

Sozoranga (Loja)

EC 42
4°20′S 79°48′W
EBA B20

Unprotected

Sozoranga is in extreme southern Loja, on the western slope of the Andes, and includes a number of sites, all within 5 km of the village. The Quebradas Yaguana and Suquinda (at 1,550–1,800 m) are 3 and 5 km west of Sozoranga, and an area of cloud-forest (c.40 ha) is on a ridge above these ravines (1,750–1,850 m). The vegetation grades from semi-deciduous lower montane forest (covering c.200 ha) to humid montane cloud-forest.

Leptotila ochraceiventris	1993	Recent sightings (e.g. P. Boesman *in litt.* 1994).
Hylocryptus erythrocephalus	1994	Regular sightings since June 1989 (e.g. P. Boesman *in litt.* 1994, A. Lewis *in litt.* 1995).
Syndactyla ruficollis	1991	Small numbers recorded in most months up to February 1991.
Lathrotriccus griseipectus	1994	One seen at 1,750 m in February 1991, and one tape-recorded in late 1994 (A. Lewis *in litt.* 1995).

Tambo Negro (Loja)

EC 43

Unprotected

4°24′S 79°51′W
EBA B20

This area lies c.2 km west of Tambo Negro which abuts the Peruvian border in south-west Loja. It extends from the Río Macara valley-bottom at 600 m to the ridge-tops on the south side of the valley at 1,000 m. The ridge is almost entirely covered by *Ceiba*-dominated deciduous forest extending to at least 1,500 ha. This is the largest patch of continuous forest in western Loja and El Oro provinces.

Leptotila ochraceiventris	1991	Sightings in September 1989 and birds heard in March 1991.
Synallaxis tithys	1991	Fairly common in January–February 1991, but apparently less so in March that year.
Hylocryptus erythrocephalus	1991	Several sightings, birds heard constantly in February–March 1991.
Syndactyla ruficollis	1991	Recorded only rarely during 1989 and 1991.
Myrmeciza griseiceps	1991	Two seen, February 1991.
Lathrotriccus griseipectus	1991	Four or five seen, February–March 1991.

Angashcola (Loja)

EC 44

Unprotected

4°34′S 79°22′W
EBAs B20, B21, B60

The area of interest lies near Amaluza, with Lomo Angashcola extending from 2,500 to 3,100 m and harbouring a humid montane cloud-forest block of 200–300 ha, part of an unbroken tract (several thousand hectares) stretching over several valleys (R. S. R. Williams *in litt*. 1995). Hunting, especially of *Penelope barbata*, is a problem (R. S. R. Williams *in litt*. 1995).

Penelope barbata	1990	A large population found July–August 1990.
Leptosittaca branickii	1991	Ten seen feeding in August 1990, and up to 19 in July 1991 (Williams and Tobias 1994).
Metallura odomae	1990	Seen 2–3 times during August–September 1990.
Syndactyla ruficollis	1991	One or two seen.

Lagunillas (Zamora-Chinchipe)

EC 45

Unprotected

4°47′S 79°22′W
EBA B18

Lagunillas is in southernmost Zamora-Chinchipe in the Cordillera Lagunillas, south of Jimbura and north of the Río Isimanchi (c.6 km north-west of San Andrés). On the eastern slope from 2,075 to 2,250 m, vegetation comprises tall, subtropical/lower temperate forest, with temperate forest extending up to the treeline at c.3,050 m (M. B. Robbins *in litt*. 1994). In 1992 the forest near the treeline was relatively undisturbed although cattle were present, and the area had been periodically burnt, but lower on the slopes (below 2,100 m) large areas had been cleared for cattle (M. B. Robbins *in litt*. 1994). Also present in this area was the recently described Chestnut-bellied Cotinga *Doliornis remseni* (Robbins *et al*. 1994, see 'Recent changes to the threatened list', above), and populations of the threatened mountain tapir *Tapirus pinchaque* and spectacled bear *Tremarctos ornatus* (Groombridge 1993, M. B. Robbins *in litt*. 1994).

Penelope barbata	1992	Three seen at 2,250 m on the eastern slope and a pair at 2,650 m on the western slope (M. B. Robbins *in litt.* 1994).
Leptosittaca branickii	1992	Flocks (typically of 15 but up to 30) seen daily in November 1992, with a presumed pair found excavating a nest-hole at 2,950 m (M. B. Robbins *in litt.* 1994).
Hapalopsittaca pyrrhops	1992	A pair seen at c.3,000 m in November 1992 (M. B. Robbins *in litt.* 1994).

Zumba (Zamora-Chinchipe)

EC 46
4°53′S 79°10′W
EBA B18

Unprotected

Zumba lies within the Río Chinchipe valley, in southernmost Zamora-Chinchipe, where to the north of the town some forest apparently still exists at c.1,200 m. Isolated patches of forest were also present on the slopes south of Zumba (along the trail to Chito) in 1986 (M. Pearman *in litt.* 1995).

Touit stictoptera	1989	A flock of 18–20.
Acestura bombus	1986	A male seen south of Zumba on the trail to Chito (M. Pearman *in litt.* 1995).

Podocarpus (Loja/Zamora-Chinchipe)

EC 47
4°08′S 78°58′W
EBAs B18, B21, B60

Podocarpus National Park (IUCN category II), 146,280 ha

The national park lies south and east of Loja, and south of Zamora. It has a very irregular topography covering altitudes from 950 to 3,700 m, i.e. from upper tropical forest in the east through subtropical, temperate and elfin forest and páramo and down again through elfin and temperate forest on the western side of the main (eastern) Andean cordillera which traverses the park (Rasmussen and Rahbek 1994). The area is situated in the only part of southern Ecuador which still retains large tracts of undisturbed forest, continuous from upper tropical to temperate zones.

Penelope barbata	1992	Possibly up to 1,000 pairs within the park.
Leptosittaca branickii	1991	Flocks of 4–20 recorded in 1989; six seen September 1991 (Williams and Tobias 1994).
Pyrrhura albipectus	1994	Small flocks of 5–8 present all year, with 30 birds seen in November 1991 (Williams and Tobias 1994), 12 on the Sabanilla trail in December 1993 (P. Boesman *in litt.* 1994), and 10 seen daily at Bombuscaro in late 1994 (A. Lewis *in litt.* 1995).
Hapalopsittaca pyrrhops	1994	Up to four pairs found breeding, November 1994 (E. P. Toyne *in litt.* 1994).
Metallura odomae	1994	Found regularly above Cajanuma (e.g. R. S. R. Williams *in litt.* 1995).
Acestrura bombus	pre-1953	Recorded from Zamora (Chapman 1926) and Sabanilla (Zimmer 1953).
Galbula pastazae	1994	Population of 4–6 pairs recorded in the Río Bombuscaro area; at least six birds at Bombuscaro in late 1994 (A. Lewis *in litt.* 1995); other records from just north of the park (Poulsen and Wege 1994).
Agriornis andicola	1965	One collected at 2,400 m.
Buthraupis wetmorei	1991	Fairly common in the correct habitat at Cajanuma (4–5 birds on 2 km of trail).

Chinapinza (Zamora-Chinchipe)

EC 48
4°00′S 78°34′W
EBA B24

Unprotected

This area embraces the ridge between La Punta and Chinapinza, near the top of the central Cordillera del Condor. The ridge reaches an elevation of c.1,700 m and supports humid cloud- and elfin forest.

Pyrrhura albipectus	1990	A flock of c.5, c.1988; three specimens collected in 1989–1990.
Hemitriccus cinnamomeipectus	1990	Three specimens collected at 1,700 m, September 1990.

Shaime (Zamora-Chinchipe)

EC 49

Unprotected

4°22'S 78°40'W
EBA B24

Adjacent to the Peruvian border at 1,000 m in the Cordillera del Condor, Shaime lies along the upper Río Nangaritza, a tributary of the upper Río Zamora. The vegetation is humid montane forest.

Wetmorethraupis sterrhopteron	1990	Four specimens collected July–August 1990.

Islas Galápagos (Territorio de Galápagos)

EC 50

Galápagos National Park (IUCN category II), 727,800 ha

0°30'S 90°30'W
EBA B16

The Galápagos comprises 13 islands of more than 10 km^2, six smaller islands, and over 40 islets with official names, the land area totals c.8,000 km^2. The archipelago straddles the Equator at c.90°W, c.960 km west of mainland Ecuador, and is spread over 45,000 km^2 of sea. Twenty-three species are endemic to the islands, and three further (seabird) species are essentially endemic breeders (Stattersfield *et al.* in prep.). Combined with the five threatened species listed below, this makes the national park one of the most important protected areas in South America (see 'Recent changes to the threatened list', above).

Pterodroma phaeopygia	1992	The breeding population in 1980 was estimated at 9,000 pairs on Santa Cruz and 1,000 pairs on Floreana, although there has been a rapid decline since then.
Nannopterum harrisi	1992	Restricted to the coast of Fernandina and Isabela; in 1986 the adult population was estimated at 1,000 birds.
Buteo galapagoensis	1992	Up to 130 breeding territories (representing c.400 birds) have been found on a number of islands, with c.50 territories on Santiago.
Nesomimus trifasciatus	1992	Confined to Gardner island (200–300 birds) and Champion island (c.50 birds).
Camarhynchus heliobates	1995	Restricted to dense mangrove swamps on eastern Fernandina and western Isabela (totalling c.500 ha). Unrecorded for over 20 years until a young bird was seen being fed north of Punta Tortuga, western Isabela, in March 1995 (R. Naveen verbally 1995).

THE GUIANAS

Rufous-sided Pygmy-tyrant *Euscarthmus rufomarginatus*

THE GUIANAS are here defined by the three contiguous entities of Guyana, Surinam and French Guiana (and should not be confused with the larger floristic region of Guayana: Gillespie and Funk 1993). The smallest of the three, French Guiana, is known to support an avifauna of c.800 species (Tostain *et al.* 1992), while Surinam has 668 species (Haverschmidt and Mees 1994) and Guyana c.740 species (M. Johnston *in litt.* 1993). Combined, they probably support over 900 species of bird, of which just one is endemic to a country (French Guiana), 27 have restricted ranges (Stattersfield *et al.* in prep.), and five are threatened (Collar *et al.* 1992). This analysis has identified four Key Areas for the threatened birds in the Guianas (see '*Key Areas*: the book', p. 11, for criteria).

THREATENED BIRDS

Five species now known to occur in the Guianas were considered by Collar *et al.* (1992) to be at risk of global extinction (Table 1); one of these (*Caprimulgus maculosus*) is confined to French Guiana and two others (*Synallaxis kollari* and *Cercomacra carbonaria*) are recent additions to the Guianan avifauna (see 'Recent changes to the threatened list', below). All five are exclusively lowland species, depending variously on gallery and seasonally flooded riverine forest (*Synallaxis kollari* and *Cercomacra carbonaria*), shrubby grasslands and savanna (*Euscarthmus rufomarginatus* and possibly *Caprimulgus maculosus*) and marshes or wet savanna (*Coturnicops notatus*). With the exception of *Euscarthmus rufomarginatus*, which is threatened by habitat loss elsewhere within its range, the Guianan threatened species are generally considered such owing to their rarity and the fact that they are so little known (see 'Old records and little-known birds', below), rather than because of the existence of any tangible threat. The distributions of the five threatened birds and their relationship to Endemic Bird Areas are shown in Figure 1.

KEY AREAS

The four Guianan Key Areas would, if adequately protected, help ensure the conservation of four of the five threatened species (Table 1). No Key Area was selected for *Coturnicops notatus* owing to the age of the single record (see 'Old records and little-known birds'), and the numerous more recent and regular records from elsewhere within its extraordinary range. However, for the four remaining Guianan threatened species, the four Key Areas embrace all the known

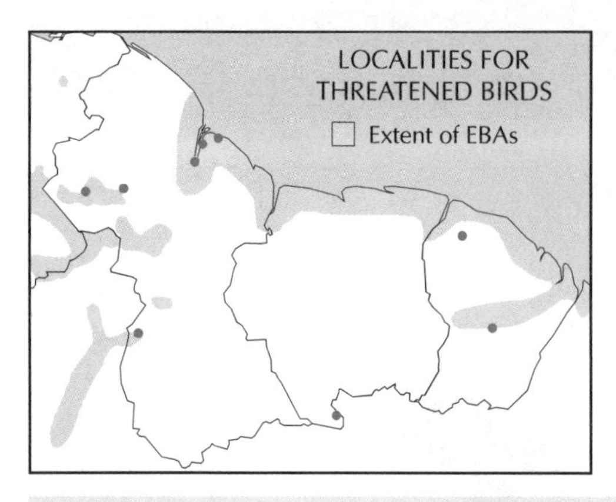

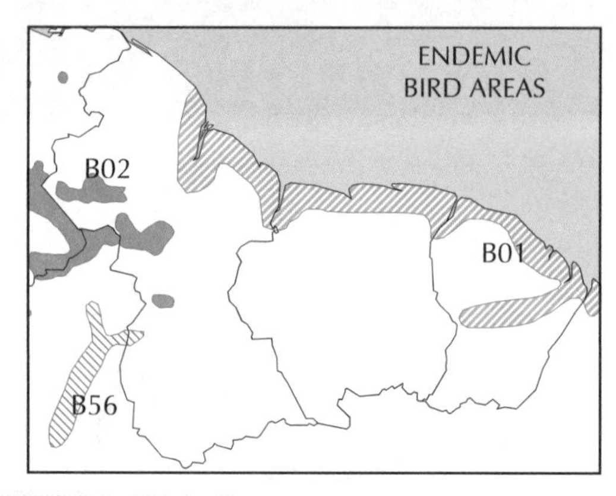

Figure 1. The localities where threatened birds have been recorded in the Guianas and their relationship to Endemic Bird Areas (EBAs).

- Three of the Guianan threatened species have restricted ranges and occur in two different EBAs as listed below (figures are numbers of these species in each EBA).

B01 Guiana shield (1) B02 Tepuís (0) B56 Upper Rio Branco (2)

localities, and are thus vitally important for the conservation of these species and for the habitats that the areas support. Just one threatened species, *Caprimulgus maculosus*, is endemic to the Guianas (French Guiana), and may be totally reliant for its survival on the integrity of habitat in the two Key Areas within which it is thought to occur (see 'Outlook', below).

KEY AREA PROTECTION

Of the four Key Areas in this region, just one, the Sipaliwini savanna (SR 01), is formally protected—which leaves the Key Areas for the French Guiana endemic and the two species confined to the upper Rio Branco EBA totally unprotected.

RECENT CHANGES TO THE THREATENED LIST

The revision by Collar *et al.* (1994) of the world list of threatened species did not bring about any changes which affect the three Guianan countries, although new distributional information has added *Synallaxis kollari* and *Cercomacra carbonaria*, both previously thought endemic to Brazil, to the list of species recorded from Guyana (Forrester 1995).

OLD RECORDS AND LITTLE-KNOWN BIRDS

Our knowledge of threatened birds in the Guianas suffers from the general paucity of information characteristic of so many species and countries within the Neotropics.

Coturnicops notatus is known in the Guianas from a single record along the Abary river, Guyana, in September 1907.

Caprimulgus maculosus is known with certainty from just one French Guianan specimen collected in 1917, with one bird (almost certainly this species) mist-netted at a second locality in 1982.

Synallaxis kollari and *Cercomacra carbonaria* both have single Guianan records from one locality (in Guyana) in January 1993.

Euscarthmus rufomarginatus is known in the Guianas

Table 1. Coverage of threatened species by Key Area. Areas in bold currently have some form of protected status.

	Key Areas occupied	No. of Key Areas protected	Total nos. of Key Areas	
			Guianas	Neotropics
Speckled Crake *Coturnicops notatus*	—	0	0	8
Cayenne Nightjar *Caprimulgus maculosus* [E]	GF 01, GF 02	0	2	2
Hoary-throated Spinetail *Synallaxis kollari*	GY 01	0	1	3
Rio Branco Antbird *Cercomacra carbonaria*	GY 01	0	1	3
Rufous-sided Pygmy-tyrant *Euscarthmus rufomarginatus*	**SR 01**	1	1	9

[E] Endemic to the Guianas

from only one area in southern Surinam where it has not been recorded since its discovery in the 1960s.

OUTLOOK

All four Key Areas are vitally important for the conservation of threatened birds within the Guianas. However, as a national endemic, *Caprimulgus maculosus* is a clear priority for attention, both in terms of Key Area protection but more urgently for surveys to determine its current status at the two known localities (i.e. the two Key Areas) and potentially elsewhere. The two Rio Branco endemics occur together at Pirara (GY 01), which is the only Key Area for either species outside Brazil and therefore very important for the long-term conservation of both birds. Although the protection of this area should be the ultimate goal, surveys are needed to clarify the status of the two species and the distribution of appropriate habit. *Euscarthmus rufomarginatus* is known from Key Areas in Bolivia and Brazil, but the Sipaliwini savanna (SR 01) is the only one for the disjunct *savannophilus* population—which has not, however, been recorded (or presumably looked for) since the mid-1960s, suggesting that species-specific surveys are needed within the nature reserve.

DATA SOURCES

The above introductory text and the Site Inventory (below) were compiled from information supplied by M. Johnston and D. Finch, as well as from the following references.

COLLAR, N. J., GONZAGA, L. P., KRABBE, N., MADROÑO NIETO, A., NARANJO, L. G., PARKER, T. A. AND WEGE, D. C. (1992) *Threatened birds of the Americas: the ICBP/IUCN Red Data Book* (Third edition, part 2). Cambridge, U.K.: International Council for Bird Preservation.

COLLAR N. J., CROSBY, M. J. AND STATTERSFIELD, A. J. (1994) *Birds to watch 2: the world list of threatened birds*. Cambridge, U.K.: BirdLife International (BirdLife Conservation Series no.4).

FORRESTER, B. C. (1995) Brazil's northern frontier sites: in search of two Rio Branco endemics. *Cotinga* 3: 51–53.

GILLESPIE, L. J. AND FUNK, V. A. (1993) Biodiversity and conservation in Guyana: a new centre for the study of biodiversity. *Global Biodiversity* 3(1): 7–11.

HAVERSCHMIDT, F. AND MEES, G. F. (1994) *Birds of Suriname*. Paramaribo: VACO Uitgeversmaatschappij.

IUCN (1992) *Protected areas of the world: a review of national systems. Volume 4: Nearctic and Neotropical.* Gland, Switzerland and Cambridge, U.K.: International Union for Conservation of Nature and Natural Resources.

STATTERSFIELD, A. J., CROSBY, M. J., LONG, A. J. AND WEGE, D. C. (in prep.) *A global directory of Endemic Bird Areas*. Cambridge, U.K.: BirdLife International (BirdLife Conservation Series).

STEPHENS, L. AND TRAYLOR, M. A. (1985) *Ornithological gazetteer of the Guianas*. Cambridge, Mass.: Museum of Comparative Zoology.

TOSTAIN, O., DUJARDIN, J.-L., ERARD, C. AND THIOLLAY, J.-M. (1992) *Oiseaux de Guyane*. Brunoy: Société d'Etudes Ornithologiques.

SITE INVENTORY

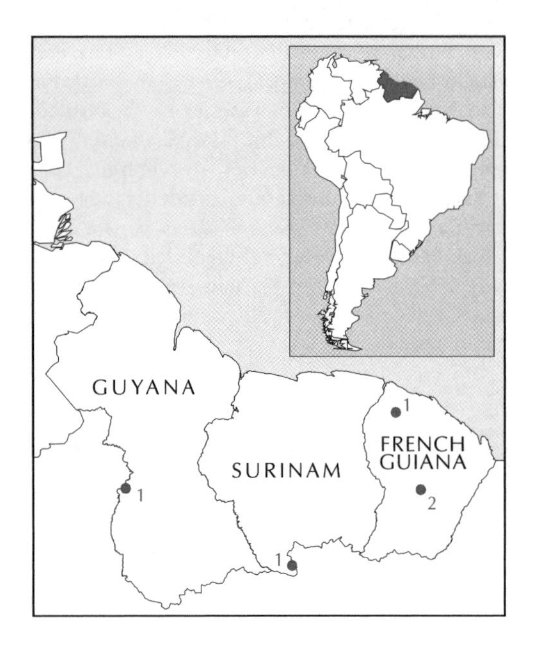

Figure 2. Key Areas in the Guianas.

Surinam
SR 01 Sipaliwini savanna

French Guiana
GF 01 Saut Tamanoir
GF 02 Saül

Guyana
GY 01 Pirara

SURINAM

Sipaliwini savanna

SR 01
2°03′N 56°01′W

Sipaliwini Savanna Nature Reserve (IUCN category IV), 100,000 ha

This area of southern Surinam, on the border with Brazil, is a region of savanna, gallery forest and isolated forest patches, between c.200 and 500 m. It is continuous with the extensive Paru savanna of Brazil, parts of which should be surveyed for the presence of the species below.

Euscarthmus rufomarginatus	c.1965	A population of unknown size representing a new subspecies *savannophilus* was discovered in the 1960s.

FRENCH GUIANA

Saut Tamanoir

GF 01
5°09′N 53°45′W
EBA B01

Unprotected

Saut Tamanoir is in the north-west of the country and on the Fleuve Mana c.10 km above the confluence with the Rivière Cokioco. It is an area of boulder-strewn rapids bordered by closed-canopy forest and more open areas with rare savanna-like clearings. Specific efforts should be made to mist-net nightjars at this locality with the aim of confirming the continued existence and taxonomic validity of *Caprimulgus maculosus*.

Caprimulgus maculosus	1917	No records here since the type- and only specimen (a male) was taken.

Saül

GF 02

Unprotected

3°35'N 53°12'W
EBA B01

This area of cleared ground and lowland forest in the centre of the country lies around the Saül airstrip. Specific efforts should be made to mist-net nightjars at this locality with the aim of confirming the continued existence and taxonomic validity of *Caprimulgus maculosus*.

Caprimulgus maculosus	1982	A female trapped was thought almost certainly to be this species; it would represent only the second known record and locality.

GUYANA

Pirara

GY 01

Unprotected

3°37'N 59°40'W
EBA B56

In the south-west of the country, this area embraces the Río Takutu and Río Ireng (which form the Guyana–Brazil border) and the Río Pirara, along which there are areas of gallery forest and seasonally flooded riverine forest with a dense understorey.

Synallaxis kollari	1993	Two seen and two heard in January (D. Finch *in litt.* 1993).
Cercomacra carbonaria	1993	One seen and one heard in January (D. Finch *in litt.* 1993).

■ MEXICO

Eared Quetzal *Euptilotis neoxenus*

MEXICO holds between 1,025 and 1,050 species (depending on which taxonomic treatment is followed), of which 769 (75%) breed, 225 (22%) are migrants, 125 (12%) are endemic to the country (Escalante Pliego *et al.* 1993, Howell and Webb 1995), 100 (10%) have restricted ranges (Stattersfield *et al.* in prep.) and 29 (3%) are threatened (Collar *et al.* 1992). This analysis has identified 58 Key Areas for the threatened birds in Mexico (see '*Key Areas*: the book', p. 11 for criteria).

THREATENED BIRDS

Of the 29 Mexican species which Collar *et al.* (1992) considered at risk of extinction, 20 are confined to the country (see Table 1). Mexico is sixth in rank in the Americas for numbers of threatened birds, but fourth for threatened country endemics (Collar *et al.* in press). The 29 threatened species rely primarily on wet forest (61% of species), but also important is dry forest (25%). There is no predominant altitudinal zone for them, comparable proportions being found in the highland temperate zone above 2,000 m (36% of species), subtropical submontane (39%), lowlands below 500 m (21%) and islands (25%). The distributions of threatened birds and their relationship to Endemic Bird Areas are shown in Figure 1.

KEY AREAS

The 58 Key Areas that have been identified in Mexico would, if adequately protected, help to ensure the conservation of 27 (96%) of its threatened species, including all those that are endemic to the country (always accepting that important new populations and areas may yet be found; but see below).

Of the 58 areas, only 20 are important for two or more (up to four) threatened species (see 'Outlook' below), and harbour potential sanctuaries for 20 threatened species, 69% of the total number. However, as vital as these areas are for the conservation of Mexican threatened species, they should not detract from the significance of the remaining Key Areas, as these are of major importance for the remaining eight species (six of which are confined to the country) whose habitat would otherwise be unrepresented. This emphasizes that in these cases targeted single-species site-based conservation is needed.

From Tables 1 and 2 it can be seen that seven species occur in just one Mexican Key Area: two of

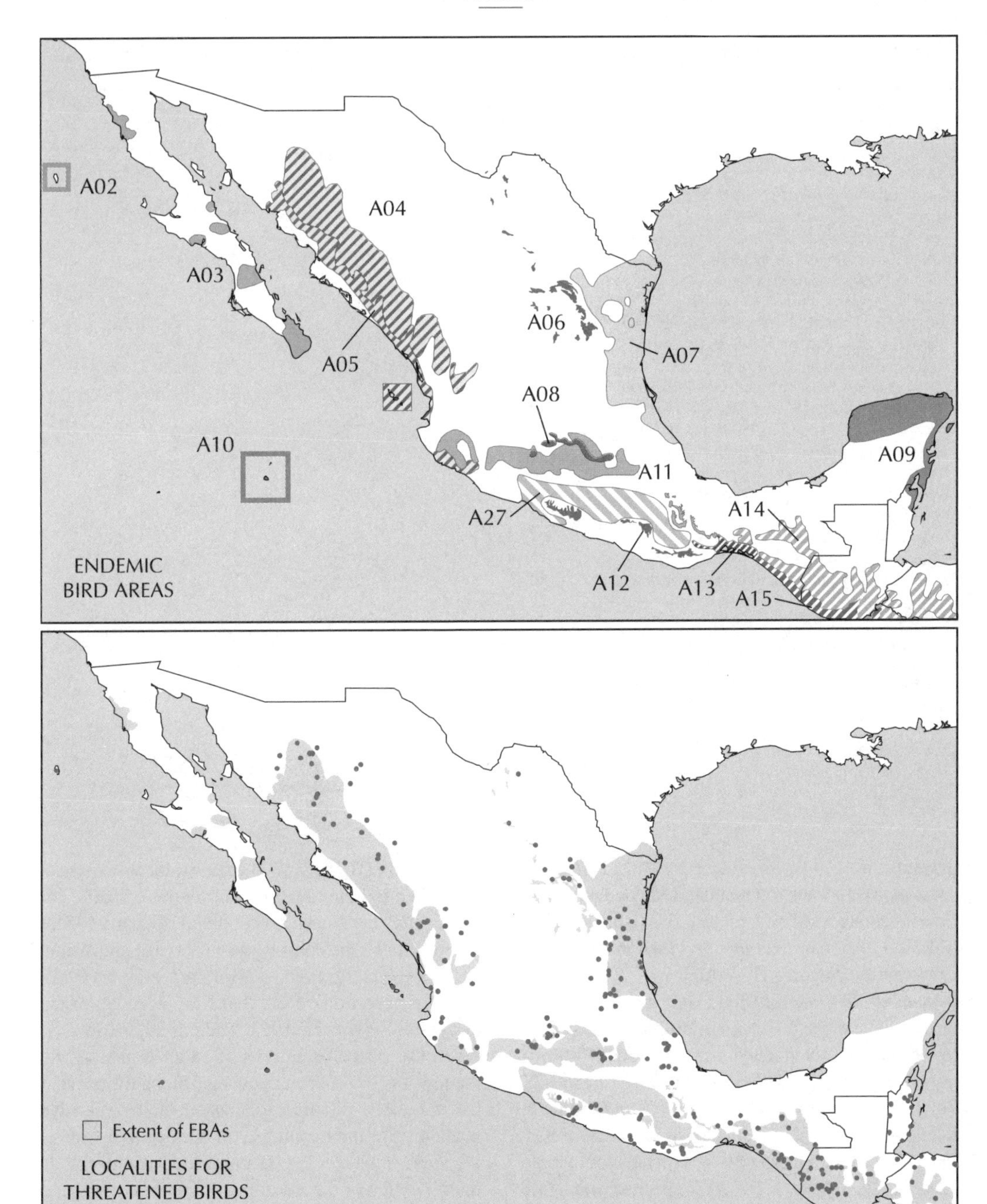

Figure 1. The localities where threatened birds have been recorded in Mexico and their relationship to Endemic Bird Areas (EBAs).

- As almost two-thirds of the threatened birds in Mexico have restricted ranges, they occur in various combinations within EBAs, which are listed below (figures are numbers of these species in each EBA).

A02 Guadalupe Island (2)
A03 Baja California (0)
A04 Sierra Madre Occidental (3)
A05 North-west Mexican Pacific slope (1)
A06 Sierra Madre Oriental (1)
A07 North-east Mexican Gulf slope (1)
A08 Central Mexican marshes (1)
A09 Yucatán peninsula (0)
A10 Socorro Island (3)
A11 Central Mexican highlands (3)
A12 Sierra Madre del Sur (4)
A13 Isthmus de Tehuantepec (0)
A14 North Central American highlands (2)
A15 North Central American Pacific slope (0)
A27 Balsas drainage (0)

Table 1. Coverage of threatened species by Key Area. Areas in bold currently have some form of protected status.

	Key Areas occupied	No. of Key Areas protected	Total nos. of Key Areas	
			Mexico	Neotropics
Townsend's Shearwater *Puffinus auricularis*	**28**,**29**,**30**	3	3	3
Guadalupe Storm-petrel *Oceanodroma macrodactyla*	**01**	1	1	1
Horned Guan *Oreophasis derbianus*	**48**,**56**,01	2	3	8
Bearded Wood-partridge *Dendrortyx barbatus*[E]	19,20,39	0	3	3
Piping Plover *Charadrius melodus*	—	0	0	0
Socorro Dove *Zenaida graysoni*[E]	**29**	1	1	1
Socorro Parakeet *Aratinga brevipes*[E]	**29**	1	1	1
Thick-billed Parrot *Rhynchopsitta pachyrhyncha*[E]	02,04,05,06,08,24,**27**	1	7	7
Maroon-fronted Parrot *Rhynchopsitta terrisi*[E]	12,**13**,14	1	3	3
Yellow-headed Amazon *Amazona oratrix*	15,**16**,17,18,21,**25**,34	2	7	8
Green-cheeked Amazon *Amazona viridigenalis*[E]	15,**16**,17,18	1	4	4
Short-crested Coquette *Lophornis brachylopha*[E]	43	0	1	1
Mexican Woodnymph *Thalurania ridgwayi*[E]	22,**23**,**26**,**27**	3	4	4
Oaxaca Hummingbird *Eupherusa cyanophrys*[E]	50,51	0	2	2
White-tailed Hummingbird *Eupherusa poliocerca*[E]	**42**,43,46	1	3	3
Eared Quetzal *Euptilotis neoxenus*	02,03,04,06,**07**,08,24	1	7	7
Keel-billed Motmot *Electron carinatum*	**48**	1	1	7
Imperial Woodpecker *Campephilus imperialis*[E]	04,05,06,24	0	4	4
Navas's Wren *Hylorchilus navai*[E]	41,**53**	1	2	2
Sumichrast's Wren *Hylorchilus sumichrasti*[E]	40,45	0	2	2
Socorro Mockingbird *Mimodes graysoni*[E]	**29**	1	1	1
Guadalupe Junco *Junco insularis*[E]	**01**	1	1	1
Sierra Madre Sparrow *Xenospiza baileyi*[E]	**36**,37,38	1	3	3
Azure-rumped Tanager *Tangara cabanisi*	**56**,57,58	1	3	4
Golden-cheeked Warbler *Dendroica chrysoparia*	52,**54**,**55**	2	3	8
Black-polled Yellowthroat *Geothlypis speciosa*[E]	31,32,33,35	0	4	4
Black-capped Vireo *Vireo atricapillus*	09,10,11,22,**25**,**26**,**27**	3	7	7
White-throated Jay *Cyanolyca mirabilis*[E]	**42**,43,46,51	1	4	4
Dwarf Jay *Cyanolyca nana*[E]	44,**47**,49	1	3	3

[E] Endemic to Mexico

these species (*Lophornis brachylopha* and *Electron carinatum*) are found on mainland Mexico while the remainder are confined to just two islands, three (*Zenaida graysoni*, *Aratinga brevipes* and *Mimodes graysoni*) on Socorro (Key Area MX 29) and two (*Oceanodroma macrodactyla* and *Junco insularis*) on Guadalupe (MX 01). In addition, at least four threatened Mexican endemics (namely *Puffinus auricularis* in MX 27, *Dendrortyx barbatus* in MX 36, *Xenospiza baileyi* in MX 33 and *Cyanolyca nana* in MX 45), although recorded from a number of Key Areas, appear to be effectively dependent on single Key Areas. Depressingly, two species which are listed as having occurred in Key Areas (*Oceanodroma macrodactyla* and *Zenaida graysoni*) are thought to be extinct (although the latter survives in captivity), and *Campephilus imperialis* is almost certainly extinct in the Key Areas for which it is listed (see 'Old records and little-known birds', below).

Not represented within the Mexican Key Area analysis is *Charadrius melodus*, which appears to be a regular passage migrant and winter visitor to the Gulf coast of eastern Mexico from Tamaulipas to central Veracruz and along the north coast of the Yucatán peninsula, with some records from the north Pacific coast (Howell 1993). Specific localities where more than 10 individuals (but never more than 27) of *C. melodus* have been noted include Laguna Madre (Tamaulipas), the coast between Tuxpan and Antón Lizardo (Veracruz) and Isla Holbox (Quintana Roo); it was also recorded from San Blas (MX 22) in the early 1980s (Howell 1993, Sada 1994). However, given the extensive amount of suitable habitat and the low level of observer coverage along the coast, it has not been possible to make a comprehensive evaluation of the importance of Mexico as a wintering area for the species, which would have to include identification of the most important sites.

KEY AREA PROTECTION

Currently, 19 (33%) of Mexico's Key Areas have some form of protected status (i.e. as national or state parks, or Biosphere or private reserves); 14 (25% of the total) are national parks or Biosphere Reserves (IUCN categories II and IX). The 39 unprotected Key Areas (67% of the total) require attention in the form of appropriate conservation measures if the populations of their threatened species are to survive. However, even the formally protected areas remain under threat and, in many, habitat degradation and uncontrolled hunting continue unchecked; effective management is thus required of activities undertaken within them.

With so many Key Areas lacking any form of protection, the long-term survival of a number of Mexican threatened species must be questioned. Six are not currently recorded from any protected Key Area (see Table 1): *Dendrortyx barbatus*, *Eupherusa cyanophrys*, *Lophornis brachylopha*, *Hylorchilus sumichrasti*, *Xenospiza baileyi* and *Geothlypis speciosa*. All six are endemic to Mexico, and Key Areas supporting populations of any of them should be regarded as high conservation priorities. Nearly all the remaining threatened Mexican endemics are known from only one protected Key Area (Table 1), which in each case will potentially be of major importance for the continued survival of the species in question (see 'Outlook', below); only *Amazona oratrix* and *Thalurania ridgwayi* have populations in two or more protected areas in Mexico.

RECENT CHANGES TO THE THREATENED LIST

With the publication of Collar *et al.* (1994), one of the 29 threatened species (*Electron carinatum*) was dropped from the Mexican threatened list, and nine birds added: Black-vented Shearwater *Puffinus opisthomelas*, Mountain Plover *Charadrius montanus*, Military Macaw *Ara militaris*, Veracruz Quail-dove *Geotrygon carrikeri*, White-fronted Swift *Cypseloides storeri*, Sinaloa Martin *Progne sinaloae*, Clarión Wren *Troglodytes tanneri*, Worthen's Sparrow *Spizella wortheni* and Belding's Yellowthroat *Geothlypis beldingi*; these additional species have not, however, been included in the Site Inventory (see '*Key Areas*: the book', p. 12).

These changes represent a high percentage difference (35%) in the composition of the threatened species list, and have mostly come about because of the new IUCN criteria applied in Collar *et al.* (1994). *Troglodytes tanneri*, for instance, is now listed because of its small range (endemic to Isla Clarión, MX 30), and the poorly known *Cypseloides storeri* and *Progne sinaloae* have been added as Data Deficient species. However, data subsequent to Collar *et al.* (1992) has prompted the upgrading of species such as *Charadrius montanus* (Knopf and Miller 1994), *Geothlypis beldingi* (Curson 1994, P. Escalante verbally 1994) and *Spizella wortheni* (Wege *et al.* 1993).

Taxonomic changes since Collar *et al.* (1992) have also led to the addition of species. Recent fieldwork has demonstrated that the two subspecies of *Hylorchilus sumichrasti* show marked differences in song and plumage and are best treated as allospecies (Atkinson *et al.* 1993). This viewpoint is followed by Howell and Webb (1995) and the 1994 revised threatened list (Collar *et al.* 1994)—as it is here. A further change involves *Geotrygon carrikeri*, which Peterson (1993) split from Purplish-backed Quail-dove *G. lawrencii*, and is confined to two mountains in the Los Tuxtlas region of south-east Veracruz.

The recent additions to the Mexican threatened species list will have some impact on the Key Area analysis, as many of the species do not occur sympatrically with other threatened birds. *Puffinus opisthomelas* breeds on Isla Guadalupe (MX 01) but its main breeding colony is on Isla Natividad off the Mexican Pacific coast. Priority sites need to be identified in northern Mexico for wintering *Charadrius montanus*, remaining areas in Baja California for *Geothlypis beldingi*, in Coahuila and Nuevo Léon for *Spizella wortheni*, and on the coastal slopes of north-west and north-east Mexico for *Ara militaris*. Identifying Key Areas for *Cypseloides storeri* and *Progne sinaloae* is not really possible as their breeding grounds are unknown.

OLD RECORDS AND LITTLE-KNOWN BIRDS

A number of threatened species in Mexico suffer from a general paucity of recent records from somewhere within their range, but recent records from other areas mean that we are able to make informed recommendations for their conservation. The exceptions are species known only from old specimen records or chance observations (from which there can be no guarantees of a current viable population); these include the following.

Oceanodroma macrodactyla was abundant at its only known breeding colony on Isla Guadalupe (MX 02) in 1906 but had apparently gone by 1922. There does remain, however, a remote possibility that small numbers survive (Collar *et al.* 1992, 1994).

Campephilus imperialis, originally distributed throughout the Sierra Madre Occidental (historical records from Key Areas MX 02, 04, 06, 23 and 25), has not certainly been seen since 1958. The first comprehensive search in 25 years, currently in progress, was initiated in November 1994 (Lammertink and Rojas Tomé 1995a,b).

Zenaida graysoni was fairly common in the forests of Socorro, but was last seen in 1958 and is classified as Extinct in the Wild (Collar *et al.* 1994). However, over 200 birds are held in captivity in the U.S.A. and Europe, and reintroductions are planned.

Table 2. Matrix of threatened species by Key Area.

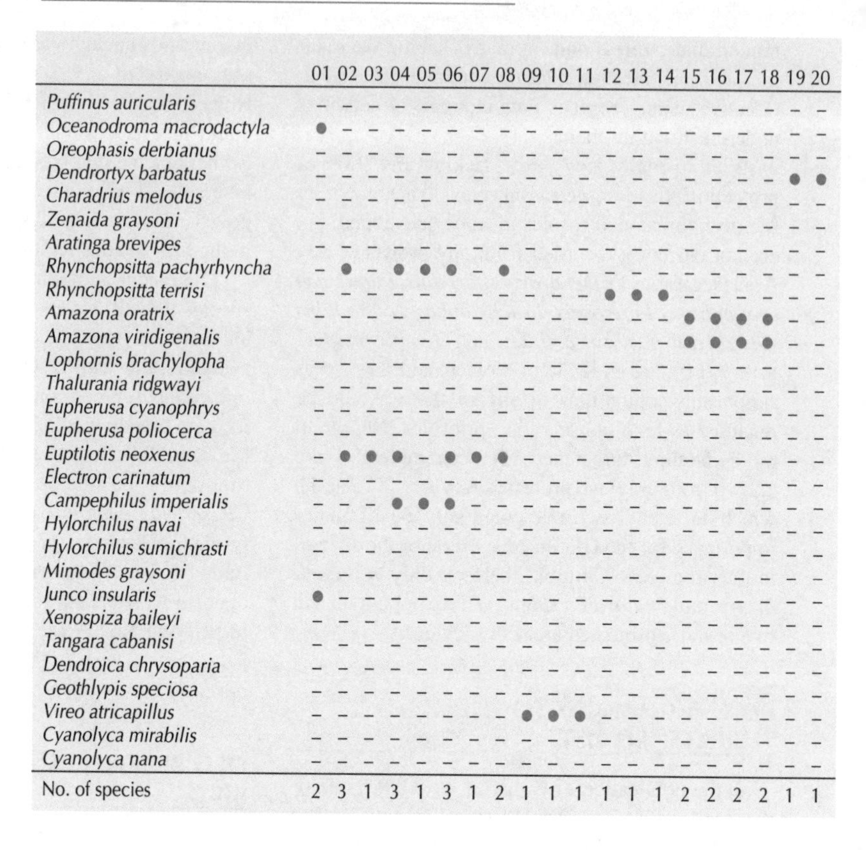

	01	02	03	04	05	06	07	08	09	10	11	12	13	14	15	16	17	18	19	20
Puffinus auricularis	–	–	–	–	–	–	–	–	–	–	–	–	–	–	–	–	–	–	–	–
Oceanodroma macrodactyla	●	–	–	–	–	–	–	–	–	–	–	–	–	–	–	–	–	–	–	–
Oreophasis derbianus	–	–	–	–	–	–	–	–	–	–	–	–	–	–	–	–	–	–	–	–
Dendrortyx barbatus	–	–	–	–	–	–	–	–	–	–	–	–	–	–	–	–	–	–	●	●
Charadrius melodus	–	–	–	–	–	–	–	–	–	–	–	–	–	–	–	–	–	–	–	–
Zenaida graysoni	–	–	–	–	–	–	–	–	–	–	–	–	–	–	–	–	–	–	–	–
Aratinga brevipes	–	–	–	–	–	–	–	–	–	–	–	–	–	–	–	–	–	–	–	–
Rhynchopsitta pachyrhyncha	–	●	–	●	●	●	–	●	–	–	–	–	–	–	–	–	–	–	–	–
Rhynchopsitta terrisi	–	–	–	–	–	–	–	–	–	–	–	●	●	●	–	–	–	–	–	–
Amazona oratrix	–	–	–	–	–	–	–	–	–	–	–	–	–	–	●	●	●	●	–	–
Amazona viridigenalis	–	–	–	–	–	–	–	–	–	–	–	–	–	–	●	●	●	●	–	–
Lophornis brachylopha	–	–	–	–	–	–	–	–	–	–	–	–	–	–	–	–	–	–	–	–
Thalurania ridgwayi	–	–	–	–	–	–	–	–	–	–	–	–	–	–	–	–	–	–	–	–
Eupherusa cyanophrys	–	–	–	–	–	–	–	–	–	–	–	–	–	–	–	–	–	–	–	–
Eupherusa poliocerca	–	–	–	–	–	–	–	–	–	–	–	–	–	–	–	–	–	–	–	–
Euptilotis neoxenus	–	●	●	●	–	●	●	●	–	–	–	–	–	–	–	–	–	–	–	–
Electron carinatum	–	–	–	–	–	–	–	–	–	–	–	–	–	–	–	–	–	–	–	–
Campephilus imperialis	–	–	–	●	●	●	–	–	–	–	–	–	–	–	–	–	–	–	–	–
Hylorchilus navai	–	–	–	–	–	–	–	–	–	–	–	–	–	–	–	–	–	–	–	–
Hylorchilus sumichrasti	–	–	–	–	–	–	–	–	–	–	–	–	–	–	–	–	–	–	–	–
Mimodes graysoni	–	–	–	–	–	–	–	–	–	–	–	–	–	–	–	–	–	–	–	–
Junco insularis	●	–	–	–	–	–	–	–	–	–	–	–	–	–	–	–	–	–	–	–
Xenospiza baileyi	–	–	–	–	–	–	–	–	–	–	–	–	–	–	–	–	–	–	–	–
Tangara cabanisi	–	–	–	–	–	–	–	–	–	–	–	–	–	–	–	–	–	–	–	–
Dendroica chrysoparia	–	–	–	–	–	–	–	–	–	–	–	–	–	–	–	–	–	–	–	–
Geothlypis speciosa	–	–	–	–	–	–	–	–	–	–	–	–	–	–	–	–	–	–	–	–
Vireo atricapillus	–	–	–	–	–	–	–	–	●	●	●	–	–	–	–	–	–	–	–	–
Cyanolyca mirabilis	–	–	–	–	–	–	–	–	–	–	–	–	–	–	–	–	–	–	–	–
Cyanolyca nana	–	–	–	–	–	–	–	–	–	–	–	–	–	–	–	–	–	–	–	–
No. of species	2	3	1	3	1	3	1	2	1	1	1	1	1	1	2	2	2	2	1	1

Table 3. Top Key Areas in Mexico. Those with the area number in bold currently have some form of protected status.

Key Area		No. of threatened spp.	Comments
01	Isla de Guadalupe	2	This Endemic Bird Area holds two single-island endemics: *Oceanodroma macrodactyla* is probably extinct and *Junco insularis* numbers as few as 100 birds.
02	Mesa de Huaracán	2	An important breeding area for *Rhynchopsitta pachyrhyncha; Euptilotis neoxenus* also occurs.
13	Cumbres de Monterrey	1	An important protected area for *Rhynchopsitta terrisi.*
16	Gómez Farías	2	An important protected area for two threatened *Amazona* parrots.
21	Islas Marías	1	These islands possess a significant population of an endemic subspecies of *Amazona oratrix*.
24	El Carricito del Huichol	2	The largest remaining tract of old-growth forest in the Sierra de Huicholes with important populations of *Euptilotis neoxenus* and wintering *Rhynchopsitta pachyrhyncha.*
27	Nevados de Colima	3	An important wintering area for *Rhynchopsitta pachyrhyncha* and possibly *Vireo atricapillus; Thalurania ridgwayi* is also present.
35	Upper Río Lerma	1	The stronghold for *Geothlypis speciosa.*
29	Socorro Island	4	This Endemic Bird Area holds three threatened single-island endemics and the principal breeding population of *Puffinus auricularis.*
37	El Capulin–La Cima	1	The only site where *Xenospiza baileyi* is regularly seen.
39	Coatepec	1	Supports *Dendrortyx barbatus.*
43	Atoyac de Alvarez–Teotepec road	3	The only known area for *Lophornis brachylopha*; important populations of *Eupherusa poliocerca* and *Cyanolyca mirabilis* are also present.
47	Cerro San Felipe	1	The only currently known area for *Cyanolyca nana.*
56	El Triunfo	2	This reserve holds substantial populations of *Tangara cabanisi* and is also important for *Oreophasis derbianus.*

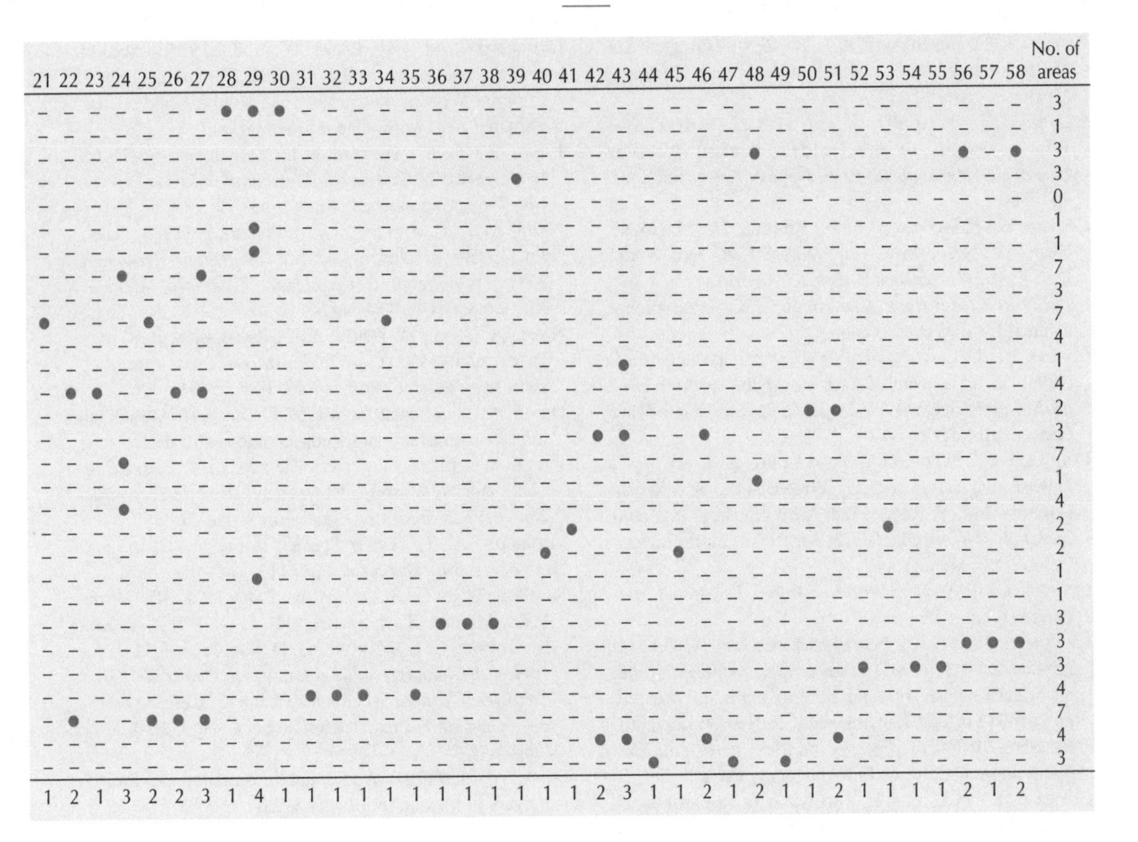

21	22	23	24	25	26	27	28	29	30	31	32	33	34	35	36	37	38	39	40	41	42	43	44	45	46	47	48	49	50	51	52	53	54	55	56	57	58	No. of areas
–	–	–	–	–	–	–	●	●	●	–	–	–	–	–	–	–	–	–	–	–	–	–	–	–	–	–	–	–	–	–	–	–	–	–	–	–	–	3
–	–	–	–	–	–	–	–	–	–	–	–	–	–	–	–	–	–	–	–	–	–	–	–	–	–	–	–	–	–	–	–	–	–	–	–	–	–	1
–	–	–	–	–	–	–	–	–	–	–	–	–	–	–	–	–	–	–	–	–	–	–	–	–	–	–	●	–	–	–	–	–	–	–	●	–	●	3
–	–	–	–	–	–	–	–	–	–	–	–	–	–	–	–	–	–	●	–	–	–	–	–	–	–	–	–	–	–	–	–	–	–	–	–	–	–	3
–	–	–	–	–	–	–	–	–	–	–	–	–	–	–	–	–	–	–	–	–	–	–	–	–	–	–	–	–	–	–	–	–	–	–	–	–	–	0
–	–	–	–	–	–	–	–	●	–	–	–	–	–	–	–	–	–	–	–	–	–	–	–	–	–	–	–	–	–	–	–	–	–	–	–	–	–	1
–	–	–	–	–	–	–	–	●	–	–	–	–	–	–	–	–	–	–	–	–	–	–	–	–	–	–	–	–	–	–	–	–	–	–	–	–	–	1
–	–	–	●	–	–	●	–	–	–	–	–	–	–	–	–	–	–	–	–	–	–	–	–	–	–	–	–	–	–	–	–	–	–	–	–	–	–	7
–	–	–	–	–	–	–	–	–	–	–	–	–	–	–	–	–	–	–	–	–	–	–	–	–	–	–	–	–	–	–	–	–	–	–	–	–	–	3
●	–	–	–	●	–	–	–	–	–	–	–	–	●	–	–	–	–	–	–	–	–	–	–	–	–	–	–	–	–	–	–	–	–	–	–	–	–	7
–	–	–	–	–	–	–	–	–	–	–	–	–	–	–	–	–	–	–	–	–	–	–	–	–	–	–	–	–	–	–	–	–	–	–	–	–	–	4
–	–	–	–	–	–	–	–	–	–	–	–	–	–	–	–	–	–	–	–	–	–	●	–	–	–	–	–	–	–	–	–	–	–	–	–	–	–	1
–	●	●	–	–	●	●	–	–	–	–	–	–	–	–	–	–	–	–	–	–	–	–	–	–	–	–	–	–	–	–	–	–	–	–	–	–	–	4
–	–	–	–	–	–	–	–	–	–	–	–	–	–	–	–	–	–	–	–	–	–	–	–	–	–	–	–	–	●	●	–	–	–	–	–	–	–	2
–	–	–	–	–	–	–	–	–	–	–	–	–	–	–	–	–	–	–	–	–	●	●	–	–	●	–	–	–	–	–	–	–	–	–	–	–	–	3
–	–	–	●	–	–	–	–	–	–	–	–	–	–	–	–	–	–	–	–	–	–	–	–	–	–	–	–	–	–	–	–	–	–	–	–	–	–	7
–	–	–	–	–	–	–	–	–	–	–	–	–	–	–	–	–	–	–	–	–	–	–	–	–	–	–	●	–	–	–	–	–	–	–	–	–	–	1
–	–	–	●	–	–	–	–	–	–	–	–	–	–	–	–	–	–	–	–	–	–	–	–	–	–	–	–	–	–	–	–	–	–	–	–	–	–	4
–	–	–	–	–	–	–	–	–	–	–	–	–	–	–	–	–	–	–	–	●	–	–	–	–	–	–	–	–	–	–	–	●	–	–	–	–	–	2
–	–	–	–	–	–	–	–	–	–	–	–	–	–	–	–	–	–	–	●	–	–	–	–	●	–	–	–	–	–	–	–	–	–	–	–	–	–	2
–	–	–	–	–	–	–	–	●	–	–	–	–	–	–	–	–	–	–	–	–	–	–	–	–	–	–	–	–	–	–	–	–	–	–	–	–	–	1
–	–	–	–	–	–	–	–	–	–	–	–	–	–	–	–	–	–	–	–	–	–	–	–	–	–	–	–	–	–	–	–	–	–	–	–	–	–	1
–	–	–	–	–	–	–	–	–	–	–	–	–	–	–	●	●	●	–	–	–	–	–	–	–	–	–	–	–	–	–	–	–	–	–	–	–	–	3
–	–	–	–	–	–	–	–	–	–	–	–	–	–	–	–	–	–	–	–	–	–	–	–	–	–	–	–	–	–	–	–	–	–	–	●	●	●	3
–	–	–	–	–	–	–	–	–	–	–	–	–	–	–	–	–	–	–	–	–	–	–	–	–	–	–	–	–	–	–	●	–	●	●	–	–	–	3
–	–	–	–	–	–	–	–	–	–	●	●	●	–	●	–	–	–	–	–	–	–	–	–	–	–	–	–	–	–	–	–	–	–	–	–	–	–	4
–	●	–	–	●	●	●	–	–	–	–	–	–	–	–	–	–	–	–	–	–	–	–	–	–	–	–	–	–	–	–	–	–	–	–	–	–	–	7
–	–	–	–	–	–	–	–	–	–	–	–	–	–	–	–	–	–	–	–	–	●	●	–	–	●	–	–	–	–	●	–	–	–	–	–	–	–	4
–	–	–	–	–	–	–	–	–	–	–	–	–	–	–	–	–	–	–	–	–	–	–	●	–	–	●	–	●	–	–	–	–	–	–	–	–	–	3
1	2	1	3	2	2	3	1	4	1	1	1	1	1	1	1	1	1	1	1	1	2	3	1	1	2	1	2	1	1	2	1	1	1	1	2	1	2	

OUTLOOK

This analysis has identified 58 Key Areas, each of which is of major importance for the conservation of threatened birds in Mexico. However, certain Key Areas stand out as top priorities (Table 3). These and the remaining areas (which are individually extremely important for many threatened Mexican endemics: see 'Key Areas' and 'Key Area protection', above) all need some degree of conservation action if the populations of their threatened birds are to remain viable. For many areas there is a requirement to confirm the continued existence or assess the population of a threatened species (*Dendrortyx barbatus* in MX 19 and 20). For others, the need is for locating and defining the extent of suitable habitat in the vicinity of the Key Area. For example, although *Lophornis brachylopha* is currently known only from the Atoyac de Alvarez-Teotepec road (MX 41) in the Sierra Madre del Sur of Guerrero, it very probably occurs throughout the Pacific slope of this mountain range, which remains largely unexplored and still has intact forest; this is similarly the case for *Eupherusa poliocerca* and *Cyanolyca mirabilis* in the Sierra de Yucuyacua (San Andrés Chicahuaxtla, MX 46) and for *E. cyanophrys* and *C. mirabilis* in the Sierra de Miahuatlán (MX 50 and 51).

DATA SOURCES

The above introductory text and the Site Inventory were compiled from information supplied by M. C. Arizmendi, E. Enkerlin, P. B. Escalante, J. Estudillo, S. N. G. Howell, M. Lammertink, A. G. Navarro, J. F. Ornelas, A. T. Peterson, K. Renton, H. Gómez de Silva, G. Spinks and R. G. Wilson, as well as from the following references.

Arizmendi, M. C., Berlanga, H., Márquez-Valdelamar, L., Navarijo, L. and Ornelas, F., eds. (1990) *Avifauna de la región de Chamela, Jalisco*. México: Instituto de Biología, UNAM (Cuadernos 4).

Atkinson, P. W., Whittingham, M. J., Gómez de Silva Garza, H., Kent, A. M. and Maier, R. T. (1993) Notes on the ecology, conservation and taxonomic status of *Hylorchilus* wrens. *Bird Conserv. Internatn.* 3: 75–85.

Balvanera, P., González, S., Jenkins, P., Nelson, G., Benítez, R., Flores, C., Galindo, C., Bunnell, F. and Valencia, S. (1992) Estudio de espécies amenazadas y en peligro de extinción. Centro de Ecología, UNAM. Unpublished report.

BENSON, R. H. AND BENSON, K. L. P. (1990) Estimated size of Black-capped Vireo population in northern Coahuila, Mexico. *Condor* 92: 777–779.

BRAUN, M. J., BRAUN, D. D. AND TERRILL, S. B. (1986) Winter records of the Golden-cheeked Warbler (*Dendroica chrysoparia*) from Mexico. *Amer. Birds* 40: 564–566.

COLLAR, N. J., GONZAGA, L. P., KRABBE, N., MADROÑO NIETO, A., NARANJO, L. G., PARKER, T. A. AND WEGE, D. C. (1992) *Threatened birds of the Americas: the ICBP/IUCN Red Data Book*. Cambridge, U.K.: International Council for Bird Preservation.

COLLAR, N. J., CROSBY, M. J. AND STATTERSFIELD, A. J. (1994) *Birds to watch 2: the world list of threatened birds*. Cambridge, U.K.: BirdLife International (BirdLife Conservation Series no. 4).

COLLAR, N. J., WEGE, D. C. AND LONG, A. J. (in press) Patterns and causes of endangerment in the New World avifauna. In J. V. Remsen, ed. *Natural history and conservation of Neotropical birds*. American Ornithologists' Union (Orn. Monogr.).

CURSON, J. (1994) *New World warblers*. London: Christopher Helm.

ESCALANTE PLIEGO, P., NAVARRO SIGÜENZA, A. G. AND PETERSON, A. T. (1993) A geographic, ecological, and historical analysis of land bird diversity in Mexico. Pp.281–317 in T. P. Ramamoorthy, R. Bye, A. Lot and J. Fa, eds. *Biological diversity of Mexico: origins and distribution*. Oxford: Oxford University Press.

GÓMEZ DE SILVA GARZA, H. AND AGUILAR RODRÍGUEZ, S. (1994) The Bearded Wood-Partridge in central Veracruz and suggestions for finding and conserving the species. *Euphonia* 3: 8–12.

GRZYBOWSKI, J. A. (1991) *Black-capped Vireo (Vireo atricapillus) recovery plan*. Austin, Texas: U.S. Fish and Wildlife Service.

HOWELL, S. N. G. AND WEBB, S. (1992) A little-known cloud forest in Hidalgo, México. *Euphonia* 1: 7–11.

HOWELL, S. N. G. (1993) Status of the Piping Plover in Mexico. *Euphonia* 2: 51–54.

HOWELL, S. N. G. AND WEBB, S. (1995) *A guide to the birds of Mexico and northern central America*. Oxford: Oxford University Press.

IUCN (1992) *Protected areas of the world: a review of national systems, 4: Nearctic and Neotropical*. Gland, Switzerland and Cambridge, U.K.: International Union for Conservation of Nature and Natural Resources.

KNOPF, F. L. AND MILLER, B. J. (1994) *Charadrius montanus*—montane, grassland, or bare-ground plover? *Auk* 111: 504–506.

LAMMERTINK, M. AND ROJAS TOMÉ, J. (1995a) Preliminary report on the Sierra de Huicholes and northern Nayarit field work of the Mexican Mountain Forest—Imperial Woodpecker project. Unpublished report.

LAMMERTINK, M. AND ROJAS TOMÉ, J. (1995b) Second preliminary report on the Mexican Mountain Forest - Imperial Woodpecker project: the southern and central Durango field work. Unpublished report.

LONG, A. J. (1987) The birds of La Yerbabuena. Pp.55–66 in *University of East Anglia Mexican Rainforest Expedition 1986*. Unpublished report.

NAVARRO, A. G. AND ESCALANTE PLIEGO, P. (1993) Aves. Pp.443–501 in *Historia natural del Parque Ecologico Estatal Omiltemi, Chilpancingo, Guerrero, México*. México: CONABIO-UNAM.

NAVARRO, A. G., PETERSON, A. T., ESCALANTE, P. B. AND BENÍTEZ, H. (1992) *Cypseloides storeri*, a new species of swift from Mexico. *Wilson Bull.* 104: 55–64.

ORNELAS, J. F. AND ARIZMENDI, M. C. (in press) Altitudinal migration: implications for conservation of the neotropical migrant avifauna of western Mexico. In A. Estrada, S. Sader and M. Wilson, eds. *Neotropical migratory birds conservation in Mexico*. Studies in Avian Biology.

PETERSON, A. T. (1993) Species status of *Geotrygon carrikeri*. *Bull. Brit. Orn. Club* 113: 166–168.

RAMÍREZ, B. P., DE SUCRE, M. A., PAREDES, Z. R., ORTIZ, P. D., MEDINA, T. I., SÁNCHEZ, R. D. I., PÉREZ, V. M., MONTÁÑEZ, G. L., VARONA, G. D. AND FLORES, H. J. A. (1993) *Estimación avifaunistica en Cerro de Oro, Tuxtepec, Oaxaca*. Sección Mexicana del Consejo Internacional para la Preservación de las Aves (CIPAMEX) (abstract).

SADA, A. M. (1994) Additional information on Piping Plovers in Mexico. *Euphonia* 3: 21.

SALAS, S. H., RAMÍREZ, G., SCHIBILI, I., DE AVILA, A. AND AGUILAR, R. (1994) Analysis of the vegetation and current land use in the state of Oaxaca II: Valles Centrales, Sierra Norte and Papaloapan region. Sociedad para el estudio de los recursos bióticos de Oaxaca, Asociación Civil (SERBO, AC). Unpublished report.

STATTERSFIELD, A. J., CROSBY, M. J., LONG, A. J. AND WEGE, D. C. (in prep.) *Global directory of Endemic Bird Areas*. Cambridge, U.K.: BirdLife International (BirdLife Conservation Series).

VÁZQUEZ, E., CEBALLOS, G., GARCÍA, A. AND RODRÍGUEZ, P. (1994) The Chamela-Cuixmala Biosphere Reserve in western Mexico. Pp.160 in *1994 international meeting of the Society for Conservation Biology and the Association for Tropical Biology*. Guadalajara: Universidad de Guadalajara (abstract).

VIDAL, R. M., MACÍAS-CABALLERO, C. AND DUNCAN, C. D. (1994) The occurrence and ecology of the Golden-cheeked Warbler in the highlands of northern Chiapas, Mexico. *Condor* 96: 684–691.

WEGE, D. C., HOWELL, S. N. G. AND SADA, A. M. (1993) The distribution and status of Worthen's Sparrow *Spizella wortheni*: a review. *Bird Conserv. Internatn.* 3: 211–220.

SITE INVENTORY

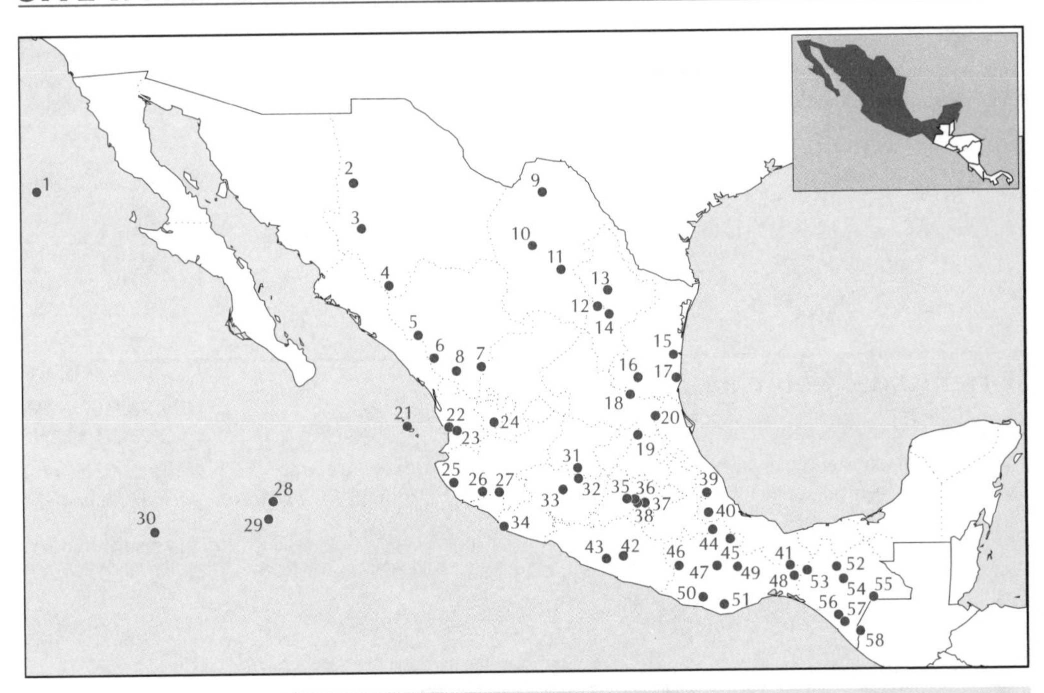

Figure 2. Key Areas in Mexico

01 Isla de Guadalupe
02 Mesa de Huaracán
03 Barranca del Cobre
04 Cerro Mohinoro
05 Las Bufas and San Blas
06 Mexiquillo
07 La Michilía
08 Monte Oscuro
09 Serranías del Burro
10 Sierra Madera
11 Sierra San Marcos
12 San Antonio de las Alazanas
13. Cumbres de Monterrey
14 Cerro del Potosí
15 Soto La Marina–La Pesca
16 Gómez Farías
17 Los Colorados Ranch
18 Río el Naranjo
19 Xilitla
20 Tlanchinol
21 Islas Marías
22 San Blas
23 Cerro San Juan
24 El Carricito del Huichol
25 Chamela–Cuixmala
26 Manantlán
27 Nevados de Colima
28 Isla San Benedicto
29 Isla Socorro
30 Isla Clarión
31 Lago Yuriria
32 Lago Cuitzeo
33 Lago Pátzcuaro
34 La Placita
35 Upper Río Lerma
36 Los Dinamos
37 El Capulín–La Cima
38 Santa Ana Tlacotenco
39 Coatepec
40 Córdoba
41 Uxpanapa
42 Omiltemi
43 Atoyac de Alvarez–Teotepec road
44 Cuasimulco
45 Cerro Oro
46 San Andrés Chicahuaxtla
47 Cerro San Felipe
48 Chimalapas
49 Cerro Zempoaltepec
50 Puerto Escondido–Oaxaca City road
51 Puerto Angel–Oaxaca City road
52 Selva Negra
53 Río la Venta–El Ocote
54 San Cristóbal de las Casas
55 Lagunas de Montebello
56 El Triunfo
57 Monte Ovando
58 Volcán Tacaná

Isla de Guadalupe (Baja California)

MX 01

29°00′N 118°20′W
EBA A02

Isla Guadalupe Special Biosphere Reserve (IUCN category I), 25,000 ha

This volcanic Pacific island of 282 km² lies c.280 km west of northern Baja California. The flora and fauna have suffered terribly at man's hand, most notably through the introduction of cats and goats at the turn of the nineteenth century. The Guadalupe Caracara *Polyborus lutosus* was extinct by the end of 1900.

Oceanodroma macrodactyla	1912	Probably extinct, but a thorough survey of all likely breeding sites at the appropriate season has not been made since 1906.
Junco insularis	1989	The population in 1989 was estimated at c.100 birds, scattered throughout the northern half of the island where adequate vegetation persists.

Mesa de Huaracán (Chihuahua)

MX 02

Unprotected

29°10′N 108°10′W
EBA A04

This area embraces the pine forests on the Mesa del Huaracán around Candelaria and Pito Real in Madera municipality, situated in the highlands of the Sierra Madre Occidental of west-central Chihuahua. Some of these forests have been damaged by fire (Balvanera *et al.* 1992).

Rhynchopsitta pachyrhyncha	1992	A group of 14 birds seen during fieldwork in June–August 1992 (Balvanera *et al.* 1992); 13 pairs bred near Madera in 1979.
Euptilotis neoxenus	1992	Seen twice during fieldwork in June–August 1992 (Balvanera *et al.* 1992).

Barranca del Cobre (Chihuahua)

MX 03

Unprotected

27°45′N 107°55′W
EBA A04

This large canyon is at the heart of the Sierra Madre Occidental in central-south-west Chihuahua, and supports extensive tracts of pine forest. The area should also hold *Rhynchopsitta pachyrhyncha*.

Euptilotis neoxenus	1990s	Regularly recorded near Barranca del Cobre, with sightings especially from the canyon leading to Cascada de Cusarare, south-west of Creel (S. N. G. Howell *in litt.* 1994).

Cerro Mohinoro (Chihuahua)

MX 04

Unprotected

25°58′N 107°03′W
EBA A04

The area ranges from 1,000 to 3,350 m in the highlands of the Sierra Madre Occidental, southern Chihuahua. The main vegetation is pine and fir forest with tropical deciduous forest at lower elevations. Some areas of the pine forest have had dead trees removed and have suffered hurricane damage (Balvanera *et al.* 1992).

Rhynchopsitta pachyrhyncha	1992	A group of six seen during fieldwork in June–August 1992 (Balvanera *et al.* 1992).
Euptilotis neoxenus	1992	Seen twice during fieldwork in June–August 1992 (Balvanera *et al.* 1992).
Campephilus imperialis	1937	The most recent record is of a male collected at 3,050 m in May 1937.

Las Bufas and San Blas (Durango)

MX 05

Unprotected

24°26′N 106°09′W
EBA A04

This area is in the highlands of the Sierra Madre Occidental in western Durango. It comprises dry open pine–oak forest and some of the region's last-remaining old-growth forests, mainly fir with some pine (Lammertink and Rojas Tomé 1995b).

Rhynchopsitta pachyrhyncha	1990s	Lammertink and Rojas Tomé (1995b) found a group of c.40 near San Blas village in March 1995; local people reported May–June breeding in the higher parts of the sierra at nearby Las Bufas.
Campephilus imperialis	1940s	From field surveys and interviews in March 1995, Lammertink and Rojas Tomé (1995b) concluded that it is now extinct in the area.

Mexiquillo (Durango)

MX 06

Unprotected

23°42′N 105°40′W
EBA A04

This canyon area lies in the Sierra Madre Occidental only a few kilometres south of the main road (route 40) near La Ciudad and between Durango and Mazatlán in south-west Durango. It supports c.1,200 ha of pristine pine–oak forest, which is currently safeguarded by a development plan to create a tourist centre there (Lammertink and Rojas Tomé 1995b).

Rhynchopsitta pachyrhyncha	1990s?	Although Lammertink and Rojas Tomé (1995b) did not record it during surveys in March 1995, it was reliably reported by local people as passing through in winter.
Euptilotis neoxenus	1995	A group of 10 seen at 2,200 m in open pine–oak forest in March 1995 (Lammertink and Rojas Tomé 1995b).
Campephilus imperialis	1882	From field surveys and interviews in March 1995, Lammertink and Rojas Tomé (1995b) concluded that it is now extinct in the area.

La Michilía (Durango)

MX 07

La Michilía Biosphere Reserve (IUCN category IX), 35,000 ha

23°25′N 104°10′W
EBA A04

This reserve is on the interior slope of the Sierra Madre Occidental around San Juan de Michilía in south-east Durango. Land varies in elevation from 1,700 to 2,950 m, and the topography is rugged with deep canyons and valleys, and the principal habitat is dry forest of pine and oak.

Euptilotis neoxenus	1980	Two pairs recorded in July 1980.

Monte Oscuro (Durango)

MX 08

Unprotected

23°18′N 104°58′W
EBA A04

This canyon area lies at and above 2,000 m in the Sierra Madre Occidental, c.80 km south-west of Mezquital in southern Durango. It holds c.1,100 ha of pristine pine–oak forest (and a population of Spotted Owl *Strix occidentalis*), which is currently safeguarded as logging is uneconomical (Lammertink and Rojas Tomé 1995b). Proposals suggesting the area be designated a sanctuary have been sent to the Instituto Nacional de Ecología (INE) for consideration (Lammertink and Rojas Tomé 1995b).

Rhynchopsitta pachyrhyncha	1995	A group of c.32 was seen at 2,300 m in January 1995 but did not roost in the canyon (Lammertink and Rojas Tomé 1995b).
Euptilotis neoxenus	1995	A group of c.16 seen at 2,050 m in lush riparian forest, January 1995 (Lammertink and Rojas Tomé 1995b).

Serranías del Burro (Coahuila)

MX 09

Unprotected

28°50′N 102°10′W
EBA A06

This mountain-range stretches c.150 km and rises to over 2,000 m in northern Coahuila, close to the border with the U.S.A. The species below occupies rich, dense, desert scrub, principally on the lower northern slopes of the mountain and along the river valleys to the canyon entrances.

Vireo atricapillus	1989	A major breeding area for the species: surveys in 1989 found 28 pairs (Benson and Benson 1990).

Sierra Madera (Coahuila)

MX 10

Unprotected

27°10′N 102°30′W
EBA A06

The Sierra Madera is a small range in central Coahuila, c.30 km south of Ocampo and c.20 km west of Cuatro Ciénagas de Carranza. The species below occupies rich, dense, desert scrub, mainly at the base of the northern slopes of the sierra and along the valleys of northward-flowing rivers.

Vireo atricapillus	1980s	A survey in the 1980s confirmed breeding (Grzybowski 1991).

Sierra San Marcos (Coahuila)

MX 11

Unprotected

26°25′N 101°35′W
EBA A06

This mountain range is located c.60 km south-west of Monclova in east-central Coahuila. The species below occupies the rich, dense, desert scrub found at the base of the northern slopes of the sierra and along the valleys of northward-flowing rivers.

Vireo atricapillus	1980s	Found breeding in this sierra, probably at the southernmost limit of the species' breeding range (Grzybowski 1991).

San Antonio de las Alazanas (Coahuila)

MX 12

Unprotected

25°16′N 100°25′W
EBA A06

This area embraces the pine forests near San Antonio de las Alazanas, c.60 km south-east of Saltillo in south-east Coahuila near the border with Nuevo León.

Rhynchopsitta terrisi	1990s	Recorded from several sites such as 15 km north of Los Lirios, Mesa de las Tablas and Ciruela.

Cumbres de Monterrey (Nuevo León)

MX 13

Cumbres de Monterrey National Park (IUCN category II), 246,500 ha

25°45′N 101°30′W
EBA A06

This area embraces the highland pine forests in the Sierra Madre Oriental to the east of Monterrey City. The national park is one of the largest designated protected areas in Mexico, but unfortunately is poorly protected. A working group is investigating how to give effective protection to the park, and an initial proposal includes reducing its size to c.150,000 ha in order to focus on the area of highest conservation value (E. Enkerlin *in litt.* 1995).

Rhynchopsitta terrisi	1994	The largest single observation since 1978, and perhaps ever, was made in October 1994 when 1,480 were seen (E. Enkerlin *in litt.* 1995). A significant portion of this species' range is covered by the national park.

MX 14

Cerro del Potosí (Nuevo León)

25°02′N 100°03′W
EBA A06

Unprotected

This area comprises the pine forests around Potosí in central Nuevo León. Much of the available habitat was damaged by fire in 1978 and logging mills have operated in the vicinity. Land above 2,000 m on Cerro del Potosí, which covers c.6,000 ha, is soon to be declared a protected area (E. Enkerlin *in litt.* 1995).

Rhynchopsitta terrisi	1990s	The species' most southerly known breeding area.

MX 15

Soto La Marina–La Pesca (Tamaulipas)

23°45′N 98°00′W
EBA A07

Unprotected

Lying at sea-level on the coastal plain of eastern Tamaulipas, c.150 km east of Ciudad Victoria, this area comprises tropical dry deciduous forest. There is some potential for a protected area to be established in the region (E. Enkerlin *in litt.* 1995).

Amazona oratrix	1990s	Small but stable population.
Amazona viridigenalis	1990s	Small but stable population.

MX 16

Gómez Farías (Tamaulipas)

23°03′N 99°09′W
EBA A07

El Cielo Biosphere Reserve (IUCN category IX), 144,530 ha

This area, which includes the Río Sabinas valley (22°59′N 98°58′W) and specifically Rancho Rinconada and the Río Frío district, is located on the lower Atlantic slope of the Sierra Madre Oriental of southern Tamaulipas. Principal habitats are semi-deciduous gallery forest (along the valleys), deciduous forest on slopes and in canyons, and dry open pine–oak forest on ridges above 1,000 m. The reserve holds one of the most extensive tracts of pristine forest on the Atlantic slope of north-east Mexico.

Amazona oratrix	1990s	Recorded occasionally on Christmas bird counts, always in small numbers.
Amazona viridigenalis	1990s	No details on current populations, but recorded almost annually on Christmas bird counts.

MX 17

Los Colorados Ranch (Tamaulipas)

23°02′N 97°55′W
EBA A07

Unprotected

Los Colorados is a privately owned, 800-ha ranch on the coastal plain of southern Tamaulipas, east-north-east of Aldama, where the main habitat is tropical dry deciduous forest which exists in patches among cattle pasture (E. Enkerlin *in litt.* 1995). Although 80-85% of the land is cleared for cattle pasture many large trees remain and both threatened *Amazona* species breed, with more than 100 nests found there during a recently completed five-year study (E. Enkerlin *in litt.* 1995).

Amazona oratrix	1990s	Small numbers breed annually.
Amazona viridigenalis	1990s	Small numbers breed annually.

Río el Naranjo (San Luis Potosí)

MX 18

Unprotected

22°30′N 99°24′W
EBA A07

Río el Naranjo is centred on Las Abritas village in north-central San Luis Potosí, close to the border with Tamaulipas. The vegetation consists mainly of humid oak–sweetgum forest, brushland, arid upland grassland and dry oak forest.

Amazona oratrix	1990	Uncommon resident, seen only in small groups or pairs.
Amazona viridigenalis	1990s	Nests in oakwoods in April, when it is quite common; scarce in winter, especially December–January (S. N. G. Howell *in litt.* 1995).

Xilitla (San Luis Potosí)

MX 19

Unprotected

21°15′N 99°10′W
EBA A11

An area including the peaks of Cerro Miramar, Cerro San Antonio (just west of Xilitla) and Cerro Conejo (west of Cerro San Antonio) in extreme south-east San Luis Potosí. The main habitat is subtropical humid evergreen forest (including cloud-forest), although these have been extensively cut and cleared such that identifying and surveying the largest surviving tracts is an important priority.

Dendrortyx barbatus	1951	Several collected. Current status needs confirmation.

Tlanchinol (Hidalgo)

MX 20

Unprotected

21°05′N 98°36′W
EBA A11

This area encompasses the montane forests located 4 km and 8 km north of Tlanchinol town along route 105. The forest appears to be one of the largest remaining tracts of cloud-forest in Hidalgo (Howell and Webb 1992).

Dendrortyx barbatus	1986	One seen in 1986, but not recorded during a three-day visit in June 1990 (Howell and Webb 1992) or during fieldwork in 1989–1993 (B. E. Hernández-Baños *per* A. T. Peterson and A. G. Navarro *in litt.* 1995).

Islas Marías (Nayarit)

MX 21

Unprotected

21°35′N 106°33′W
EBA A05

The four islands, lying c.150 km off the coast of Nayarit, are San Juanito, María Madre, María Magdalena and María Cleofas. They support a number of endemic taxa.

Amazona oratrix	1990s	The endemic subspecies *tresmariae* is nomadic within and between islands so no discrete site can be identified for it, although in April 1983 hundreds, possibly thousands, gathered daily on fruit trees in the main port of Balleto on María Madre. However, transects on the island in 1984 calculated a mean of four per 8 km and estimated the population at fewer than 800.

San Blas (Nayarit)

MX 22

21°33′N 105°13′W
EBA A05

Unprotected

Inland from the coastal town of San Blas in central Nayarit lies an area of foothills and humid canyons on the Pacific slope. The main habitats are tropical semi-deciduous forest, brushy thickets and shaded coffee plantations.

Thalurania ridgwayi	1990s	Inhabits wooded gullies and coffee plantations above La Bajada village, near San Blas.
Vireo atricapillus	1990s	Small numbers winter in varied habitats including brushy thickets, tangles associated with palm forest, overgrown fields and coffee plantations, but the bird is easily overlooked (S. N. G. Howell *in litt.* 1995).

Cerro San Juan (Nayarit)

MX 23

21°26′N 104°58′W
EBA A05

Cerro San Juan Special Biosphere Reserve (IUCN category IV), 27,000 ha

This reserve is located in coastal foothills on the Pacific slope, c.25 km south-west of Tepic in south-west Nayarit. The vegetation is tropical semi-evergreen forest with some montane forest, and is still in excellent condition (P. B. Escalante *in litt.* 1995).

Thalurania ridgwayi	1990s	Recorded, but no details available (P. B. Escalante *in litt.* 1995).

El Carricito del Huichol (Jalisco)

MX 24

21°41′N 103°47′W
EBA A04

Unprotected

This area embraces the highest part (c.2,500 m) of the south-east portion of the Sierra de Huicholes. It comprises four patches of mature pine–oak forest, totalling 2,100 ha, among drier and secondary pine–oak forest. Lammertink and Rojas Tomé (1995a) found the area to be the most important remaining old-growth zone within the entire Sierra de Huicholes and mountains of northern Nayarit, and recommended that a protected area of 11,000 ha be established. *Xenospiza baileyi* (the type-series) was collected only 20 km to the east at Bolaños in 1889.

Rhynchopsitta pachyrhyncha	1990s	Not recorded by Lammertink and Rojas Tomé (1995a) during surveys in November–December 1994, but reliably reported as a winter visitor (January–March), in flocks of 30–100.
Euptilotis neoxenus	1994	Seen several times at different sites within the proposed reserve (Lammertink and Rojas Tomé 1995a).
Campephilus imperialis	1940s	Extinct in the area, although the oldest Huichol indians remember it (Lammertink and Rojas Tomé 1995a).

Chamela–Cuixmala (Jalisco)

MX 25

Chamela–Cuixmala Biosphere Reserve (IUCN category IX), 13,142 ha

19°50′N 105°03′W
EBA A05

The ecology of this recently decreed reserve, comprising an area from Arroyo Chamela to the Río Cuitzmala on the coastal plain of south-east Jalisco, has been studied extensively at the Chamela Research Station, which lies within the reserve. The main vegetation type is tropical dry deciduous forest, although the reserve also has semi-deciduous forest, riparian vegetation, mangrove swamps, lagoons and coastal dunes. Importantly, Chamela-Cuixmala is one of the only protected areas of tropical deciduous forest in Mexico (Vázquez *et al.* 1994).

Amazona oratrix	c.1995	Irregular; occurs locally in small numbers in lowland semi-deciduous forest (K. Renton *in litt.* 1995).
Vireo atricapillus	1980s	Listed for the reserve (Arizmendi *et al.* 1990).

Manantlán (Jalisco/Colima)

MX 26

Sierra de Manantlán Biosphere Reserve (IUCN category IX), 139,577 ha

19°33′N 104°10′W
EBA A05, A11

Lying at 400–2,860 m in the Sierra de Manantlán of south-west Jalisco, with a small part in neighbouring Colima, the reserve supports a variety of vegetation types including tropical dry forest, tropical deciduous forest, oak, pine, fir and cloud-forest. There are historical records of *Euptilotis neoxenus* which remain unconfirmed (Ornelas and Arizmendi in press).

Thalurania ridgwayi	1992	Common at Puerto Los Mazos on the lower slope of the reserve. Several were trapped at Las Joyas field station in August–September 1992 at 1,990 m, indicating that it is an altitudinal migrant (Ornelas and Arizmendi in press).
Vireo atricapillus	1992	Rare: two trapped at Las Joyas field station in riparian growth in November 1992 (J. F. Ornelas *in litt.* 1995).

Nevados de Colima (Jalisco/Colima)

MX 27

Nevados de Colima National Park (IUCN category II), 22,200 ha

19°31′N 103°38′W
EBA A05, A11

The park embraces Volcán de Fuego de Colima (3,820 m) and Volcán Nevado de Colima (4,240 m), which support extensive areas of forest, most notably pine and pine–oak, but also semi-deciduous forest, coffee plantations and fields on the lower slopes.

Rhynchopsitta pachyrhyncha	1990s	This area appears to hold one of the major wintering populations.
Thalurania ridgwayi	1990s	Present on the volcano's slopes (A. T. Peterson and A. G. Navarro *in litt.* 1995).
Vireo atricapillus	1990s	Found to be fairly common on the lower south-west slopes of Volcán de Fuego at 1,000–2,000 m, inhabiting wooded gullies, coffee plantations and even overgrown fields (S. N. G. Howell *in litt.* 1995).

Isla San Benedicto (Colima)

MX 28

Revillagigedo Islands Biosphere Reserve (IUCN category IX), 636,685 ha (in part)

19°18′N 110°49′W

San Benedicto, c.400 km south of the tip of Baja California, is one of the four oceanic, volcanic islands comprising the Revillagigedo group. An eruption in 1952 covered the 20-km² island in ash and pumice and devastated its seabird colony. Some low vegetation has regenerated, mostly in the northern part.

Puffinus auricularis	1952?	The colony is thought to have been wiped out by the eruption, but sightings of the species a few kilometres north of the island suggest that it might recolonize.

MX 29

Isla Socorro (Colima)

18°45′N 110°58′W
EBA A10

Revillagigedo Islands Biosphere Reserve (IUCN category IX), 636,685 ha (in part)

Socorro (150 km²) is the largest of four oceanic islands comprising the Revillagigedo group. It consists mainly of undulating hills, but there are steep slopes and dry canyons with the highest point on Cerro Evermann (1,130 m). The natural vegetation is scrub below 700 m and broadleaved forest at higher elevations and on the northern slope.

Puffinus auricularis	1990s	The stronghold for the species, though the population is threatened by feral cats; breeding is mainly in forest above 550 m.
Zenaida graysoni	1958	This Socorro endemic is now known only from a (relatively healthy) captive population. A conservation programme plans eventually to reintroduce it.
Aratinga brevipes	1990s	An estimated 400–500 birds in 1990–1991, most commonly found in forest at 350–850 m.
Mimodes graysoni	1990s	Work from June 1993 to June 1994 produced a population estimate of 300–400 birds (J. E. Martínez-Gómez and R. L. Curry in prep.).

MX 30

Isla Clarión (Colima)

18°22′N 114°40′W

Revillagigedo Islands Biosphere Reserve (IUCN category IX), 636,685 ha (in part)

Clarión (28.4 km²), though part of the Revillagigedo group, lies c.370 km west of the other islands. Most of the terrain is gently sloping, rising to 335 m. It is arid, and rabbits and feral pigs have caused serious damage to the vegetation.

Puffinus auricularis	1986	All burrows found during a survey in 1990 were old, so its current status is unclear.

MX 31

Lago Yuriria (Guanajuato)

20°15′N 101°06′W
EBA A08

Unprotected

Lago Yuriria is a small shallow lake (7 × 3 km) in the Río Lerma drainage, located in southern Guanajuato, c.20 km south of Salamanca and 20 km north of Lago Cuitzeo (MX 32).

Geothlypis speciosa	1991	The most recent records were on the south side of the lake in July. Common Yellowthroat *G. trichas* is more abundant, occurring at a ratio of 7:1.

MX 32

Lago Cuitzeo (Michoacán)

19°55′N 101°05′W
EBA A08

Unprotected

A long (25 km) shallow lake in the Río Lerma drainage of northern Michoacán, close to the border with Guanajuato. Lago Cuitzeo dried out at least once during the 1980s and water-levels fluctuate widely.

Geothlypis speciosa	1990s	Apparently quite abundant, outnumbering Common Yellowthroat *G. trichas* by 3:1.

Lago Pátzcuaro (Michoacán)

MX 33

Unprotected

19°35′N 101°35′W
EBA A08

A shallow lake in the central part of the state, north of Pátzcuaro and 28 km east of Morelia. Water-levels fluctuate dramatically and thus affect the lake-side marshes where the species below occurs.

Geothlypis speciosa	1980s	A small population was recorded a number of times in the 1980s. Common Yellowthroat *G. trichas* is more abundant, occurring at a ratio of 7:1.

La Placita (Michoacán)

MX 34

Unprotected

18°27′N 103°30′W

This village lies on the east bank of the Río Maquili c.1 km from the sea and c.20 km south-east of the Coahuayana estuary. Surveys need to be conducted throughout the coastal zone of Michoacán to assess the current status of this threatened parrot.

Amazona oratrix	1950	Formerly fairly common in coconuts and small trees in the vicinity of the village, but there is no recent information.

Upper Río Lerma (México)

MX 35

Unprotected

19°17′N 99°32′W
EBA A08

This area embraces the headwaters of the Río Lerma west of Mexico City and east of Toluca and includes the marshes around Lerma da Villada, San Mateo Atenco and San Pedro Techuchuco. Much of the area has been drained for agriculture with the few existing marshes still totally unprotected. The area embraced much of the range of the extinct Slender-billed Grackle *Quiscalis palustris*.

Geothlypis speciosa	1990s	Found to be common in several areas, most notably just north of Lerma de Villada, and between Almoloya and Texcalyacac.

Los Dinamos (Distrito Federal)

MX 36

Cañada Los Dinamos (IUCN category V), size unknown

19°16′N 99°17′W
EBA A11

The vegetation of this canyon and adjacent mountains immediately south-west of Mexico City is mainly humid pine forest but there are also open areas of bunch-grass on the highest parts.

Xenospiza baileyi	1993	One bird in September 1993 (P. B. Escalante *in litt.* 1995); one of the only recent records of the species away from Key Area MX 37.

MX 37

El Capulín–La Cima (Distrito Federal/Morelos)

19°20′N 99°22′W
EBA A11

Unprotected

This is an area of bunch-grass (zacatón) interspersed with pine and oak forest near the highest point on the old highway (route 95) between Mexico City and Cuernavaca and immediately south-east of Volcán El Pelado. Agriculture is increasing in the area and the grasslands have been burnt indiscriminately in the past. There is a field station at El Capulín which gives some protection to the habitat (P. B. Escalante *in litt.* 1995).

Xenospiza baileyi	1994	A major stronghold for the species, and the only area where it is seen regularly throughout the year.

MX 38

Santa Ana Tlacotenco (Distrito Federal)

19°09′N 98°58′W
EBA A11

Unprotected

This area comprises open pine–oak and pine woodland with bunch-grass (zacatón) understorey near the road to Ozumba, c.30 km south-east of Mexico City.

Xenospiza baileyi	1990	A sighting of 2–3 birds in January 1990 is the only record despite casual searches (R. G. Wilson *in litt.* 1994)—but the species is easy to overlook except when breeding.

MX 39

Coatepec (Veracruz)

19°27′N 96°58′W
EBA A11

Unprotected

Located 15 km south-west of Jalapa, this area encompasses the humid region between 1,000 and 2,500 m, west of Coatepec and east of the 4,282-m-high Cofre de Perote volcano. The Cofre de Perote National Park (11,700 ha) is centred on the summit of the volcano, and habitat for the threatened species possibly lies on its eastern slopes and thus within the park boundary.

Dendrortyx barbatus	1994	Seen in July 1994 near Coatepec; the first record from the area in 25 years (Gómez de Silva Garza and Aguilar Rodríguez 1994).

MX 40

Córdoba (Veracruz)

18°50′N 96°55′W

Unprotected

Limestone (karst) outcrops covered in semi-deciduous and evergreen forest characterize this area, and are all located near Córdoba, the main area being on the outskirts of Amatlán town. Although these steep rocky limestone areas are generally poor for cultivation, extensive coffee plantations and some limestone quarrying are destroying the forest.

Hylorchilus sumichrasti	1995	Seen regularly in the forest on the limestone outcrops (e.g. G. Spinks *in litt.* 1995).

Uxpanapa (Veracruz/Oaxaca)

MX 41

Unprotected

17°10′N 94°10′W

This humid forest area is in the northern part of the Uxpanapa region, mainly in the Coatzacoalcos drainage of eastern Veracruz, with a small part in neighbouring Oaxaca. Much of the forest has been cleared for agriculture, leaving isolated fragments among the limestone outcrops. Uxpanapa is receiving attention from several conservation organizations including a proposed Los Chimalapas–Uxpanapa Biosphere Reserve which would cover 800,000 ha of this and the neighbouring Chimalapas region (MX 48).

Hylorchilus navai	1995	Locally common in the Veracruz part of Uxpanapa where there are numerous limestone outcrops. Found for the first time in the Oaxaca part of Uxpanapa in April 1995 (H. Gómez de Silva *in litt.* 1995).

Omiltemi (Guerrero)

MX 42

Omiltemi State Ecological Park (IUCN category II), 9,600 ha

17°30′N 99°40′W
EBA A12

This state park is c.20 km west of Chilpancingo in the Sierra Madre del Sur of southern Guerrero. The area is rugged with heavily vegetated humid canyons holding the best habitat for the species mentioned below.

Eupherusa poliocerca	1990s	Favours heavily vegetated humid canyons in the park.
Cyanolyca mirabilis	1980s	Recorded regularly in the 1980s, especially in the humid canyons and montane forests (Navarro and Escalante Pliego 1993).

Atoyac de Alvarez–Teotepec road (Guerrero)

MX 43

Unprotected

17°25′N 100°13′W
EBA A12

This area embraces the forest between Atoyac de Alvarez and Teotepec, along the only road that crosses the rugged and remote Sierra Madre del Sur of southern Guerrero. The 40-km stretch of road varies in elevation from 900 m, where there are patches of semi-evergreen forest, up to pristine cloud-forest at 1,800 m, although coffee cultivation is widespread in the intervening zones and forest clearance is increasing throughout the area. This is the type-locality for the recently described White-fronted Swift *Cypseloides storeri* (Navarro *et al.* 1992).

Lophornis brachylopha	1995	Has been seen along the road near the villages of Arroyo Grande, Paraíso and Nueva Delhi. This is the only area within its tiny range where it is regularly recorded (e.g. G. Spinks *in litt.* 1995).
Eupherusa poliocerca	1995	One of the commonest hummingbirds in these montane forests (e.g. G. Spinks *in litt.* 1995).
Cyanolyca mirabilis	1995	Most sightings come from cloud-forest, especially near Cerro Teotepec (e.g. G. Spinks *in litt.* 1995).

Cuasimulco (Oaxaca)

MX 44
18°17′N 96°48′W
EBA A11

Unprotected

The Cuasimulco area lies in the central part of the Sierra de Juárez (Sierra Norte), an isolated range abutting the Atlantic lowlands of northern Oaxaca. This is one of the most extensive areas of cloud-forest left in Mexico, and also holds large tracts of pristine humid pine–oak forest (Salas *et al.* 1994).

Cyanolyca nana	1968	Collected at Santos Reyes Pábalo, the highest pass in the Sierra de Juárez on the road from Oaxaca City to Tuxtepec (route 75).

Cerro Oro (Oaxaca)

MX 45
18°00′N 96°15′W

Unprotected

Cerro Oro lies between Tuxtepec and the Presidente Miguel de la Madrid reservoir in northern Oaxaca. The area comprises humid evergreen forest on limestone karst, although this is rapidly being cleared for agriculture.

Hylorchilus sumichrasti	1990s	Locally common (Ramírez *et al.* 1993, H. Gómez de Silva *in litt.* 1994).

San Andrés Chicahuaxtla (Oaxaca)

MX 46
17°11′N 97°53′W
EBA A12

Unprotected

San Andrés Chicahuaxtla is near the highest point along the paved highway (route 125) from Putla (de Guerrero) to Laguna which passes over the Sierra de Yucuyacua, a high range adjacent to the Pacific lowlands and close to the border with Guerrero. The habitat along this road is semi-deciduous forest and montane forest, although the traditional collecting localities north of Putla are now heavily deforested and further degraded by goats (S. N. G. Howell *in litt.* 1995).

Eupherusa poliocerca	1987	All sightings are from localities along the main road north of Putla.
Cyanolyca mirabilis	1964	The only observations are from San Andrés Chicahuaxtla. Short visits to the area in recent years failed to find it (A. T. Peterson and A. Navarro *in litt.* 1995).

Cerro San Felipe (Oaxaca)

MX 47
17°10′N 96°40′W
EBA A11

Benito Juárez National Park (IUCN category II), 2,737 ha

Cerro San Felipe is a high mountain (3,100 m) in the Sierra Aloapaneca (Sierra de San Felipe), 13 km north-east of Oaxaca City. The main habitat of the area is humid pine–oak forest and cloud-forest. The southern slope of the mountain lies within Benito Juárez National Park although the extent of overlap is difficult to gauge as the park's boundaries were never demarcated (Salas *et al.* 1994).

Cyanolyca nana	1995	Seen regularly (e.g. G. Spinks *in litt.* 1995).

Chimalapas (Oaxaca)

MX 48
17°10′N 94°20′W

Chimalapas Ecological and Indigenous Reserve, c.20,000 ha
Los Chimalapas–Uxpanapa Biosphere Reserve (proposed), 800,000 ha

The Chimalapas is a mountainous region located at the western end of the Sierra Madre de Chiapas in the municipalities of San Miguel Chimalapa and Santa María Chimalapa in eastern Oaxaca. The area holds some of the most rugged and isolated terrain in Mexico and a wide range of forest types including rainforest, pine–oak forest and montane forest (including cloud- and elfin forest).

Oreophasis derbianus	1980s	Status uncertain: there are unconfirmed reports, and some birds in a private zoo were apparently collected from these mountains (J. Estudillo *in litt.* 1994).
Electron carinatum	1995	One record from near San Isidro la Gringa (A. T. Peterson and A. Navarro *in litt.* 1995).

Cerro Zempoaltepec (Oaxaca)

MX 49
17°08′N 96°01′W
EBA A11

Unprotected

This mountain (3,400 m) is in the Sierra de Zempoaltepec, a small isolated range 100 km west of Oaxaca City. The main habitat is humid pine–oak and montane forest (including cloud-forest).

Cyanolyca nana	1942	This is the south-easternmost area in the species' range. The lack of records in spite of some recent surveys indicates that the bird may be extinct in the area (A. T. Peterson and A. Navarro *in litt.* 1995).

Puerto Escondido–Oaxaca City road (Oaxaca)

MX 50
16°12′N 97°07′W
EBA A12

Unprotected

The road from Puerto Escondido to Oaxaca City (route 131) crosses the central portion of the Sierra de Miahuatlán, an isolated coastal range. The forest between San Pedro Juchatengo and San Gabriel Mixtepec defines the area whose habitat includes the upper reaches of tropical semi-deciduous forest and cloud-forest.

Eupherusa cyanophrys	1990s	Records come from the roadside forest between San Pedro Juchatengo and San Gabriel Mixtepec, with good populations at Santa Rosa–Río Salado (A. T. Peterson and A. Navarro *in litt.* 1995).

Puerto Angel–Oaxaca City road (Oaxaca)

MX 51
15°58′N 96°27′W
EBA A12

Unprotected

The road between the coastal resort of Puerto Angel and Oaxaca City (route 175) crosses the Sierra de Miahuatlán through an area embracing the tropical, semi-deciduous and cloud-forest, though much of these are now heavily degraded.

Eupherusa cyanophrys	1990s	Many sightings from roadside forest, especially between Pluma Hidalgo and La Soledad.
Cyanolyca mirabilis	1990s	Occasional sightings near San Miguel Suchixtepec at its intersection with the Puerto Angel road.

MX 52

Selva Negra (Chiapas)

17°07′N 92°51′W
EBA A14

Unprotected

This area embraces the Selva Negra region around Pueblo Nuevo Solistihuacán in the northern part of the Central Highlands of northern Chiapas. The main habitat is pine–oak–*Liquidambar* forest with montane and cloud-forest on ridges above 2,200 m.

Dendroica chrysoparia	1986	Fieldwork at La Yerbabuena, July–September 1986, first recorded the species on 16 August, with a steady increase in numbers peaking at 40 birds on 5 September (Long 1987). The area should be surveyed in December–February to see if it winters, since it was recently found in San Cristóbal de las Casas (MX 51).

MX 53

Río la Venta–El Ocote (Chiapas)

17°01′N 93°47′W

El Ocote Special Biosphere Reserve (IUCN category IV), 48,140 ha

This area is located at the mouth of the Río la Venta where it enters Laguna Nezahualcóyotl in western Chiapas, but close to the border with Oaxaca and Veracruz states. Limestone outcrops covered by humid evergreen forest are the main habitat.

Hylorchilus navai	1992	An estimate of 10–25 birds per km^2 was made in 1992 (Atkinson *et al.* 1993).

MX 54

San Cristóbal de las Casas (Chiapas)

16°44′N 92°38′W
EBA A14

Cerro Huitepec Reserve (private), 230 ha

This town is located at 2,100 m in the Central Highlands of north-central Chiapas. The main habitat in the vicinity is various types of pine–oak forest, with some montane forest, although a high proportion of the land is cultivated.

Dendroica chrysoparia	1992	Winters in small numbers: 46 sightings, 1990–1992 (Vidal *et al.* 1994).

MX 55

Lagunas de Montebello (Chiapas)

16°10′N 91°40′W
EBA A14

Lagunas de Montebello National Park (IUCN category II), 6,022 ha

The park is in the Central Highlands c.40 km southwest of Comitán in east-central Chiapas, near the Guatemala border. It supports pine–oak–*Liquidambar* forests with montane and cloud-forest in the more humid gullies.

Dendroica chrysoparia	1978	A sighting in January 1978 suggests the species might winter in small numbers within the park (Braun *et al.* 1986).

El Triunfo (Chiapas)

MX 56

El Triunfo Biosphere Reserve (IUCN category I), 119,177 ha

15°37′N 92°48′W
EBA A14

This area is located in the mountains of the Sierra Madre de Chiapas, just within the northern limits of the Soconusco region. There are five core areas totalling 30,000 ha, which embrace the higher mountain slopes and peaks, with the remainder of the reserve comprising a buffer zone. The main habitat above 1,500 m is montane rainforest and cloudforest dominated by oaks, with subtropical forest confined to the lower slopes.

Oreophasis derbianus	1994	Found on a number of the reserve's mountain peaks (e.g. R. Sólis *in litt.* 1994).
Tangara cabanisi	1994	The species' main stronghold, and all recent Mexican records are from several sites within the reserve (e.g. R. Sólis *in litt.* 1994).

Monte Ovando (Chiapas)

MX 57

Unprotected

15°24′N 92°36′W
EBA A14

This mountain is (predominantly) on the Pacific slope of the Sierra Madre de Chiapas, just outside the south-eastern limits of the El Triunfo Biosphere Reserve. The forests are broadleaf evergreen but much of the land below 1,200 m has been cleared for coffee.

Tangara cabanisi	1937	One specimen.

Volcán Tacaná (Chiapas)

MX 58

Unprotected

15°07′N 92°06′W
EBA A14

Volcán Tacaná (4,092 m) lies on the Guatemala border in the Sierra Madre de Chiapas. Habitat destruction has been severe on the Mexican side, with little forest remaining below 1,900 m.

Oreophasis derbianus	1989	The Miguel del Alvarez Toro Zoo received five young birds probably captured on Volcán Tacaná in 1989.
Tangara cabanisi	c.1965	One specimen.

NORTH CENTRAL AMERICA

Golden-cheeked Warbler *Dendroica chrysoparia* (with Hermit Warbler *D. occidentalis*)

IN COMBINATION, the five contiguous countries defined here as north Central America probably support over 750 species of bird: Belize c.519, El Salvador 445, Guatemala c.667, Honduras c.663 and Nicaragua c.614 (Monroe 1968, Land 1970, Steffee 1972, Thurber *et al.* 1987, Miller and Miller 1992). Of these, c.170 are migratory species that breed to the north of the region (Monroe 1968, Howell and Webb 1995), one is a national endemic, 34 (4.5%) have restricted ranges (Stattersfield *et al.* in prep.) and six are threatened (Collar *et al.* 1992). This analysis has identified 14 Key Areas for threatened birds in north Central America (see '*Key Areas*: the book', p. 11, for criteria).

THREATENED BIRDS

Six north Central American species were considered at risk of extinction by Collar *et al.* (1992), one of which, *Amazilia luciae*, is a national (Honduran) endemic (Tables 1 and 2). These threatened birds are split evenly between the tropical/lower subtropical zone (up to c.1,000 m) and the subtropical/temperate zone (1,000 m and above), with four species relying on wet forest (including pine–oak associations), and two (*Amazona oratrix* and *Amazilia luciae*) primarily dependent on dry forest (Collar *et al.* in press). All six species are threatened by the loss of their respective habitats, although for *A. oratrix* this is combined with the devastating effects of trade, with *Oreophasis derbianus* also under pressure from hunting (Collar *et al.* in press). The distributions of these threatened birds and their relationship to Endemic Bird Areas are shown in Figure 1.

KEY AREAS

The 14 north Central American Key Areas would, if adequately protected, help ensure the conservation of all six threatened species in the region—always accepting that important new populations or areas may yet be found. Three of these areas support two threatened species, with the remainder important for their populations of single threatened birds (Table 3). *Amazilia luciae* is totally reliant for its survival on the integrity of habitat in the two Key Areas from which it is known (see 'Outlook', below). Although represented in a number of Key Areas in Mexico,

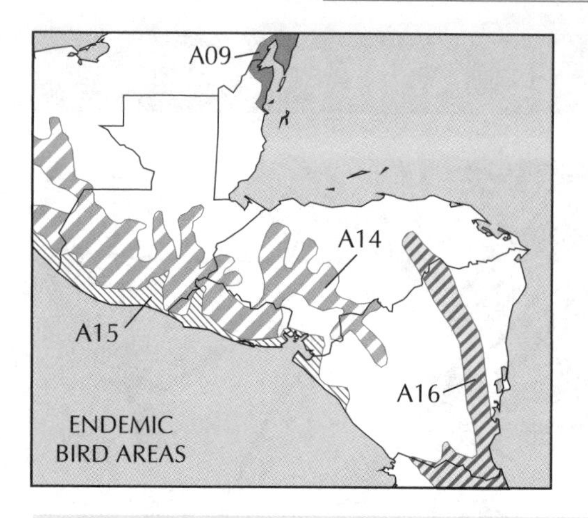

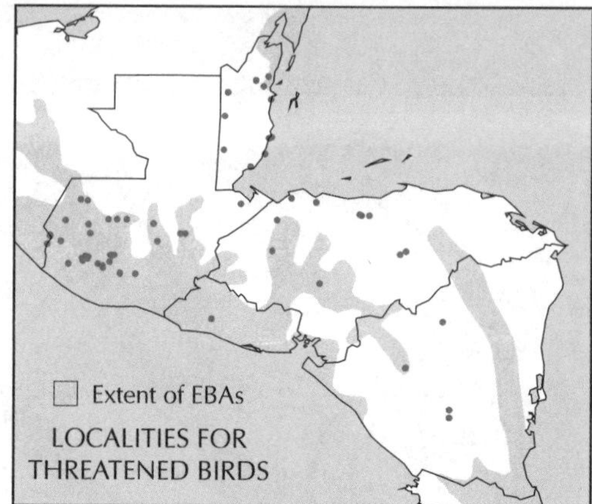

Figure 1. The localities where threatened birds have been recorded in northern Central America and their relationship to Endemic Bird Areas (EBAs).

- Four northern Central American threatened birds have restricted ranges, three of which occur in one of the region's EBAs as listed below (figures are numbers of these species in each EBA). The fourth species, *Amazilia luciae*, however, is confined to a small and dwindling area of arid thorn-forest within the interior lowlands of Honduras, where it is under intense pressure from habitat destruction (in close proximity to, but not within the Central American Caribbean slope EBA, A16).

A14 North Central American highlands (3) A15 North Central American Pacific slope (0) A16 Central American Caribbean slope (0)

only single areas have been identified for *Amazona oratrix* and *Tangara cabanisi* in north Central America; the integrity of these Key Areas is essential for the continued survival of both species in this region. *Dendroica chrysoparia* winters in southernmost Mexico (Chiapas), but the centre of its wintering range appears to be in Guatemala and Honduras (see 'Old records and little-known birds' and 'Outlook', below). Similarly, the centre of abundance for *Electron carinatum* lies within Honduras (and to a lesser extent neighbouring Guatemala and Belize), and thus its survival may well depend on the adequate conservation of the three Key Areas from which it is known in north Central America.

KEY AREA PROTECTION

Of the region's 14 Key Areas, six (43%) currently have formal protected status, although only two (15%) fall within IUCN categories I or II. The eight Key Areas (57% of the total) that are currently unprotected require attention in the form of appropriate conservation measures if the populations of their threatened species are to survive. While most of the threatened species are reasonably well represented, even the formally protected areas remain under threat and, in many, habitat degradation and hunting continue unchecked, suggesting the need for practical—and effective—management of these activities. Most urgent, however, is the need to protect the two known Key Areas for *Amazilia luciae*, which

Table 1. Coverage of threatened species by Key Area. Areas in bold currently have some form of protected status.

	Key Areas occupied	No. of Key Areas protected	Total nos. of Key Areas	
			N.C. America	Neotropics
Horned Guan *Oreophasis derbianus*	**GT 01**, GT 02, GT 03, GT 04, **GT 05**	2	5	8
Yellow-headed Amazon *Amazona oratrix*	**BZ 01**	1	1	8
Honduran Emerald *Amazilia luciae* [E]	HN 02, HN 03	0	2	2
Keel-billed Motmot *Electron carinatum*	**BZ 02**, GT 06, **HN 01**	2	3	7
Azure-rumped Tanager *Tangara cabanisi*	**GT 01**	1	1	4
Golden-cheeked Warbler *Dendroica chrysoparia*	**GT 05**, GT 06, HN 04, HN 05, SV 01	1	5	8

[E] Endemic to Honduras

	Belize	Guatemala	Honduras	Nicaragua	El Salvador
Oreophasis derbianus	—	•	?[N]	—	—
Amazona oratrix	•	•[N]	•[N]	—	—
Amazilia luciae	—	—	•	—	—
Electron carinatum	•	•	•	•[N]	—
Tangara cabanisi	—	•	—	—	—
Dendroica chrysoparia	—	•	•	•[N]	•

[N] No Key Areas identified

Table 2. National distribution of threatened species.

is increasingly threatened from grazing pressure and agricultural expansion; the continued survival of this species is entirely dependent on the integrity of the arid thorn-forest at Coyoles (HN 02) and Olanchito (HN 03) (see 'Outlook', below).

RECENT CHANGES TO THE THREATENED LIST

With the publication of Collar *et al.* (1994), two species—Military Macaw *Ara militaris* and Three-wattled Bellbird *Procnias tricarunculata*—were added to the north Central American threatened species list (but are not included within the Site Inventory; see '*Key Areas*: the book', p.12), with *Electron carinatum* being relegated to Near Threatened status; all these changes resulted from the use of new criteria and/or new data. Recent reports of *Dendroica chrysoparia* from Volcán de San Salvador (SV 01) in El Salvador (*Field Notes* 1994, 48: 871–872) have added this country to the range states of the species, the first globally threatened bird recorded from El Salvador, and confirmation that *Amazona oratrix* is present in Guatemala (in the central Petén and the mangroves of the coastal north-west) and Honduras (in the north-west, to the Sula valley) (Collar *et al.* 1994, Howell and Webb 1995) has added this bird to the threatened species list of both nations. *Procnias tricarunculata* breeds in humid evergreen forest from 1,000 to 1,800 m in north-central Nicaragua, wintering on the Caribbean slope of eastern Honduras (Olancho) and northern Nicaragua from 300 to 700 m (Howell and Webb 1995) where it is broadly sympatric with *Ara militaris* (Monroe 1968); neither species occurs within Key Areas identified in this analysis, but both should be considered in any future conservation strategies or initiatives, and areas for their conservation need to be identified as a high priority.

OLD RECORDS AND LITTLE-KNOWN BIRDS

Most of the threatened species in north Central America are reasonably well known and their presence has been recorded in recent years, at least in their respective Key Areas. The exception to this, however, is *Dendroica chrysoparia*: surprisingly very little is known of its true distribution or requirements in its north Central American winter quarters. Pulich (1976), contrary to subsequent representations of his maps and data (e.g. Braun *et al.* 1986), documents *D. chrysoparia* as follows: Guatemala in just two areas, Usumatlán (two localities 11 km apart, with records from the 1950s) and Tactic (the type-locality, with no records since 1859); Honduras in just two areas, La Esperanza (with records from the 1960s) and Cantoral (records during the 1930s); and Nicaragua, where the bird is known only from two specimens collected at Matagala in 1891. Recent records amount to one bird seen on Cerro San Gil (GT 06) in 1991, 12 seen near La Esperanza (HN 04) in 1975, and the reports from Volcán de San Salvador (SV 01) in 1993 and 1994.

The current status and distribution of all six of the region's threatened species away from their respective Key Areas is little known and in each case warrants attention (see 'Outlook', below).

	BZ		GT						HN					SV	No. of
	01	02	01	02	03	04	05	06	01	02	03	04	05	01	areas
Oreophasis derbianus	–	–	•	•	•	•	•	–	–	–	–	–	–	–	5
Amazona oratrix	•	–	–	–	–	–	–	–	–	–	–	–	–	–	1
Amazilia luciae	–	–	–	–	–	–	–	–	–	•	•	–	–	–	2
Electron carinatum	–	•	–	–	–	–	–	•	•	–	–	–	–	–	3
Tangara cabanisi	–	–	•	–	–	–	–	–	–	–	–	–	–	–	1
Dendroica chrysoparia	–	–	–	–	–	–	•	•	–	–	–	•	•	•	5
No. of species	1	1	2	1	1	1	2	2	1	1	1	1	1	1	

Table 3. Matrix of threatened species by Key Area.

OUTLOOK

Although each Key Area is of major importance, the most urgent requirement for threatened bird conservation in north Central America is the protection and guaranteed integrity of the two Key Areas identified for *Amazilia luciae*; without such action, this Honduran endemic will soon become extinct. Field surveys are needed to determine the status and abundance of the threatened species within each Key Area, but such surveys should also focus on identifying any additional areas that might exist for these same birds. This is particularly true for *Amazona oratrix* (in Guatemala and Honduras), *Dendroica chrysoparia* and *Tangara cabanisi*. Consideration should also be given to reports seemingly referring to *Oreophasis derbianus* ('large birds with single red horns') in the Cerro Volcán Pacayita Biological Reserve of Honduras.

DATA SOURCES

The above introductory text and the Site Inventory (below) were compiled from information supplied by Fundacíon Defensores de la Naturaleza, C. Ladd and D. S. Weber, as well as from the following references.

BRAUN, M. J., BRAUN, D. D. AND TERRILL, S. B. (1986) Winter records of the Golden-cheeked Warbler (*Dendroica chrysoparia*) from Mexico. *Amer. Birds* 40: 564–566.

COLLAR, N. J., GONZAGA, L. P., KRABBE, N., MADROÑO NIETO, A., NARANJO, L. G., PARKER, T. A. AND WEGE, D. C. (1992) *Threatened birds of the Americas: the ICBP/IUCN Red Data Book* (Third edition, part 2). Cambridge, U.K.: International Council for Bird Preservation.

COLLAR N. J., CROSBY, M. J. AND STATTERSFIELD, A. J. (1994) *Birds to watch 2: the world list of threatened birds*. Cambridge, U.K.: BirdLife International (BirdLife Conservation Series no.4).

COLLAR, N. J., WEGE, D. C. AND LONG, A. J. (in press) Patterns and causes of endangerment in the New World avifauna. In J. V. Remsen, ed. *Natural history and conservation of Neotropical birds*. American Ornithologists' Union (Orn. Monogr.).

HOWELL, S. N. G. AND WEBB, S. (1992) New and noteworthy bird records from Guatemala and Honduras. *Bull. Brit. Orn. Club* 112: 42–49.

HOWELL, S. N. G. AND WEBB, S. (1995) *A guide to the birds of Mexico and northern Central America*. Oxford, U.K.: Oxford University Press.

IUCN (1992) *Protected areas of the world: a review of national systems. Volume 4: Nearctic and Neotropical.* Gland, Switzerland and Cambridge, U.K.: International Union for Conservation of Nature and Natural Resources.

KROLL, J. C. (1980) Habitat requirements of the Golden-cheeked Warbler: management implications. *J. Range Management* 33: 60–65.

LAND, H. C. (1970) *Birds of Guatemala*. Wynnewood, Penn.: Livingston Publishing Company, for International Committee for Bird Preservation Pan-American Section.

LYONS, J. (1990) Winter habitat survey of the Golden-cheeked Warbler (*Dendroica chrysoparia*) in Guatemala. Unpublished report to Resources Protection Division, Texas Parks and Wildlife Department, Austin, Texas.

MILLER, B. W. AND MILLER, C. M. (1988) Mussel Creek Drainage: a report with recommendations resulting from 1987 surveys as requested by the Belize Heritage Society. Unpublished report.

MILLER, B. W. AND MILLER, C. M. (1992) Checklist of birds of Belize: field checklist. Draft manuscript.

MONROE, B. L. (1968) *A distributional survey of the birds of Honduras*. American Ornithologists' Union (Orn. Monogr. 7).

PARKER, T. A., HOLST, B. K., EMMONS, L. H. AND MEYER, J. R. (1993) *A biological assessment of the Columbia River Forest Reserve, Toledo District, Belize (Rapid Assessment Program)*. Washington, D.C.: Conservation International.

PULICH, W. M. (1976) *The Golden-cheeked Warbler: a bioecological study*. Austin, Texas: Texas Parks and Wildlife Department.

SCOTT, D. A. AND CARBONELL, M. (1986) *A directory of Neotropical wetlands*. Cambridge, U.K.: International Union for Conservation of Nature and Natural Resources, and Slimbridge, U.K.: International Waterfowl Research Bureau.

STATTERSFIELD, A. J., CROSBY, M. J., LONG, A. J. AND WEGE, D. C. (in prep.) *A global directory of Endemic Bird Areas*. Cambridge, U.K.: BirdLife International (BirdLife Conservation Series).

STEFFEE, N. D. (1972) *Field checklist of the birds of Nicaragua*. Kissimmee, Florida: Flying Carpet Tours, Inc.

THURBER, W. A., SERRANO, J. F., SERMEÑO, A. AND BENITEZ, M. (1987) Status of uncommon and previously unreported birds of El Salvador. *Proc. West. Found. Vert. Zool.* 3: 109–293.

VANNINI, J. P. AND MORALES, C. L. (1990) Evaluation of populations of Highland Guans (*Penelopina nigra*) in plantation and secondary forests in Guatemala. Unpublished report.

SITE INVENTORY

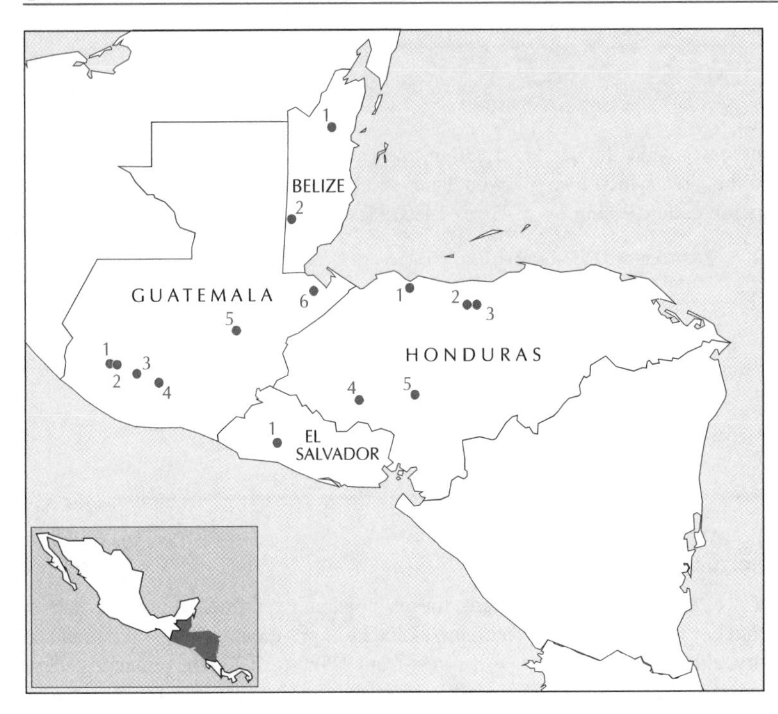

Figure 2. Key Areas in north Central America.

Belize
BZ 01 Crooked Tree–Bermudean Landing
BZ 02 Chiquibul

El Salvador
SV 01 Volcán de San Salvador

Guatemala
GT 01 Volcán Santa María
GT 02 Volcán Zunil and Zunil ridge
GT 03 Volcán Atitlán and Tolimán
GT 04 Volcán Acatenango and Volcán de Fuego
GT 05 Sierra de las Minas
GT 06 Cerro San Gil

Honduras
HN 01 Tela–Lancetilla
HN 02 Coyoles
HN 03 Olanchito
HN 04 La Esperanza
HN 05 Cerro Cantoral

BELIZE

Crooked Tree–Bermudean Landing (Orange Walk/Belize)

BZ 01
17°45′N 88°32′W

Crooked Tree Wildlife Sanctuary (IUCN category IV), 1,470 ha
Bermudean Landing Community Baboon Sanctuary (private), 4,680 ha

This area 45 km north-west of Belize City embraces the complex of lagoons along the length of Black Creek to the Belize River, but also includes the northern portion of the Mussell Creek drainage and the contiguous Baboon Sanctuary. The area comprises swamp forest, humid tropical forest, and freshwater lagoons and marshes, and is extremely important for concentrations of waterfowl (Scott and Carbonell 1986, Miller and Miller 1988).

Amazona oratrix	1992	A roost of 70–80 found in the baboon sanctuary in February 1992; 'considerable numbers' in the Mussell Creek area during February 1988 (Miller and Miller 1988).

Chiquibul (Cayo/Toledo)

BZ 02
16°35′N 89°05′W

Chiquibul National Park (IUCN category II), 107,607 ha
Bladen Branch Nature Reserve (IUCN category I), 39,256 ha
Caracol Archaeological Reserve (IUCN category IV), 20,000 ha

The Chiquibul area covers the Vaca plateau and the southern portion of the Maya mountains (including Little Quartz Ridge and the Columbia River Forest Reserve) in southern Belize.

Electron carinatum	1992	At least 17–24 birds have been found in the vicinity of Caracol; locally fairly common elsewhere on the Vaca plateau (Parker *et al.* 1993).

EL SALVADOR

Volcán de San Salvador (San Salvador)

SV 01
13°44′N 89°17′W

Unprotected

El Picacho, the highest point on Volcán de San Salvador, forms the centre of this area which also includes the adjacent El Boqueron crater. Rising to 1,830 m, the slopes of the volcano are covered in woodland, scrub and coffee plantations (*American Birds* 1980, 34: 676).

Dendroica chrysoparia	1994	First recorded in December 1993 (*Field Notes* 1994, 48: 871–872), with further reports the following year.

GUATEMALA

Volcán Santa María (Quetzaltenango)

GT 01
14°45′N 91°33′W
EBA A14

Finca El Faro Reserve (private), 670 ha

Situated in the mountains of western Guatemala, Volcán Santa María has suffered considerable habitat loss through volcanic activity, although two adjacent fincas (El Faro and El Patzulin) on the south-west slopes retain significant tracts of forest. Finca El Faro, for example, an experimental private reserve, comprises 300 ha of pre-montane moist and montane wet forest from 1,500 to 2,200 m, remnant gallery forest in river canyons below 1,500 m, and various plantation crops (Vannini and Morales 1990).

Oreophasis derbianus	c.1960	Recorded from the volcano, but no records during one year of fieldwork in the mid-1980s.
Tangara cabanisi	1990	Mostly recorded at 1,280–1,450 m, the fincas on the south-west slopes are regular localities.

Volcán Zunil and Zunil ridge (Quetzaltenango)

GT 02
14°44′N 91°27′W
EBA A14

Unprotected

This area includes the locality Fuentes Georgina, north-west of the Atitlán complex, and aerial surveys in 1970 indicated substantial tracts of forest on the slopes of Volcán Zunil. A sighting of a 'probable' *Dendroica chrysoparia* in 1975 (Lyons 1990) requires verification, but suggests that this species should be considered during field surveys in the area.

Oreophasis derbianus	1990	Regularly recorded from Fuentes Georgina; five seen together in July 1990.

Volcán Atitlán and Tolimán (Suchitepequez)

GT 03

Atitlán National Park (IUCN category VIII), 54,773 ha

14°37′N 91°11′W
EBA A14

This area embraces the three volcanoes (Tolimán, Atitlán and San Pedro) to the south and west of Lago de Atitlán. Aerial surveys in 1970 demonstrated substantial tracts of forest remaining on Volcanes Tolimán and Atitlán and some on San Pedro, although military operations in this area have added to the effects of farming and hunting in reducing numbers of *Oreophasis derbianus*. Finca Mocca is a private reserve on the southern slopes of this area which protects a 400-ha block of pristine cloud-forest known to support this species (Vannini and Morales 1990).

Oreophasis derbianus	1989	Records from all three volcanoes within the park, and from Finca Mocca (Vannini and Morales 1990).

Volcán Acatenango and Volcán de Fuego (Escuintla/Sacatepequez)

GT 04

Unknown

14°30′N 90°53′W
EBA A14

The southern and eastern slopes of this volcano massif retain considerable forest cover between 400 and 3,975 m. However, forest on the western slopes has been devastated by volcanic activity from Volcán de Fuego. Much of the cloud-forest is now apparently part of a 'national park' (Vannini and Morales 1990), the details of which are unknown.

Oreophasis derbianus	1989	Two birds shot by hunters confirm the presence of this species, which is reported to be common (Vannini and Morales 1990).

Sierra de las Minas (Alta Verapaz/Zacapa)

GT 05

Sierra de las Minas Biosphere Reserve (IUCN category I), 105,700 ha

15°10′N 89°50′W
EBA A14

This large, remote mountain range (c.4,000 km^2) in eastern Guatemala, north of the Río Motagua and south-west of Lago de Izabal, was designated a Biosphere Reserve in 1990. The sierra rises to c.3,000 m and supports cloud-forest, lower montane and pre-montane forest, with c.1,300 km^2 of cloud-forest (65% thought to be virgin), this is probably the largest unbroken tract of this habitat in Central America (Fundación Defensores de la Naturaleza *in litt.* 1993). However, agricultural expansion, logging, hunting and colonization all threaten the area.

Oreophasis derbianus	1988	One shot at 2,500 m, c.20 km north-west of Río Hondo; reportedly quite common in this remote area (Howell and Webb 1992).
Dendroica chrysoparia	1958	Specimens collected c.10 km north-west of Usumatlán (c.15°00′N 89°50′W) in December 1958; previous records in late August (Pulich 1976).

Cerro San Gil (Izabal)

GT 06

15°40′N 88°47′W

Cerro San Gil Proposed Ecological Reserve, 47,400 ha

Lying in easternmost Guatemala, effectively at the easternmost end of the Sierra de las Minas, Cerro San Gil is proposed as an ecological reserve (by FUNDAECO). Within the core area of the proposed reserve are large undisturbed tracts of subtropical and tropical wet forest which protect an important watershed for local villages and larger, more distant urban areas (D. S. Weber *in litt.* 1992).

Electron carinatum	1991	Up to eight noted at 200–700 m during February 1991, when one was seen near a probable nest-burrow (Howell and Webb 1992).
Dendroica chrysoparia	1991	A male seen in humid evergreen forest at 900 m during February 1991 (Howell and Webb 1992).

HONDURAS

Tela–Lancetilla (Atlántida)

HN 01

15°42′N 87°28′W

Lancetilla Biological Reserve (IUCN category IV), 1,681 ha
Jardín Botánico de Lancetilla Protected Area (IUCN category IV), 1,253 ha

Tela lies on the north-west coast, with forest (including that of Lancetilla) found c.3–11 km inland of Tela along the Río Tela. The two reserves contain much untouched forest, and although large numbers of people visit the botanic gardens, and a well-used trail runs through the forest reserve, disturbance in most areas appears to be minimal.

Electron carinatum	1991	A series of recent records have shown that at least six birds are present.

Coyoles (Yoro)

HN 02

15°29′N 86°41′W

Unprotected

The town of Coyoles is in the upper Río Aguán valley in north-central Honduras (and just 16 km west of Olanchito, HN 03). An area of arid thorn-forest 6 km to the west-north-west holds *Amazilia luciae* but suffers from heavy grazing of the understorey and continuing conversion of forest to pineapple plantations.

Amazilia luciae	1988	Common.

Olanchito (Yoro)

HN 03

15°29′N 86°33′W

Unprotected

Olanchito is in the upper Río Aguán valley (16 km east of Coyoles, HN 02). A large tract of arid thorn-forest and scrub 4–5 km west of town is frequented by *Amazilia luciae* but has been cut over, is heavily grazed, and suffers from continuing conversion of forest to pineapple plantations.

Amazilia luciae	1991	A sizeable population (probably breeding) was recorded during March 1991, when 22–28 birds were found in 250 ha of habitat.

HN 04

La Esperanza (Intibucá)

14°16′N 88°10′W
EBA A14

Unprotected

La Esperanza, the departmental capital of Intibucá, lies c.24 km south-west of Jesús de Otoro. In 1963, the species below was found in an area of broad-leaf oak forest with a few scattered pines c.7 km south-east of town at c.1,600–1,700 m (Monroe 1968, Pulich 1976). In 1975 birds were seen in an area of ocote pine *Pinus oocarpa*, feeding in the shrubby understorey dominated by oaks *Quercus* and sweet-gum *Liquidambar*, c.4 km south-west of town at c.1,500 m, the area was then (1975) being managed by the Corporación de Hondureña Forestal (Kroll 1980).

Dendroica chrysoparia	1975	Three collected and two additional males seen in January 1963 (Monroe 1968); 12 birds seen in March 1975 (Kroll 1980).

HN 05

Cerro Cantoral (D.C.)

14°20′N 87°24′W
EBA A14

Unprotected

The mountain, c.8 km north-west of Archaga, rises to over 1,800 m and supports pine–oak forest. Surveys are urgently needed to determine the current status of the forest and of the species below.

Dendroica chrysoparia	1936	Four collected, November–February 1931–1936 (Monroe 1968).

PANAMA

Bare-necked Umbrellabird *Cephalopterus glabricollis*

PANAMA is a land-bridge where the faunas of North and South America meet and intermingle. One consequence is that Panama's avifauna is exceptionally large, with a total of at least 929 species in just 75,500 km^2; 122 of these occur only as long-distance migrants, primarily from breeding areas in North America (Ridgely and Gwynne 1989). Eleven species are endemic to Panama, 93 have restricted ranges (Stattersfield *et al.* in prep.) and five are threatened (Collar *et al.* 1992). This analysis has identified 11 Key Areas for the threatened birds in Panama (see '*Key Areas*: the book', p. 11, for criteria).

THREATENED BIRDS

Five Panamanian species were considered at risk of extinction by Collar *et al.* (1992), one of which, *Selasphorus ardens*, is confined to the country (see Table 1). All five species are dependent on wet forest (although *S. ardens* also occurs above the tree-line and *Carpodectes antoniae* relies primarily on mangroves, being present in adjacent wet forest only seasonally), and all species are distributed in the tropical and subtropical zones (up to 2,000 m), where they are threatened by loss of habitat and, in two cases, by virtue of their restricted ranges (Collar *et al.* in press). The distributions of these threatened birds and their relationship to Endemic Bird Areas are shown in Figure 1.

KEY AREAS

The 11 Panamanian Key Areas would, if adequately protected, help ensure the conservation of all five of the country's threatened species—always accepting that important new populations and areas may yet be found. Of the 11 Key Areas only one is important for more than a single threatened species. Just one threatened species, *Selasphorus ardens*, is endemic to Panama, and is thus totally reliant for its survival on the integrity of habitat in the three Key Areas from which it is known (see 'Outlook', below). Both *Carpodectes antoniae* and *Cephalopterus glabricollis* are poorly known within Panama, but are well represented in Key Areas in Costa Rica (see p. 155). However, the two east Panama species, *Crypturellus kerriae* and *Xenornis setifrons*, are known outside the country only by old records from a single Colom-

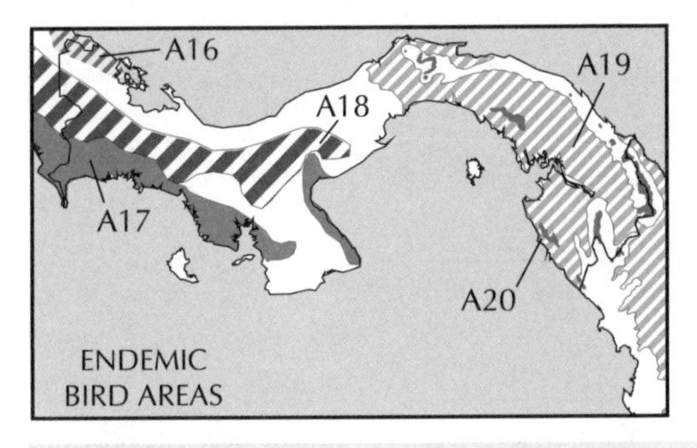

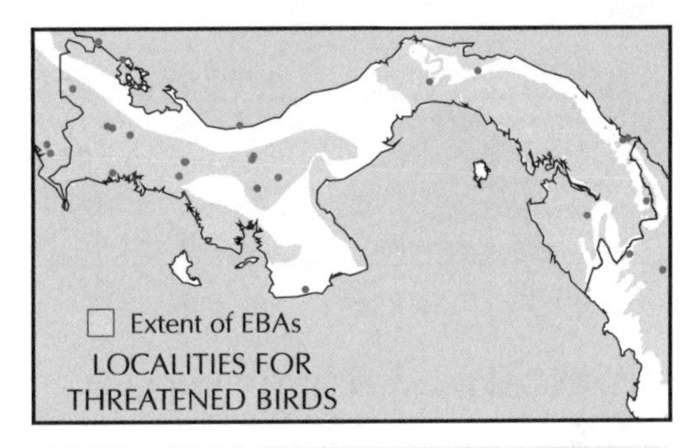

Figure 1. The localities where threatened birds have been recorded in Panama and their relationship to Endemic Bird Areas (EBAs).

- All the Panamanian threatened species have restricted ranges and thus occur within various EBAs, which are listed below (figures are numbers of these species in each EBA).

A16 Central American Caribbean slope (1)
A17 South Central American Pacific slope (2)
A18 Costa Rica and Panama highlands (2)
A19 Darién and Urabá lowlands (2)
A20 East Panama and Darién highlands (0)

bian Key Area (CO 17 Serranía del Baudó, p. 139), and thus appear to be dependent on the integrity of the relevant Key Areas in Panama.

KEY AREA PROTECTION

A satisfactory attribute of Key Areas in Panama is the fact that eight (73%) currently have some form of protected status, six (55% of the total) as national parks (IUCN category II). The three Key Areas that are currently unprotected urgently require attention in the form of appropriate conservation measures if the populations of their threatened species are to survive. This is especially important for *Carpodectes antoniae*, which in Panama is known only from one unprotected Key Area (Río Corotú–Puerto Armuelles, PA 01), and *Selasphorus ardens*, known from only one protected Key Area. Key Areas harbouring the other three threatened species are all protected. However, even these formally protected areas remain under threat, and habitat degradation in many continues unchecked; effective management of activities undertaken within them is thus required.

RECENT CHANGES TO THE THREATENED LIST

With the publication of Collar *et al.* (1994), four species (Military Macaw *Ara militaris*, Turquoise Cotinga *Cotinga ridgwayi*, Three-wattled Bellbird *Procnias tricarunculata* and Yellow-green Finch *Pselliophorus luteoviridis*) were added to the Panamanian threatened species list, resulting primarily from the adoption of new criteria; these species have not, however, been included in the Site Inventory (see '*Key Areas*: the book', p. 12). Each is sympatric with one of the previously listed threatened birds, and their addition would thus have no major impact on the Key Area analysis. This is well demonstrated by *Pselliophorus luteoviridis*, the entire range of which is embraced by the Cerro Colorado (PA 04) and Cerro Tute Key Areas (PA 05) where it occurs together with *Selasphorus ardens* (Ridgely and Gwynne 1989). Nonetheless, each of the four recently added species should be considered in any future conservation strategies or initiatives.

Table 1. Coverage of threatened species by Key Area. Areas in bold currently have some form of protected status.

	Key Areas occupied	No. of Key Areas protected	Total nos. of Key Areas	
			Panama	Neotropics
Chocó Tinamou *Crypturellus kerriae*	**10,11**	2	2	3
Glow-throated Hummingbird *Selasphorus ardens* [E]	04,05,**06**	1	3	3
Speckled Antshrike *Xenornis setifrons*	**07,08,09,10**	4	4	5
Yellow-billed Cotinga *Carpodectes antoniae*	01	0	1	9
Bare-necked Umbrellabird *Cephalopterus glabricollis*	**02,03**	2	2	6

[E] Endemic to Panama

	01	02	03	04	05	06	07	08	09	10	11	No. of areas
Crypturellus kerriae	–	–	–	–	–	–	–	–	–	●	●	2
Selasphorus ardens	–	–	–	●	●	●	–	–	–	–	–	3
Xenornis setifrons	–	–	–	–	–	–	●	●	●	●	–	4
Carpodectes antoniae	●	–	–	–	–	–	–	–	–	–	–	1
Cephalopterus glabricollis	–	●	●	–	–	–	–	–	–	–	–	2
No. of species	1	1	1	1	1	1	1	1	1	2	1	

Table 2. Matrix of threatened species by Key Area.

OLD RECORDS AND LITTLE-KNOWN BIRDS

There have been records of all five threatened species during the 1980s and 1990s, although there is a general lack of detailed information available for each of the birds, and a number of the Key Areas have been defined on the basis of old records of species which, it is to be hoped, still exist there. This is the case for the Key Areas of La Amistad and Volcán Barú (PA 02), Cerro Tacarcuna (PA 09) and Cerros de Quía (PA 11). Among the most poorly known species in Panama is *Carpodectes antoniae*, which in recent decades has been recorded only from the Burica peninsula. However, the seasonal movements of *Cephalopterus glabricollis* remain virtually unknown, and thus it is represented in Key Areas solely on its breeding grounds, and *Crypturellus kerriae* has been recorded from only a single Key Area during the past 25 years.

OUTLOOK

With 11 Key Areas in the country, and just one where two threatened species occur together, each area is a site of major importance for the conservation of threatened birds. However, top Key Areas for protection include Río Corotú (PA 01), the only Key Area in Panama for *Carpodectes antoniae*, and Cerro Colorado (PA 04) and Cerro Tute (PA 05), two of only three areas currently known for *Selasphorus ardens*. These then are perhaps Panama's top Key Areas for threatened birds. For the areas identified on the basis of old records—such as La Amistad and Volcán Barú (PA 02), Cerro Tacarcuna (PA 09) and Cerros de Quía (PA 11)—there is a need to locate and define the extent of suitable habitat within or near the Key Areas, and, for the species which define these areas, surveys should aim to confirm their continued existence or assess the status of their populations.

DATA SOURCES

The above introductory text and the Site Inventory (below) were compiled from information supplied by W. Adsett, G. Angehr and P. Boesman as well as from the following references.

COLLAR, N. J., GONZAGA, L. P., KRABBE, N., MADROÑO NIETO, A., NARANJO, L. G., PARKER, T. A. AND WEGE, D. C. (1992) *Threatened birds of the Americas: the ICBP/IUCN Red Data Book* (Third edition, part 2). Cambridge, U.K.: International Council for Bird Preservation.

COLLAR N. J., CROSBY, M. J. AND STATTERSFIELD, A. J. (1994) *Birds to watch 2: the world list of threatened birds*. Cambridge, U.K.: BirdLife International (BirdLife Conservation Series no.4).

COLLAR, N. J., WEGE, D. C. AND LONG, A. J. (in press) Patterns and causes of endangerment in the New World avifauna. In J. V. Remsen, ed. *Natural history and conservation of Neotropical birds*. American Ornithologists' Union (Orn. Monogr.).

ENGLEMAN, D. (1994) The field editor's report. *Toucan* 20(7): 4–5.

ENGLEMAN, D. AND ENGLEMAN L. (1992) Birding spots in Panama: the Fortuna reserve. *Toucan* 18(7): 5–7.

IUCN (1992) *Protected areas of the world: a review of national systems. Volume 4: Nearctic and Neotropical.* Gland, Switzerland and Cambridge, U.K.: International Union for Conservation of Nature and Natural Resources.

RIDGELY, R. S. AND GWYNNE, J. A. (1989) *A guide to the birds of Panama with Costa Rica, Nicaragua, and Honduras*. Second edition. Princeton: Princeton University Press.

STATTERSFIELD, A. J., CROSBY, M. J., LONG, A. J. AND WEGE, D. C. (in prep.) *A global directory of Endemic Bird Areas*. Cambridge, U.K.: BirdLife International (BirdLife Conservation Series).

SITE INVENTORY

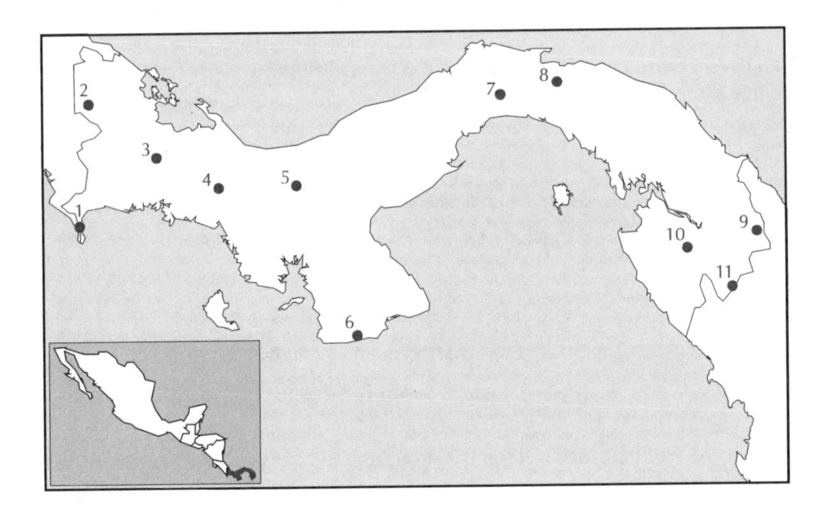

Figure 2. Key Areas in Panama.

01 Río Corotú
02 La Amistad and Volcán Barú
03 Fortuna
04 Cerro Colorado
05 Cerro Tute
06 Cerro Hoya
07 Cerro Azul–Cerro Jefe
08 Nusugandi
09 Cerro Tacarcuna
10 Cerro Pirre
11 Cerros de Quía

Río Corotú (Chiriquí)

PA 01

8°10′N 82°54′W
EBA A17

Unprotected

This area includes the Río Corotú (up to the headwater region), inland from Puerto Armuelles on the Península de Burica in western Chiriquí. Habitat destruction in western Chiriquí (including this area) has been extensive, with remaining habitat (both wet forest and mangrove) in urgent need of protection.

Carpodectes antoniae 1982 Various sight records suggest a remnant population.

La Amistad and Volcán Barú (Bocas del Toro/Chiriquí)

PA 02

9°09′N 82°49′W
EBA A18

La Amistad National Park (IUCN category II), 207,000 ha
Volcán Barú National Park (IUCN category II), 14,000 ha

These two contiguous parks in westernmost Panama represent an extensive area of highland forest which, however, is poorly known and its importance for the species mentioned below can only be guessed at. Adjacent tracts of lowland and foothill forest on the Caribbean slope should be considered in any conservation initiatives concerning this area.

Cephalopterus glabricollis 1960 A breeding population almost certainly persists.

Fortuna (Chiriquí/Bocas del Toro)

PA 03

8°43′N 82°16′W
EBA A18

La Fortuna Water Production Reserve (IUCN category VIII), 15,000 ha
Fortuna Forest Reserve (status unknown), c.100,000 ha

The Fortuna Water Production Reserve protects the watershed of the Fortuna reservoir, on the continental divide between Gualaca and Chiriquí Grande, and comprises almost intact montane forest (Engleman and Engleman 1992). Like La Amistad (PA 02), adjacent tracts of lowland and foothill forest on the Caribbean slope need to be considered in any conservation initiatives concerning this area.

Cephalopterus glabricollis 1992 Regularly encountered at various places in this area, March–September (Engleman and Engleman 1992).

Cerro Colorado (Chiriquí)

PA 04

Unprotected

8°28′N 81°45′W
EBA A18

Cerro Colorado (including the adjacent Cerro Florés) sits astride the continental divide (and therefore on the border with Bocas del Toro) in the Serranía de Tabasará in eastern Chiriquí. The foothill and lower highland forests, and alpine meadows in this area are largely untouched, but remain totally unprotected and deserve more attention.

Selasphorus ardens	1995	Found regularly above Cerro Colorado but considered uncommon, although 12 were seen in ericaceous scrub in early 1995 (G. Angehr verbally 1995).

Cerro Tute (Veraguas)

PA 05

Unprotected

8°29′N 81°06′W
EBA A18

Cerro Tute is above Santa Fé in the Serranía de Tabasará, and sits astride the continental divide. The foothill and lower highland forests in this area are largely untouched, but remain totally unprotected, poorly studied ornithologically, and deserving of more attention.

Selasphorus ardens	1984	Recorded from a number of localities but apparently rare.

Cerro Hoya (Veraguas)

PA 06

Cerro Hoya National Park (IUCN category II), 32,557 ha

7°16′N 80°36′W
EBA A17

Cerro Hoya, which rises to c.2,100 m, is at the southern end of the Azuero peninsula, and lies on the peninsula divide near the border with Los Santos. Little is known of the status of the forest, or of its avifauna, although the area is clearly important and harbours a number of isolated populations (e.g. *Selasphorus ardens* and Painted Parakeet *Pyrrhura picta*).

Selasphorus ardens	1994	Discovered in March 1994, this population is the only one outside the central highlands, and the only one protected (Engleman 1994).

Cerro Azul–Cerro Jefe (Panama)

PA 07

Chagres National Park (IUCN category II), 129,000 ha

9°12′N 79°24′W
EBA A19

The Cerro Azul–Cerro Jefe region (which includes Altos de Pacora) is in the Cordillera de San Blas, on the continental divide, c.50 km north-east of Panama city. Although protected within the national park, there has been some habitat loss on the lower slopes which has opened up the montane forests to the damaging effects of wind (W. Adsett verbally 1994).

Xenornis setifrons	1994	Never abundant but seen regularly since 1992 on a trail between the two mountains (*Toucan* 1992–1994, 18–20).

PA 08

Nusugandi (San Blas/Panamá)

9°18′N 78°56′W

Comarca Kuna Yala Indigenous Reserve (IUCN category VII), 320,000 ha

EBA A19

Nusugandi is on the continental divide (and thus on the border between San Blas and Panamá provinces) between Golfo de San Blas and El Llano. It is the base for the forest reserve (PEMASKY) headquarters.

Xenornis setifrons	1995	Never abundant but regularly recorded since 1992 on one trail (*Toucan* 1992–1994, 18, 19, 20, P. Boesman *in litt.* 1995).

PA 09

Cerro Tacarcuna (Darién)

8°05′N 77°17′W

Darién National Park (IUCN category II), 579,000 ha

EBA A19

This mountain lies at the southern end of the Serranía del Darién, near the border with Colombia. It is the highest in the range and is just part of the large, relatively untouched Darién National Park. The area has been poorly studied, and may also be important for *Crypturellus kerriae* (as well as for a number of other east Panama endemics).

Xenornis setifrons	1964	A number of records but no recent surveys.

PA 10

Cerro Pirre (Darién)

7°57′N 77°52′W

Darién National Park (IUCN category II), 579,000 ha

EBA A19

Cerro Pirre is at the northern end of the Serranía de Pirre in eastern Darién, just to the north of the Alturas de Nique which forms the border with Colombia. Like Cerro Tacarcuna (PA 09), Cerro Pirre is relatively inaccessible, and consequently understudied (although access is possible via Cana on the eastern slope and Pirre Station in the lowlands on the western slope). The area is protected within the Darién National Park, is essentially undisturbed, and is of key importance for a number of east Panama lowland and highland endemics (Ridgely and Gwynne 1989).

Crypturellus kerriae	1995	A number reports of this species having been heard (also *Toucan* 1993, 19(9): 5). Tape recordings were made of birds calling in the foothills around Cana in April 1995 (P. Boesman *in litt.* 1995).
Xenornis setifrons	1993	Discovered in the area during 1993 (*Toucan* 1993, 19(9): 5).

PA 11

Cerros de Quía (Darién)

7°35′N 77°27′W

Darién National Park (IUCN category II), 579,000 ha

EBA A19

Cerros de Quía form the central part of the border with Colombia, and lie within the Darién National Park. The area is relatively inaccessible and therefore under-studied, but maintains essentially pristine forest and should be surveyed for the presence of both the species mentioned below and for *Xenornis setifrons* which also probably occurs.

Crypturellus kerriae	1970	Found to be fairly common on the steep slopes of higher ridges, but no recent surveys.

PARAGUAY

Helmeted Woodpecker *Dryocopus galeatus*

PARAGUAY is home to about 645 bird species, of which 576 (69%) are presumed to breed, 80 (12%) are austral migrants, 34 (5%) are Nearctic migrants (Hayes 1995), 14 (2%) have restricted ranges (though none is endemic to Paraguay: Stattersfield *et al.* in prep.) and 24 (4%) are threatened (Collar *et al.* 1992). This analysis has identified 23 Key Areas for the threatened birds of Paraguay (see '*Key Areas*: the book', p. 11, for criteria).

THREATENED BIRDS

The 24 Paraguayan threatened species which Collar *et al.* (1992) listed rely primarily on grasslands and chaco (39% of the species) and wet forest (35%), and all are threatened by habitat loss (Collar *et al.* in press). The distribution of threatened birds is shown in Figure 1.

KEY AREAS

The 23 Paraguayan Key Areas that have been identified would, if adequately protected, help ensure the conservation of 20 (83%) of the threatened species in the country (always accepting that important new populations and areas may yet be found). Of these areas, 15 are important for two or more (up to nine) threatened species, and these are therefore perhaps the most efficient areas currently known in which to conserve Paraguay's threatened birds (see 'Outlook', below). The eight areas which each harbour three or more threatened species (see Table 3) together represent potential sanctuaries for 14 threatened species, 58% of the total number. However, vital as these areas are for the conservation of Paraguayan threatened species, they should not detract from the significance of the other Key Areas, as these are of importance for the remaining six species whose habitat would otherwise be unrepresented.

From Tables 1 and 2 it can be seen that six species (*Mergus octosetaceus*, *Coturnicops notatus*, *Claravis godefrida*, *Amazona pretrei*, *Euscarthmus rufomarginatus* and *Sporophila falcirostris*) occur in just one Paraguayan Key Area, although several of the species are known in Paraguay only from single records (see 'Old records and little-known birds', below).

Not represented within the Paraguayan Key Area

Figure 1. The localities where threatened birds have been recorded in Paraguay and their relationship to Endemic Bird Areas (EBAs).

- The threatened species are not distributed evenly throughout Paraguay, the majority being found in the Oriente region (Paraguayan territory east of the Río Paraguay), especially in the humid forests and savannas of the Alto Paraná (the westernmost extension of the south-east Brazilian lowland and foothills, EBA B52) and also in the wet grasslands and savannas of Ñeembucú and the dry forests and savannas of the Campo Cerrado region. The only threatened species within the Chaco region (Paraguayan territory west of the Río Paraguay), are restricted to the wet palm savannas of the Bajo Chaco near the Río Paraguay.

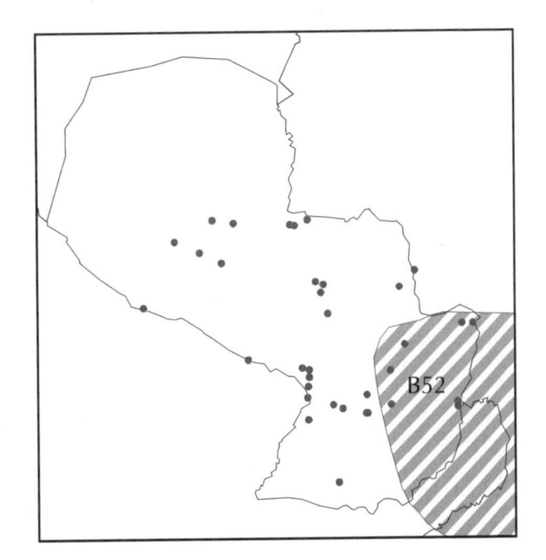

analysis are four of the country's threatened species, as follows.

Anodorhynchus glaucus is known from Paraguay only by c.10 skins and two testimonies from the eighteenth century (Collar *et al.* 1992).

Eleothreptus anomalus has been recorded only by a handful of specimens and a couple of possible sightings, all east of the Río Paraguay. Although recent fieldwork (in Argentina) has improved our knowledge of the species' habits, its voice is still unknown so it is likely to go unnoticed in many parts of its range (Pearman and Abadie 1995).

Gubernatrix cristata is known in Paraguay only from a handful of specimens, the only locality information being for two taken at 'Villa Rica' (Villarrica) in 1905 (Collar *et al.* 1992).

Sporophila frontalis is known in Paraguay only from specimens which were collected around the turn of the century but are otherwise undated.

Table 1. Coverage of threatened species by Key Area. Areas in bold currently have some form of protected status.

	Key Areas occupied	No. of Key Areas protected	Total nos. of Key Areas	
			Paraguay	Neotropics
Brazilian Merganser *Mergus octosetaceus*	07	0	1	7
Crowned Eagle *Harpyhaliaetus coronatus*	01,**06**,**16**,19,**21**	3	5	18
Black-fronted Piping-guan *Pipile jacutinga*	**06**,**09**,**10**,**11**,**12**,**13**,**14**,**16**	8	8	29
Rufous-faced Crake *Laterallus xenopterus*	04,08	0	2	3
Speckled Crake *Coturnicops notatus*	04	0	1	8
Purple-winged Ground-dove *Claravis godefrida*	**09**	1	1	9
Glaucous Macaw *Anodorhynchus glaucus*	—	0	0	0
Hyacinth Macaw *Anodorhynchus hyacinthinus*	01,**03**,**06**	2	3	15
Red-spectacled Amazon *Amazona pretrei*	**09**	1	1	14
Vinaceous Amazon *Amazona vinacea*	01,**03**,**06**,07,**09**,**11**,**12**,**14**	6	8	35
Sickle-winged Nightjar *Eleothreptus anomalus*	—	0	0	9
Helmeted Woodpecker *Dryocopus galeatus*	**06**,08,**09**,**10**,**12**,**13**	5	6	15
Dinelli's Doradito *Pseudocolopteryx dinellianus*	**13**,19	1	2	5
Rufous-sided Pygmy-tyrant *Euscarthmus rufomarginatus*	02	0	1	9
São Paulo Tyrannulet *Phylloscartes paulistus*	**06**,08,**09**,**16**	3	4	17
Russet-winged Spadebill *Platyrinchus leucoryphus*	**06**,08,**17**	2	3	19
Strange-tailed Tyrant *Yetapa risora*	19,**20**,22	1	3	9
Ochre-breasted Pipit *Anthus nattereri*	18,23	0	2	6
Temminck's Seedeater *Sporophila falcirostris*	05	0	1	14
Buffy-throated Seedeater *Sporophila frontalis*	—	0	0	22
Rufous-rumped Seedeater *Sporophila hypochroma*	**06**,19,22	1	3	15
Marsh Seedeater *Sporophila palustris*	**06**,08,22	1	3	15
Yellow Cardinal *Gubernatrix cristata*	—	0	0	11
Saffron-cowled Blackbird *Xanthopsar flavus*	**09**,15	1	2	12

Table 2. Matrix of threatened species by Key Area.

	01	02	03	04	05	06	07	08	09	10	11	12	13	14	15	16	17	18	19	20	21	22	23	No. of areas
Mergus octosetaceus	–	–	–	–	–	–	●	–	–	–	–	–	–	–	–	–	–	–	–	–	–	–	–	1
Harpyhaliaetus coronatus	●	–	–	–	–	●	–	–	–	–	–	–	–	–	–	●	–	–	●	–	●	–	–	5
Pipile jacutinga	–	–	–	–	–	●	–	–	●	●	●	●	●	●	–	●	–	–	–	–	–	–	–	8
Laterallus xenopterus	–	–	–	●	–	–	–	●	–	–	–	–	–	–	–	–	–	–	–	–	–	–	–	2
Coturnicops notatus	–	–	–	●	–	–	–	–	–	–	–	–	–	–	–	–	–	–	–	–	–	–	–	1
Claravis godefrida	–	–	–	–	–	–	–	–	●	–	–	–	–	–	–	–	–	–	–	–	–	–	–	1
Anodorhynchus glaucus	–	–	–	–	–	–	–	–	–	–	–	–	–	–	–	–	–	–	–	–	–	–	–	0
Anodorhynchus hyacinthinus	●	–	●	–	–	●	–	–	–	–	–	–	–	–	–	–	–	–	–	–	–	–	–	3
Amazona pretrei	–	–	–	–	–	–	–	–	●	–	–	–	–	–	–	–	–	–	–	–	–	–	–	1
Amazona vinacea	●	–	●	–	–	●	●	–	●	–	●	●	–	●	–	–	–	–	–	–	–	–	–	8
Eleothreptus anomalus	–	–	–	–	–	–	–	–	–	–	–	–	–	–	–	–	–	–	–	–	–	–	–	0
Dryocopus galeatus	–	–	–	–	–	●	–	●	●	●	–	●	●	–	–	–	–	–	–	–	–	–	–	6
Pseudocolopteryx dinellianus	–	–	–	–	–	–	–	–	–	–	–	–	●	–	–	–	–	–	●	–	–	–	–	2
Euscarthmus rufomarginatus	–	●	–	–	–	–	–	–	–	–	–	–	–	–	–	–	–	–	–	–	–	–	–	1
Phylloscartes paulistus	–	–	–	–	–	●	–	●	●	–	–	–	–	–	–	●	–	–	–	–	–	–	–	4
Platyrinchus leucoryphus	–	–	–	–	–	●	–	●	–	–	–	–	–	–	–	–	●	–	–	–	–	–	–	3
Yetapa risora	–	–	–	–	–	–	–	–	–	–	–	–	–	–	–	–	–	–	●	●	–	●	–	3
Anthus nattereri	–	–	–	–	–	–	–	–	–	–	–	–	–	–	–	–	–	●	–	–	–	–	●	2
Sporophila falcirostris	–	–	–	–	●	–	–	–	–	–	–	–	–	–	–	–	–	–	–	–	–	–	–	1
Sporophila frontalis	–	–	–	–	–	–	–	–	–	–	–	–	–	–	–	–	–	–	–	–	–	–	–	0
Sporophila hypochroma	–	–	–	–	–	●	–	–	–	–	–	–	–	–	–	–	–	–	●	–	–	●	–	3
Sporophila palustris	–	–	–	–	–	●	–	●	–	–	–	–	–	–	–	–	–	–	–	–	–	●	–	3
Gubernatrix cristata	–	–	–	–	–	–	–	–	–	–	–	–	–	–	–	–	–	–	–	–	–	–	–	0
Xanthopsar flavus	–	–	–	–	–	–	–	–	●	–	–	–	–	–	●	–	–	–	–	–	–	–	–	2
No. of species	3	1	2	2	1	9	2	5	7	2	2	3	3	2	1	3	1	1	4	1	1	3	1	

KEY AREA PROTECTION

A total of 12 (52%) of Paraguay's Key Areas currently have some form of protected status (i.e. as national parks, biological, ecological or private reserves, or natural monuments), but only one is a national park (IUCN category II). The 11 unprotected Key Areas (48% of the total) require attention in the form of appropriate conservation measures if the populations of their threatened species are to survive. However, even the formally protected areas remain under threat and, in many, habitat degradation continue unchecked; effective management of activities undertaken within them is thus required.

Five Paraguayan threatened species are not currently recorded from any protected Key Area: *Mergus octosetaceus*, *Coturnicops notatus*, *Laterallus xenopterus*, *Euscarthmus rufomarginatus* and *Sporophila falcirostris* (Table 1). All of these except *L. xenopterus* are known from just one Paraguayan Key Area but their current status in Paraguay is also uncertain (see 'Old records and little-known birds'). A further eight Paraguayan threatened species are known from only one protected Key Area (Table 1) which in each case is potentially important for the continued survival of the species in question (see 'Outlook', below). Species in this category include *Claravis godefrida*, *Amazona pretrei*, *Pseudocolopteryx dinellianus*, *Yetapa risora*, *Sporophila hypochroma*, *S. palustris* and *Xanthopsar flavus*, though all the protected-area records of *A. pretrei*, *P. dinellianus*, *Y. risora* and *X. flavus* were single observations, probably of vagrant or wandering individuals (Brooks *et al.* 1993).

RECENT CHANGES TO THE THREATENED LIST

With the publication of Collar *et al.* (1994), one of the 24 threatened species (*Eleothreptus anomalus*) was dropped from the Paraguayan threatened list, and three birds were added: Blue-winged Macaw *Ara maracana*, Black-and-white Monjita *Heteroxolmis dominicana* and Black-masked Finch *Coryphaspiza melanotis* (but these have not been included in the Site Inventory; see '*Key Areas*: the book', p. 12).

All three of the species added to the threatened list previously had Near Threatened status in Collar *et al.* (1992), with the new criteria and/or new data having promoted their upgrading.

The recent additions will not have any major distributional impact on the Key Area analysis, as two of them (*Heteroxolmis dominicana* and *Coryphaspiza melanotis*) are known from just one or two old records, and there only a handful of records of *Ara maracana*, including recent sightings of a bird at Itabó (PY 12), a pair at Estancia Primavera (Concepción) and also a 1930 specimen from Serranía San Luis (PY 03) (López 1992, Brooks *et al.* 1993).

Table 3. Top Key Areas in Paraguay. Those with the area number in bold currently have some form of protected status.

Key Area	No. of threatened spp.	Comments
06 Mbaracayú	9	The largest tract of humid forest remaining in Paraguay and a key site for several of the country's threatened species.
08 Estancia La Fortuna	5	Located close to PY 06 and important for some threatened forest birds but also one of the only sites where *Laterallus xenopterus* has been recorded.
09 Estancia Itabó	5	A private reserve with an important wintering population of *Amazona vinacea*; the site of Paraguay's only confirmed sighting of *Claravis godefrida*.

OLD RECORDS AND LITTLE-KNOWN BIRDS

Our knowledge of threatened birds in Paraguay has increased significantly through fieldwork over the last ten years (see Hayes 1995 for a comprehensive review). Many species are, however, still little-known, though recent records mean that we are able to make informed recommendations for their conservation. The exceptions are species known only from old specimen records or chance observations (from which there can be no guarantees of a current viable population), and in Paraguay these include the following.

Mergus octosetaceus is thought probably to be extinct in Paraguay (Hayes 1995), having been seen with certainty only once, in 1984 (PY 07), but not since, despite searches.

Coturnicops notatus has been recorded in Paraguay just three times, all before 1945.

Amazona pretrei is poorly documented in Paraguay and, indeed, Hayes (1995) lists the species only as hypothetical for the country. The single record covered by the Site Inventory (PY 09) was of one bird, thought to be an escaped cagebird or a vagrant, seen by a single observer (Brooks *et al.* 1993).

Euscarthmus rufomarginatus is known in Paraguay from an undocumented report and a single specimen from Zanja Morotí (PY 02) collected in 1940 (Olrog 1979, Hayes 1995).

Sporophila falcirostris has been recorded just once near Salto de Guairá (PY 05).

OUTLOOK

This analysis has identified 23 Key Areas, each of which is important for the conservation of threatened birds in Paraguay. However, three Key Areas stand out as top priorities (Table 3): they are the most important areas for at least six threatened species, and 12 (50%) of the Paraguayan threatened birds have been recorded from them. These and the remaining Key Areas (which are individually extremely important for many threatened Paraguayan endemics: see 'Key Areas' and 'Key Area protection', above) all need some degree of conservation action, often in the form of protection, if Paraguay is to retain populations of each of its threatened birds and prevent future extinctions.

DATA SOURCES

The above introductory text and the Site Inventory were compiled from information supplied by J. C. Almada, T. M. Brooks, L. Carlo, O. Carrill, R. P. Clay, E. Z. Esquivel, F. E. Hayes, J. C. Lowen, A. Madroño N., M. Pearman, D. M. Pullan and R. S. Ridgely, as well as from the following references.

Acevedo, C., Fox, J., Gauto, R., Granizo, T., Keel, S., Pinazzo, J., Spinzi, L., Sosa, W. and Vera, V. (1990) *Areas prioritarias para la conservación en la región oriental del Paraguay*. Asunción: Centro de Datos para la Conservación.

Brooks, T. M., Barnes, R., Bartrina, L., Butchart, S. H. M., Clay, R. P., Esquivel, E. Z., Etcheverry, N. I., Lowen, J. C. and Vincent, J. (1993) *Bird surveys and conservation in the Paraguayan Atlantic Forest: Project CANOPY '92 final report*. Cambridge, U.K.: BirdLife International (Study Report 57).

Collar, N. J., Gonzaga, L. P., Krabbe, N., Madroño Nieto, A., Naranjo, L. G., Parker, T. A. and Wege, D. C. (1992) *Threatened birds of the Americas: the ICBP/IUCN Red Data Book*. Cambridge, U.K.: International Council for Bird Preservation.

Collar, N. J., Crosby, M. J. and Stattersfield, A. J. (1994) *Birds to watch 2: the world list of threatened birds*. Cambridge, U.K.: BirdLife International (BirdLife Conservation Series no. 4).

Collar, N. J., Wege, D. C. and Long, A. J. (in press) Patterns and causes of endangerment in the New World avifauna. In J. V. Remsen, ed. *Natural history and conservation of Neotropical birds*. American Ornithologists' Union (Orn. Monogr.).

DPNVS (1993) *Plan maestro del Sistema Nacional de Areas Silvestres Protegidas del Paraguay (SINASIP).* Asunción, Paraguay. Unpublished report.

HAYES, F. E. (1995) *Status, distribution and biogeogarphy of the birds of Paraguay.* Colorado Springs: American Birding Association (Monogr. in Field Orn. 1).

IUCN (1992) *Protected areas of the world: a review of national systems,* 4: *Nearctic and Neotropical.* Gland, Switzerland and Cambridge, U.K.: International Union for Conservation of Nature and Natural Resources.

LÓPEZ, N. E. (1992) Observaciones sobre la distribución de psitacidos en el Departamento de Concepción, Paraguay. *Bol. Mus. Nac. Hist. Nat. Paraguay.* 11: 2–25.

MADROÑO, A. AND ESQUIVEL, E. Z. (1994) Investigación y monitoreo de los recursos naturales de la Reserva Natural del Bosque Mbaracayú (plan no 1: ornitología). Fundación Moisés Bertoni (FMB). Unpublished report.

OLROG, C. C. (1979) Notas ornitológicas, XI: sobre la colección del Instituto Miguel Lillo. *Acta Zool. Lilloana* 33: 5–7.

PAYNTER, R. A. (1989) *Ornithological gazetteer of Paraguay.* Second edition. Cambridge, Mass.: Museum of Comparative Zoology.

PEARMAN, M. AND ABADIE, E. (1995) Field identification, ecology and status of the Sickle-winged Nightjar *Eleothreptus anomalus. Cotinga* 3: 12–14.

STATTERSFIELD, A. J., CROSBY, M. J., LONG, A. J. AND WEGE, D. C. (in prep.) *Global directory of Endemic Bird Areas.* Cambridge, U.K.: BirdLife International (BirdLife Conservation Series).

SITE INVENTORY

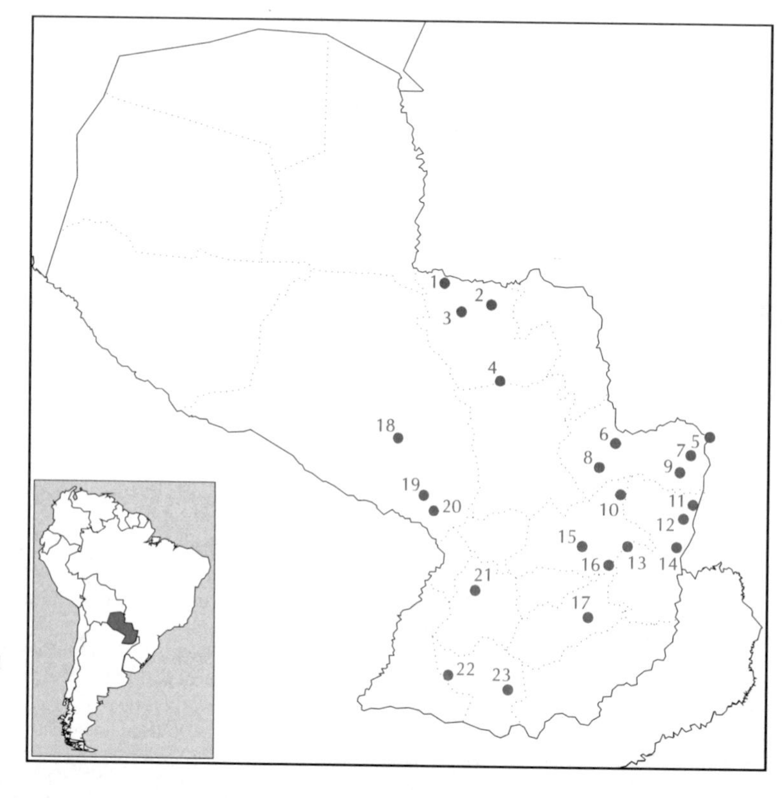

Figure 2. Key Areas in Paraguay.

01 Estancia Centurión
02 Zanja Morotí
03 Serranía San Luis
04 Horqueta
05 Salto de Guairá
06 Mbaracayú
07 Río Carapá
08 Estancia La Fortuna
09 Estancia Itabó
10 Estancia La Golondrina
11 Limoy
12 Itabó
13 Estancia San Antonio
14 Tatí Yupí
15 Encarnación
16 Estancia La Golondrina
17 Caaguazu
18 Estancia Leopoldina
19 Río Confuso–Trans-Chaco Highway
20 Estancia La Golondrina
21 Cerro Acahay
22 Santa Elisa–San Juan Bautista
23 San Patricio

Estancia Centurión (Concepción)

PY 01

22°15′S 57°35′W

Estrella Scientific Reserve (proposed), 50,000 ha

The estancia is c.10 km south of the Río Apa and 33 km north-east of Saladero Risso in north-west Concepción. The area comprises mainly cerrado with some sub-humid forest.

Harpyhaliaetus coronatus	1989	One observation near the estancia.
Anodorhynchus hyacinthinus	1989	One observation (López 1992).
Amazona vinacea	1989	Recorded, but status unknown (López 1992).

Zanja Morotí (Concepción)

PY 02

22°30′S 57°00′W

Unprotected

Zanja Morití is a hilly region at c.330 m in the vicinity of the Río Apa in east-central Concepción. More surveys are needed in the cerrado habitat of this area.

Euscarthmus rufomarginatus	1944	One collected in December 1944 (Olrog 1979). The only record for Paraguay.

Serranía San Luis (Concepción)

PY 03

22°35′S 57°22′W

Serranía San Luis Ecological Reserve (IUCN category V), 10,273 ha

The area is c.40 km east of Río Paraguay and 30 km south of Río Apa in north-west Concepcíon; Estancia San Luis, an old collecting locality, is adjacent to the north, and Estancia Santa María is adjacent to the south. The rocky outcrop of the serranía stands in an otherwise flat landscape and is a rugged terrain covered in dense forest with some trees in excess of 20 m; in its flatter areas the forest is more open and mixed with grassland (Acevedo *et al.* 1990).

Anodorhynchus hyacinthinus	1988	A pair seen at Estancia San Luis in December 1988 (López 1992).
Amazona vinacea	1988	Recorded at Estancia Santa María de la Sierra (López 1992).

Horqueta (Concepción)

PY 04

23°24′S 56°53′W

Unprotected

The town of Horqueta, located in southern Concepción, is an old site for two threatened species of rail, both of which are known from only a handful of localities. A search of marshes in the vicinity of the town could find populations of these rare and little-known species, which, however, may be—or may have been—present only seasonally.

Laterallus xenopterus	1933	The type-locality for the species.
Coturnicops notatus	1938	A female collected 7 km east of Horqueta.

Salto de Guairá (Canindeyú)

PY 05

Unprotected

24°02′S 54°16′W
EBA B52

Along the road south-west of Salto de Guairá, which lies on the upper Río Paraná on the border with Brazil, is an area of humid forest dominated by bamboo. Between visits in 1977 and 1991 forest cover had, at least near the road, much diminished (R. S. Ridgely *in litt.* 1994).

Sporophila falcirostris	1977	Many present in bamboo stands in 1977, but none seen during a visit in 1991.

Mbaracayú (Canindeyú)

PY 06

Mbaracayú Forest Nature Reserve (IUCN category V), 62,979 ha

24°07′S 55°26′W
EBA B52

This area, located in the upper Jejui basin close to the border with Brazil, is the largest block of humid subtropical forest left in Paraguay. The area received official protection in 1991 and is now being managed by Fundación Moisés Bertoni (FMB).

Harpyhaliaetus coronatus	1992	A pair observed in cerrado bordered by transitional forest near Lagunita (Brooks *et al.* 1993).
Pipile jacutinga	1995	A stronghold for the species. Recorded regularly at several localities during survey work in recent years (Brooks *et al.* 1993, Madroño and Esquivel 1994).
Anodorhynchus hyacinthinus	1994	Three sightings by three different park guards in 1982, 1993 and 1994 (Madroño and Esquivel 1994, A. Madroño *in litt.* 1995).
Amazona vinacea	1994	Present in small numbers (Madroño and Esquivel 1994) but recorded only infrequently, so its status is uncertain.
Dryocopus galeatus	1994	First recorded in 1994 and at least two pairs bred (raising one and two young) in October (Madroño and Esquivel 1994, R. P. Clay *in litt.* 1994, D. M. Pullan *in litt.* 1995).
Phylloscartes paulistus	1994	A number of records throughout 1994 suggest the site to be a stronghold (Madroño and Esquivel 1994, R. P. Clay *in litt.* 1994, D. M. Pullan *in litt.* 1995).
Platyrinchus leucoryphus	1994	A handful of observations in 1992 and 1994 indicate that this area is probably a stronghold for the species in Paraguay (Brooks *et al.* 1993, Madroño and Esquivel 1994, R. P. Clay *in litt.* 1994).
Sporophila hypochroma	1994	Three males seen in a large flock of *Sporophila* spp. in grasslands at Lagunita in November 1994 (Madroño and Esquivel 1994, R. P. Clay *in litt.* 1994, D. M. Pullan *in litt.* 1995).
Sporophila palustris	1994	Two males seen in a large flock of *Sporophila* spp. in grasslands at Lagunita in November 1994 (Madroño and Esquivel 1994, R. P. Clay *in litt.* 1994, D. M. Pullan *in litt.* 1995).

Río Carapá (Canindeyú)

PY 07

Unprotected

24°15′S 54°30′W
EBA B52

The middle and lower sections of the Río Carapá stretch across eastern Canindeyú where it enters the northern end of the Itaipú dam lake. The Key Area embraces the river and humid forests along the middle part of the river including the forests 16 km south-south-east of Mbaracayú (24°16′S 54°35′W) and around Catueté (24°11′S 54°41′W).

Mergus octosetaceus	1984	The only recent record for Paraguay.
Amazona vinacea	1989	Reported to be common in July 1989 around Catueté.

PY 08

Estancia La Fortuna (Canindeyú)

24°24′S 55°38′W
EBA B52

Unprotected

Estancia La Fortuna embraces the humid subtropical forests and marshes, with dense bunch-grass-like vegetation 1.5–2 m in height, north of Curuguaty.

Laterallus xenopterus	1979	Two collected 6.3 km by road north-east of Curuguaty in July 1976 and July 1979, and two collected 13.3 km by road north-east of Curuguaty in August 1978 and July 1979. The only recent records in Paraguay.
Dryocopus galeatus	1989	There are records 13 km north of Curuguaty (July 1979), 10 km north (September 1989), and a probable sighting at Estancia La Fortuna in June 1991.
Phylloscartes paulistus	1991	One sighting at Estancia La Fortuna in June.
Platyrinchus leucoryphus	1978	Three collected in August 1978.
Sporophila palustris	1990	One sight record.

PY 09

Estancia Itabó (Canindeyú)

24°27′S 54°38′W
EBA B52

Itabó Nature Reserve (private), 8,000 ha

This estancia is located 26 km west of route 4, the nearest town being Puerto Kyha in eastern Alto Paraná. About 8,000 ha of the estancia comprises humid forest, which is being managed primarily for heart of palm *Euterpe edulis* (Brooks *et al.* 1993). The palms, and therefore the forest, are protected by guards employed by the estancia (Brooks *et al.* 1993).

Pipile jacutinga	1992	Three records of four birds during 16 days of fieldwork in August (Brooks *et al.* 1993).
Claravis godefrida	1994	A female seen in December was the first Paraguayan record since 1893 (R. P. Clay *in litt.* 1994).
Amazona pretrei	1992	A single sighting is the sole record for Paraguay. Possibly a vagrant or an escaped cagebird, as the site is well outside the currently known range (Brooks *et al.* 1993).
Amazona vinacea	1994	The commonest amazon in winter, with a minimum of 50 birds. Less common in December 1994, suggesting that the summer (and presumably breeding) population is lower (Brooks *et al.* 1993, R. P. Clay *in litt.* 1994).
Dryocopus galeatus	1994	Recorded several times during fieldwork in August 1992, probably involving at least five individuals; sightings also in 1991 and 1994 (Brooks *et al.* 1993, R. P. Clay *in litt.* 1994).
Phylloscartes paulistus	1992	Probably an uncommon resident (Brooks *et al.* 1993, R. P. Clay *in litt.* 1994).
Xanthopsar flavus	1992	One seen in open grassland (Brooks *et al.* 1993).

PY 10

Estancia La Golondrina (Canindeyú/Caaguazú)

24°43′S 55°22′W
EBA B52

Golondrina II Nature Reserve (private), 20,000 ha
Acaray-mí Ecological Reserve (proposed), 25,000 ha

Ríos Aracay-mí and Piratí-y define most of the boundary of this large estancia, 35 km east of Curuguaty. Much of the land is used for cattle-ranching (Brooks *et al.* 1993). The private reserve is in the north of the estancia, and in spite of continued logging some large trees still remain (Brooks *et al.* 1993).

Pipile jacutinga	1992	Three sightings of two birds in disturbed, selectively logged forest (Brooks *et al.* 1993).
Dryocopus galeatus	1992	Sightings were made during fieldwork in 1991 and 1992 but the population may be only very small (Brooks *et al.* 1993).

Limoy (Alto Paraná)

PY 11

Limoy Biological Reserve (IUCN category V), 14,332 ha

24°50'S 54°28'W
EBA B52

The reserve is adjacent to the Itaipú dam lake with its eastern boundary alongside the main lake, its southern border along the Río Limoy and its northern border along the Río Itambey. Tall and relatively pristine humid forest is the main habitat.

Pipile jacutinga	1988?	Listed as rare; no further information.
Amazona vinacea	1990	Apparently good numbers in and around the reserve, e.g. 70 birds seen 2 km south of the reserve in May 1990 (López 1992).

Itabó (Alto Paraná)

PY 12

Itabó Biological Reserve (IUCN category V), 11,260 ha

25°00'S 54°35'W
EBA B52

The reserve is located c.50 km east of Itakyry with its southern boundary following the shoreline of the Itaipú dam lake in eastern Alto Paraná. The main habitat is tall humid forest.

Pipile jacutinga	c.1988	Listed as rare; no further information.
Amazona vinacea	c.1988	Listed as common; no further information.
Dryocopus galeatus	c.1988	Listed as rare; no further information.

Estancia San Antonio (Alto Paraná)

PY 13

San Antonio Nature Reserve (private), 1,000 ha

25°18'S 55°20'W
EBA B52

This estancia is c.36 km west of Santa Rita in southern Alto Paraná. Much of the area remains forested, but land is also set aside for cattle-ranching and agriculture. The private reserve in the north-west corner of the estancia comprises 1,000 ha of humid forest.

Pipile jacutinga	1992	Two sightings (probably of one bird) during 12 days of fieldwork in July—August 1992 (Brooks *et al.* 1993).
Dryocopus galeatus	1992	Several sightings involving a minimum of five individuals during fieldwork in July—August 1992 (Brooks *et al.* 1993).
Pseudocolopteryx dinellianus	1992	A bird seen twice in July 1992 appears to have been a vagrant, as the record is north of the currently known range (Brooks *et al.* 1993).

Tatí Yupí (Alto Paraná)

PY 14

Tatí Yupí Biological Refuge (IUCN category V), 2,245 ha

25°20'S 54°40'W
EBA B52

The reserve lies on the western shore of the Itaipú dam lake at its southern end immediately north-east of Hernandarias. It includes areas of tall humid forest, and cerrado, but has suffered from logging and much of the forest is secondary.

Pipile jacutinga	c.1988	Listed as rare; no further information.
Amazona vinacea	c.1988	Listed as common; no further information.

Encarnación (Itapua)

PY 15

27°20′S 55°50′W

Unprotected

The general area lies west of Encarnación immediately north of the Río Paraná in southern Itapua, and the habitat consists mainly of seasonally flooded grassland.

Xanthopsar flavus	1989	A group of five seen 10 km east of San Cosmé and Damián.

Estancia La Golondrina (Caazapá)

PY 16

25°33′S 55°30′W
EBA B52

Golondrina III Nature Reserve (private), 15,477 ha

This estancia is in the eastern corner of Caazapá, adjacent to the southern banks of the Río Monday. Over half of it comprises humid forest, but most has been selectively logged. Much of the rest of the land is turned over to cattle-ranching and cotton, soya and maize cultivation, although there are 2,000 ha of wetlands.

Harpyhaliaetus coronatus	1992	An immature seen over cotton fields at the edge of the forest in July (Brooks *et al.* 1993).
Pipile jacutinga	c.1992	Reliably reported by local indians (Brooks *et al.* 1993).
Phylloscartes paulistus	1992	Recorded twice during 16 days fieldwork in July 1992 (Brooks *et al.* 1993).

Caaguazu (Caazapá)

PY 17

26°10′S 55°45′W
EBA B52

Caaguazu Ecological Reserve (IUCN category II), 16,000 ha

This area is located between San Juan Nepomuceno and San Agustín in central Caazapá, and comprises principally humid forest (PROVEPA *per* R. P. Clay *in litt.* 1994).

Platyrinchus leucoryphus	1993	One collected in October (PROVEPA *per* R. P. Clay *in litt.* 1994).

Estancia Leopoldina (Presidente Hayes)

PY 18

24°05′S 58°10′W

Unprotected

This area is located at about Km 187 on the Trans-Chaco Highway c.150 km north-west of Asunción. The habitat is humid natural grassland (M. Pearman *in litt.* 1995).

Anthus nattereri	1989	One singing in May 1989 (M. Pearman *in litt.* 1995); this is the locality given as near Monte Lindo (23°57′S 57°12′W) in Collar *et al.* (1992).

Río Confuso–Trans-Chaco Highway (Presidente Hayes)

PY 19

24°45′S 57°50′W

Tacuara National Park (proposed), 150,000 ha

This ill-defined Key Area lies 80 km north-west of Asuncíon in south-east Presidente Hayes, between the Río Confuso and Km 75–100 on the Trans-Chaco Highway. The area is within the lower humid chaco, which holds large areas of seasonally flooded wet palm savanna.

Harpyhaliaetus coronatus	1989	One seen at Estancia San José near Río Confuso in July.
Pseudocolopteryx dinellianus	1990	Two seen in May 1990 at Km 79, and another at Km 100 in June 1990.
Yetapa risora	1989	Seen at Km 103 on the Trans-Chaco Highway and near Estancia San José in July.
Sporophila hypochroma	1991	Two males (and an unidentified female *Sporophila*) seen near Km 100 on the Trans-Chaco Highway in February 1991.

Estancia La Golondrina (Presidente Hayes)

PY 20

24°52′S 57°10′W

Golondrina I Nature Reserve (private), 5,497 ha

The reserve is in the lower chaco of south-east Presidente Hayes near the lower Río Confuso, 40 km north of Asunción. The area holds extensive wet grasslands with streams and seasonally flooded palm savanna. Much of the land is grazed by cattle, and some of the streams have been dammed to provide permanent water for livestock.

Yetapa risora	1994	At least eight seen on six dates in November 1994 (D. M. Pullan *in litt.* 1995).

Cerro Acahay (Paraguarí)

PY 21

24°56′S 57°42′W

Acahay Massif Natural Monument (IUCN category unknown), 2,500 ha

A small isolated massif, located between the towns of Quindy and Colonia La Colmena in eastern Paraguarí. The terrain is steep and rocky, and cactus scrub is the main habitat.

Harpyhaliaetus coronatus	1980s	Reported by Acevedo *et al.* (1990).

Santa Elisa–San Juan Bautista (Misiones)

PY 22

25°00′S 57°30′W
EBA B52

Unprotected

The area of north-central Misiones comprises seasonally inundated wet grassland with some land under cattle-ranching. *Anthus nattereri* should be looked for here.

Yetapa risora	1994	At least nine seen in November 1994 c.3 km east of Estancia Santa Elisa (D. M. Pullan *in litt.* 1995); 47 seen along c.10 km of road in the same area in June 1991.
Sporophila hypochroma	1994	Three males seen in November 1994, c.3 km east of Estancia Santa Elisa (D. M. Pullan *in litt.* 1995).
Sporophila palustris	1991	One seen in March 1991.

PY 23

San Patricio (Misiones)

27°02′S 56°45′W

Unprotected

The area is south of San Patricio which is located c.30 km south-east of San Ignacio in southern Misiones. The principal habitat is presumably natural rolling grassland.

Anthus nattereri	1977	Birds seen displaying in August 1977.

■

PERU

Golden-backed Mountain-tanager *Buthraupis aureodorsalis*

PERU harbours one of the most species-rich avifaunas of any country in the world. More than 1,678 species of resident and migrant bird are recorded (Parker *et al.* 1982), of which about 113 (7%) are endemic to the country, 216 (13%) have restricted ranges (Stattersfield *et al.* in prep.) and 64 (4%) are threatened (Collar *et al.* 1992). This analysis has identified 89 Key Areas for threatened birds in Peru (see '*Key Areas*: the book', p. 11, for criteria).

THREATENED BIRDS

Of the 64 Peruvian species which Collar *et al.* (1992) considered to be at risk of extinction, 31 (48.5%) are confined to the country (Table 1), making Peru second in rank in the Americas for numbers of threatened birds (Collar *et al.* in press). The majority of these threatened species are distributed in the subtropical and temperate zones, with 36% confined to the temperate zone above 2,000 m (Collar *et al.* in press). Wet forest habitats are essential to 53% of the threatened birds, 20% are reliant on dry forests and the remainder are split between páramo grasslands, wetlands and riverine areas; the loss or alteration of all these habitats is a threat to over 73% of the species, 14% are affected by hunting and 17% are at risk simply because of the small area of their own ranges (Collar *et al.* in press). The distributions of these threatened birds and their relationship to Endemic Bird Areas are shown in Figure 1.

KEY AREAS

The 89 Peruvian Key Areas would, if adequately protected, help ensure the conservation of 62 (97%) of the threatened species (but see 'Old records and little-known birds', below). Of these areas, 46 are important for two or more (up to 11) threatened species, and are therefore the most efficient areas (currently known) in which to conserve Peru's threatened birds (see 'Outlook', below). The six areas which each harbour five or more threatened species together represent potential populations of 29 threatened species (45% of the total number); this is increased to 43 species (67%) if the 16 Key Areas with four or more threatened species are considered

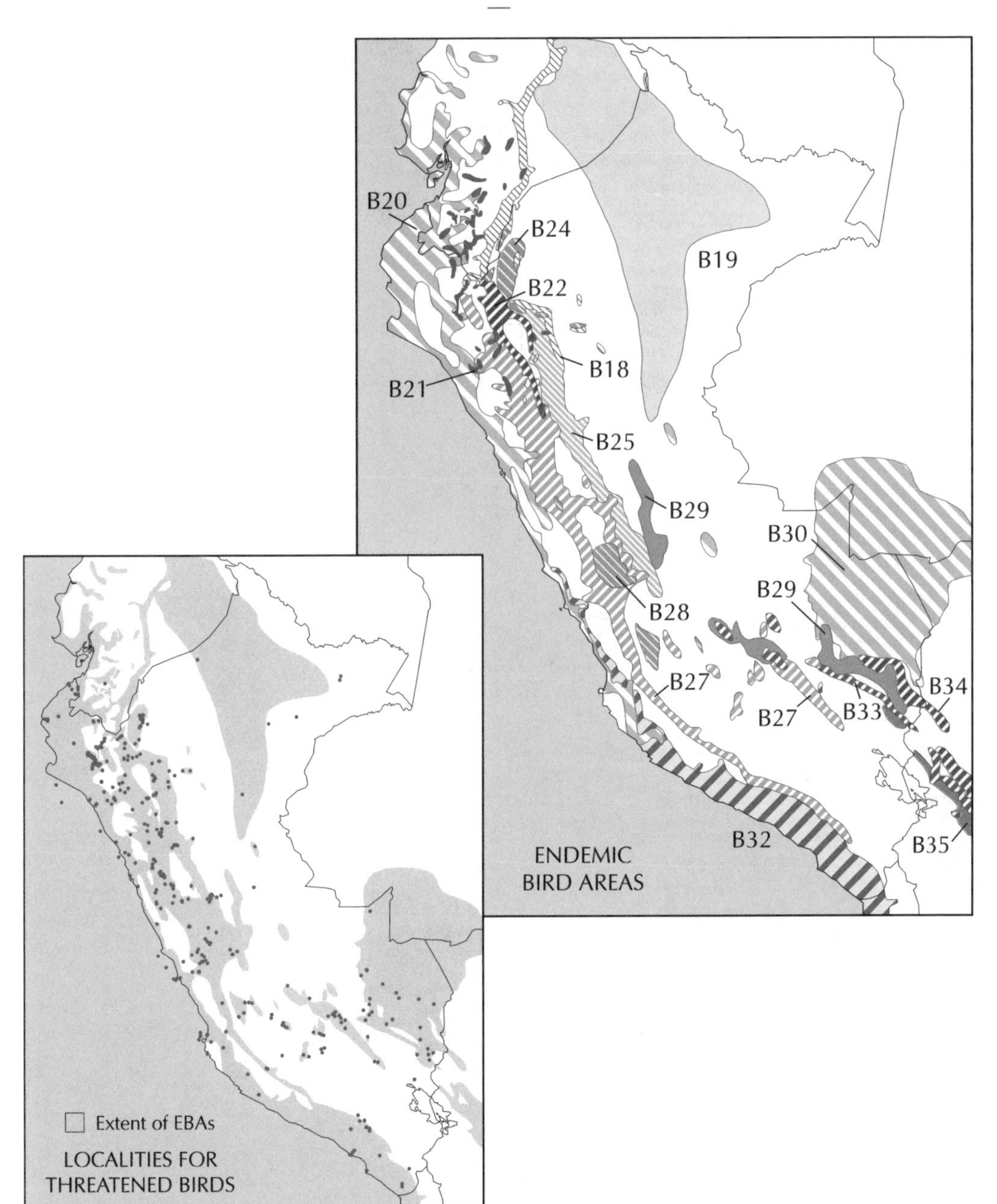

Figure 1. The localities where threatened birds have been recorded in Peru and their relationship to Endemic Bird Areas (EBAs).

- Most of the threatened birds in Peru have restricted ranges and thus occur together in various combinations within EBAs, which are listed below (figures are numbers of these species in each EBA). The distribution of threatened birds in Peru is dominated by the Andes, but of particular importance are EBAs in the Tumbesian region of north-west Peru (B20), the high Peruvian Andes (B27) and the dry Marañón valley (B22).

B18 East Andes of Ecuador and northern Peru (0)
B19 Napo and upper Amazon lowlands (1)
B20 Tumbesian western Ecuador and Peru (12)
B21 South/Central Andean forests (2)
B22 Marañón valley (7)
B24 Sub-Andean ridge-top forests of south-east Ecuador and northern Peru (5)
B25 North-east Peruvian cordilleras (4)
B27 High Peruvian Andes (12)
B28 Junín puna (3)
B29 East Andean foothills of Peru (1)
B30 South-east Peruvian lowlands (2)
B32 South Peruvian and north Chilean Pacific slope (3)
B34 Lower Bolivian yungas (2)
B60 Central Andean páramo (2)

Table 1. Coverage of threatened species by Key Areas. Areas in bold currently have some form of protected status.

	Key Areas occupied	No. of Key Areas protected	Total nos. of Key Areas	
			Peru	Neotropics
Black Tinamou *Tinamus osgoodi*	78,**79**,**81**	2	3	4
Kalinowski's Tinamou *Nothoprocta kalinowskii*[E]	34	0	1	1
Taczanowski's Tinamou *Nothoprocta taczanowskii*[E]	70,**71**,73,77,85	1	5	5
Junín Grebe *Podiceps taczanowskii*[E]	**59**	1	1	1
Peruvian Diving-petrel *Pelecanoides garnotii*	**64**	1	1	3
Grey-backed Hawk *Leucopternis occidentalis*	**01**,07	1	2	20
Lesser Collared Forest-falcon *Micrastur buckleyi*	13,53,**79**	1	3	5
White-winged Guan *Penelope albipennis*[E]	24,25,26,27	0	4	4
Bearded Guan *Penelope barbata*	03,05,06,10,11,28	0	6	12
Southern Helmeted Curassow *Pauxi unicornis*	52,**81**	1	2	4
Wattled Curassow *Crax globulosa*	80	0	1	4
Junín Rail *Laterallus tuerosi*[E]	**59**	1	1	1
Peruvian Pigeon *Columba oenops*[E]	16,17,21,27,33	0	5	5
Ochre-bellied Dove *Leptotila ochraceiventris*	**01**,03,27	1	3	14
Golden-plumed Parakeet *Leptosittaca branickii*	16,17,**32**,35,36,50,**79**	2	7	22
Yellow-faced Parrotlet *Forpus xanthops*[E]	15,16,17,21,33	0	5	5
Spot-winged Parrotlet *Touit stictoptera*	07,23	0	2	8
Red-faced Parrot *Hapalopsittaca pyrrhops*	06	0	1	7
Long-whiskered Owlet *Xenoglaux loweryi*[E]	16,22	0	2	2
Purple-backed Sunbeam *Aglaeactis aliciae*[E]	31,33	0	2	2
Royal Sunangel *Heliangelus regalis*[E]	07,17,23	0	3	3
Neblina Metaltail *Metallura odomae*	06	0	1	3
Grey-bellied Comet *Taphrolesbia griseiventris*[E]	11,12,46,47	0	4	4
Marvellous Spatuletail *Loddigesia mirabilis*[E]	18,19,20,23	0	4	4
Chilean Woodstar *Eulidia yarrellii*	89	0	1	5
Little Woodstar *Acestrura bombus*	28,33,50	0	3	15
Yellow-browed Toucanet *Aulacorhynchus huallagae*[E]	**32**,36	1	2	2
Royal Cinclodes *Cinclodes aricomae*	72,73	0	2	2
White-bellied Cinclodes *Cinclodes palliatus*[E]	**59**,60,61,62	1	4	4
White-browed Tit-spinetail *Leptasthenura xenothorax*[E]	72,73,74,75,76	0	5	5
Chestnut-throated Spinetail *Synallaxis cherriei*	—	0	0	4
Apurímac Spinetail *Synallaxis courseni*[E]	**71**	1	1	1
Blackish-headed Spinetail *Synallaxis tithys*	**01**	1	1	7
Russet-bellied Spinetail *Synallaxis zimmeri*[E]	41,42,43	0	3	3
Pale-tailed Canastero *Asthenes huancavelicae*[E]	38,39,40,**44**,48, 62,66,68,69	1	9	9
Russet-mantled Softtail *Thripophaga berlepschi*[E]	06,16,18,20,35	0	5	5
Henna-hooded Foliage-gleaner *Hylocryptus erythrocephalus*	**01**,25,27	1	3	12
Rufous-necked Foliage-gleaner *Syndactyla ruficollis*	**01**,**02**,03,05,10,11,28	2	7	15
Ash-throated Antwren *Herpsilochmus parkeri*[E]	23	0	1	1
Yellow-rumped Antwren *Terenura sharpei*	82	0	1	3
Grey-headed Antbird *Myrmeciza griseiceps*	**01**,03,05	1	3	7
White-masked Antbird *Pithys castanea*[E]	—	0	0	0
Rufous-fronted Antthrush *Formicarius rufifrons*[E]	**79**,80,**81**	2	3	3
White-cheeked Cotinga *Zaratornis stresemanni*[E]	38,**44**,55,56,57,58,**67**	2	7	7
Slaty Becard *Pachyramphus spodiurus*	**01**,08,15	1	3	9
Ash-breasted Tit-tyrant *Anairetes alpinus*	38,**44**,45,72,73,75	1	6	7
Cinnamon-breasted Tody-tyrant *Hemitriccus cinnamomeipectus*	07,16,22	0	3	4
Pacific Royal Flycatcher *Onychorhynchus occidentalis*	**01**	1	1	10
Grey-breasted Flycatcher *Lathrotriccus griseipectus*	**01**,**02**,07,08	2	4	20
White-tailed Shrike-tyrant *Agriornis andicola*	10,34,**44**,50,77,**86**	2	6	11
Ochraceous Attila *Attila torridus*	**01**	1	1	11
Peruvian Plantcutter *Phytotoma raimondii*[E]	29,30,37	0	3	3
Slender-billed Finch *Xenospingus concolor*	63,65,**87**	1	3	7
Grey-winged Inca-finch *Incaspiza ortizi*[E]	04,09,21	0	3	3
Plain-tailed Warbling-finch *Poospiza alticola*[E]	21,**32**,35,38,**44**,45	2	6	6
Rufous-breasted Warbling-finch *Poospiza rubecula*[E]	12,40,42,**44**,56,57	1	6	6
Golden-backed Mountain-tanager *Buthraupis aureodorsalis*[E]	**32**,35,49,51	1	4	4
Masked Mountain-tanager *Buthraupis wetmorei*	06	0	1	7

cont.

Table 1 (cont.)

	Key Areas occupied	No. of Key Areas protected	Total nos. of Key Areas	
			Peru	Neotropics
Orange-throated Tanager *Wetmorethraupis sterrhopteron*	13,14	0	2	3
Green-capped Tanager *Tangara meyerdeschauenseei*[E]	83,84	0	2	2
Sira Tanager *Tangara phillipsi*[E]	52	0	1	1
Tamarugo Conebill *Conirostrum tamarugense*	**86**,88	1	2	8
Selva Cacique *Cacicus koepckeae*[E]	54	0	1	1
Saffron Siskin *Carduelis siemiradzkii*	**01**	1	1	4

[E] Endemic to Peru

(see Table 3). However, as vital as these areas with multiple threatened species are, they should not detract from the significance of the remaining Key Areas, each of which is of major importance for the adequate conservation of Peru's threatened birds (see 'Outlook').

From Tables 1 and 2 it can be seen that 18 species occur in just one Peruvian Key Area; seven of these species (*Nothoprocta kalinowskii*, *Podiceps taczanowskii*, *Laterallus tuerosi*, *Synallaxis courseni*, *Herpsilochmus parkeri*, *Tangara phillipsi* and *Cacicus koepckeae*) occur only in Peru, and are thus—on present evidence—reliant for their survival on the integrity of habitat in their respective single Key Areas (see 'Outlook'). Indeed, *Cacicus koepckeae* occurs in just one area and is not sympatric with other threatened birds, suggesting that, at least in this case, targeted single-species conservation is a necessity. An extension to this list should perhaps be made to include *Xenoglaux loweryi*, *Aglaeactis aliciae*, *Aulacorhynchus huallagae* and *Tangara meyerdeschauenseei*, all endemic to Peru but each known from just two Key Areas. All of the species mentioned above are represented within the Key Areas listed in Table 3 (see 'Outlook').

Two Peruvian threatened species are not represented within the Key Area analysis for the country, as follows.

Synallaxis cherriei is known in Peru from four localities, but better defined areas with more recent records have been documented within Brazil and Ecuador.

Pithys castanea is known from a single 1937 specimen collected from a locality ('Andoas') in the Amazonian lowlands of either Loreto in Peru or Oriente in Ecuador. Until the species is found again, or until the actual type-locality is known, site-based conservation measures will remain inappropriate.

KEY AREA PROTECTION

Only 11 (c.12%) of Peru's Key Areas currently have some form of protected status, just three of them as national parks (IUCN category II). The 78 Key Areas (c.88% of the total) that are currently unprotected urgently require attention in the form of appropriate conservation measures if the populations of their threatened species are to survive. However, even the few formally protected areas remain under threat and, in many, habitat degradation and uncontrolled hunting continue unchecked; effective management is thus required of activities undertaken within them. Perhaps more important than the level of Key Area protection is the fact that as many as 32 (50%) of Peru's threatened species are not currently known from any protected Key Area; 19 of these species (*Nothoprocta kalinowskii*, *Penelope albipennis*, *Columba oenops*, *Forpus xanthops*, *Xenoglaux loweryi*, *Aglaeactis aliciae*, *Heliangelus regalis*, *Taphrolesbia griseiventris*, *Loddigesia mirabilis*, *Leptasthenura xenothorax*, *Synallaxis zimmeri*, *Thripophaga berlepschi*, *Herpsilochmus parkeri*, *Pithys castanea*, *Phytotoma raimondii*, *Incaspiza ortizi*, *Tangara meyerdeschauenseei*, *Tangara phillipsi* and *Cacicus koepckeae*) are endemic to the country (and three known only from single Key Areas: see Table 1) and therefore totally reliant on the integrity of the habitat within their respective Key Areas if their continued survival is to be guaranteed. Key Areas supporting populations of any of these unprotected Peruvian endemics should be regarded as high conservation priorities.

RECENT CHANGES TO THE THREATENED LIST

With the publication of Collar *et al.* (1994), 10 of the 64 threatened species (*Micrastur buckleyi*, *Xenoglaux loweryi*, *Metallura odomae*, *Aulacorhynchus huallagae*, *Synallaxis cherriei*, *Thripophaga berlepschi*, *Pachyramphus spodiurus*, *Hemitriccus cinnamomeipectus*, *Incaspiza ortizi* and *Tangara phillipsi*) were dropped from the Peruvian threatened species list

Table 2. Matrix of threatened species by Key Area. Totals of areas and species refer to the whole of Peru. Two threatened Peruvian species, *Synallaxis cherriei* and *Pithys castanea,* do not occur in any Key Area and are omitted here.

	01	02	03	04	05	06	07	08	09	10	11	12	13	14	15	16	17	18
Nothoprocta kalinowskii	–	–	–	–	–	–	–	–	–	–	–	–	–	–	–	–	–	–
Leucopternis occidentalis	●	–	–	–	–	–	●	–	–	–	–	–	–	–	–	–	–	–
Micrastur buckleyi	–	–	–	–	–	–	–	–	–	–	–	–	●	–	–	–	–	–
Penelope albipennis	–	–	–	–	–	–	–	–	–	–	–	–	–	–	–	–	–	–
Penelope barbata	–	–	●	–	●	●	–	–	–	●	●	–	–	–	–	–	–	–
Columba oenops	–	–	–	–	–	–	–	–	–	–	–	–	–	–	–	●	●	–
Leptotila ochraceiventris	●	–	●	–	–	–	–	–	–	–	–	–	–	–	–	–	–	–
Leptosittaca branickii	–	–	–	–	–	–	–	–	–	–	–	–	–	–	–	●	●	–
Forpus xanthops	–	–	–	–	–	–	–	–	–	–	–	–	–	–	●	●	●	–
Touit stictoptera	–	–	–	–	–	–	●	–	–	–	–	–	–	–	–	–	–	–
Hapalopsittaca pyrrhops	–	–	–	–	–	●	–	–	–	–	–	–	–	–	–	–	–	–
Xenoglaux loweryi	–	–	–	–	–	–	–	–	–	–	–	–	–	–	–	●	–	–
Aglaeactis aliciae	–	–	–	–	–	–	–	–	–	–	–	–	–	–	–	–	–	–
Heliangelus regalis	–	–	–	–	–	–	●	–	–	–	–	–	–	–	–	–	●	–
Metallura odomae	–	–	–	–	–	●	–	–	–	–	–	–	–	–	–	–	–	–
Taphrolesbia griseiventris	–	–	–	–	–	–	–	–	–	–	●	●	–	–	–	–	–	–
Loddigesia mirabilis	–	–	–	–	–	–	–	–	–	–	–	–	–	–	–	–	–	●
Acestrura bombus	–	–	–	–	–	–	–	–	–	–	–	–	–	–	–	–	–	–
Aulacorhynchus huallagae	–	–	–	–	–	–	–	–	–	–	–	–	–	–	–	–	–	–
Synallaxis tithys	●	–	–	–	–	–	–	–	–	–	–	–	–	–	–	–	–	–
Thripophaga berlepschi	–	–	–	–	–	●	–	–	–	–	–	–	–	–	–	●	●	–
Hylocryptus erythrocephalus	●	–	–	–	–	–	–	–	–	–	–	–	–	–	–	–	–	–
Syndactyla ruficollis	●	●	●	–	●	–	–	–	–	●	●	–	–	–	–	–	–	–
Herpsilochmus parkeri	–	–	–	–	–	–	–	–	–	–	–	–	–	–	–	–	–	–
Myrmeciza griseiceps	●	–	●	–	●	–	–	–	–	–	–	–	–	–	–	–	–	–
Pachyramphus spodiurus	●	–	–	–	–	–	–	●	–	–	–	–	–	–	●	–	–	–
Hemitriccus cinnamomeipectus	–	–	–	–	–	–	●	–	–	–	–	–	–	–	–	●	–	–
Onychorhynchus occidentalis	●	–	–	–	–	–	–	–	–	–	–	–	–	–	–	–	–	–
Lathrotriccus griseipectus	●	●	–	–	–	–	●	●	–	–	–	–	–	–	–	–	–	–
Agriornis andicola	–	–	–	–	–	–	–	–	–	●	–	–	–	–	–	–	–	–
Attila torridus	●	–	–	–	–	–	–	–	–	–	–	–	–	–	–	–	–	–
Phytotoma raimondii	–	–	–	–	–	–	–	–	–	–	–	–	–	–	–	–	–	–
Incaspiza ortizi	–	–	–	●	–	–	–	–	●	–	–	–	–	–	–	–	–	–
Poospiza alticola	–	–	–	–	–	–	–	–	–	–	–	–	–	–	–	–	–	–
Poospiza rubecula	–	–	–	–	–	–	–	–	–	–	–	●	–	–	–	–	–	–
Buthraupis aureodorsalis	–	–	–	–	–	–	–	–	–	–	–	–	–	–	–	–	–	–
Buthraupis wetmorei	–	–	–	–	–	●	–	–	–	–	–	–	–	–	–	–	–	–
Wetmorethraupis sterrhopteron	–	–	–	–	–	–	–	–	–	–	–	–	●	●	–	–	–	–
Carduelis siemiradskii	●	–	–	–	–	–	–	–	–	–	–	–	–	–	–	–	–	–
No. of species	11	2	4	1	3	5	5	2	1	3	3	2	2	1	2	6	5	1

	38	39	40	41	42	43	44	45	46	47	48	49	50	51	52	53	54	55
Tinamus osgoodi	–	–	–	–	–	–	–	–	–	–	–	–	–	–	–	–	–	–
Nothoprocta taczanowskii	–	–	–	–	–	–	–	–	–	–	–	–	–	–	–	–	–	–
Podiceps taczanowskii	–	–	–	–	–	–	–	–	–	–	–	–	–	–	–	–	–	–
Pelecanoides garnotii	–	–	–	–	–	–	–	–	–	–	–	–	–	–	–	–	–	–
Micrastur buckleyi	–	–	–	–	–	–	–	–	–	–	–	–	–	–	–	●	–	–
Pauxi unicornis	–	–	–	–	–	–	–	–	–	–	–	–	–	–	●	–	–	–
Crax globulosa	–	–	–	–	–	–	–	–	–	–	–	–	–	–	–	–	–	–
Laterallus tuerosi	–	–	–	–	–	–	–	–	–	–	–	–	–	–	–	–	–	–
Leptosittaca branickii	–	–	–	–	–	–	–	–	–	–	–	–	●	–	–	–	–	–
Taphrolesbia griseiventris	–	–	–	–	–	–	–	–	●	●	–	–	–	–	–	–	–	–
Eulidia yarrellii	–	–	–	–	–	–	–	–	–	–	–	–	–	–	–	–	–	–
Acestrura bombus	–	–	–	–	–	–	–	–	–	–	–	–	●	–	–	–	–	–
Cinclodes aricomae	–	–	–	–	–	–	–	–	–	–	–	–	–	–	–	–	–	–
Cinclodes palliatus	–	–	–	–	–	–	–	–	–	–	–	–	–	–	–	–	–	–
Leptasthenura xenothorax	–	–	–	–	–	–	–	–	–	–	–	–	–	–	–	–	–	–
Synallaxis courseni	–	–	–	–	–	–	–	–	–	–	–	–	–	–	–	–	–	–
Synallaxis zimmeri	–	–	–	●	●	●	–	–	–	–	–	–	–	–	–	–	–	–
Asthenes huancavelicae	●	●	●	–	–	–	●	–	–	–	●	–	–	–	–	–	–	–
Terenura sharpei	–	–	–	–	–	–	–	–	–	–	–	–	–	–	–	–	–	–
Formicarius rufifrons	–	–	–	–	–	–	–	–	–	–	–	–	–	–	–	–	–	–
Zaratornis stresemanni	●	–	–	–	–	–	●	–	–	–	–	–	–	–	–	–	–	●
Anairetes alpinus	●	–	–	–	–	–	●	–	–	–	–	–	–	–	–	–	–	–
Agriornis andicola	–	–	–	–	–	–	●	–	–	–	–	–	●	–	–	–	–	–
Xenospingus concolor	–	–	–	–	–	–	–	–	–	–	–	–	–	–	–	–	–	–
Poospiza alticola	●	–	–	–	–	–	●	–	–	–	–	–	–	–	–	–	–	–
Poospiza rubecula	–	–	●	–	●	–	●	–	–	–	–	–	–	–	–	–	–	–
Buthraupis aureodorsalis	–	–	–	–	–	–	–	–	–	–	–	●	–	●	–	–	–	–
Tangara meyerdeschauenseei	–	–	–	–	–	–	–	–	–	–	–	–	–	–	–	–	–	–
Tangara phillipsi	–	–	–	–	–	–	–	–	–	–	–	–	–	–	●	–	–	–
Conirostrum tamarugense	–	–	–	–	–	–	–	–	–	–	–	–	–	–	–	–	–	–
Cacicus koepckeae	–	–	–	–	–	–	–	–	–	–	–	–	–	–	–	–	●	–
No. of species	4	1	2	1	2	1	6	2	1	1	1	1	3	1	2	1	1	1

19	20	21	22	23	24	25	26	27	28	29	30	31	32	33	34	35	36	37	No. of areas
–	–	–	–	–	–	–	–	–	–	–	–	–	–	–	●	–	–	–	1
–	–	–	–	–	–	–	–	–	–	–	–	–	–	–	–	–	–	–	2
–	–	–	–	–	–	–	–	–	–	–	–	–	–	–	–	–	–	–	3
–	–	–	–	–	●	●	●	●	–	–	–	–	–	–	–	–	–	–	4
–	–	–	–	–	–	–	–	–	●	–	–	–	–	–	–	–	–	–	6
–	–	●	–	–	–	–	–	●	–	–	–	–	–	●	–	–	–	–	5
–	–	–	–	–	–	–	–	●	–	–	–	–	–	–	–	–	–	–	3
–	–	–	–	–	–	–	–	–	–	–	–	–	●	–	–	●	●	–	7
–	–	●	–	–	–	–	–	–	–	–	–	–	–	●	–	–	–	–	5
–	–	–	–	●	–	–	–	–	–	–	–	–	–	–	–	–	–	–	2
–	–	–	–	–	–	–	–	–	–	–	–	–	–	–	–	–	–	–	1
–	–	–	●	–	–	–	–	–	–	–	–	–	–	–	–	–	–	–	2
–	–	–	–	–	–	–	–	–	–	–	–	●	–	●	–	–	–	–	2
–	–	–	–	●	–	–	–	–	–	–	–	–	–	–	–	–	–	–	3
–	–	–	–	–	–	–	–	–	–	–	–	–	–	–	–	–	–	–	1
–	–	–	–	–	–	–	–	–	–	–	–	–	–	–	–	–	–	–	4
●	●	–	–	●	–	–	–	–	–	–	–	–	–	–	–	–	–	–	4
–	–	–	–	–	–	–	–	–	●	–	–	–	–	●	–	–	–	–	3
–	–	–	–	–	–	–	–	–	–	–	–	–	●	–	–	–	●	–	2
–	–	–	–	–	–	–	–	–	–	–	–	–	–	–	–	–	–	–	1
–	●	–	–	–	–	–	–	–	–	–	–	–	–	–	–	●	–	–	5
–	–	–	–	–	–	●	–	●	–	–	–	–	–	–	–	–	–	–	3
–	–	–	–	–	–	–	–	–	●	–	–	–	–	–	–	–	–	–	7
–	–	–	–	●	–	–	–	–	–	–	–	–	–	–	–	–	–	–	1
–	–	–	–	–	–	–	–	–	–	–	–	–	–	–	–	–	–	–	3
–	–	–	–	–	–	–	–	–	–	–	–	–	–	–	–	–	–	–	3
–	–	–	●	–	–	–	–	–	–	–	–	–	–	–	–	–	–	–	3
–	–	–	–	–	–	–	–	–	–	–	–	–	–	–	–	–	–	–	1
–	–	–	–	–	–	–	–	–	–	–	–	–	–	–	–	–	–	–	4
–	–	–	–	–	–	–	–	–	–	–	–	–	–	–	●	–	–	–	6
–	–	–	–	–	–	–	–	–	–	–	–	–	–	–	–	–	–	–	1
–	–	–	–	–	–	–	–	–	–	●	●	–	–	–	–	–	–	●	3
–	–	●	–	–	–	–	–	–	–	–	–	–	–	–	–	–	–	–	3
–	–	●	–	–	–	–	–	–	–	–	–	–	●	–	–	●	–	–	6
–	–	–	–	–	–	–	–	–	–	–	–	–	–	–	–	–	–	–	6
–	–	–	–	–	–	–	–	–	–	–	–	–	●	–	–	●	–	–	4
–	–	–	–	–	–	–	–	–	–	–	–	–	–	–	–	–	–	–	1
–	–	–	–	–	–	–	–	–	–	–	–	–	–	–	–	–	–	–	2
–	–	–	–	–	–	–	–	–	–	–	–	–	–	–	–	–	–	–	1
1	2	4	2	4	1	2	1	4	3	1	1	1	4	4	2	4	2	1	

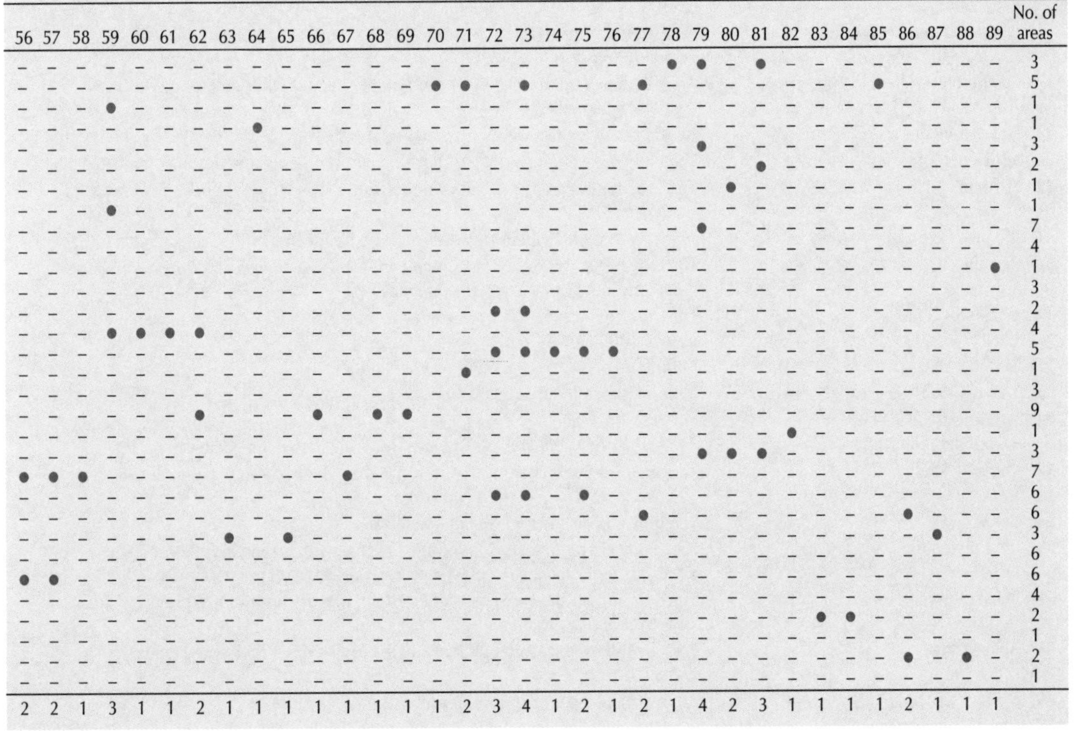

56	57	58	59	60	61	62	63	64	65	66	67	68	69	70	71	72	73	74	75	76	77	78	79	80	81	82	83	84	85	86	87	88	89	No. of areas
–	–	–	–	–	–	–	–	–	–	–	–	–	–	–	–	–	–	–	–	–	–	●	●	–	●	–	–	–	–	–	–	–	–	3
–	–	–	–	–	–	–	–	–	–	–	–	–	–	●	●	–	●	–	–	–	●	–	–	–	–	–	–	–	●	–	–	–	–	5
–	–	–	●	–	–	–	–	–	–	–	–	–	–	–	–	–	–	–	–	–	–	–	–	–	–	–	–	–	–	–	–	–	–	1
–	–	–	–	–	–	–	–	●	–	–	–	–	–	–	–	–	–	–	–	–	–	–	–	–	–	–	–	–	–	–	–	–	–	1
–	–	–	–	–	–	–	–	–	–	–	–	–	–	–	–	–	–	–	–	–	–	–	●	–	–	–	–	–	–	–	–	–	–	3
–	–	–	–	–	–	–	–	–	–	–	–	–	–	–	–	–	–	–	–	–	–	–	–	–	●	–	–	–	–	–	–	–	–	2
–	–	–	–	–	–	–	–	–	–	–	–	–	–	–	–	–	–	–	–	–	–	–	–	●	–	–	–	–	–	–	–	–	–	1
–	–	–	●	–	–	–	–	–	–	–	–	–	–	–	–	–	–	–	–	–	–	–	–	–	–	–	–	–	–	–	–	–	–	1
–	–	–	–	–	–	–	–	–	–	–	–	–	–	–	–	–	–	–	–	–	–	–	●	–	–	–	–	–	–	–	–	–	–	7
–	–	–	–	–	–	–	–	–	–	–	–	–	–	–	–	–	–	–	–	–	–	–	–	–	–	–	–	–	–	–	–	–	–	4
–	–	–	–	–	–	–	–	–	–	–	–	–	–	–	–	–	–	–	–	–	–	–	–	–	–	–	–	–	–	–	–	–	●	1
–	–	–	–	–	–	–	–	–	–	–	–	–	–	–	–	–	–	–	–	–	–	–	–	–	–	–	–	–	–	–	–	–	–	3
–	–	–	–	–	–	–	–	–	–	–	–	–	–	–	–	●	●	–	–	–	–	–	–	–	–	–	–	–	–	–	–	–	–	2
–	–	–	●	●	●	●	–	–	–	–	–	–	–	–	–	–	–	–	–	–	–	–	–	–	–	–	–	–	–	–	–	–	–	4
–	–	–	–	–	–	–	–	–	–	–	–	–	–	–	–	●	●	●	●	●	–	–	–	–	–	–	–	–	–	–	–	–	–	5
–	–	–	–	–	–	–	–	–	–	–	–	–	–	–	●	–	–	–	–	–	–	–	–	–	–	–	–	–	–	–	–	–	–	1
–	–	–	–	–	–	–	–	–	–	–	–	–	–	–	–	–	–	–	–	–	–	–	–	–	–	–	–	–	–	–	–	–	–	3
–	–	–	–	–	–	●	–	–	–	●	–	●	●	–	–	–	–	–	–	–	–	–	–	–	–	–	–	–	–	–	–	–	–	9
–	–	–	–	–	–	–	–	–	–	–	–	–	–	–	–	–	–	–	–	–	–	–	–	–	–	●	–	–	–	–	–	–	–	1
–	–	–	–	–	–	–	–	–	–	–	–	–	–	–	–	–	–	–	–	–	–	–	●	●	●	–	–	–	–	–	–	–	–	3
●	●	●	–	–	–	–	–	–	–	–	●	–	–	–	–	–	–	–	–	–	–	–	–	–	–	–	–	–	–	–	–	–	–	7
–	–	–	–	–	–	–	–	–	–	–	–	–	–	–	–	●	●	–	●	–	–	–	–	–	–	–	–	–	–	–	–	–	–	6
–	–	–	–	–	–	–	–	–	–	–	–	–	–	–	–	–	–	–	–	–	●	–	–	–	–	–	–	–	–	●	–	–	–	6
–	–	–	–	–	–	–	●	–	●	–	–	–	–	–	–	–	–	–	–	–	–	–	–	–	–	–	–	–	–	–	●	–	–	3
–	–	–	–	–	–	–	–	–	–	–	–	–	–	–	–	–	–	–	–	–	–	–	–	–	–	–	–	–	–	–	–	–	–	6
●	●	–	–	–	–	–	–	–	–	–	–	–	–	–	–	–	–	–	–	–	–	–	–	–	–	–	–	–	–	–	–	–	–	6
–	–	–	–	–	–	–	–	–	–	–	–	–	–	–	–	–	–	–	–	–	–	–	–	–	–	–	–	–	–	–	–	–	–	4
–	–	–	–	–	–	–	–	–	–	–	–	–	–	–	–	–	–	–	–	–	–	–	–	–	–	–	●	●	–	–	–	–	–	2
–	–	–	–	–	–	–	–	–	–	–	–	–	–	–	–	–	–	–	–	–	–	–	–	–	–	–	–	–	–	–	–	–	–	1
–	–	–	–	–	–	–	–	–	–	–	–	–	–	–	–	–	–	–	–	–	–	–	–	–	–	–	–	–	–	●	–	●	–	2
–	–	–	–	–	–	–	–	–	–	–	–	–	–	–	–	–	–	–	–	–	–	–	–	–	–	–	–	–	–	–	–	–	–	1
2	2	1	3	1	1	2	1	1	1	1	1	1	1	1	2	3	4	1	2	1	2	1	4	2	3	1	1	1	1	2	1	1	1	

(Collar *et al.* 1992), with the following 10 added: Ringed Storm-petrel *Oceanodroma hornbyi*, Markham's Storm-petrel *O. markhami* (endemic), Andean Flamingo *Phoenicopterus andinus*, Puna Flamingo *Phoenicopterus jamesi*, Rufous-headed Chachalaca *Ortalis erythroptera*, Brown Wood-rail *Aramides wolfi*, Military Macaw *Ara militaris*, Black-tailed Antbird *Myrmoborus melanurus* (endemic), Chestnut-bellied Cotinga *Doliornis remseni* and Black-masked Finch *Coryphaspiza melanotis*; these additional species have not, however, been included in the Site Inventory (see '*Key Areas*: the book', p. 12).

The changes represent a high percentage difference (31%) brought about primarily by the new criteria and data only recently made available, although *Doliornis remseni* was described as new to science since the 1992 listing. These recent additions will not have any major distributional impact on the Key Area analysis, as many of the species occur sympatrically with at least one other Peruvian threatened bird. For example, *Coryphaspiza melanotis* is present in the Pamapas del Heath National Sanctuary which is contiguous with the Tambopata–Candamo Reserved Zone (PE 81), these two areas being proposed as the basis of a single strict protected area, the Tambopata–Heath (or Bahuaja–Sonene) National Park (Foster *et al.* 1994). Similarly, *Myrmoborus melanurus* occurs at Lago Yarinacocha (PE 53: Ridgely and Tudor 1994), *Ara militaris* can still be found in at least the southern Cordillera del Colán (PE 17: Barnes *et al.* 1995), and the only known breeding locality for *Oceanodroma markhami* is within the Paracas National Reserve (PE 64: Collar *et al.* 1994). However, each of these new additions needs to be considered in any future priority-setting exercise, and additional areas will need to be identified for some of the species (e.g. the two flamingos).

Table 3. Top Key Areas in Peru. Those with the area number in bold currently have some form of protected status.

Key Area	No. of threatened spp.	Comments
01 Tumbes National Forest	11	The only protected area for most of the species present, and the only Key Area for *Synallaxis tithys, Onychorhynchus occidentalis, Attila torridus* and *Carduelis siemiradzkii.*
06 Cerro Chinguela	5	The only Key Area for *Hapalopsittaca pyrrhops, Metallura odomae* and *Buthraupis wetmorei.*
07 San José de Lourdes	5	An important area for all five species, though all can be found in other Peruvian Key Areas.
16 Northern Cordillera del Colán	6	An extremely important area for each of the species, and one of only two sites known for *Xenoglaux loweryi.*
17 Southern Cordillera del Colán	5	An extremely important area for each of the species, whose presence was confirmed as recently as 1994.
21 Balsas–Celedín	4	An important area for all four threatened species.
23 Jesús del Monte–Jirillo	4	The entire known range of *Herpsilochmus parkeri* lies within this important area.
25 El Tocto	2	This area holds the largest known population of the Peruvian endemic *Penelope albipennis.*
31 Molino	1	The only Key Area where *Aglaeactis aliciae* has been recorded recently.
32 Río Abiseo	4	The only protected Key Area for *Aulacorhynchus huallagae* and *Buthraupis aureodorsalis,* both Peruvian endemics.
34 Hacienda Tulpo	2	The only Key Area for *Nothoprocta kalinowskii,* though it has not been recorded since 1900.
35 Mashua–La Caldera	4	An important area for three Peruvian endemics.
38 Yanac–Quebrada Tutapac	4	Important for all four of its threatened species.
44 Huascarán–Cordillera Blanca	6	The only protected Key Area for *Anairetes alpinus, Asthenes huancavelicae* and *Poospiza rubecula,* the last two species being Peruvian endemics.
52 Cerros del Sira	2	*Tangara phillipsi* is endemic to this Key Area, which is also one of only two Peruvian sites for *Pauxi unicornis.*
54 Balta	1	The only known site for *Cacicus koepckeae.*
59 Lago de Junín	3	Both *Podiceps taczanowskii* and *Laterallus tuerosi* are endemic to this Key Area—the only protected area from which *Cinclodes palliatus* has been recorded.
71 Bosque Ampay	2	The only known site for *Synallaxis courseni,* and the only known protected area for *Nothoprocta taczanowskii.*
73 Abra Málaga	4	Possibly the most important Peruvian area for the three *Polylepis–Gynoxys* specialists.
79 Manu	4	An important area for all four of its threatened species.

OLD RECORDS AND LITTLE-KNOWN BIRDS

Many Peruvian threatened species are little known, although our knowledge is based on records from the 1970s or early 1980s (e.g. *Xenoglaux loweryi*, *Aglaeactis aliciae*, *Wetmorethraupis sterrhopteron* and *Tangara meyerdeschauenseei*). However, these records do provide sufficient information to allow informed recommendations to be made for the species' conservation through the identification of Key Areas. Exceptions to this are species known from old specimen records or chance observations (from which there can be no guarantee of a current viable population), which in Peru include the following.

Nothoprocta kalinowskii is known only from two Peruvian localities, over 900 km apart, at which the only two specimens were collected in and before 1900. One of the two localities has not been precisely traced (but is near Cuzco).

Crax globulosa is little known throughout its range, with local people in Peru suggesting that it has disappeared almost totally within the last 30 years. The few recent Peruvian records have added little to our knowledge of its status or ecology.

Synallaxis cherriei (see 'Key Areas', above).

Terenura sharpei, in Peru, is known only from a single specimen collected in 1900, although there are relatively recent records from adjacent Bolivia.

Pithys castanea (see 'Key Areas', above).

Tangara phillipsi is known only from several sightings and four specimens collected in July 1969 on the isolated Cerros del Sira, there being no subsequent records of the species or information about this poorly known area.

Cacicus koepckeae is known with certainty only from two specimens collected at a single locality in 1963 and 1965; it is virtually unknown in life. Reports of the species from Manu are now widely refuted (S. K. Robinson verbally 1994).

OUTLOOK

This analysis has identified 89 Key Areas, each of which is of major importance for the conservation of threatened birds in Peru. However, Key Areas standing out as top priorities (Table 3) comprise those which play host to populations of five or more threatened species, and areas with four threatened species if these add significantly to the coverage of species. Also listed are the areas for species endemic to Peru and (effectively) known from just one Key Area, and areas supporting the primary population of a species.

The 20 Key Areas in Table 3 are the most important for a number of threatened species, and represent populations of 50 Peruvian threatened birds, or 78% of the total. These and the remaining Key Areas (which are individually extremely important for many threatened Peruvian endemics: see 'Key Areas' and 'Key Area protection', above) all need some degree of conservation action if the populations of their threatened birds are to remain viable. Owing to the general paucity of recent information on Peruvian threatened birds, there is a requirement for fieldwork to target those species (and their respective Key Areas) listed under 'Old records and little-known birds' (above), with the aim of locating and defining the extent of suitable habitat in the Key Area and its vicinity. For many areas there is a need to confirm the continued existence or to assess the population of the threatened species for which it was defined.

DATA SOURCES

The above introductory text and the Site Inventory (below) were compiled from information supplied by the late T. A. Andrews, S. Butchart, G. Engblom, I. Franke, M. Pearman, S. K. Robinson and T. S. Schulenberg, as well as from the following references.

Barnes, R., Butchart, S., Clay, R., Davies, C. and Seddon, N. (1995) The conservation status of the Cordillera de Colán, northern Peru. *Cotinga* 3: 6–7.

Barrio, J. (1995) The Rufous-breasted Warbling-finch *Poospiza rubecula* in Bosque Zárate, Peru. *Cotinga* 3: 56–57.

Best, B. J. and Clarke, C. T., eds. (1991) *The threatened birds of the Sozoranga region, southwest Ecuador*. Cambridge, U.K.: International Council for Bird Preservation (Study Report 44).

Collar, N. J., Gonzaga, L. P., Krabbe, N., Madroño Nieto, A., Naranjo, L. G., Parker, T. A. and Wege, D. C. (1992) *Threatened birds of the Americas: the ICBP/IUCN Red Data Book* (Third edition, part 2). Cambridge, U.K.: International Council for Bird Preservation.

Collar N. J., Crosby, M. J. and Stattersfield, A. J. (1994) *Birds to watch 2: the world list of threatened birds*. Cambridge, U.K.: BirdLife International (BirdLife Conservation Series no.4).

Collar, N. J., Wege, D. C. and Long, A. J. (in press) Patterns and causes of endangerment in the New World avifauna. In J. V. Remsen, ed. *Natural history and conservation of Neotropical birds*. American Ornithologists' Union (Orn. Monogr.).

Davies, C. W. N., Barnes, R., Butchart, S. H. M., Fernández, M. and Seddon, N. (1994) The conservation status of the Cordillera de Colán: a report based on bird and mammal surveys in 1994. Unpublished report.

DAVIS, T. J. (1986) Distribution and natural history of some birds from the Departments of San Martín and Amazonas, northern Peru. *Condor* 88: 50–56.

DAVIS, T. J. AND O'NEILL, J. P. (1986) A new species of antwren (Formicariidae: *Herpsilochmus*) from Peru, with comments on the systematics of other members of the genus. *Wilson Bull.* 98: 337–352.

FOSTER, R. B., CARR, J. L. AND FORSYTH, A. B., EDS. (1994) *The Tambopata–Candamo Reserved Zone of southeastern Perú: a biological assessment (Rapid Assessment Program).* Washington, D.C.: Conservation International.

GARDNER, N. (1986) A birder's guide to travel in Peru. Unpublished report.

HARRIS, M. P. (1980) Avifauna del Lago de Junín, Peru. *Publ. Museo de Hist. Nat. Javier Prado* 27: 1–14.

IUCN (1992) *Protected areas of the world: a review of national systems. Volume 4: Nearctic and Neotropical.* Gland, Switzerland and Cambridge, U.K.: International Union for Conservation of Nature and Natural Resources.

KRATTER, A. W. (1995) Status, habitat, and conservation of the Rufous-throated Antthrush *Formicarius rufifrons. Bird Conserv. Internatn.* 5(2–3) (in press).

LEO, M., ORTÍZ, E. AND RODRÍGUEZ, L. (1988) Results of 1988 fieldwork faunal inventory Río Abiseo National Park, Peru. Unpublished report.

PARKER, T. A., PARKER, S. A. AND PLENGE, M. A. (1982) *An annotated checklist of Peruvian birds.* Vermillion, South Dakota: Buteo Books.

PARKER, T. A., SCHULENBERG, T. S., GRAVES, G. R. AND BRAUN, M. J. (1985) The avifauna of the Huancabamba region, northern Peru. Pp.169–197 in P. A. Buckley, M. S. Foster, E. S. Morton, R. S. Ridgely and F. G. Buckley, eds. *Neotropical ornithology.* Washington, D.C.: American Ornithologists' Union (Orn. Monogr. 36).

PARKER, T. A., SCHULENBERG, T. S., KESSLER, M. AND WUST, W. H. (1995) Natural history and conservation of the endemic avifauna in north-west Peru. *Bird Conserv. Internatn.* 5(2–3) (in press).

RIDGELY, R. S. AND TUDOR, G. (1994) *The birds of South America*, 2. Austin: University of Texas Press.

STATTERSFIELD, A. J., CROSBY, M. J., LONG, A. J. AND WEGE, D. C. (in prep.) *A global directory of Endemic Bird Areas.* Cambridge, U.K.: BirdLife International (BirdLife Conservation Series).

STEPHENS, L. AND TRAYLOR, M. A. (1983) *Ornithological gazetteer of Peru.* Cambridge, Mass.: Museum of Comparative Zoology.

VALQUI, T. (1994) The extinction of the Junín Flightless Grebe? *Cotinga* 1: 42–44.

SITE INVENTORY

Tumbes National Forest (Tumbes)

PE 01

Tumbes National Forest (IUCN category VIII), 75,102 ha

3°49′S 80°17′W
EBA B20

Part of the North-west Peru Biosphere Reserve, Tumbes National Forest is situated in eastern Tumbes department, the eastern and southern boundaries being formed by the border with Ecuador. Most records come from the Cordillera Larga in the east of the park. This reserve, El Angolo (see PE 02), and the contiguous Cerros de Amotape National Park, encompass the largest remaining tract of deciduous and moist forest west of the Andes, but receive only meagre protection.

Leucopternis occidentalis	1988	Uncommon or rare; at least 2–3 pairs in the east of the forest (also Parker *et al.* 1995).
Leptotila ochraceiventris	1988	Common at Campo Verde in February 1986, but less so in July 1988, suggesting seasonal movements (Parker *et al.* 1995).
Synallaxis tithys	1988	Common in 1979, with 20 collected; fairly common in 1988 (Parker *et al.* 1995).
Hylocryptus erythrocephalus	1988	At least seasonally fairly common, with 12 found along 1 km of trail in July 1988 (also Parker *et al.* 1995).
Syndactyla ruficollis	1988	Inconspicuous and apparently uncommon to rare (also Parker *et al.* 1995).
Myrmeciza griseiceps	1986	The reserve is at the lower elevational range for this species. Last recorded in 1986 by M. Kessler (T. S. Schulenberg *in litt.* 1994).
Pachyramphus spodiurus	1988	Rare at two localities during 1979, with several heard and a male seen in July 1988 by T. A. Parker and W. Wust (T. S. Schulenberg *in litt.* 1994).
Onychorhynchus occidentalis	1988	Only a few recorded during surveys in 1979, and singles recorded in 1988 (T. S. Schulenberg *in litt.* 1994, B. P. Walker *in litt.* 1995).
Lathrotriccus griseipectus	1988	Uncommon at two localities (also Parker *et al.* 1995).
Attila torridus	1988	Only two records from this reserve (also Parker *et al.* 1995).
Carduelis siemiradzkii	1988	Rare, with three flocks of 4–8 seen near El Caucho in 1988 (also Parker *et al.* 1995).

El Angolo (Tumbes/Piura)

PE 02

El Angolo Hunting Reserve (IUCN category VIII), 65,000 ha

4°28'S 80°48'W
EBA B20

El Angolo lies at 700 m within the Cerros de Amotape, on the border of Tumbes and Piura departments, and south-west of the Cordillera Larga (which is probably part of the same geological formation). It forms part of the North-west Peru Biosphere Reserve (see PE 01), and, given its contiguity with Tumbes (PE 01), it is likely that the threatened species complement will be higher once fuller surveys have been conducted.

Syndactyla ruficollis	1972	One collected in November 1972.
Lathrotriccus griseipectus	1972	Two collected in November 1972.

Cerro Chacas (Piura)

PE 03

Unprotected

4°36'S 79°44'W
EBAs B20, B21

The mountain rises to c.3,000 m and lies c.5 km north of Ayabaca in northern Piura, c.15 km from the Ecuadorean border. Montane cloud-forest is present at 2,600 m but was recently being actively felled (Best and Clarke 1991).

Penelope barbata	1989	Three seen and several heard, with a population estimated to be a few tens of birds (Best and Clarke 1991).
Leptotila ochraceiventris	1989	One seen in cloud-forest at 2,625 m in September 1989 (Best and Clarke 1991).
Syndactyla ruficollis	1989	Common in cloud-forest, September 1989 (Best and Clarke 1991).
Myrmeciza griseiceps	1989	One seen in dense bamboo understorey of cloud-forest at 2,625 m, September 1989 (Best and Clarke 1991).

Huancabamba (Piura)

PE 04

Unprotected

5°14'S 79°28'W
EBA B22

Huancabamba lies in a narrow rain-shadow valley at c.1,925 m in eastern Piura department. The area comprises a single rocky hilltop c.2 km north-east of town along the road to Sapalache. The hilltop is covered in dense herbaceous scrub with small *Acacia* and cacti, though other areas near Huancabamba have been totally cleared for cultivation and agriculture.

Incaspiza ortizi	1989	Uncommon at 2,125 m in 1980 (when six were collected), with up to eight seen subsequently. This represents the northernmost population of the species.

Cruz Blanca (Piura)

PE 05

Unprotected

5°20'S 79°32'W
EBAs B20, B21

Cruz Blanca lies at the crest of the western cordillera along the Canchaque–Huancabamba road in eastern Piura department. The western slope of the ridge was (in 1980) covered by a mixed evergreen forest from 3,050 down to 2,150 m, below which only scattered patches of forest survive owing to human (and associated livestock) pressure (Parker *et al.* 1985).

Penelope barbata	1987	Uncommon (and probably decreasing) during surveys up to 1980; three seen in 1987 (B. P. Walker *in litt.* 1995).
Syndactyla ruficollis	1980	Small numbers recorded during surveys up to 1980.
Myrmeciza griseiceps	1980	Found to be rare during surveys up to 1980.

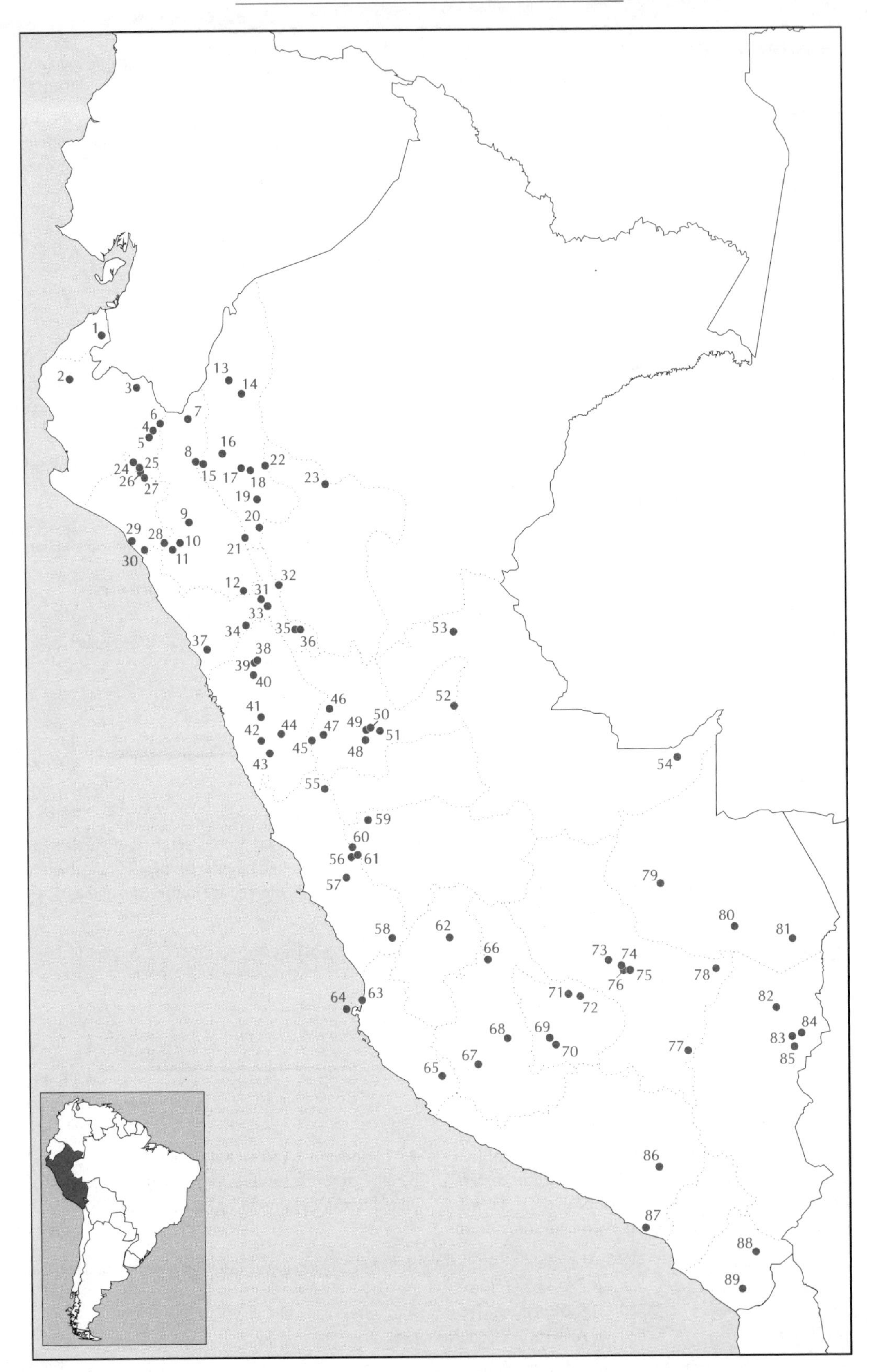
1
2
3
4
5
6
7
8
9
10
11
12
13
14
15
16
17
18
19
20
21
22
23
24
25
26
27
28
29
30
31
32
33
34
35
36
37
38
39
40
41
42
43
44
45
46
47
48
49
50
51
52
53
54
55
56
57
58
59
60
61
62
63
64
65
66
67
68
69
70
71
72
73
74
75
76
77
78
79
80
81
82
83
84
85
86
87
88
89

Cerro Chinguela (Piura/Cajamarca)

PE 06

Unprotected

5°07'S 79°23'W
EBAs B21, B60

Rising to 3,715 m, Cerro Chinguela forms part of the cordillera north and east of Huancabamba (5 km north-east of Sapalache) in north-east Piura/north-westernmost Cajamarca. The páramo zone is grazed by a small number of cattle and is burnt annually by local herders. However, cloud-forest persists on both slopes, to 2,500 m on the west side and 1,500 m on the east (Parker *et al.* 1985).

Penelope barbata	1980	Uncommon (and probably decreasing) on both slopes at 2,400–2,900 m.
Hapalopsittaca pyrrhops	1989	Pairs recorded at 2,500–3,000 m in 1977 and 1978; one bird seen in 1989.
Metallura odomae	1989	Fairly common in its páramo habitat above 2,800 m.
Thripophaga berlepschi	1987	Two found on the upper slopes of the west side of the mountain (M. Pearman *in litt.* 1995).
Buthraupis wetmorei	1980	Found at 2,900 m in 1980, but not seen since, despite searches in 1989.

San José de Lourdes (Cajamarca)

PE 07

Unprotected

5°04'S 78°54'W
EBA B24

The town of San José de Lourdes is on the left bank of the Río Chinchipe in northern Cajamarca department. Above town are elfin forested ridges at c.2,000 m, while evergreen moist forest occurs down to 800 m below town. Deforestation was proceeding rapidly in 1984.

Leucopternis occidentalis	c.1976	This record is the only one east of the Andes and is now doubted by some authorities.
Touit stictoptera	1976	One collected above town in 1976.
Heliangelus regalis	1983	Common at 1,800–2,200 m in 1975 (when the species was discovered) and in 1976, still present in numbers at 1,700–1,850 m in 1983 (B. P. Walker *in litt.* 1995).
Hemitriccus cinnamomeipectus	1976	Two (including the type) collected at 1,800–2,200 m in 1976, the species then being considered rare.
Lathrotriccus griseipectus	1976	Very common 2 km north of town (at c.830 m) in August 1976.

Figure 2 (opposite). Key Areas in Peru.

01 Tumbes National Forest
02 El Angolo
03 Cerro Chacas
04 Huancabamba
05 Cruz Blanca
06 Cerro Chinguela
07 San José de Lourdes
08 Jaén
09 La Esperanza
10 Taulís
11 Paucal
12 Cajabamba
13 Río Comaina and Río Cenepa
14 Urakusa
15 Corral Quemado
16 Northern Cordillera del Colán
17 Southern Cordillera del Colán
18 Florida and Laguna Pomacochas
19 Levanto
20 Leimebamba
21 Balsas–Celedín
22 Abra Patricia
23 Jesús del Monte
24 Hacienda Boca Chica
25 El Tocto
26 Hacienda Recalí
27 Km 21, Olmos–Choloque road
28 Seques
29 Reque
30 Río de Saña and Rafan
31 Molino
32 Río Abiseo
33 Soquián and Chagual
34 Hacienda Tulpo
35 Mashua and La Caldera
36 Cumpang and Utcubamba
37 Hacienda Buenavista
38 Yanac and Quebrada Tutapac
39 Yuracmarca–Yanac
40 Huaylas–Quitacocha
41 Chacchan
42 Bosque San Damián
43 Bosque de Noquo
44 Huascarán
45 Cerro Huansala
46 Cullcui
47 Huánaco–La Unión road
48 Santa María del Valle
49 Quilluacocha and Bosque Unchog
50 Bosque Zapatogocha
51 Sariapunta
52 Cerros del Sira
53 Lago Yarinacocha
54 Balta
55 Pueblo Quichas
56 Upper Santa Eulalia valley
57 Bosque Zárate
58 Hortigal
59 Lago de Junín
60 Marcapomacocha
61 Pampa Pucacocha
62 Yauli
63 Pisco
64 Isla San Gallán
65 Nazca
66 Ayacucho
67 Pampa Galeras
68 Río Mayobamba
69 Mutca
70 Bosque de Naupallagta
71 Bosque Ampay
72 Cerro Runtacocha–Morococha
73 Abra Málaga
74 Nevada Chaiñapuerto
75 Yanacocha lakes
76 Urubamba
77 Abra La Raya
78 Quincemil
79 Manu
80 Río Colorado
81 Tambopata–Candamo
82 Inca Mine
83 Sandia
84 Maruncunca
85 Valcón
86 Salinas y Aguada Blanca
87 Río Tambo and Lagunas de Mejía
88 North-east of Tarata
89 Tacna

Jaén (Cajamarca)

PE 08

Unprotected

5°42′S 78°47′W
EBA B22

Jaén is on the left bank of the Marañón drainage in eastern Cajamarca. Any remnant forest areas either north or south of town are likely to be of importance for the two west Ecuador/north-west Peru endemics listed below.

Pachyramphus spodiurus	1983	Recorded 2 km north of town at 350 m, and 2 km east of town at 750 m (B. P. Walker *in litt.* 1995).
Lathrotriccus griseipectus	1968	One collected south of town at 900 m, July 1968.

La Esperanza (Cajamarca)

PE 09

Unprotected

6°36′S 78°54′W
EBA B27

La Esperanza is at 1,700 m on the upper Río Chancay, on the western slope of the West Andes, c.5 km north-east of Santa Cruz. The area is (or was) presumably covered in dense, arid, montane scrub.

Incaspiza ortizi	1951	One (the type-specimen) collected at 1,800 m.

Taulís (Cajamarca)

PE 10

Unprotected

6°54′S 79°03′W
EBAs B20, B21

Taulís is on the upper Río de Saña which flows down the Pacific slope of the West Andes in west-central Cajamarca. The area includes the nearby Haciendas Montseco and Udima, and thus spans altitudes from c.1,150 to 3,400 m. Large areas of montane forest apparently still survive.

Penelope barbata	1987	Most records of what may have been a sizeable population come from the 1950s; the 1987 record is a single specimen.
Syndactyla ruficollis	1926	Recorded from 2,700 m in June–July 1926.
Agriornis andicola	1952	One collected at 3,400 m.

Paucal (Cajamarca)

PE 11

Unprotected

7°00′S 79°10′W
EBAs B20, B21, B27

Paucal is on the uppermost Río de Saña drainage, high on the Pacific slope of the West Andes, and appears not to have been surveyed for almost 70 years.

Penelope barbata	c.1926	A collecting locality mentioned by I. Franke (verbally 1991).
Taphrolesbia griseiventris	1874	The type-locality, from where, however, there are no subsequent records.
Syndactyla ruficollis	c.1926	Mentioned for this locality.

Cajabamba (Cajamarca)

PE 12
7°37′S 78°03′W
EBA B27

Unprotected

Cajabamba is at 2,655 m on the left bank of the Marañón drainage in south-east Cajamarca department. In 1894, the surrounding hills were covered in small brushwood, with only the canyons containing some small trees; similar habitat remained in 1995, but neither of the species mentioned below was seen during a brief search (B. P. Walker *in litt.* 1995).

Taphrolesbia griseiventris	1894	A number of specimens and sightings from around town.
Poospiza rubecula	1894	Two collected at 2,750 m.

Río Comaina and Río Cenepa (Amazonas)

PE 13
4°27–33′S 78°12–17′W
EBA B24

Unprotected

The Ríos Comaina and Cenepa are left-bank tributaries of the middle Marañón, in northern Amazonas department, their confluence is near Chávez Valdivia, west of which are the foothills (200–800 m, covered in humid montane forest) of the Cordillera del Condor, with the various collecting localities lying within Aguaruna Indian territory. Recent (1994) surveys away from Indian settlements and near the headwaters of the Río Comaina at 1,200–1,500 m have confirmed the presence of intact forest and a population of *Galbula pastazae* (T. S. Schulenberg *in litt.* 1994), suggesting that the entire watershed is of key importance.

Micrastur buckleyi	1978	One collected in 1978 near Kusú, with another taken the previous year at 200 m on the Río Cenepa.
Wetmorethraupis sterrhopteron	1964	Common at a number of localities above 600 m in 1964.

Urakusa (Amazonas)

PE 14
4°42′S 78°03′W
EBA B24

Unprotected

Urakusa (or Oracuza) is on the right bank of the middle Río Marañón in north-east Amazonas department. About 3 km west of this locality are the foothills at the northern end of Cordillera del Colán, the forested slopes of which are important for the species mentioned below.

Wetmorethraupis sterrhopteron	1978	Common above 600 m in 1964, but uncommon along the road to Urakusa in 1978 (T. S. Schulenberg *in litt.* 1994).

Corral Quemado (Amazonas)

PE 15
5°44′S 78°40′W
EBA B22

Unprotected

Corral Quemado lies within the arid tropical zone on the Río Marañón (8 km south of Bellavista), where the road from Olmos to Bagua Chica crosses the river.

Forpus xanthops	c.1979	Recorded on a number of occasions.
Pachyramphus spodiurus	c.1978	Sight records during the late 1970s almost certainly refer to this species.

Northern Cordillera del Colán (Amazonas)

PE 16

Unprotected

5°35′S 78°22′W
EBAs B22, B24, B25

As defined here, this area embraces the northern end of the Cordillera del Colán (in the Río Utcubamba drainage), especially the localities east of Bagua along the La Peca Nueva trail, from 2,500 m (where ridge-top elfin or cloud-forest persists) down to c.700 m at (for example) Hacienda Morerilla. Recent reports from local people suggest that little forest remains in this part of the cordillera owing primarily to forest clearance for cattle grazing and lucrative drug cash crops (Davies *et al.* 1994). A metalled road has recently been built up to a microwave station above La Peca Nueva (Davies *et al.* 1994), and what little forest remains is presumably in urgent need of protection.

Columba oenops	1955	Recorded at a number of localities in this area.
Leptosittaca branickii	1978	Local and uncommon.
Forpus xanthops	1955	Recorded near Bagua in the low-altitude, more arid vegetation.
Xenoglaux loweryi	1978	Recorded east of Bagua, when two were mist-netted at 2,165 m in 1978 (T. S. Schulenberg *in litt.* 1994).
Thripophaga berlepschi	1978	A number collected at 2,500 m in 1978.
Hemitriccus cinnamomeipectus	1978	One specimen.

Southern Cordillera del Colán (Amazonas)

PE 17

Unprotected

5°48′S 78°02–06′W
EBAs B22, B24, B25

This area, at the southern end of the Cordillera del Colán, includes the Río Cristobal and Río Comboca catchments on the northern bank of the Utcubamba, c.30–40 km east of Bagua Grande (Davies *et al.* 1994). It holds the largest remaining tract of humid cloud- and elfin forest on the cordillera, and was recently found to support a population of the threatened yellow-tailed woolly monkey *Lagothrix flavicauda* (Davies *et al.* 1994). However, the progress of deforestation is alarming, and this area, which is extremely important for numerous restricted-range birds as well as for the threatened species mentioned below, is in urgent need of protection (Barnes *et al.* 1995).

Columba oenops	1994	Singles or pairs seen nine times around Comboca during two days in 1994 suggest that the species is common (Davies *et al.* 1994).
Leptosittaca branickii	1994	One recorded above San Cristobal at 1,800 m (Davies *et al.* 1994).
Forpus xanthops	1994	A pair and a group of four seen over cultivated land at 500 m, c.10 km east of Bagua Grande (i.e. just west of this area) (Davies *et al.* 1994).
Heliangelus regalis	1994	Common within its restricted habitat above San Cristobal at 1,600–1,950 m in 1994 (Davies *et al.* 1994). This is only the third known locality for the species.
Thripophaga berlepschi	1994	Recorded six times at 1,800–1,950 m above San Cristobal in August 1994 (Davies *et al.* 1994).

Florida and Laguna Pomacochas (Amazonas)

PE 18

Unprotected

5°50′S 77°55′W
EBA B25

Florida is at the southernmost end of the Cordillera del Colán (east-south-east of PE 17) on the right bank of the Utcubamba in Bongara province, where the Bagua Grande–Rioja road crosses the cordillera at c.2,150 m. The area east of Laguna Pomacochas and Florida (e.g. along the Río Chido) supports secondary growth and some forest remnants (also Gardner 1986), although the area is under severe pressure from slash-and-burn farming (M. Pearman *in litt.* 1995).

Loddigesia mirabilis	1990	Most recent records are June 1987 (a female seen daily: M. Pearman *in litt.* 1995) and May 1990 (T. A. Andrews *in litt.* 1990).

Levanto (Amazonas)

PE 19

Unprotected

6°16′S 77°49′W
EBA B25

Levanto is in the Utcubamba valley at 2,400 m, c.5 km south-east of Chachapoyas. Habitat suitable for the species mentioned below was apparently present above town in 1932, since when, however, there appear not to have been any surveys.

Loddigesia mirabilis	c.1932	Recorded between 2,590 and 2,745 m.

Leimebamba (Amazonas)

PE 20

Unprotected

6°41′S 77°47′W
EBA B25

Situated on the upper Utcubamba, Leimebamba here includes the slopes up to the páramo above town. Much of the land in the páramo zone is currently cultivated or used by roaming cattle, making the easily approached elfin forests very vulnerable. Likewise, the forests and woodlots on the lower slopes are also under heavy pressure from cultivation.

Loddigesia mirabilis	1977	A male seen in May 1977 at 2,200 m.
Thripophaga berlepschi	1933	'Not rare' at two localities south-west of Leimebamba, 1932–1933.

Balsas–Celedín (Amazonas/Cajamarca)

PE 21

Unprotected

6°50′S 78°01′W
EBAs B22, B27

The town of Balsas lies in the arid zone at c.850 m on the right bank of the Marañón, Amazonas. The riparian woods and nearby open dry forest are suffering from gradual degradation, but are important for two of the species mentioned below. The other species rely on the area west of Balsas on the road to Celedín (on the left bank of the Marañón), where the Andean slopes rise to c.2,900 m and comprise open *Acacia* woodland, grass and thorny scrub, and ultimately *Alnus* thickets and *Gynoxys* shrubs.

Columba oenops	1975	Groups of 3–6 seen in riparian woods and adjacent forest in 1975.
Forpus xanthops	1983	A pair at 1,300 m (B. P. Walker *in litt.* 1995).
Incaspiza ortizi	1983	Twelve collected at Hacienda Limón in 1975 and three at the same locality in 1983 (B. P. Walker *in litt.* 1995).
Poospiza alticola	c.1975	Recorded from a shrubby patch at 2,900 m, east of Celedín.

Abra Patricia (San Martín)

PE 22

Unprotected

5°46′S 77°41′W
EBA B24

Abra Patricia (or Pardo de Miguel) is at a pass on the Amazonas–San Martín border, on the Rioja–Ingeno road. An area of ridge-top elfin and cloud-forest lies at c.1,900 m, 10 km north-east of the town, in the Río Mayo drainage. This forest appears to be relatively inaccessible and so far untouched.

Xenoglaux loweryi	1976	Records in the immediate vicinity of this, the type-locality, suggest that it may be locally not uncommon.
Hemitriccus cinnamomeipectus	1989	Records suggest that it is not uncommon.

Jesús del Monte (San Martín)

PE 23
6°03′S 76°44′W
EBA B24

Unprotected

Jesús del Monte is c.15 km north-east of Jirillo (east-south-east of Moyobamba) on an old mule trail to the town of Balsapuerto, some distance to the north (Davis 1986). The trail runs over a low, isolated mountain ridge (c.1,500 m) supporting a very heterogeneous habitat ranging from savanna-like vegetation and semi-stunted forest to tall cloud-forest (Davis 1986, Davis and O'Neill 1986), and near-threatened species such as *Campylopterus villaviscencio* and *Henicorhina leucoptera* are present (Davis 1986, T. A. Andrews *in litt.* 1990, M. Pearman *in litt.* 1995). Lowland areas in the Mayo valley (Huallaga drainage) west of the ridge are almost entirely deforested, and forest clearings are gradually encroaching further up into this mountain area (also T. S. Schulenberg *in litt.* 1994).

Touit stictoptera	1983	Flocks of 5–25 seen daily during October–November 1983.
Heliangelus regalis	1986	Only 2–3 were recorded in a month's survey, October–November 1983, with just one male seen during a short visit in 1986 (T. S. Schulenberg *in litt.* 1994).
Loddigesia mirabilis	1987	Recorded along this trail in June 1987 (M. Pearman *in litt.* 1995).
Herpsilochmus parkeri	1990	The area (at c.1,350 m) along this trail embraces the entire known range of this species which is, however, fairly common; last recorded in April–May 1990 near Jesús del Monte (T. A. Andrews *in litt.* 1990).

Hacienda Boca Chica (Lambayeque)

PE 24
5°42′S 79°48′W
EBA B20

Unprotected

Hacienda Boca Chica is on the Pacific-slope foothills of the Andes in northern Lambayeque department, and includes the dry wooded valleys of Quebrada de Pavas and Quebrada Mugo Mugo, in both of which the species below has been recorded.

Penelope albipennis	c.1980	A population of 5–6 pairs is thought to exist, possibly supplemented by locally dispersing birds.

El Tocto (Lambayeque)

PE 25
5°47′S 79°42′W
EBA B20

Unprotected

El Tocto is a quebrada (in the Río Cascajal drainage) north-east of Olmos that flows down the Pacific slope of the Andean foothills (from at least 1,200 m) of northernmost Lambayeque department, and includes the dry wooded side valleys of Quebrada la Pachinga, two Quebradas Paltorán, Quebrada Cachaco–Quebrada Rosas, Quebrada Caballito, Quebrada Granada, Quebrada Peña Blanca and Quebrada Pomapara.

Penelope albipennis	1990	This is the largest population of this species, with at least 12 pairs present, although 12 birds were estimated in Quebrada Caballito alone in August 1989; last recorded May 1990 (T. A. Andrews *in litt.* 1990).
Hylocryptus erythrocephalus	1986	One seen at 500 m on the Quebrada Caballito in February 1986.

Hacienda Recalí (Lambayeque)

PE 26

5°51'S 79°41'W
EBA B20

Unprotected

Hacienda Recalí is in the Cascajal drainage on the Pacific slope of the Andean foothills in northern Lambayeque, and includes the dry wooded valleys of Quebrada El Algodonal, Quebrada Oberito, Quebrada Las Torcazas and Quebrada El Barranco.

Penelope albipennis	c.1980	There is a population of at least six pairs.

Km 21, Olmos–Choloque road (Lambayeque)

PE 27

5°56'S 79°37'W
EBAs B20, B22

Unprotected

This network of trails is c.21 km east-north-east of Olmos in the Río Olmos drainage, on the Pacific slope of the Andean foothills in northern Lambayeque (M. Pearman *in litt.* 1995). The trails penetrate an area of dry forest which in 1987 appeared largely intact, and supported numerous restricted-range species from the Tumbesian Endemic Bird Area (B20) (M. Pearman *in litt.* 1995).

Penelope albipennis	1987	Reliable reports by local people appear to refer to the presence of at least one pair (M. Pearman *in litt.* 1995).
Columba oenops	1987	One seen June 1987 (M. Pearman *in litt.* 1995).
Leptotila ochraceiventris	1987	Several recorded in June 1987 (M. Pearman *in litt.* 1995).
Hylocryptus erythrocephalus	1990	One seen in dry forest in June 1987, with others seen in May 1990 (T. A. Andrews *in litt.* 1990, M. Pearman *in litt.* 1995).

Seques (Lambayeque)

PE 28

6°54'S 79°18'W
EBAs B20, B21

Unprotected

Lying on the upper Río de Saña which flows down the Pacific slope of the West Andes, Seques is further west than the Taulís area (PE 10) and is centred around 1,500 m where there is humid evergreen and lower montane forest.

Penelope barbata	c.1926	A collecting locality reported by I. Franke (verbally 1991).
Acestrura bombus	c.1926	Mentioned for this locality.
Syndactyla ruficollis	1926	A record at 1,500 m.

Reque (Lambayeque)

PE 29

6°52'S 79°50'W
EBA B20

Unprotected

Reque is c.10 km south of Chiclayo in southern Lambayeque department, an area of barren coastal dunes within which a remnant 50-ha patch of *Prosopis*, *Acacia*, *Capparis*, etc., was found. However, this patch of habitat had been heavily disturbed, and was subject to intensive understorey grazing.

Phytotoma raimondii	1989	Twenty birds found in August 1989.

Río de Saña and Rafan (Lambayeque)

PE 30
7°00′S 79°38′W
EBA B20

Unprotected

Habitat suitable for the species below, namely isolated riparian thickets, *Acacia* and associated desert scrub, exists in southern Lambayeque around the village of Rafan (10 km west of Mocupé: M. Pearman *in litt.* 1995) and c.5 km north-north-east of Rafan near the Río de Saña.

Phytotoma raimondii	1990	Collected near the Río de Saña in 1978, three males and three females seen around Rafan in 1987, and others recorded in 1990 (T. A. Andrews *in litt.* 1990).

Molino (La Libertad)

PE 31
7°45′S 77°46′W
EBA B22

Unprotected

Molino is a village in a heavily populated area c.10 km north-west of Aricapampa on the road from Trujillo to the Río Marañón in eastern La Libertad department. This area is in the temperate zone (c.3,000 m) with the vegetation comprising shrubs and *Alnus* and *Eucalyptus* trees.

Aglaeactis aliciae	1995	Up to seven seen at 3,095 m (B. P. Walker *in litt.* 1995).

Río Abiseo (La Libertad/San Martín)

PE 32
7°32′S 77°29′W
EBAs B25, B27

Río Abiseo National Park (IUCN category II), 274,520 ha

The National Park is in the north-east Andes, primarily in San Martín department, but bordering eastern La Libertad, north-east of Pataz (which is situated in the Río Marañón valley). The park protects an important area comprising five life-zones including páramo, tropical montane forest and extensive undisturbed cloud-forest (Leo *et al.* 1988).

Leptosittaca branickii	1981	Three collected in August 1981.
Aulacorhynchus huallagae	1989	One collected at 2,500 m.
Poospiza alticola	1981	Up to four seen during July–August 1981.
Buthraupis aureodorsalis	1981	Seen three times near the treeline in August 1981.

Soquián and Chagual (La Libertad)

PE 33
7°51′S 77°40′W
EBA B22

Unprotected

These two localities (Soquián being a hacienda) are on the left bank of the Marañón valley, in an area of arid subtropical desert scrub and cultivated plants between c.1,000 and 2,000 m, but with montane shrubbery at around 3,000 m near Succha.

Columba oenops	1979	Uncommon in August 1979 when two were collected at 1,050 m (T. S. Schulenberg *in litt.* 1994).
Forpus xanthops	1979	Fairly common in 1979 (T. S. Schulenberg *in litt.* 1994).
Aglaeactis aliciae	1932	Seven collected (probably near Succha) in June 1932.
Acestrura bombus	1932	Recorded from this locality.

Hacienda Tulpo (La Libertad)

PE 34
8°08′S 78°01′W
EBA B27

Unprotected

Hacienda Tulpo is situated at c.3,000 m c.19 km east of Santiago de Chuco, in the Pacific drainage of south-west La Libertad. The area comprised pastures, potato and barley fields in 1900 but appears not to have been surveyed since.

Nothoprocta kalinowskii	1900	One specimen (of only two known).
Agriornis andicola	1900	Two collected May 1900.

Mashua and La Caldera (La Libertad)

PE 35
8°12′S 77°14′W
EBAs B25, B27

Unprotected

Mashua and La Caldera lie either side of the crest of the East Andes in south-east La Libertad. Both are on the Tayabamba–Ongón trail, with Mashua at the treeline east of the crest (an area of wet temperate forest), and La Caldera on the western slope just below the crest (in an area of shrubby temperate forest), both between 3,300 and 3,500 m.

Leptosittaca branickii	1979	One specimen at 3,350 m at Mashua.
Thripophaga berlepschi	1979	Rather uncommon during fieldwork in September 1979 (at Mashua), with 2–3 seen on three occasions and one collected (also T. S. Schulenberg *in litt.* 1994).
Poospiza alticola	1979	One specimen collected at 3,500 m near La Caldera.
Buthraupis aureodorsalis	1979	Recorded between 3,150 and 3,350 m at Mashua.

Cumpang and Utcubamba (La Libertad)

PE 36
8°12′S 77°09′W
EBA B25

Unprotected

Cumpang and Utcubamba are in close proximity on the eastern slope of the East Andes in the upper Huallaga drainage (near the Río Mishollo), in south-east La Libertad. Both localities are on the trail between Tayabamba and Ongón, and between 1,800 and 2,500 m they are within the subtropical to temperate cloud-forest zone.

Leptosittaca branickii	1900	Recorded from 2,400 m, August 1900.
Aulacorhynchus huallagae	1979	Four collected at Cumpang in October 1979, the type was taken from a small group seen near Utcubamba in 1932.

Hacienda Buenavista (La Libertad)

PE 37
8°29′S 78°38′W
EBA B20

Unprotected

Hacienda Buenavista is situated in the Río Chao valley at c.200–300 m on the Pacific slope of southern La Libertad department. The status of habitat suitable for the species below is unknown.

Phytotoma raimondii	1975	One collected.

Yanac and Quebrada Tutapac (Ancash)

PE 38

Unprotected

8°39′S 77°50′W
EBA B27

Yanac is at 2,860 m in the Río Santa drainage, at the northern end of the Cordillera Blanca in northern Ancash department. Most records of the species mentioned below come from areas to the south of (i.e. above) Yanac between 3,500 and 4,500 m, where the Quebrada Tutapac is found.

Asthenes huancavelicae	1979	A small population found below Yanac at 2,700 m.
Zaratornis stresemanni	1976	A number of specimens come from areas south of Yanac.
Anairetes alpinus	1976	Reportedly not rare above Yanac in 1976 when three were collected.
Poospiza alticola	c.1976	Recent sightings come from Quebrada Tutapac at 3,660 m.

Yuracmarca–Yanac (Ancash)

PE 39

Unprotected

8°41′S 77°53′W
EBA B27

In the Río Santa drainage, the road between Yanac and Yuracmarca drops from c.2,900 m to 1,400 m. A small population of the species mentioned below is found in dry shrubby habitat at c.1,800–1,900 m, between 10 km south-west of Yanac and c.20 km from Yuracmarca.

Asthenes huancavelicae	1979	A number have been collected along this road.

Huaylas–Quitacocha (Ancash)

PE 40

Unprotected

8°52′S 77°54′W
EBA B27

Huaylas is on the left bank of the Río Santa, in the foothills at the northern end of the Cordillera Negra in northern Ancash. Lying between 2,700 and 3,400 m (at Quitacocha) this is an area of dense bushes and small trees in a rocky landscape.

Asthenes huancavelicae	1979	A small breeding population was found.
Poospiza rubecula	1979	One specimen (of a pair) was collected.

Chacchan (Ancash)

PE 41

Unprotected

9°30′S 77°47′W
EBA B27

The town of Chacchan lies on the Pacific slope of the Cordillera Negra on the right bank of the Río Casma valley. Above town between 2,800 and 2,900 m is a 1-ha patch of habitat in which the species below has been found, though it was not seen there in 1988 (B. P. Walker *in litt.* 1995).

Synallaxis zimmeri	1983	Seven found in August 1983.

Bosque San Damián (Ancash)

PE 42
9°51′S 77°47′W
EBA B27

Unprotected

Bosque San Damián is above the town of San Damián (between 1,800 and 2,400 m), on the Pacific slope of the southern Cordillera Negra, along the right bank of the Río Huarmey valley.

Synallaxis zimmeri	1985	Four collected in 1980 and two seen in 1985.
Poospiza rubecula	1985	Two seen in October 1985.

Bosque de Noquo (Ancash)

PE 43
10°02′S 77°39′W
EBA B27

Unprotected

Bosque de Noquo is near Pararin on the Pacific slope at the southern end of the Cordillera Negra in southern Ancash department. The area presumably contains dry thorny scrub or shrubby forest favoured by the species below.

Synallaxis zimmeri	1988	Six collected at 2,850 m in May 1988, suggesting that the species is fairly common.

Huascarán (Ancash)

PE 44
9°45′S 77°28′W
EBA B27

Huascarán National Park (IUCN category II), 340,000 ha

The National Park embraces much of the central part of the Cordillera Blanca and Nevado de Huascarán in central Ancash department, from c.8°45′S to 10°05′S. Most of the localities where the threatened species have been recorded are concentrated in the north (especially around Llanganuco lake), and in the Nevado de Huascarán, although this almost certainly reflects observer bias. The *Polylepis–Gynoxys* woodlands, even within the national park, are dwindling owing to the activities of man, although some large areas still exist (e.g. around Llanganuco) (S. Butchart *in litt.* 1994).

Asthenes huancavelicae	1988	Two seen on the eastern slope at 4,000–4,200 m.
Zaratornis stresemanni	1994	Recorded from a number of localities on both slopes at 3,400–4,400 m, including two at Llanganuco in 1994 (S. Butchart *in litt.* 1994).
Anairetes alpinus	1987	Recorded at 4,350 m in August 1987.
Agriornis andicola	1987	One seen on the eastern slope at 4,250 m in February 1987.
Poospiza alticola	1994	Recorded from a number of localities on the eastern slope of the cordillera, especially near Llanganuco (e.g. S. Butchart *in litt.* 1994, B. P. Walker *in litt.* 1995).
Poospiza rubecula	1988	An adult seen at 3,700 m on the eastern slope, possibly just outside the park.

Cerro Huansala (Ancash)

PE 45
9°51′S 76°59′W
EBA B27

Unprotected

Cerro Huansala is at the south-east end of the Cordillera Blanca in south-east Ancash department. The vegetation at c.3,700 m comprises *Polylepis–Gynoxys* woodland.

Anairetes alpinus	c.1987	Records come from 3,700 m.
Poospiza alticola	c.1987	Recent sightings come from 3,700 m.

Cullcui (Huánaco)

PE 46

Unprotected

9°23′S 76°42′W
EBA B27

Cullcui is a small settlement on the right bank of the upper Río Marañón which is bounded by steep, dry shrub-covered hills in the arid subtropical zone (c.3,200 m).

Taphrolesbia griseiventris	1983	At least one seen in 1983 (T. S. Schulenberg *in litt.* 1994), this being the first record here since a specimen from 1922.

Huánaco–La Unión road (Huánaco)

PE 47

Unprotected

9°46′S 76°48′W
EBA B27

This area is in the upper Río Marañón valley, where the Huánaco–La Unión road crosses the Marañón. At this point there are steep slopes covered in shrubs, cacti and *Puya* and other bromeliads.

Taphrolesbia griseiventris	1975	Three or more seen in May 1975.

Santa María del Valle (Huánaco)

PE 48

Unprotected

9°51′S 76°08′W
EBA B27

Santa María del Valle is c.12 km downriver from Huánaco town, on the right bank of the Río Huallaga, where there are arid slopes (with cacti) above town at 2,000–2,100 m.

Asthenes huancavelicae	c.1989	An undescribed subspecies is reported to have a breeding population.

Quilluacocha and Bosque Unchog (Huánaco)

PE 49

Unprotected

9°42′S 76°07′W
EBA B25

These two localities are in the Cordillera Carpish region of central Huánaco: Quilluacocha is a swampy lake at the headwaters of a stream that flows past Acomayo, with Bosque Unchog above it at 3,600 m comprising a tract of elfin forest (and mossy brushland) on the pass between Churubamba and Hacienda Paty, c.14 km north-north-west of Acomayo.

Buthraupis aureodorsalis	1988	About 25 specimens taken in this area, mostly 1973–1975, fairly common at 3,325 m in 1988 (B. P. Walker *in litt.* 1995).

Bosque Zapatogocha (Huánaco)

PE 50

Unprotected

9°40′S 76°03′W
EBA B25

Bosque Zapatogocha is an area of subtropical forest at 2,600–3,000 m, c.6 km north-east of Acomayo in the Cordillera Carpish. Nearby but slightly lower (at 2,000 m) is Chinchao, on the upper Río Chinchao (part of the Huallaga drainage).

Leptosittaca branickii	c.1975	Recorded from this locality.
Acestrura bombus	1922	Recorded from Chinchao.
Agriornis andicola	1975	One collected at c.2,600 m.

Sariapunta (Huánuco)

PE 51

9°43′S 75°54′W
EBA B25

Unprotected

The town of Sariapunta is near Pillao, at 3,000 m on the Cordillera de Carpish in central Huánaco department. In 1973 there were areas of elfin forest above town, surrounded by grassland.

Buthraupis aureodorsalis	1973	Two collected in October 1973.

Cerros del Sira (Huánuco)

PE 52

9°21′S 74°43′W

Unprotected

The Cerros del Sira (or Sira Pico) are a series of isolated peaks between the Ríos Pachitea and Ucayali which reach c.2,500 m and are connected by a low ridge to the East Andes. In 1969, the peaks were covered with undisturbed humid cloud-forest.

Pauxi unicornis	1969	The only two specimens of the race *koepckeae* were collected on the south-west slopes.
Tangara phillipsi	1969	This is the only known locality for the species, which, however, was fairly common in 1969.

Lago Yarinacocha (Ucayali)

PE 53

8°15′S 74°43′W
EBA B19

Unprotected

Lago Yarinacocha is a large oxbow formed by the Río Ucayali, c.5 km north-north-west of Pucallpa in northern Ucayali department (formerly Loreto).

Micrastur buckleyi	1946	One collected.

Balta (Ucayali)

PE 54

10°08′S 71°13′W
EBA B30

Unprotected

Balta is a Cashinahua Indian village on the Río Curanja, a tributary of the Río Alto Purús in the lowlands of easternmost Ucayali department, and only c.30 km from the Brazilian border.

Cacicus koepckeae	1965	The only confirmed site for the species: the type was collected from a group of six seen in 1963, another being taken in 1965.

Pueblo Quichas (Lima)

PE 55

10°34′S 76°47′W
EBA B27

Unprotected

Pueblo Quichas (formerly Hacienda Quichas) is north of Oyón on the Pacific slope of the Cordillera Huayhuash in north-east Lima department.

Zaratornis stresemanni	1987	An estimated 500 birds (the largest known concentration) at c.4,000–4,200 m.

Upper Santa Eulalia valley (Lima)

PE 56

Unprotected

11°35′S 76°22′W
EBA B27

The Santa Eulalia, in east-central Lima department, is a tributary of the Rímac. In its upper reaches, 13 km west of Milloc, between 3,600 and 4,200 m, are areas of *Polylepis* and mixed *Polylepis–Gynoxys* woodland, the lower slopes comprising composite scrub and woodland.

Zaratornis stresemanni	c.1987	Fairly common at 3,600–4,200 m.
Poospiza rubecula	1985	Small numbers at 2,400–3,650 m.

Bosque Zárate (Lima)

PE 57

Unprotected

11°53′S 76°27′W
EBA B27

The Bosque Zárate is a relict Andean forest at 2,700–3,300 m on the south-facing slope of a ravine (the Río San Bartolomé) that descends to the Río Rimac, on the West Andean cordillera behind Surco (Barrio 1995). The habitat comprises mixed woodland with bushy undergrowth at its edges, and is dominated by *Oreopanax*, *Myrcianthes* and *Escallonia* (Barrio 1995). Forested habitat covers only c.120 ha, but with the inclusion of shrubby areas this increases to c.360 ha (Barrio 1995).

Zaratornis stresemanni	1953	The type-locality (after which the genus takes its name) with records and specimens from 2,700–2,900 m.
Poospiza rubecula	1994	Previous records come from 2,600–2,900 m, but most recently groups of up to six (including singing birds) were recorded at 2,850–2,900 m in March–April 1994, in an area of 40 ha (Barrio 1995).

Hortigal (Lima)

PE 58

Unprotected

12°47′S 75°44′W
EBA B28

Hortigal is in southernmost Lima department on the Cordillera de Turpicotay. Above (or south-west) of the village are areas of *Polylepis* and mixed *Polylepis–Gynoxys* woodland.

Zaratornis stresemanni	1987	Fairly common at 3,800–4,350 m.

Lago de Junín (Junín)

PE 59

Junín National Reserve (IUCN category VIII), 53,000 ha

11°02′S 76°06′W
EBA B28

The lake is situated at 4,080 m in north-west Junín department and covers 14,320 ha with an additional 11,900 ha of temporarily flooded meadows and marshland. It has deteriorated greatly owing to pollution and water-level fluctuations.

Podiceps taczanowskii	1993	The species is confined to this lake, and there are now just 50 birds (Valqui 1994).
Laterallus tuerosi	c.1983	The species is known only from marshes on the south-west shore.
Cinclodes palliatus	1978	One seen near the lake (Harris 1980).

Marcapomacocha (Junín)

PE 60
11°26′S 76°21′W
EBA B28

Unprotected

Marcapomacocha is a town at 4,415 m on the shores of Lago Marcapomacocha in the Río Mantaro watershed. Above town are puna grasslands with some areas of boggy terrain, the habitat to which the species below is adapted.

Cinclodes palliatus	1989	Records of up to five in one cushion-plant bog (B. P. Walker *in litt.* 1995).

Pampa Pucacocha (Junín/Lima)

PE 61
11°33′S 76°16′W
EBA B28

Unprotected

Pampa Pucacocha is situated at 4,400 m on the south-east base of Nevada Raujunte. The area stretches to the west, and therefore just into Lima department. This area is a mineral-rich, well-watered cushion-plant bog, situated below a glacier, with rocky outcrops and stony slopes nearby.

Cinclodes palliatus	1985	A population of 3–4 pairs has been recorded.

Yauli (Huancavelica)

PE 62
12°47′S 74°49′W
EBA B28

Unprotected

The town of Yauli is c.10 km east of Huancavelica at 3,400 m on the Río Mantaro drainage. Records of the species mentioned below come from areas above the town to the south-west (towards Huancavelica).

Cinclodes palliatus	1947	One collected at 4,940 m, south-west of Yauli.
Asthenes huancavelicae	1979	A number have been collected at 3,500–3,700 m.

Pisco (Ica)

PE 63
13°42′S 76°13′W
EBA B32

Unprotected

Pisco is at the mouth of the Río Pisco, on the coast of northern Ica department, just north of the Paracas National Reserve boundary (PE 64). The area around the mouth of the river, and on higher ground to the north, appears to be particularly suitable for the species mentioned below.

Xenospingus concolor	1983	Many recent sightings and a number of specimens come from this general area.

Isla San Gallán (Ica)

PE 64
13°50′S 76°28′W

Paracas National Reserve (IUCN category VIII), 335,000 ha

This dry, barren island with high hills is situated 5 km off the Paracas peninsula, on the coast of northern Ica department. Introduced dogs, exploitation for food and increased tourism all threaten the species below.

Pelecanoides garnotii	c.1991	Recent studies found 100 colonies totalling c.1,200 nests, one of the largest remaining populations in Peru.

Nazca (Ica)

PE 65

Unprotected

14°50′S 74°57′W
EBA B32

Nazca is at c.600 m on the Río Blanco (part of the Río Grande drainage), c.55 km from the Pacific coast of Ica department. The status of the riparian shrubbery on which the species below primarily depends is unknown.

Xenospingus concolor	1982	A number have been collected, three recently seen in hotel grounds, April 1982.

Ayacucho (Ayacucho)

PE 66

Unprotected

13°07′S 74°13′W
EBA B27

The city of Ayacucho lies at 2,750 m on the Río Mantaro drainage in northern Ayacucho department. In the surrounding areas, and along the road heading west towards the coast, arid habitat suitable for the species mentioned below persists.

Asthenes huancavelicae	1988	Abundant.

Pampa Galeras (Ayacucho)

PE 67

Pampa Galeras National Reserve (IUCN category VIII), 6,500 ha

14°40′S 74°23′W
EBA B28

This protected area of puna and *Polylepis–Gynoxys* woods lies between Nazca and Puquio, on the Pacific slope of south-west Ayacucho department. This is one of only two protected areas of suitable habitat within the range of the species below.

Zaratornis stresemanni	1987	Reported to be fairly common, although most recently only a few birds have been recorded.

Río Mayobamba (Ayacucho)

PE 68

Unprotected

14°14–22′S 73°52–57′W
EBA B27

The river is a tributary of the Río Pampas, south of Ayacucho town. The stretch between Chipao and Santa Ana de Huaycahnacho is arid and harbours a large population of the species mentioned below.

Asthenes huancavelicae	c.1990	40–50 pairs (of the rare subspecies *usheri*) are thought to occur along this stretch of the river.

Mutca (Apurímac)

PE 69

Unprotected

14°17′S 73°15′W
EBA B27

Mutca is on the Río Chalhuanca, a left-bank tributary of the Apurímac, and c.15 km north-west of Chalhuanca. The habitat in this area is a mosaic of stony ground and semi-arid montane scrub-forest.

Asthenes huancavelicae	1995	One (of the rare subspecies *usheri*) was collected 1987; fairly common at 2,600 m in 1995 (B. P. Walker *in litt.* 1995).

Bosque de Naupallagta (Apurímac)

PE 70

Unprotected

14°23′S 73°09′W
EBA B27

This site, while not precisely located by Stephens and Traylor (1983), is most probably east of Caraybamba on the road to Antabamba in southern Apurímac department. The precise location and current status of this forest (and the avifauna supported by it) are in urgent need of elucidation.

Nothoprocta taczanowskii	1977	A specimen taken at 3,650 m in July 1977 is one of the few known records of this species.

Bosque Ampay (Apurímac)

PE 71

Ampay National Sanctuary (IUCN category V), 3,635 ha

13°38′S 72°57′W
EBA B27

Bosque Ampay is a large *Podocarpus* forest and brush area on the southern slope of Nevada Ampay, north-west of Abancay. Other areas on the Nevada (such as Cerro Turronmocco) are equally important and support a mosaic habitat with small tuber fields and heavily grazed patches interspersed with copses, small woods and scrub.

Nothoprocta taczanowskii	1989	Known from several recent sightings, although apparently uncommon.
Synallaxis courseni	1989	The only known area for this species; population estimated at 300–400 pairs.

Cerro Runtacocha–Morococha (Apurímac)

PE 72

Unprotected

13°40–41′S 72°46–47′W
EBA B27

The Cerro Runtacocha–Morococha massif is south-east of Abancay in north-central Apurímac department, and supports c.10 isolated, semi-humid, mixed *Polylepis–Gynoxys* woodlands (each 1–4 ha). Also in this area but further east is Cerro Balcón, where a 75-ha *Polylepis–Gynoxys* woodland exists.

Cinclodes aricomae	1989	This entire area is thought to support a population estimated at c.40 pairs.
Leptasthenura xenothorax	1989	Only 25 pairs were estimated to occur in the area.
Anairetes alpinus	c.1989	Perhaps as few as 30 pairs survive in the area's 10 (or more) woodland patches.

Abra Málaga (Cuzco)

PE 73

Unprotected

13°08′S 72°19′W
EBA B27

Abra Málaga is in the Cordillera Vilcanota at a high pass (on the Cuzco–Quillabamba road) on the ridge that separates the Urubamba and Santa María drainages. Between c.4,000 and 4,200 m there are a number of small, widely separated but extremely important *Polylepis–Gynoxys* woods: 2–3 km south-west of Abra Málaga, an area of 12–15 ha with other much smaller patches nearby; Canchaillo, c.10 km north-west of Abra Málaga; and an area 10 km north-north-east of Abra Málaga.

Nothoprocta taczanowskii	c.1985	Seen at 3,500 m near Canchaillo.
Cinclodes aricomae	1993	A small population (c.1 pair/ha) survives (G. Engblom *in litt.* 1993).
Leptasthenura xenothorax	1993	At least 15 families (or c.50 birds) survive in the area south of Abra Málaga (G. Engblom *in litt.* 1993).
Anairetes alpinus	1993	A good population survives (G. Engblom *in litt.* 1993).

Nevada Chaiñapuerto (Cuzco)

PE 74
13°13′S 72°07′W
EBA B27

Unprotected

This area lies c.35 km south-east of Abra Málaga in the valley ending at Urubamba town, and lies c.5 km east-north-east of the town. At 3,800–4,500 m is a *Polylepis–Gynoxys* wood in a semi-humid glacial cirque valley, with trees and bushes continuing up ravines and crevices on the surrounding rocky slopes.

Leptasthenura xenothorax	1987	10 family groups (c.35 birds) estimated to inhabit the woodland.

Yanacocha lakes (Cuzco)

PE 75
13°17′S 71°59′W
EBA B27

Unprotected

The lakes are at the head of the Huayocari valley, on the slope of Nevada Chicon in Cuzco department. The area holds a mixed *Polylepis–Gynoxys* wood at 3,700–3,800 m.

Leptasthenura xenothorax	1989	Recorded only in 1989.
Anairetes alpinus	1989	Recorded only in 1989.

Urubamba (Cuzco)

PE 76
13°17′S 72°05′W
EBA B27

Unprotected

Urubamba is in the Cordillera Vilcanota, a patch of *Polylepis–Gynoxys* woodland being present c.5 km east-north-east of town, up the valley between 3,800 and 4,500 m (c.35 km south-east of Abra Málaga).

Leptasthenura xenothorax	c.1987	Recently recorded from this locality.

Abra La Raya (Cuzco)

PE 77
14°29′S 71°05′W
EBA B27

Unprotected

Abra La Raya is a pass at 4,315 m on the divide between Lago Titicaca and the Amazonian drainage, thus on the border of Cuzco and Puno departments. The origin of the Río Urubamba is within marshes and lagoons immediately below the pass.

Nothoprocta taczanowskii	1983	Recorded December 1983 (N. Krabbe *in litt.* 1984).
Agriornis andicola	1917	Six collected in April 1917.

Quincemil (Cuzco)

PE 78
13°16′S 70°38′W
EBA B29

Unprotected

This area is c.15 km east of Quincemil in the Marcapata valley, easternmost Cuzco department, along the road to Puerto Maldonado. The species mentioned below was recorded from an area of low ridges (800–900 m) which are part of a semi-isolated mountain supporting epiphyte-laden forest on the upper slopes and in ravines.

Tinamus osgoodi	1974	One seen.

Manu (Madre de Dios/Cuzco)

PE 79
12°00'S 71°30'W
EBAs B29, B30

Manu National Park (IUCN category II), 1,532,806 ha
Manu Reserved Zone (IUCN category VI), 257,000 ha

The National Park and Reserved Zone (parts of which form a Biosphere Reserve and World Heritage site) extend from the lowlands and Andean foothills in western Madre de Dios up to the puna zone (above 4,000 m) in north-central Cuzco. Covering all of the major ecosystems from puna to lowland tropical forest, this is probably one of the most biodiverse protected areas in the world, and the presence of four threatened birds emphasizes its importance. Records of *Cacicus koepckeae* from this area are now widely refuted (S. K. Robinson verbally 1994).

Tinamus osgoodi	1985	Recorded from three localities within c.5 km of the National Park boundary, all on ridges that extend into the park.
Micrastur buckleyi	1989	A juvenile recorded in July 1989 within the Reserved Zone.
Leptosittaca branickii	1989	A flock of 25 seen between Shintuya and Paucartambo, just outside the eastern boundary of the park.
Formicarius rufifrons	1990	An estimated 50–250 pairs; combined with numbers at Tambopata (PE 81), this represents 20% of the world population (Kratter 1995).

Río Colorado (Madre de Dios)

PE 80
12°39'S 70°20'W
EBA B30

Unprotected

The two species mentioned below have been recorded at the mouth of this river where it flows into the Río Madre de Dios. The area has not been explored ornithologically in recent years (Kratter 1995).

Crax globulosa	c.1942	Recorded from this locality.
Formicarius rufifrons	1958	The type- and two other specimens originate from this locality.

Tambopata–Candamo (Madre de Dios/Puno)

PE 81
12°50'S 69°25'W
EBA B30

Tambopata–Candamo Reserved Zone (IUCN category III), 1,478,942 ha

The Reserved Zone covers a large area of land south of the Río Madre de Dios between the Ríos Heath and Tambopata in south-east Madre de Dios and north-east Puno departments. Records of the species mentioned below come from the south-western part of the reserve, on the Cerros del Távara (300–900 m) and where the Río Távara joins the Tambopata near Astillero (Foster *et al.* 1994).

Tinamus osgoodi	1992	Rare on the Cerros del Távara during May–June 1992 (Foster *et al.* 1994).
Pauxi unicornis	1992	One recorded in the Cerros del Távara.
Formicarius rufifrons	1993	Recent records come from near the mouth of the Távara, and from Ccolpa de Guacamayos, on the west bank of the Tambopata (Foster *et al.* 1994), with the population here estimated at 100–500 pairs (Kratter 1995: see comments for this species under PE 79).

Inca Mine (Puno)

PE 82
13°51′S 69°41′W
EBA B34

Unprotected

Inca Mine is a general gold-mining area near the town of Santo Domingo, on the Río Inambari along the Limbani–Astillero road, c.19 km north of Oroya in northern Puno department.

Terenura sharpei	1900	The only record from Peru.

Sandia (Puno)

PE 83
14°17′S 69°26′W
EBA B34

Unprotected

Sandia is on the Río Sandia, a dry inter-montane valley that feeds into the Inambari drainage in north-east Puno department. The area between Sandia (2,180 m) and Azalaya (1,750 m), which is 12 km downstream (north-north-east), has been identified as important for the species mentioned below.

Tangara meyerdeschauenseei	1980	Fairly common along this valley in November 1980.

Maruncunca (Puno)

PE 84
14°14′S 69°17′W
EBA B34

Unprotected

Abra de Maruncunca is c.20 km east of Sandia, on a ridge separating the Huari Huari and Tambopata drainages, the west side of which supports tall, humid evergreen forest at c.2,000 m. This area is within the upper Inambari drainage in north-east Puno department.

Tangara meyerdeschauenseei	1990	Fairly common in 1980; two seen in 1990 (B. P. Walker *in litt.* 1995).

Valcón (Puno)

PE 85
14°26′S 69°24′W
EBA B27

Unprotected

Valcón is on the eastern slope of the Andes, c.5 km north-north-west of Quiaca, in the temperate zone where humid evergreen forest (restricted owing to human disturbance and the steepness of the terrain) borders puna (largely grazed).

Nothoprocta taczanowskii	1980	Two specimens (including a chick) collected at c.3,000 m in October 1980.

Salinas y Aguada Blanca (Arequipa)

PE 86
16°12′S 71°33′W
EBA B32

Salinas y Aguada Blanca National Reserve (IUCN category VIII), 366,936 ha

The National Reserve embraces a large area of the cordilleras east and north of the town of Arequipa. The most important areas known for the two species mentioned below appear to be the Nevado Chachani (c.25 km north of Arequipa), Chiguata and Cerro Pichupichu (both 15–30 km east of Arequipa).

Agriornis andicola	1952	One collected on the slopes of Nevado Chachani.
Conirostrum tamarugense	1990	Flocks of 12 or more recently recorded at c.3,900 m on Nevado Chachani; a flock of 5–6 north-east of Chiguata in 1990 (B. P. Walker *in litt.* 1995).

Río Tambo and Lagunas de Mejía (Arequipa)

PE 87

Laguna de Mejía National Sanctuary (IUCN category IV), 691 ha

17°06′S 71°46′W
EBA B32

The Río Tambo enters the Pacific near the town of Punta de Bombón in southern Arequipa. The stretch of river between El Fiscal (where the Pan-American Highway crosses it) and the coast, including the coastal lagoons to the north, is extremely important for the species mentioned below, being one of the few areas where it has been described as common.

Xenospingus concolor	1987	Locally common along the Río Tambo, but rare at Lagunas de Mejía.

North-east of Tarata (Moquegua)

PE 88

Unprotected

17°28′S 70°02′W
EBA B32

Tarata is on the upper Río Sama, on the Pacific slope of the Andes in southernmost Peru. Some 2–25 km north-east of Tarata, at c.4,050 m on the Pacna–Puno road, an area of low, open *Polylepis* woodland was found to support the species mentioned below.

Conirostrum tamarugense	1983	Small numbers seen.

Tacna (Tacna)

PE 89

Unprotected

18°01′S 70°15′W
EBA B32

Gardens within the city of Tacna, and vegetation around the airport to the south, support (at least seasonally) a population of the species below.

Eulidia yarrellii	1984	Four recorded during December 1977 and a female seen in 1984 (B. P. Walker *in litt.* 1995).

URUGUAY

URUGUAY has just over 400 species of bird (Cuello 1985) of which c.204 breed (Gore and Gepp 1978), four (1%) have restricted ranges (Stattersfield *et al.* in prep.) and 11 (<3%) are threatened (Collar *et al.* 1992). This analysis has identified seven Key Areas for the threatened birds in Uruguay (see '*Key Areas*: the book', p. 11, for criteria).

THREATENED BIRDS

Of the 11 Uruguayan species which Collar *et al.* (1992) considered to be at risk of extinction, none are endemic to the country (see Table 1). These threatened birds rely primarily on grasslands (seven species) and wetlands (four). The main threat is habitat loss (affecting eight of the species), but also significant for *Gubernatrix cristata* and *Sporophila palustris* is capture for the cage-bird trade. The distributions of the species are shown in Figure 1.

KEY AREAS

Uruguay's seven Key Areas would, if adequately protected, help to ensure the conservation of six (55%) of its threatened species (always accepting that important new populations and areas may yet be found). Of these areas, two are important for 2–3 threatened species, and the rest each harbour one species (see also 'Outlook', below). From Tables 1 and 2 it can be seen that two species (*Coturnicops notatus* and *Yetapa risora*) each occur in just one Uruguayan Key Area but their survival in the country cannot be guaranteed even by the integrity of habitat within their respective Key Area (see 'Old records and little-known birds', below).

Not represented within the Uruguayan Key Area analysis are five of the country's threatened species, as follows.

Harpyhaliaetus coronatus has not been seen in Uruguay for more than 50 years, and is likely to be extinct in the country.

Porzana spiloptera is known with certainty from only two birds: one taken on the Arroyo Pando prior to 1926 and an immature collected on the Arroyo Solís Grande (Maldonado) in 1973. As it is not clear exactly where the Arroyo Solís Grande specimen was taken, a Key Area has not been chosen.

Numenius borealis is known in the country from only two specimens, both lacking date and locality.

Larus atlanticus is a rare winter (May–November) visitor to the south coast of Uruguay and so there

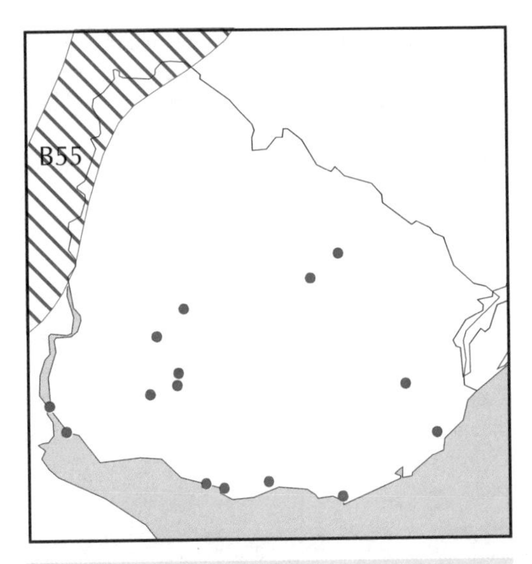

Figure 1. The localities where threatened birds have been recorded in Uruguay and their relationship to Endemic Bird Areas (EBAs).

- Of the Uruguayan threatened species only *Sporophila palustris* has a restricted range. It occurs in the Entre Ríos wet grasslands EBA (B55).

are no particularly important sites for it, although up to six individuals were seen on Laguna Rocha in 1993–1994 (PROBIDES *in litt.* 1995).

Anodorhynchus glaucus is considered extinct by Collar *et al.* (1994).

KEY AREA PROTECTION

Just two of Uruguay's Key Areas currently have some form of protected status, but none are strict nature reserves, national parks or Biosphere Reserves (IUCN categories I, II or IX). The five unprotected Key Areas require attention in the form of appropriate conservation measures if the populations of their threatened species are to survive. Two Uruguayan threatened species (*Coturnicops notatus* and *Yetapa risora*) are not currently recorded from any protected Key Area (Table 1).

RECENT CHANGES TO THE THREATENED LIST

With the publication of Collar *et al.* (1994), none of the threatened species listed in 1992 was dropped from the Uruguayan threatened list, but one—Black-and-white Monjita *Heteroxolmis dominicana*—was added (though it has not been included within the Site Inventory: see '*Key Areas*: the book', p. 12). *H. dominicana* formerly occurred in many areas in Uruguay, but by the 1970s was considered a scarce resident (Gore and Gepp 1978); a careful revision of its current distribution is needed to determine appropriate Key Areas.

OLD RECORDS AND LITTLE-KNOWN BIRDS

Our knowledge of threatened birds in Uruguay suffers from a general paucity of recent records for a number of species, but recent records from other areas mean that we are able to make informed recommendations for their conservation. However, in addition to the species listed under 'Key Areas' (above), *Yetapa risora* is known in Uruguay only from a few and mainly old records, and there can be no guarantees of a current viable population; the bird seen in 1986 at Playa Pascual (UY 06: see Site Inventory) may have been no more than a wandering individual.

OUTLOOK

This analysis has identified seven Key Areas, each of which is of importance for the conservation of threat-

Table 1. Coverage of threatened species by Key Area. Areas in bold currently have some form of protected status.

	Key Areas occupied	No. of Key Areas protected	Total nos. of Key Areas	
			Uruguay	Neotropics
Crowned Eagle *Harpyhaliaetus coronatus*	—	0	0	18
Dot-winged Crake *Porzana spiloptera*	—	0	0	4
Speckled Crake *Coturnicops notatus*	07	0	1	8
Eskimo Curlew *Numenius borealis*	—	0	0	0
Olrog's Gull *Larus atlanticus*	—	0	0	5
Glaucous Macaw *Anodorhynchus glaucus*	—	0	0	0
Strange-tailed Tyrant *Yetapa risora*	06	0	1	9
Marsh Seedeater *Sporophila palustris*	04, **05**	1	2	15
Yellow Cardinal *Gubernatrix cristata*	02, **03**	1	2	11
Saffron-cowled Blackbird *Xanthopsar flavus*	04, **05**	1	2	12
Pampas Meadowlark *Sturnella militaris*	01, **05**	1	2	4

Table 2. Matrix of threatened species by Key Area.

	01	02	03	04	05	06	07	No. of areas
Harpyhaliaetus coronatus	–	–	–	–	–	–	–	0
Porzana spiloptera	–	–	–	–	–	–	–	0
Coturnicops notatus	–	–	–	–	–	–	●	1
Numenius borealis	–	–	–	–	–	–	–	0
Larus atlanticus	–	–	–	–	–	–	–	0
Anodorhynchus glaucus	–	–	–	–	–	–	–	0
Yetapa risora	–	–	–	–	–	●	–	1
Sporophila palustris	–	–	–	●	●	–	–	2
Gubernatrix cristata	–	●	●	–	–	–	–	2
Xanthopsar flavus	–	–	–	●	●	–	–	2
Sturnella militaris	●	–	–	–	●	–	–	2
No. of species	1	1	1	2	3	1	1	

ened birds in Uruguay. The wetland and marshes in eastern Uruguay such as Bañados de India Muerta (UY 04) and Bañados de Santa Teresa and los Indios (UY 05) stand out as priorities, and together they support populations of six of Uruguay's 11 threatened species. These and the remaining Key Areas (which are individually important for many threatened species: see 'Key Areas' and 'Key Area protection', above) all need some degree of conservation action if the populations of their threatened birds are to remain viable. For some areas such as Playa Pascual (UY 06) and Juan L. Lacaze (UY 07) there is a requirement to confirm the continued existence or locating (and defining the extent of) suitable habitat in the vicinity of the Key Area.

DATA SOURCES

The above introductory text and the Site Inventory were compiled from information supplied by Programa de Conservación de la Biodiversidad y Desarollo Sustenable en los Humedales del Este (PROBIDES), Forestal Oriental SA, M. Retamosa, J. C. Rudolf, M. Santos and S. Umpiérrez, as well as from the following references.

ARBALLO, E. AND GAMBAROTA, J. C. (1987) Registro de la Tijereta de las Pajas para el Uruguay. *Nuestras Aves* 13: 16–17.

COLLAR, N. J., GONZAGA, L. P., KRABBE, N., MADROÑO NIETO, A., NARANJO, L. G., PARKER, T. A. AND WEGE, D. C. (1992) *Threatened birds of the Americas: the ICBP/IUCN Red Data Book.* Cambridge, U.K.: International Council for Bird Preservation.

COLLAR, N. J., CROSBY, M. J. AND STATTERSFIELD, A. J. (1994) *Birds to watch 2: the world list of threatened birds.* Cambridge, U.K.: BirdLife International (BirdLife Conservation Series no. 4).

CUELLO, J. P. (1985) Lista de referencia y bibliografía de las aves uruguayas. *Mus. Dámaso Antonio Larrañaga Ser. Divulg.* no. 1.

GORE, M. E. J. AND GEPP, A. R. M. (1978) *Las aves del Uruguay.* Montevideo: Mosca Hermanos.

IUCN (1992) *Protected areas of the world: a review of national systems*, 4: *Nearctic and Neotropical.* Gland, Switzerland, and Cambridge, U.K.: International Union for Conservation of Nature and Natural Resources.

PAYNTER, R. A. (1994) *Ornithological gazetteer of Uruguay.* Second edition. Cambridge, Mass.: Museum of Comparative Zoology.

SCOTT, D. A. AND CARBONELL, M. (1986) *A directory of Neotropical wetlands.* Cambridge and Slimbridge, U.K.: International Union for Conservation of Nature and Natural Resources and International Waterfowl Research Bureau.

STATTERSFIELD, A. J., CROSBY, M. J., LONG, A. J. AND WEGE, D. C. (in prep.) *Global directory of Endemic Bird Areas.* Cambridge, U.K.: BirdLife International (BirdLife Conservation Series).

SITE INVENTORY

Río Arapey (Salto)

UY 01

Unprotected

31°20'S 56°55'W

This area is located near to where the Ríos Arapey and Sosas meet, c.90 km west of Tacuarembó and c.80 km east of Salto in east-central Salto.

Sturnella militaris	1993	An important wintering population of c.100 birds was found in July (*Cotinga* 1995, 2: 30).

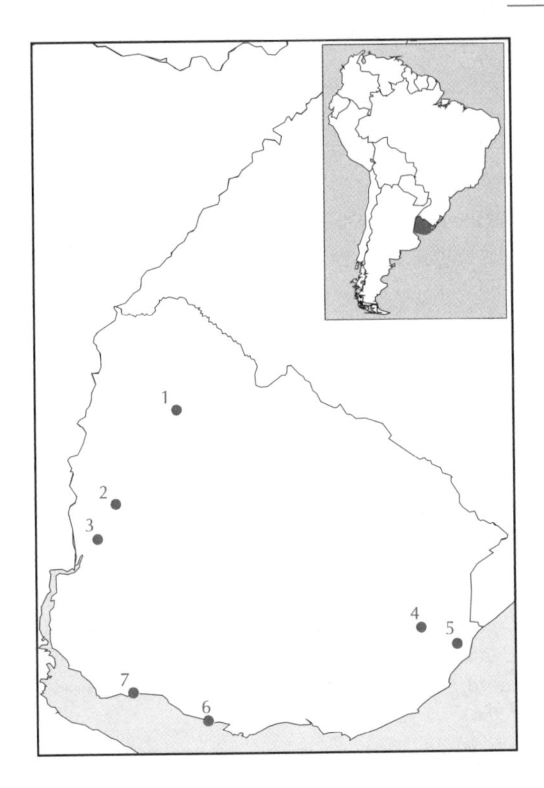

Figure 2. Key Areas in Uruguay.

01 Río Arapey
02 Piedras Coloradas
03 El Rosario
04 Bañados de India Muerta
05 Bañados de Santa Teresa and los Indios
06 Playa Pascual
07 Juan L. Lacaze

Piedras Coloradas (Paysandú)

UY 02
32°21′S 57°38′W

Unprotected

An area of hilly country lying c.50 km of Paysandú city in western Paysandú. The principal habitats are grassland with rocky outcrops, and woodland.

Gubernatrix cristata	1992	Six seen near El Refugio village (J. C. Rudolf *in litt.* 1995).

El Rosario (Río Negro)

UY 03
32°45′S 57°51′W

El Rosario (IUCN category IV), 313 ha

Located c.60 km north-east of Fray Bentos in south-west Río Negro, this reserve is owned by Forestal Oriental SA and comprises parkland, grassland and woodland along streams (J. C. Rudolf *in litt.* 1995).

Gubernatrix cristata	1992	Two seen and filmed in parkland (J. C. Rudolf *in litt.* 1995).

Bañados de India Muerta (Rocha)

UY 04
33°45′S 54°05′W

Unprotected

The freshwater marshes and seasonally flooded grassland of Bañados de India Muerta are located 80 km north of Rocha in north-central Rocha.

Sporophila palustris	1987	Three records of two birds: October 1985, November 1986 and December 1987.
Xanthopsar flavus	1987	A flock of 32 birds was seen in October 1987 and one of 28 in December 1987.

Bañados de Santa Teresa and los Indios (Rocha)

UY 05
33°56′S 53°40′W

El Potrerillo de Santa Teresa Reserve, 715 ha
Bañado los Indios Reserve, 2,487 ha

These marshes lie on the northern side of Laguna Negra in east-central Rocha. El Potrerillo de Santa Teresa is owned by the Ministerio de Vivienda, Ordenamiento Territorial y Medio Ambiente and managed by them in collaboration with PROBIDES (PROBIDES *in litt.* 1995); it comprises part of the shoreline of Laguna Negra, monte woodland, marshes and grassland with palms *Butia capitata* (S. Umpiérrez and M. Santos *in litt.* 1995). Bañados los Indios reserve is managed by PROBIDES and comprises seasonally inundated and permanent marshes and wetlands (PROBIDES *in litt.* 1995). The area, together with Laguna Negra and the extensive Bañados San Miguel and Laguna Merin, was designated a Ramsar site in 1984 (Scott and Carbonell 1986).

Sporophila palustris	1961	Six or seven seen daily in February 1961 north-west of Laguna Negra.
Xanthopsar flavus	1995	Found recently at Los Indios reserve, highest count being 69 in March 1995, the largest group recorded in Rocha department (S. Umpiérrez and M. Santos *in litt.* 1995). There is a breeding population at El Potrerillo de Santa Teresa Reserve, where the largest number observed was 44 in October 1994 (M. Santos and M. Retamosa *in litt.* 1995).
Sturnella militaris	1994	One seen in grassland at El Potrerillo de Santa Teresa in November 1994 (M. Santos and M. Retamosa *in litt.* 1995).

Playa Pascual (San José)

UY 06
34°45′S 56°35′W

Unprotected

This site is c.500 m from the Río de la Plata near Playa Pascual town and only 31 km west of Montevideo in south-east San José. The habitat is inundated grassland with some open water and pasture (Arballo and Gambarota 1987).

Yetapa risora	1986	One photographed in October 1986.

Juan L. Lacaze (Colonia)

UY 07
34°26′S 57°27′W

Unprotected

This town is on the Río de la Plata, 46 km east of Colonia del Sacramento in southern Colonia. A search of marshes in the vicinity could turn up populations of *Coturnicops notatus*.

Coturnicops notatus	1985	A brood of three found (one captured alive) on a farm near Juan L. Lacaze, December 1985.

▪ VENEZUELA

Rusty-faced Parrot *Hapalopsittaca amazonina*

VENEZUELA is home to approximately 1,300 species of resident and migrant bird, of which about 46 (c.3.5%) are endemic to the country (Meyer de Schauensee and Phelps 1978), 117 (c.9%) have restricted ranges (Stattersfield *et al.* in prep.) and 20 (1.5%) are threatened (Collar *et al.* 1992). This analysis has identified 26 Key Areas for the threatened birds in Venezuela (see '*Key Areas*: the book', p. 11, for criteria).

THREATENED BIRDS

Of the 20 Venezuelan species which Collar *et al.* (1992) considered at risk of extinction, 11 (55%) are confined to the country (Table 1), making Venezuela ninth in rank in the Americas for numbers of threatened birds (Collar *et al.* in press). Around 80–90% of the threatened species rely on wet forest in the tropical and subtropical zones (i.e. up to 2,000 m), and are almost exclusively threatened by habitat loss or disturbance (Collar *et al.* in press). The distributions of these threatened birds and their relationship to Endemic Bird Areas are shown in Figure 1.

KEY AREAS

The 26 Venezuelan Key Areas would, if adequately protected, help ensure the conservation of all 20 threatened species in the country—always accepting that important new populations and areas may yet be found (but see 'Old records and little-known birds', below). Of the 26 Key Areas, 16 are important for two or more (up to five) threatened species, and these are therefore the most efficient areas currently known in which to conserve Venezuela's threatened birds (see 'Outlook', below). The eight areas which each harbour three or more threatened species (Table 3) together represent potential sanctuaries for 17 threatened species, 85% of the total number. However, as vital as these areas are for the conservation of Venezuela's threatened species, they should not detract from the significance of Key Areas holding just one or two species, as these are of major importance for the remaining six species (three of which are confined to the country).

From Tables 1 and 2 it can be seen that eight species occur in just one Venezuelan area, and four

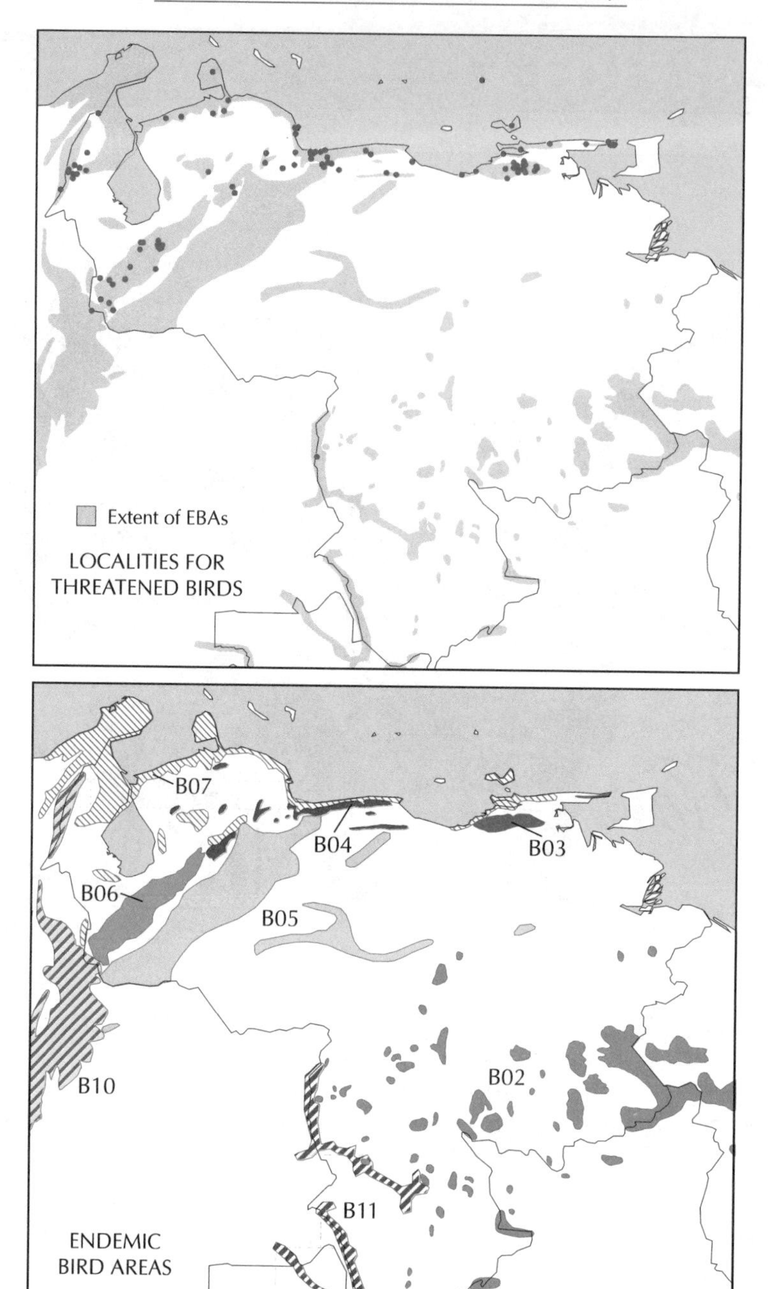

Figure 1. The localities where threatened birds have been recorded in Venezuela and their relationship to Endemic Bird Areas (EBAs).

- Most of the threatened birds in Venezuela have restricted ranges and thus occur together in various combinations within EBAs, which are listed below (figures are numbers of these species in each EBA). Threatened birds in Venezuela are almost exclusively confined to the Andes and the narrow coastal zone, but of particular importance are EBAs in the Cordillera de Caripe and Paria peninsula (B03), the Cordillera de Mérida (B06) and the páramo de Tamá of the East Andes (B10).

B02 Tepuís (0) *
B03 Cordillera de Caripe and Paria peninsula (6)
B04 Cordillera de la Costa Central (2)
B05 Llanos (0)
B06 Cordillera de Mérida (4)
B07 Caribbean dry zone (3)
B10 East Andes of Colombia (6)
B11 Upper Río Negro and Orinoco white-sand forest (1)

* See 'Recent changes to the threatened list'

Table 1. Coverage of threatened species by Key Area. Areas in bold currently have some form of protected status.

	Key Areas occupied	No. of Key Areas protected	Total nos. of Key Areas	
			Venezuela	Neotropics
Northern Helmeted Curassow *Pauxi pauxi*	**01,02,09**,10,**12,13,16**	6	7	10
Plain-flanked Rail *Rallus wetmorei* [E]	**11,12,13**	3	3	3
Rusty-flanked Crake *Laterallus levraudi* [E]	**09,11**,14,15,**17**	3	5	5
Speckled Crake *Coturnicops notatus*	**06**	1	1	8
Rusty-faced Parrot *Hapalopsittaca amazonina*	**02,03,06**	3	3	8
Yellow-shouldered Amazon *Amazona barbadensis*	**07**,08,**11**,18,**19**,20	3	6	6
White-tailed Sabrewing *Campylopterus ensipennis*	21,**22,23,25**	3	4	4
Táchira Emerald *Amazilia distans* [E]	**02**,04	1	2	2
Scissor-tailed Hummingbird *Hylonympha macrocerca* [E]	**25**	1	1	1
Orinoco Softtail *Thripophaga cherriei* [E]	26	0	1	1
White-throated Barbtail *Margarornis tatei* [E]	21,**22,23,25**	3	4	4
Recurve-billed Bushbird *Clytoctantes alixii*	**01**	1	1	4
Táchira Antpitta *Grallaria chthonia* [E]	**02**	1	1	1
Hooded Antpitta *Grallaricula cucullata*	**02**	1	1	3
Slaty-backed Hemispingus *Hemispingus goeringi* [E]	**03**,05,**06**	2	3	3
Venezuelan Flowerpiercer *Diglossa venezuelensis* [E]	**22,23**,24,**25**	3	4	4
Paria Redstart *Myioborus pariae* [E]	**25**	1	1	1
Grey-headed Warbler *Basileuterus griseiceps* [E]	21,**22,23**,24	2	4	4
Red Siskin *Carduelis cucullatus*	**13,16,23**	2	3	4
Yellow-faced Siskin *Carduelis yarrellii*	14	0	1	8

[E] Endemic to Venezuela

of them (*Hylonympha macrocerca*, *Thripophaga cherriei*, *Grallaria chthonia* and *Myioborus pariae*) are endemic to the country and thus totally reliant for their survival on the integrity of habitat in their respective Key Areas (see 'Outlook', below). Indeed, *T. cherriei* occurs in just one area and is not sympatric with other threatened birds, suggesting that, at least in this case, targeted single-species conservation is a necessity. Although occurring in a number of Key Areas (but endemic to Venezuela), *Basileuterus griseiceps* and *Diglossa venezuelensis* appear to be dependent on forests in the Cerro Negro area (VE 23), and in spite of being at the centre of its distribution and abundance in Venezuela, *Carduelis cucullatus* is confirmed as having occurred in just two (protected) Key Areas, for only one of which (Guatopo, VE 16) are there recent records (see 'Outlook'). Outside Venezuela, *Amazona barbadensis*

Table 2. Matrix of threatened species by Key Area.

	01	02	03	04	05	06	07	08	09	10	11	12	13	14	15	16	17	18	19	20	21	22	23	24	25	26	No. of areas
Pauxi pauxi	●	●	–	–	–	–	–	–	●	●	–	●	●	–	–	●	–	–	–	–	–	–	–	–	–	–	7
Rallus wetmorei	–	–	–	–	–	–	–	–	–	–	●	●	●	–	–	–	–	–	–	–	–	–	–	–	–	–	3
Laterallus levraudi	–	–	–	–	–	–	–	–	●	–	●	–	–	●	●	–	●	–	–	–	–	–	–	–	–	–	5
Coturnicops notatus	–	–	–	–	–	●	–	–	–	–	–	–	–	–	–	–	–	–	–	–	–	–	–	–	–	–	1
Hapalopsittaca amazonina	–	●	●	–	–	●	–	–	–	–	–	–	–	–	–	–	–	–	–	–	–	–	–	–	–	–	3
Amazona barbadensis	–	–	–	–	–	–	●	●	–	–	●	–	–	–	–	–	–	●	●	●	–	–	–	–	–	–	6
Campylopterus ensipennis	–	–	–	–	–	–	–	–	–	–	–	–	–	–	–	–	–	–	–	–	●	●	●	–	●	–	4
Amazilia distans	–	●	–	●	–	–	–	–	–	–	–	–	–	–	–	–	–	–	–	–	–	–	–	–	–	–	2
Hylonympha macrocerca	–	–	–	–	–	–	–	–	–	–	–	–	–	–	–	–	–	–	–	–	–	–	–	–	●	–	1
Thripophaga cherriei	–	–	–	–	–	–	–	–	–	–	–	–	–	–	–	–	–	–	–	–	–	–	–	–	–	●	1
Margarornis tatei	–	–	–	–	–	–	–	–	–	–	–	–	–	–	–	–	–	–	–	–	●	●	●	–	●	–	4
Clytoctantes alixii	●	–	–	–	–	–	–	–	–	–	–	–	–	–	–	–	–	–	–	–	–	–	–	–	–	–	1
Grallaria chthonia	–	●	–	–	–	–	–	–	–	–	–	–	–	–	–	–	–	–	–	–	–	–	–	–	–	–	1
Grallaricula cucullata	–	●	–	–	–	–	–	–	–	–	–	–	–	–	–	–	–	–	–	–	–	–	–	–	–	–	1
Hemispingus goeringi	–	–	●	–	●	●	–	–	–	–	–	–	–	–	–	–	–	–	–	–	–	–	–	–	–	–	3
Diglossa venezuelensis	–	–	–	–	–	–	–	–	–	–	–	–	–	–	–	–	–	–	–	–	–	●	●	●	●	–	4
Myioborus pariae	–	–	–	–	–	–	–	–	–	–	–	–	–	–	–	–	–	–	–	–	–	–	–	–	●	–	1
Basileuterus griseiceps	–	–	–	–	–	–	–	–	–	–	–	–	–	–	–	–	–	–	–	–	●	●	●	●	–	–	4
Carduelis cucullatus	–	–	–	–	–	–	–	–	–	–	–	–	●	–	–	●	–	–	–	–	–	–	●	–	–	–	3
Carduelis yarrellii	–	–	–	–	–	–	–	–	–	–	–	–	–	●	–	–	–	–	–	–	–	–	–	–	–	–	1
No. of species	2	5	2	1	1	3	1	1	2	1	2	2	3	2	1	2	1	1	1	1	3	4	5	2	5	1	

and *Campylopterus ensipennis* also occur on islands in the Caribbean which are, however, outside the region covered by this analysis.

KEY AREA PROTECTION

It is encouraging to find that 15 (58%) of Venezuela's Key Areas are currently under some form of protection, 13 (50% of the total) being national parks (IUCN category II). The 11 Key Areas (42% of the total) currently unprotected require attention in the form of appropriate conservation measures if the populations of their threatened species are to survive. However, even the formally protected areas remain under threat and, in many, habitat degradation and hunting continue unchecked; effective management of activities undertaken within them is thus required. All but two Venezuelan threatened species, namely *Thripophaga cherriei* and *Carduelis yarrellii*, are known from protected Key Areas. These two species occur in single unprotected Key Areas (although *C. yarrellii* occurs also outside Venezuela), and therefore warrant urgent attention which should be focused on determining their true conservation status, and ultimately on ensuring their protection (see 'Old records and little-known birds', below). The three Key Areas supporting Venezuelan endemics which occur in just one area (*Hylonympha macrocerca*, *Grallaria chthonia* and *Myioborus pariae*), are each under some form of protected status. Five other species are known from just one protected Key Area, but can be found in unprotected Venezuelan Key Areas, or in Key Areas outside the country (see Table 1).

RECENT CHANGES TO THE THREATENED LIST

With the publication of Collar *et al.* (1994), three species—Tepuí Tinamou *Crypturellus ptaritepui*, Military Macaw *Ara militaris* and Guaiquinima Whitestart *Myioborus cardonai*—were added to the Venezuelan threatened species list (though these species have not been included in the Site Inventory; see '*Key Areas*: the book', p. 12). Addition of the two Tepuí species was due to their ranges being less than 100 km^2 (Collar *et al.* 1994); they are endemic to (different) single Venezuelan mountains (Ptari-tepuí and Cerro Guaiquinima respectively) which are not covered in the present Key Area analysis but should in future be priorities for bird conservation. In Venezuela, *A. militaris* is found in the north, and in recent years has become very local and is under severe threat from habitat loss and trade (Collar *et al.* 1994); its addition will not have any major impact on the analysis, as populations occur in at least Morrocoy and Cuare (VE 11: Lentino and Goodwin 1991), Henri Pittier (VE 13: Lentino and Goodwin 1993) and Guatopo (VE 16: Lentino *et al.* 1993).

OLD RECORDS AND LITTLE-KNOWN BIRDS

Our knowledge of many Venezuelan threatened species suffers from the general paucity of recent records from all but a small number of regularly visited areas. Fortunately, these few Key Areas provide us with recent records (i.e. within the last 10–15 years) of most species. Exceptions to this are species known from old specimen records or chance observations (from which there can be no guarantees of a

Table 3. Top Key Areas in Venezuela. Those with the area number in bold currently have some form of protected status.

Key Area		No. of threatened spp.	Comments
02	El Tamá	5	The only confirmed area for *Grallaria chthonia* and *Amazilia distans*.
06	Sierras de la Culata y Nevada	3	Probably the most reliable and important known area for *Hapalopsittaca amazonina* and *Hemispingus goeringi*.
11	Morrocoy and Cuare	3	An important protected area for all three species known from it.
13	Henri Pittier	3	Probably one of the most important Key Areas for *Pauxi pauxi* and *Rallus wetmorei*.
21	Cerro Peonía	3	An old collecting locality, but one of only three areas with three of the threatened Cordillera de Caripe endemics.
22	Cerro Turumiquire	4	An old collecting locality, but one of only two areas with all four threatened Cordillera de Caripe endemics.
23	Cerro Negro and El Guácharo	5	Probably the most important known area for *Basileuterus griseiceps* and *Diglossa venezuelensis*.
25	Península de Paria	5	The only known area for *Hylonympha macrocerca* and *Myioborus pariae*, and upon which their survival is totally dependent.

current viable population), and in Venezuela these include the following.

Coturnicops notatus is represented by a single specimen collected c.1915.

Amazilia distans was first collected in 1954 since when there have been just five unconfirmed observations.

Thripophaga cherriei is known only from the type-locality on the upper Río Orinoco.

Clytoctantes alixii is known only (in Venezuela) from specimens collected in or before 1953 at a number of localities in the Sierra de Perijá.

Grallaria chthonia remains known from just the type-locality where the last specimens were collected in 1956.

Carduelis yarrellii is known only (in Venezuela) from two collecting localities on the south shore of Lago Valencia, although there is a possibility that these birds may have been escaped cage-birds (the main range is in north-east Brazil).

OUTLOOK

Each of the 26 Key Areas is a site of major importance for the conservation of threatened birds in Venezuela. These areas would, if each was adequately protected, help ensure the survival of all the country's threatened species, although the 16 areas with two or more species are perhaps the most efficient areas (currently known) in which to conserve Venezuela's threatened birds, with the eight Key Areas (listed in Table 3) embracing populations of three or more threatened species (collectively representing populations of 17 threatened birds), standing out as top priorities. These areas urgently require guaranteed protection.

The areas listed in Table 3 cover three (of the four) Venezuelan endemics known from just one Key Area; these are *Hylonympha macrocerca*, *Grallaria chthonia* and *Myioborus pariae*, the fourth being *Thripophaga cherriei*, the survival of which is dependent on the continued integrity of forest in the Río Capuana Key Area (VE 26). Although attention should be focused on all Venezuelan endemics, or species known from just one Key Area, individual species also require action. *Carduelis cucullatus*, for example, has been found in just two (protected) Key Areas, from only one of which are there recent records: as Venezuela is the centre of its distribution and abundance, other areas urgently need to be found and protected from trappers, and suggestions that the bird may occur in Terepaima and El Avila National Parks should be investigated.

All 26 Key Areas need some degree of conservation action if the populations of the threatened birds they harbour are to remain viable. For areas identified on the basis of old records (such as Sierra de Perijá, San Esteban, Lago Valencia, Cerro Peonía and Cerro Turumiquire) there is a need to locate and define the extent of suitable habitat within or in the vicinity of the Key Area, and for the species which define these areas surveys should aim to confirm their continued existence or assess the status of their populations. This is especially true in areas for *Amazilia distans*, *Grallaria chthonia* and *Thripophaga cherriei*, but also for species such as *Clytoctantes alixii* and *Carduelis yarrellii* (see 'Old records and little-known birds', above).

DATA SOURCES

The above introductory text and the Site Inventory (below) were compiled from information supplied by the late T. A. Andrews, P. Boesman, G. Kirwan, M. Lentino, A. Luy, J. L. Mateo and C. Sharpe, as well as from the following references.

Boesman, P. and Curson, J. (1995) Grey-headed Warbler *Basileuterus griseiceps* in danger of extinction? *Cotinga* 3: 35–39.

Collar, N. J., Gonzaga, L. P., Krabbe, N., Madroño Nieto, A., Naranjo, L. G., Parker, T. A. and Wege, D. C. (1992) *Threatened birds of the Americas: the ICBP/IUCN Red Data Book* (Third edition, part 2). Cambridge, U.K.: International Council for Bird Preservation.

Collar N. J., Crosby, M. J. and Stattersfield, A. J. (1994) *Birds to watch 2: the world list of threatened birds*. Cambridge, U.K.: BirdLife International (BirdLife Conservation Series no.4).

Collar, N. J., Wege, D. C. and Long, A. J. (in press) Patterns and causes of endangerment in the New World avifauna. In J. V. Remsen, ed. *Natural history and conservation of Neotropical birds*. American Ornithologists' Union (Orn. Monogr.)

Goodwin, M.-L. and Lentino, M. (1992) *Lista de las aves del Parque Nacional Yacambú, Estado Lara, Venezuela*. Caracas: Sociedad Conservacionista Audubon de Venezuela.

IUCN (1992) *Protected areas of the world: a review of national systems*, 4: *Nearctic and Neotropical*. Gland, Switzerland and Cambridge, U.K.: International Union for Conservation of Nature and Natural Resources.

Lentino, M. and Goodwin, M.-L. (1991) *Lista de las aves del Parque Nacional Morrocoy, Refugio de Fauna Silvestre de Cuare y areas aledañas, Estado Falcón, Venezuela*. Caracas: Sociedad Conservacionista Audubon de Venezuela.

Lentino, M. and Goodwin, M.-L. (1993) *Lista de las aves del Parque Nacional Henri Pittier (Rancho Grande),*

Estado Aragua, Venezuela. Caracas: Sociedad Conservacionista Audubon de Venezuela.

LENTINO, M., LUY, A. AND GOODWIN, M.-L. (1993) *Lista de las aves del Parque Nacional Guatopo, Estado Miranda, Venezuela.* Caracas: Sociedad Conservacionista Audubon de Venezuela.

MEYER DE SCHAUENSEE, R. AND PHELPS, W. (1978) *A guide to the birds of Venezuela.* Princeton: Princeton University Press.

PAYNTER, R. A., JR. (1982) *Ornithological gazetteer of Venezuela.* Second edition. Cambridge, Mass.: Museum of Comparative Zoology.

SILVA, J. L. AND STRAHL, S. D. (1991) Human impact on populations of chachalacas, guans, and curassows (Galliformes: Cracidae) in Venezuela. Pp.37–52 in J. G. Robinson and K. H. Redford, eds. *Neotropical wildlife use and conservation.* Chicago: University of Chicago Press.

STATTERSFIELD, A. J., CROSBY, M. J., LONG, A. J. AND WEGE, D. C. (in prep.) *A global directory of Endemic Bird Areas.* Cambridge, U.K.: BirdLife International (BirdLife Conservation Series).

SITE INVENTORY

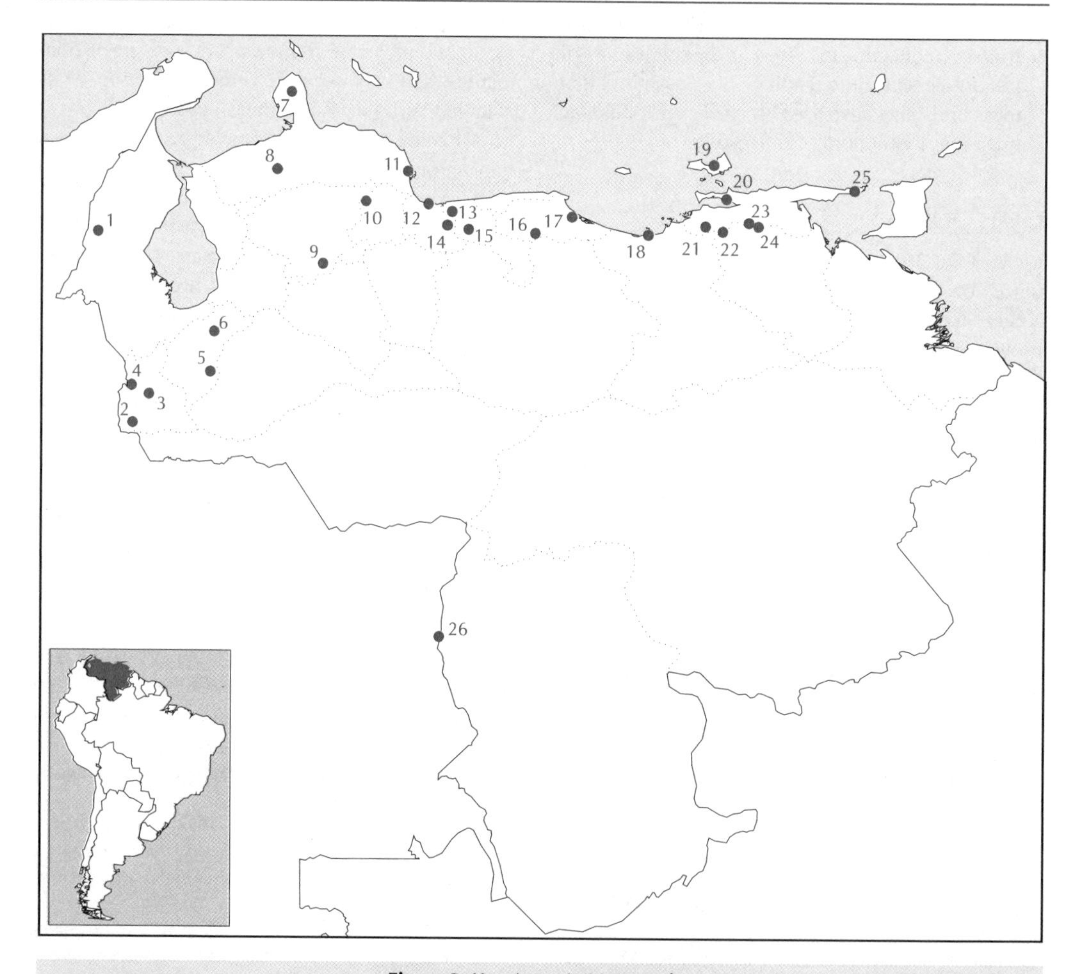

Figure 2. Key Areas in Venezuela.

01 Sierra de Perijá
02 El Tamá
03 Páramos del Batallón y La Negra
04 San Juan de Colón
05 Páramo de Aricagua
06 Sierras de la Culata y Nevada
07 Piedra Honda, Paraguaná
08 Las Veritas
09 Yacambú
10 Hato Jaguar
11 Morrocoy and Cuare
12 San Esteban
13 Henri Pittier
14 Lago de Valencia
15 Embalse de Taguaiguai
16 Guatopo
17 Laguna de Tacarigua
18 Puerto Píritu–Barcelona
19 Isla de Margarita
20 Península de Araya
21 Cerro Peonía
22 Cerro Turumiquire
23 Cerro Negro and El Guácharo
24 Teresén–Los Cumbres de San Bonifacio
25 Península de Paria
26 Río Capuana

Sierra de Perijá (Zulia)

VE 01

10°05′N 72°45′W
EBA B10

Sierra de Perijá National Park (IUCN category II), 295,288 ha

Centred at the northern end of the national park, on the eastern slope of the Sierra de Perijá between 500 and 1,500 m, this is an area of now-fragmented tropical and subtropical humid forest. The national park does not cover the lower-altitude eastern localities for the species listed below which may benefit from an extension of the park boundaries, although land clearance and cattle ranching even within the park suggest that there may be little suitable habitat to extend to (C. Sharpe *in litt.* 1995). Systematic searches are urgently required for the two threatened species in this area.

Pauxi pauxi	c.1953	Known from five localities in this area which is the main range of the subspecies *gilliardi*. Records in 1987, just over the border in Colombia, suggest the continued presence of the species which is however suffering from hunting (M. Pearman *in litt.* 1995).
Clytoctantes alixii	1953	This area embraces four of the five known Venezuelan localities.

El Tamá (Táchira)

VE 02

7°32′N 72°15′W
EBA B10

El Tamá National Park (IUCN category II), 139,000 ha

This extension of the Colombian East Andes includes localities such as Hacienda la Providencia, and areas on the periphery of the national park, thus embracing forest from tropical to temperate zones. The area is contiguous with Tamá National Park in Colombia, and in combination they represent an extremely important area. Deforestation is apparently progressing rapidly, although forest at Hacienda la Providencia (between 1,150 and 2,050 m) along the Río Chiquito was in pristine condition in 1990 (M. Pearman *in litt.* 1995).

Pauxi pauxi	1952	Mentioned for this locality.
Hapalopsittaca amazonina	1993	This is the only known area for the subspecies *amazonina* in Venezuela; most recently, 10 seen near the treeline in January 1993 (J. Curson *in litt.* 1993).
Amazilia distans	1988	Known with certainty from two specimens taken in 1954, although there have been several recent unconfirmed records from this area.
Grallaria chthonia	1956	Despite specific searches in September 1990, still known from only four specimens taken at Hacienda la Providencia in February 1955 and March 1956 (M. Pearman *in litt.* 1995).
Grallaricula cucullata	c.1988	The subspecies *venezuelana* is known only from one record at Hacienda la Providencia in 1956, and a recent sighting from elsewhere in the national park.

Páramos del Batallón y La Negra (Táchira)

VE 03

7°55′N 72°01′W
EBA B06

Páramos del Batallón y La Negra National Park (IUCN category II), 95,200 ha

Lying at the westernmost end of the Cordillera de Mérida, this area comprises upper subtropical and temperate-zone habitats and includes the Páramo Zumbador and Queniquea road. Forest in this area at altitudes suitable for the species listed here is being seriously depleted.

Hapalopsittaca amazonina	1981	A pair seen at 2,200 m along the Queniquea road.
Hemispingus goeringi	c.1989	Two recent records from the Páramo Zumbador.

San Juan de Colón (Táchira)

VE 04

Unprotected

8°02′N 72°16′W
EBA B06

This area, which is rapidly losing its native vegetation, includes the coffee plantations north of San Juan de Colón at c.500–1,000 m. It needs to be thoroughly searched for the presence of this species about which almost nothing is known. It is possible that, like other *Amazilia*, the species is not reliant on primary habitats (M. Lentino *in litt.* 1994), although this remains to be proved.

Amazilia distans	c.1990	Can apparently be found in coffee plantations at c.800 m.

Páramo de Aricagua (Mérida)

VE 05

Unprotected

8°13′N 71°08′W
EBA B06

Lying north of Aricagua, this area includes remains of temperate forest and páramo. Forest destruction, even in the temperate zone, has been widespread in this region.

Hemispingus goeringi	pre-1950	Mentioned for this locality.

Sierras de la Culata y Nevada (Mérida)

VE 06

Sierra de la Culata National Park (IUCN category II), 200,400 ha
Sierra Nevada National Park (IUCN category II), 276,446 ha

8°45′N 71°05′W
EBA B06

This area embraces the sierras north-east and south-east of Mérida, both of which are partially within national parks. Suitable habitats immediately east of Mérida remain unprotected. Habitat destruction in this region, even in the temperate zone, has been quite extensive.

Pauxi pauxi	1993	Known from along the San Isidro road within the Sierra Nevada National Park (P. Boesman *in litt.* 1995).
Coturnicops notatus	c.1915	One collected during June 1914 or 1916.
Hapalopsittaca amazonina	1993	Birds have been seen recently near La Mucuy along the 'Pico Humboldt trail', within the Sierra Nevada National Park (G. Kirwan *in litt.* 1993), and historically in the Sierra de la Culata National Park.
Hemispingus goeringi	1993	Well known and often recorded near La Mucuy along the 'Pico Humboldt trail' within the Sierra Nevada National Park (e.g. G. Kirwan *in litt.* 1993).

Piedra Honda, Paraguaná (Falcón)

VE 07

Piedra Honda (private reserve), area unknown

11°57′N 70°00′W
EBA B07

This small private reserve (the current status of which is unknown) on the Península de Paraguaná, managed (at least previously) by Fundación Bioma, comprises xerophytic scrubland dominated by cacti and thornbrush.

Amazona barbadensis	1992	30–40 birds present in the reserve.

VE 08

Las Veritas (Falcón)

10°55′N 70°12′W
EBA B07

Unprotected

Las Veritas (and environs) is south-west of Pedregal, in the area between the Ríos Pedregal and Japure. The local pet trade in this area is in urgent need of control, as this is the centre of abundance for this amazon.

Amazona barbadensis	1989	Up to 700 have been recorded in a roost near Las Veritas, and the species is apparently abundant in this area.

VE 09

Yacambú (Lara)

9°40′N 69°32′W
EBA B06

Yacambú National Park (IUCN category II), 14,580 ha

The national park is in south-east Lara at the north-east end of the Cordillera de Mérida. It is mountainous (550–2,200 m) and noted for its dense, tropical and subtropical forest (Goodwin and Lentino 1992), including a considerable area of cloud-forest (C. Sharpe *in litt.* 1995). A dam is being constructed near the park, the effects of which are as yet unknown.

Pauxi pauxi	1995	A number of recent records from the national park (e.g. H. Laidlow *per* P. Boesman *in litt.* 1995).
Laterallus levraudi	1993	Recently recorded from a man-made pool within the park (e.g. P. Boesman *in litt.* 1995).

VE 10

Hato Jaguar (Yaracuy)

c.10°30′N 68°55′W
EBA B06

Unprotected

Hato Jaguar is an active cattle ranch adjacent to the villages of Cararapito and Gusanilla in north-west Yaracuy (Silva and Strahl 1991). Primary humid cloud-forest (up to at least 850 m) covers 16,000 ha, with the remainder converted to grassland on which cattle ranching and agriculture are the principal activities (Silva and Strahl 1991). The area is sparsely populated, but both illegal sport and subsistence hunting are fairly common (Silva and Strahl 1991).

Pauxi pauxi	1985	Four birds recorded (giving a density of 8/km^2) during censuses at 650–850 m in July 1985 (Silva and Strahl 1991).

VE 11

Morrocoy and Cuare (Falcón)

10°54′N 68°19′W
EBA B07

Morrocoy National Park (IUCN category II), 32,090 ha
Cuare Faunal Refuge (IUCN category IV), 11,825 ha

These two adjacent protected areas comprise shallow coastal lagoons, marshes and mangrove swamps. Development pressures (and pollution), especially from the tourist industry, are a serious problem in this coastal area. Cuare refuge is Venezuela's only Ramsar site and the most important feeding area for Greater Flamingo *Phoenicopterus ruber* in the southern Caribbean (J. L. Mateo *in litt.* 1991). A record of *Pauxi pauxi* from this area (in c.1875) is almost certainly in error (M. Lentino *in litt.* 1994).

Rallus wetmorei	1986	Twelve specimens collected in the 1950s, but there has been just one recent record.
Laterallus levraudi	pre-1991	Recorded as 'casual' from this area (Lentino and Goodwin 1991).
Amazona barbadensis	1990	At least 35 birds are known to roost in a dry forest patch at the base of cliffs behind the village of Sanare (M. Pearman *in litt.* 1995).

San Esteban (Carabobo)

VE 12

San Esteban National Park (IUCN category II), 43,500 ha

10°28′N 68°01′W
EBAs B04,B07

San Esteban is on the coast of Carabobo, west of Caracas (and adjacent to Henri Pittier National Park, VE 13), where a number of localities east of Puerto Cabello support mangrove swamps and shallow lagoons, and inland the forest extends up to cloud-forest at 1,900 m. The mangroves and coastal wetland habitats in this region are under serious threat from development. Hunting pressure is high, as in Henri Pittier, and the level of protection is even lower than in that park (P. Boesman *in litt.* 1995).

Pauxi pauxi	c.1875	The type-locality, with just one subsequent (nineteenth century) record, but, with cloud-forest extending up to 1,900 m, the species' continued occurrence is to be expected (P. Boesman *in litt.* 1995).
Rallus wetmorei	1945	Eleven specimens collected during 1944 and 1945.

Henri Pittier (Aragua)

VE 13

Henri Pittier National Park (IUCN category II), 107,000 ha

10°22′N 67°41′W
EBAs B04,B07

Lying west of Caracas and adjacent to San Esteban National Park, this reserve embraces a number of coastal marshy areas as well as tropical and subtropical forest (and cloud-forest between 1,100 and 1,500 m: Silva and Strahl 1991). Coastal habitats in this area are under severe pressure from development projects, and the forest is subject to disturbance from adjacent roads and illegal hunting (see Silva and Strahl 1991).

Pauxi pauxi	1995	Occurs at low density (2 per km^2) in the national park, but is declining (e.g. Silva and Strahl 1991); regularly recorded (e.g. P. Boesman *in litt.* 1995).
Rallus wetmorei	1993	One pair seen in 1991 and 1993 (A. Fernandez Badillo *per* M. Lentino *in litt.* 1994).
Carduelis cucullatus	c.1990	Has been recorded in the park (Lentino and Goodwin 1993).

Lago de Valencia (Carabobo/Aragua)

VE 14

Unprotected

10°11′N 67°45′W
EBA B04

This large lake lies south of the main Cordillera de la Costa and east of Valencia. Historically, at least, it supported extensive wetlands. The lake is polluted by industrial waste, and the wetlands in irrigated areas are affected by pesticides. It is not known whether any suitable wetland habitat remains, or indeed if appropriate vegetation persists for *Carduelis yarrellii*.

Laterallus levraudi	1947	A number were collected in the vicinity, 1942–1947.
Carduelis yarrellii	1914	At least 24 were collected along the southern lake shore (El Trompillo) in 1914, although other specimens (from near Pirapira) may postdate these.

Embalse de Taguaiguai (Aragua)

VE 15

Unprotected

10°08′N 67°27′W
EBA B04

Also known as Bella Vista, this reservoir is c.15 km east of Lago de Valencia, and originally had dense fringing aquatic vegetation. Falling water-levels during the 1980s may have caused a reduction in available habitat, which in turn could be responsible for the disappearance of this threatened species (surveys are needed to confirm this).

Laterallus levraudi	1985	Up to 12 birds present throughout 1985, but not seen since.

VE 16

Guatopo (Miranda/Guárico)

10°05′N 66°30′W
EBA B04

Guatopo National Park (IUCN category II), 122,464 ha

This park, set south-east of Caracas at the easternmost end of the Cordillera de la Costa Central, covers both the tropical zone, characterized mainly by deciduous and semi-deciduous forests as well as xerophytic brush, and the subtropical zone, embracing the greater part of the park, with the dominant vegetation being high, dense evergreen rainforest (Lentino *et al.* 1993).

Pauxi pauxi	c.1988	A population is present in the national park.
Carduelis cucullatus	c.1988	A small population reputedly occurs in the national park, although its current status is unknown.

VE 17

Laguna de Tacarigua (Miranda)

10°18′N 65°59′W
EBA B04

Laguna de Tacarigua National Park (IUCN category II), 39,100 ha

A large, mangrove-fringed coastal lagoon east of Caracas, including, at the western end, San José de Río Chico. Lagoons and associated habitat are under severe pressure from tourist-related development.

Laterallus levraudi	1940	Six collected during September 1940.

VE 18

Puerto Píritu–Barcelona (Anzoátegui)

10°04′N 64°55′W
EBA B07

Unprotected

A strip of xerophytic vegetation along the coast and the road between the two towns (west of the Cordillera de Caripe). The capture of birds for the local pet trade is a serious problem which may be exacerbated by habitat degradation.

Amazona barbadensis	1991	Small numbers have repeatedly been found.

VE 19

Isla de Margarita (Nueva Esparta)

11°00′N 64°00′W
EBA B07

Cerro El Copey National Park (IUCN category II), 7,130 ha
Laguna de la Restinga National Park (IUCN category II), 18,862 ha

Covering 1,000 km^2, this Caribbean island lies c.30 km off the Araya peninsula. Localities included within this Key Area area are confined to the western third of the island. Burgeoning tourism and local trade in this amazon are both serious problems on the island.

Amazona barbadensis	1988	Up to 800 birds thought to be present on the island.

VE 20

Península de Araya (Sucre)

10°33′N 63°49′W
EBA B07

Unprotected

This peninsula, south of Isla de Margarita, parallels the coast on the northern side of the Golfo de Cariaco. The xerophytic vegetation on the peninsula is under severe pressure from overgrazing by goats and firewood-gathering by local people.

Amazona barbadensis	1986	There are few records from this area, although a population probably exists.

Cerro Peonía (Anzoátegui/Monagas/Sucre)

VE 21

Unprotected

10°11′N 64°07′W
EBA B03

Cerro Peonía is one of the highest peaks at the western end of the Cordillera de Caripe, and was historically covered in subtropical montane forest. The area is under severe human population pressure which has resulted in extensive degradation of the montane forest.

Campylopterus ensipennis	pre-1958	Mentioned for this locality.
Margarornis tatei	pre-1950	Mentioned for this locality.
Basileuterus griseiceps	1941	Two collected at 1,200 m in November 1941.

Cerro Turumiquire (Sucre/Monagas)

VE 22

Macizo Montañoso del Turumiquire Protective Zone (IUCN category V), 540,000 ha

10°07′N 63°52′W
EBA B03

At almost 2,600 m, Cerro Turumiquire is the highest mountain in the Cordillera de Caripe, and used to be covered in humid montane forest. The area includes historical collecting localities (on the northern slope) such as Carapas, Boca la Trinidad and Mirasol. Clearance for agriculture and pasture has been widespread, and little undisturbed forest remains.

Campylopterus ensipennis	c.1986	Recorded recently only from the 'hydraulic' reserve although there are numerous specimen records from the area.
Margarornis tatei	1963	The type- and another specimen were collected in 1925, with just five taken since then.
Diglossa venezuelensis	1963	Over 20 collected February–March 1963 (11 taken in 1925).
Basileuterus griseiceps	1963	Eight collected January–February 1963 (22 taken previously).

Cerro Negro and El Guácharo (Sucre/Monagas)

VE 23

El Guácharo National Park (IUCN category II), 62,700 ha

10°14′N 63°30′W
EBA B03

Cerro Negro is a high mountain within an important national park at the eastern end of the Cordillera de Caripe, 10 km north-west of Caripe. The area formerly supported extensive stands of palm-rich humid montane forest. The agricultural practices of local campesinos (forest clearance, repeated burnings and removal of understorey for coffee) have reduced suitable forest within this park to a small percentage of its former extent.

Campylopterus ensipennis	1995	Apparently one of the commoner birds, present all year (P. Boesman *in litt.* 1994, 1995).
Margarornis tatei	1994	Collected in 1943 but not seen again until August 1994, the first mainland record for 30 years (Boesman and Curson 1995).
Diglossa venezuelensis	1994	A population recently found at c.1,700 m, with one bird seen in April 1994 (Boesman and Curson 1995).
Basileuterus griseiceps	1995	A pair in 1993 was the first confirmed record since 1943; up to four seen regularly since (e.g. Boesman and Curson 1995, P. Boesman *in litt.* 1995).
Carduelis cucullatus	1898	Mentioned for this locality.

Teresén–Los Cumbres de San Bonifacio (Monagas)

VE 24
10°11′N 63°22′W
EBA B03

Unprotected

Los Cumbres de San Bonifacio is a small mountain range at the eastern end of the Cordillera de la Costa in northern Monagas state, and forms the border with Sucre. The area c.12 km east of Teresén is dominated by coffee plantations, but some partially degraded montane forest persists along tributaries of the Río Caripe (*Cotinga* 1994, 2: 30–31, also Boesman and Curson 1995). Any remnant forest patches in this area may well be important for the two species listed here and in urgent need of survey and protection.

Diglossa venezuelensis	1993	One seen on two consecutive days in December 1993 (Boesman and Curson 1995).
Basileuterus griseiceps	1993	One seen on two consecutive days in December 1993 (Boesman and Curson 1995).

Península de Paria (Sucre)

VE 25
10°40′N 62°00′W
EBA B03

Península de Paria National Park (IUCN category II), 37,500 ha

A large peninsula projecting east towards Trinidad. Montane tropical forest exists between c.400 and 1,200 m, with peaks such as Cerro Azul, Cerro Patao, Cerro El Olvido and Cerro Humo especially important. Changes in agricultural practices have led to increased degradation of forest within the park which embraces most of the forest on the peninsula. The southern slope of Cerro Humo, of critical importance for *Myioborus pariae*, currently lies outside the park but is gradually being incorporated (C. Sharpe *in litt.* 1995). A number of sources suggest that *Pauxi pauxi* may be present in the park, and this should perhaps be investigated (C. Sharpe *in litt.* 1995).

Campylopterus ensipennis	1994	Found to be common in available habitat on Cerro Humo, and at the eastern end of the peninsula (e.g. P. Boesman *in litt.* 1994).
Hylonympha macrocerca	1994	Recent records come from Cerro Humo and Cerro El Olvido where the species is locally common; up to seven per day on Cerro Humo (P. Boesman *in litt.* 1994) and 5–6 per day on Cerro El Olvido (C. Sharpe *in litt.* 1995).
Margarornis tatei	1994	Populations occur on a number of peaks; recent records come from Cerro Humo (e.g. P. Boesman *in litt.* 1994).
Diglossa venezuelensis	1989	Known in this area only from a number of recent records near Melenas.
Myioborus pariae	1994	This species reaches its maximum density on Cerro Humo, whence come the majority of records (e.g. P. Boesman *in litt.* 1994).

Río Capuana (Amazonas)

VE 26
4°42′N 67°50′W
EBA B11

Unprotected

The Río Capuana is a small tributary on the east bank of the upper Río Orinoco. The primary habitat in this area is lowland tropical forest, although there are evidently clearings along the river banks and small streams. There appears to be little habitat disturbance although the area lies just c.150 km south of Puerto Ayacucho which is rapidly developing as the trade and tourist centre for the region.

Thripophaga cherriei	1970	The type- and only known locality: two collected in February 1899 and four others in March 1970.

APPENDIX 1: Key Areas listed by species

INCLUDED here are all globally threatened species occurring in the Neotropics (as listed by Collar *et al.* 1992), with the Key Areas from which each has been recorded.

Black Tinamou *Tinamus osgoodi*
Colombia: 59.
Peru: 78, 79, 81.

Chocó Tinamou *Crypturellus kerriae*
Colombia: 17.
Panama: 10, 11.

Magdalena Tinamou *Crypturellus saltuarius*
Colombia: 05.

Kalinowski's Tinamou *Nothoprocta kalinowskii*
Peru: 34.

Taczanowski's Tinamou *Nothoprocta taczanowskii*
Peru: 70, 71, 73, 77, 85.

Lesser Nothura *Nothura minor*
Brazil: 061, 062, 096, 098, 104.

Dwarf Tinamou *Taoniscus nanus*
Brazil: 060, 062, 104.

Colombian Grebe *Podiceps andinus*
Colombia: 39, 40, 41, 44, 45.

Junín Grebe *Podiceps taczanowskii*
Peru: 59.

Townsend's Shearwater *Puffinus auricularis*
Mexico: 28, 29, 30.

Pink-footed Shearwater *Puffinus creatopus*
Chile: 13, 17.

Defilippe's Petrel *Pterodroma defilippiana*
Chile: 16, 17.

Dark-rumped Petrel *Pterodroma phaeopygia*
Ecuador: 50.

Guadalupe Storm-petrel *Oceanodroma macrodactyla*
Mexico: 01.

Peruvian Diving-petrel *Pelecanoides garnotii*
Chile: 09, 11.
Peru: 64.

Galápagos Cormorant *Nannopterum harrisi*
Ecuador: 50.

Brazilian Merganser *Mergus octosetaceus*
Argentina: 15, 16, 19.
Brazil: 058, 061, 072.
Paraguay: 07.

White-necked Hawk *Leucopternis lacernulata*
Brazil: 036, 056, 066, 068, 074, 076, 077, 078, 084, 088, 089, 092, 102, 103, 106, 113, 114, 115, 116, 120, 125, 128.

Grey-backed Hawk *Leucopternis occidentalis*
Ecuador: 03, 04, 05, 09, 11, 16, 17, 18, 19, 20, 21, 25, 32, 35, 36, 37, 38, 40.
Peru: 01, 07.

Crowned Eagle *Harpyhaliaetus coronatus*
Argentina: 03, 10, 12, 28, 37, 38.
Bolivia: 02, 11.
Brazil: 058, 061, 062, 090, 100.
Paraguay: 01, 06, 16, 19, 21.

Galápagos Hawk *Buteo galapagoensis*
Ecuador: 50.

Lesser-collared Forest-falcon *Micrastur buckleyi*
Ecuador: 12, 23.
Peru: 13, 53, 79.

Plumbeous Forest-falcon *Micrastur plumbeus*
Colombia: 53, 64, 65.
Ecuador: 02.

White-winged Guan *Penelope albipennis*
Peru: 24, 25, 26, 27.

Bearded Guan *Penelope barbata*
Ecuador: 32, 33, 34, 44, 45, 47.
Peru: 03, 05, 06, 07, 11, 28.

Chestnut-bellied Guan *Penelope ochrogaster*
Brazil: 017, 018, 057.

Cauca Guan *Penelope perspicax*
Colombia: 25, 51, 53, 54, 55.

Black-fronted Piping-guan *Pipile jacutinga*
Argentina: 15, 16, 19, 22.
Brazil: 056, 068, 074, 086, 088, 100, 105, 106, 107, 108, 109, 110, 112, 114, 116, 120, 126.
Paraguay: 06, 09, 10, 11, 12, 13, 14, 16.

Horned Guan *Oreophasis derbianus*
Guatemala: 01, 02, 03, 04, 05.
Mexico: 48, 56, 01.

Alagoas Curassow *Mitu mitu*
Brazil: 037.

Northern Helmeted Curassow *Pauxi pauxi*
Colombia: 02, 30, 38.
Venezuela: 01, 02, 06, 09, 10, 12, 13, 16.

Southern Helmeted Curassow *Pauxi unicornis*
Bolivia: 09, 17.
Peru: 52, 81.

Blue-billed Curassow *Crax alberti*
Colombia: 06, 08, 09.

Red-billed Curassow *Crax blumenbachii*
Brazil: 053, 056, 068, 074, 076, 077.

Wattled Curassow *Crax globulosa*
Brazil: 006, 007.
Colombia: 61.
Peru: 80.

Bearded Wood-partridge *Dendrortyx barbatus*
Mexico: 19, 20, 39.

Gorgeted Wood-quail *Odontophorus strophium*
Colombia: 32, 35, 42.

Austral Rail *Rallus antarcticus*
Argentina: 41.
Chile: 14.

Bogotá Rail *Rallus semiplumbeus*
Colombia: 39, 40, 41, 42, 44, 45, 47, 49.

Plain-flanked Rail *Rallus wetmorei*
Venezuela: 11, 12, 13.

Dot-winged Crake *Porzana spiloptera*
Argentina: 11, 32, 33, 34.

Rusty-flanked Crake *Laterallus levraudi*
Venezuela: 09, 11, 14, 15, 17.

Junín Rail *Laterallus tuerosi*
Peru: 59.

Rufous-faced Crake *Laterallus xenopterus*
Brazil: 062.
Paraguay: 04, 08.

Speckled Crake *Coturnicops notatus*
Argentina: 11, 27, 32.
Brazil: 101.
Colombia: 62.
Paraguay: 04.
Uruguay: 07.
Venezuela: 06.

Horned Coot *Fulica cornuta*
Argentina: 01, 02, 07.
Bolivia: 25.
Chile: 08, 10.

Piping Plover *Charadrius melodus*
No Key Areas identified.

Eskimo Curlew *Numenius borealis*
No Key Areas identified.

Olrog's Gull *Larus atlanticus*
Argentina: 11, 33, 35, 36, 43.

Peruvian Pigeon *Columba oenops*
Peru: 16, 17, 21, 27, 33.

Socorro Dove *Zenaida graysoni*
Mexico: 29.

Blue-eyed Ground-dove *Columbina cyanopis*
Brazil: 016.

Purple-winged Ground-dove *Claravis godefrida*
Argentina: 15, 16.
Brazil: 078, 086, 088, 103, 106, 126.
Paraguay: 09.

Tolima Dove *Leptotila conoveri*
Colombia: 27, 28, 58.

Ochre-bellied Dove *Leptotila ochraceiventris*
Ecuador: 19, 20, 21, 32, 36, 37, 38, 40, 41, 42, 43.
Peru: 01, 03, 27.

Blue-headed Quail-dove *Starnoenas cyanocephala*
Brazil: 016.

Glaucous Macaw *Anodorhynchus glaucus*
No Key Areas identified.

Hyacinth Macaw *Anodorhynchus hyacinthinus*
Bolivia: 05, 07.
Brazil: 010, 011, 015, 017, 019, 020, 023, 025, 057, 59.
Paraguay: 01, 03, 06.

Lear's Macaw *Anodorhynchus leari*
Brazil: 040.

Spix's Macaw *Cyanopsitta spixii*
Brazil: 039.

Blue-throated Macaw *Ara glaucogularis*
Bolivia: 03.

Red-fronted Macaw *Ara rubrogenys*
Bolivia: 09, 21, 22, 23.

Golden-capped Parakeet *Aratinga auricapilla*
Brazil: 042, 044, 051, 056, 063, 067, 068, 072.

Socorro Parakeet *Aratinga brevipes*
Mexico: 29.

Golden Parakeet *Guaruba guarouba*
Brazil: 009, 013, 022.

Golden-plumed Parakeet *Leptosittaca branickii*
Colombia: 24, 25, 26, 28, 53, 56, 59, 67, 68.
Ecuador: 06, 24, 27, 34, 45, 47.
Peru: 16, 17, 32, 35, 36, 50, 79.

Yellow-eared Parrot *Ognorhynchus icterotis*
Colombia: 28, 53, 56, 57, 59, 66.
Ecuador: 07.

Thick-billed Parrot *Rhynchopsitta pachyrhyncha*
Mexico: 02, 04, 05, 06, 08, 24, 27.

Maroon-fronted Parrot *Rhynchopsitta terrisi*
Mexico: 12, 13, 14.

White-necked Parakeet *Pyrrhura albipectus*
Ecuador: 23, 47, 48.

Flame-winged Parakeet *Pyrrhura calliptera*
Colombia: 30, 35, 38, 39, 48, 49, 50.

Blue-chested Parakeet *Pyrrhura cruentata*
Brazil: 044, 051, 054, 055, 056, 067, 068, 074, 075, 076, 077, 084, 087.

El Oro Parakeet *Pyrrhura orcesi*
Ecuador: 25, 32.

Rufous-fronted Parakeet *Bolborhynchus ferrugineifrons*
Colombia: 24, 25, 26.

Yellow-faced Parrotlet *Forpus xanthops*
Peru: 15, 16, 17, 21, 33.

Brown-backed Parrotlet *Touit melanonota*
Brazil: 084, 086, 088, 092, 095, 103, 111, 113, 115, 116.

Spot-winged Parrotlet *Touit stictoptera*
Colombia: 62.

Ecuador: 12, 14, 22, 23, 46.
Peru: 07, 23.

Golden-tailed Parrotlet *Touit surda*
Brazil: 029, 035, 036, 037, 046, 051, 054, 055, 056, 074, 076, 078, 084, 086, 087, 088, 099, 106, 107, 109, 114, 116.

Rusty-faced Parrot *Hapalopsittaca amazonina*
Colombia: 22, 48, 49, 56, 59.
Venezuela: 02, 03, 06.

Fuertes's Parrot *Hapalopsittaca fuertesi*
Colombia: 24, 26.

Red-faced Parrot *Hapalopsittaca pyrrhops*
Ecuador: 22, 27, 33, 34, 45, 47.
Peru: 06.

Yellow-shouldered Amazon *Amazona barbadensis*
Venezuela: 07, 08, 11, 18, 19, 20.

Red-tailed Amazon *Amazona brasiliensis*
Brazil: 111, 112, 113, 114, 115, 116, 120, 123, 124.

Yellow-headed Amazon *Amazona oratrix*
Belize: 01.
Mexico: 15, 16, 17, 18, 21, 25, 34.

Red-spectacled Amazon *Amazona pretrei*
Argentina: 16, 21, 26.
Brazil: 132, 133, 134, 135, 136, 137, 138, 140, 141, 142.
Paraguay: 09.

Red-browed Amazon *Amazona rhodocorytha*
Brazil: 037, 055, 056, 068, 073, 074, 075, 076, 077, 078, 084, 087, 100.

Vinaceous Amazon *Amazona vinacea*
Argentina: 15, 18, 20, 21.
Brazil: 063, 064, 067, 073, 097, 099, 109, 110, 114, 117, 118, 119, 121, 127, 129, 131, 132, 133, 134, 135, 136, 137, 138.
Paraguay: 01, 03, 06, 07, 09, 11, 12, 14.

Green-cheeked Amazon *Amazona viridigenalis*
Mexico: 15, 16, 17, 18.

Blue-bellied Parrot *Triclaria malachitacea*
Argentina: 15, 16.
Brazil: 076, 078, 080, 084, 090, 100, 106, 108, 109, 110, 111, 114, 116, 137, 139.

Banded Ground-cuckoo *Neomorphus radiolosus*
Colombia: 52, 53, 65.
Ecuador: 02.

Long-whiskered Owlet *Xenoglaux loweryi*
Peru: 16, 22.

White-winged Nightjar *Caprimulgus candicans*
Bolivia: 02.
Brazil: 061.

Cayenne Nightjar *Caprimulgus maculosus*
French Guiana: 01, 02.

Sickle-winged Nightjar *Eleothreptus anomalus*
Argentina: 14, 15, 24, 26, 30, 31.
Brazil: 062, 098, 104, 122.

White-chested Swift *Cypseloides lemosi*
Colombia: 55.

Hook-billed Hermit *Glaucis dohrnii*
Brazil: 055, 056, 074, 077.

White-tailed Sabrewing *Campylopterus ensipennis*
Venezuela: 21, 22, 23, 25.

Short-crested Coquette *Lophornis brachylopha*
Mexico: 43.

Coppery Thorntail *Popelairia letitiae*
No Key Areas identified.

Mexican Woodnymph *Thalurania ridgwayi*
Mexico: 22, 23, 26, 27.

Sapphire-bellied Hummingbird *Lepidopyga lilliae*
Colombia: 03, 04.

Mangrove Hummingbird *Amazilia boucardi*
Costa: Rica 05, 06, 07, 08, 10, 12, 13.

Chestnut-bellied Hummingbird *Amazilia castaneiventris*
Colombia: 07, 31, 33, 37.

Táchira Emerald *Amazilia distans*
Venezuela: 02, 04.

Honduran Emerald *Amazilia lucia*
Honduras: 02, 03.

Oaxaca Hummingbird *Eupherusa cyanophrys*
Mexico: 50, 51.

White-tailed Hummingbird *Eupherusa poliocerca*
Mexico: 42, 43, 46.

Scissor-tailed Hummingbird *Hylonympha macrocerca*
Venezuela: 25.

Purple-backed Sunbeam *Aglaeactis aliciae*
Peru: 31, 33.

Black Inca *Coeligena prunellei*
Colombia: 35, 36, 42, 43.

Juan Fernández Firecrown *Sephanoides fernandensis*
Chile: 17.

Royal Sunangel *Heliangelus regalis*
Peru: 07, 17, 23.

Turquoise-throated Puffleg *Eriocnemis godini*
No Key Areas identified.

Colourful Puffleg *Eriocnemis mirabilis*
Colombia: 53.

Black-breasted Puffleg *Eriocnemis nigrivestis*
Ecuador: 08, 10.

Hoary Puffleg *Haplophaedia lugens*
Colombia: 65, 66.
Ecuador: 07, 09.

Violet-throated Metaltail *Metallura baroni*
Ecuador: 27.

Neblina Metaltail *Metallura odomae*
Ecuador: 44, 47.
Peru: 06.

Grey-bellied Comet *Taphrolesbia griseiventris*
Peru: 11, 12, 46, 47.

Marvellous Spatuletail *Loddigesia mirabilis*
Peru: 18, 19, 20, 23.

Chilean Woodstar *Eulidia yarrellii*
Chile: 02, 03, 04, 05.
Peru: 89.

Esmeraldas Woodstar *Acestrura berlepschi*
Ecuador: 19

Little Woodstar *Acestrura bombus*
Ecuador: 07, 11, 12, 18, 19, 22, 29, 37, 38, 41, 46, 47.
Peru: 28, 33, 50.

Glow-throated Hummingbird *Selasphorus ardens*
Panama: 04, 05, 06.

Eared Quetzal *Euptilotis neoxenus*
Mexico: 02, 03, 04, 06, 07, 08, 24.

Keel-billed Motmot *Electron carinatum*
Costa: Rica 01, 02, 03.
Belize: 02.
Guatemala: 06.
Honduras: 01.
Mexico: 48.

Coppery-chested Jacamar *Galbula pastazae*
Colombia: 69.
Ecuador: 12, 13, 22, 23, 47.

Three-toed Jacamar *Jacamaralcyon tridactyla*
Brazil: 067, 068, 081, 082, 083, 085.

White-mantled Barbet *Capito hypoleucus*
Colombia: 07, 09, 14, 15, 23.

Yellow-browed Toucanet *Aulacorhynchus huallagae*
Peru: 32, 36.

Helmeted Woodpecker *Dryocopus galeatus*
Argentina: 15, 16, 17, 22, 23.
Brazil: 108, 109, 116, 126.
Paraguay: 06, 08, 09, 10, 12, 13.

Imperial Woodpecker *Campephilus imperialis*
Mexico: 04, 05, 06, 24.

Moustached Woodcreeper *Xiphocolaptes falcirostris*
Brazil: 025, 026, 047, 063, 064.

Royal Cinclodes *Cinclodes aricomae*
Peru: 72, 73.

White-bellied Cinclodes *Cinclodes palliatus*
Peru: 59, 60, 61, 62.

Masafuera Rayadito *Aphrastura masafuerae*
Chile: 18.

White-browed Tit-spinetail *Leptasthenura xenothorax*
Peru: 72, 73, 74, 75, 76.

Chestnut-throated Spinetail *Synallaxis cherriei*
Brazil: 010, 011, 015.
Ecuador: 15.

Apurímac Spinetail *Synallaxis courseni*
Peru: 71.

Plain Spinetail *Synallaxis infuscata*
Brazil: 032, 034, 035, 036.

Hoary-throated Spinetail *Synallaxis kollari*
Brazil: 001, 002.
Guyana: 01.

Blackish-headed Spinetail *Synallaxis tithys*
Ecuador: 19, 20, 31, 37, 39, 43.
Peru: 01

Russet-bellied Spinetail *Synallaxis zimmeri*
Peru: 41, 42, 43.

Austral Canastero *Asthenes anthoides*
Argentina: 38, 39, 42.
Chile: 12, 15.

Berlepsch's Canastero *Asthenes berlepschi*
Bolivia: 13.

Pale-tailed Canastero *Asthenes huancavelicae*
Peru: 38, 39, 40, 44, 48, 62, 66, 68, 69.

Cipó Canastero *Asthenes luizae*
Brazil: 066.

Russet-mantled Softtail *Thripophaga berlepschi*
Peru: 06, 16, 18, 20, 35.

Orinoco Softtail *Thripophaga cherriei*
Venezuela: 26.

Striated Softtail *Thripophaga macroura*
Brazil: 046, 051, 076, 084.

White-throated Barbtail *Margarornis tatei*
Venezuela: 21, 22, 23, 25.

Alagoas Foliage-gleaner *Philydor novaesi*
Brazil: 035.

Henna-hooded Foliage-gleaner *Hylocryptus erythrocephalus*
Ecuador: 19, 20, 37, 38, 39, 40, 41, 42, 43.
Peru: 01, 25, 27.

Rufous-necked Foliage-gleaner *Syndactyla ruficollis*
Ecuador: 32, 37, 38, 41, 42, 43, 44.
Peru: 01, 02, 03, 05, 10, 11, 28.

Great Xenops *Megaxenops parnaguae*
Brazil: 024, 025, 028, 033, 042, 044, 047, 049, 059, 063.

Bolivian Recurvebill *Simoxenops striatus*
Bolivia: 09, 14, 17.

White-bearded Antshrike *Biatas nigropectus*
Argentina: 15, 16, 20.
Brazil: 084, 086, 088, 109, 126.

Recurve-billed Bushbird *Clytoctantes alixii*
Colombia: 07, 08, 09.
Venezuela: 01.

Rondônia Bushbird *Clytoctantes atrogularis*
Brazil: 014.

Speckled Antshrike *Xenornis setifrons*
Colombia: 17.
Panama: 07, 08, 09, 10.

Bicoloured Antvireo *Dysithamnus occidentalis*
Colombia: 52, 53.
Ecuador: 12, 14.

Plumbeous Antvireo *Dysithamnus plumbeus*
Brazil: 067, 068, 076, 077, 078.

Rio de Janeiro Antwren *Myrmotherula fluminensis*
No Key Areas identified.

Ashy Antwren *Myrmotherula grisea*
Bolivia: 04, 09, 12, 14, 17.

Alagoas Antwren *Myrmotherula snowi*
Brazil: 035.

Ash-throated Antwren *Herpsilochmus parkeri*
Peru: 23.

Pectoral Antwren *Herpsilochmus pectoralis*
Brazil: 021, 038, 041, 044.

Black-hooded Antwren *Formicivora erythronotos*
Brazil: 094.

Narrow-billed Antwren *Formicivora iheringi*
Brazil: 048, 050, 065.

Restinga Antwren *Formicivora littoralis*
Brazil: 091, 093.

Yellow-rumped Antwren *Terenura sharpei*
Bolivia: 12, 16.
Peru: 82.

Orange-bellied Antwren *Terenura sicki*
Brazil: 030, 035, 036.

Rio Branco Antbird *Cercomacra carbonaria*
Brazil: 003, 004.
Guyana: 01.

Fringe-backed Fire-eye *Pyriglena atra*
Brazil: 043.

Slender Antbird *Rhopornis ardesiaca*
Brazil: 048, 050.

Grey-headed Antbird *Myrmeciza griseiceps*
Ecuador: 36, 37, 38, 43.
Peru: 01, 03, 05.

Scalloped Antbird *Myrmeciza ruficauda*
Brazil: 031, 032, 034, 035, 036, 075, 076.

White-masked Antbird *Pithys castanea*
No Key Areas identified.

Rufous-fronted Antthrush *Formicarius rufifrons*
Peru: 79, 80, 81.

Moustached Antpitta *Grallaria alleni*
Colombia: 26, 59.

Táchira Antpitta *Grallaria chthonia*
Venezuela: 02.

Giant Antpitta *Grallaria gigantea*
Colombia: 56.
Ecuador: 07, 08, 13.

Brown-banded Antpitta *Grallaria milleri*
Colombia: 25, 26.

Bicoloured Antpitta *Grallaria rufocinerea*
Colombia: 22, 25, 26, 28, 56.

Hooded Antpitta *Grallaricula cucullata*
Colombia: 53, 59.
Venezuela: 02

Stresemann's Bristlefront *Merulaxis stresemanni*
Brazil: 052.

Brasília Tapaculo *Scytalopus novacapitalis*
Brazil: 062, 066, 071, 072.

Bahia Tapaculo *Scytalopus psychopompus*
Brazil: 045, 052.

Shrike-like Cotinga *Laniisoma elegans*
Brazil: 051, 067, 071, 078, 084, 086, 087, 090, 092, 100, 103, 106, 107, 116.

Grey-winged Cotinga *Tijuca condita*
Brazil: 088, 090.

Black-headed Berryeater *Carpornis melanocephalus*
Brazil: 035, 051, 055, 056, 074, 076, 077, 080, 106, 108, 109, 111, 112, 113, 114, 116, 120, 123.

White-cheeked Cotinga *Zaratornis stresemanni*
Peru: 38, 44, 55, 56, 57, 58, 67.

Buff-throated Purpletuft *Iodopleura pipra*
Brazil: 029, 035, 051, 084, 086, 088, 094, 102, 103, 113.

Kinglet Cotinga *Calyptura cristata*
No Key Areas identified.

Cinnamon-vented Piha *Lipaugus lanioides*
Brazil: 051, 067, 068, 071, 076, 078, 080, 086, 088, 102, 109, 110, 114, 120.

Banded Cotinga *Cotinga maculata*
Brazil: 054, 055, 056, 068, 074, 076, 077.

White-winged Cotinga *Xipholena atropurpurea*
Brazil: 029, 031, 035, 036, 037, 053, 054, 055, 056, 074, 076, 077, 084.

Yellow-billed Cotinga *Carpodectes antoniae*
Costa: Rica 07, 08, 09, 10, 11, 12, 13, 14.
Panama: 01.

Bare-necked Umbrellabird *Cephalopterus glabricollis*
Costa: Rica 01, 02, 03, 04.
Panama: 02, 03.

Golden-crowned Manakin *Pipra vilasboasi*
Brazil: 012.

Black-capped Manakin *Piprites pileatus*
Argentina: 20.
Brazil: 086, 097, 100, 137.

Ash-breasted Tit-tyrant *Anairetes alpinus*
Bolivia: 15.
Peru: 38, 44, 45, 72, 73, 75.

Dinelli's Doradito *Pseudocolopteryx dinellianus*
Argentina: 08, 11, 13.
Paraguay: 13, 19.

Rufous-sided Pygmy-tyrant *Euscarthmus rufomarginatus*
Bolivia: 01, 05.
Brazil: 005, 012, 016, 062, 069, 096.
Paraguay: 02.
Surinam: 01.

Long-tailed Tyrannulet *Phylloscartes ceciliae*
Brazil: 035, 036.

Antioquia Bristle-tyrant *Phylloscartes lanyoni*
Colombia: 09, 15, 23.

São Paulo Tyrannulet *Phylloscartes paulistus*
Argentina: 15, 23.
Brazil: 084, 089, 102, 103, 106, 108, 109, 114, 116, 126, 128.
Paraguay: 06, 08, 09, 16.

Minas Gerais Tyrannulet *Phylloscartes roquettei*
Brazil: 064.

Cinnamon-breasted Tody-tyrant *Hemitriccus cinnamomeipectus*
Ecuador: 48.
Peru: 07, 16, 22.

Fork-tailed Pygmy-tyrant *Hemitriccus furcatus*
Brazil: 051, 084, 086, 102, 103, 105, 106.

Buff-breasted Tody-tyrant *Hemitriccus mirandae*
Brazil: 026, 027, 036.

Kaempfer's Tody-tyrant *Hemitriccus kaempferi*
Brazil: 128, 130.

Buff-cheeked Tody-flycatcher *Todirostrum senex*
Brazil: 008.

Russet-winged Spadebill *Platyrinchus leucoryphus*
Argentina: 15.
Brazil: 078, 084, 092, 094, 102, 103, 108, 109, 110, 112, 116, 120, 125, 126, 128.
Paraguay: 06, 08, 17.

Pacific Royal Flycatcher *Onychorhynchus occidentalis*
Ecuador: 11, 16, 17, 19, 20, 21, 25, 26, 32.
Peru: 01.

Grey-breasted Flycatcher *Lathrotriccus griseipectus*
Ecuador: 07, 11, 17, 19, 20, 21, 25, 26, 32, 35, 37, 38, 39, 41, 42, 43.
Peru: 01, 02, 07, 08.

White-tailed Shrike-tyrant *Agriornis andicola*
Argentina: 07.
Bolivia: 24.
Chile: 01.
Ecuador: 30, 47.
Peru: 10, 34, 44, 50, 77, 86.

Strange-tailed Tyrant *Yetapa risora*
Argentina: 14, 16, 24, 27, 28.
Paraguay: 19, 20, 22.
Uruguay: 06.

Ochraceous Attila *Attila torridus*
Ecuador: 04, 11, 16, 17, 19, 25, 32, 34, 37, 38.
Peru: 01.

Slaty Becard *Pachyramphus spodiurus*
Ecuador: 05, 11, 16, 25, 31, 35.
Peru: 01, 08, 15.

Peruvian Plantcutter *Phytotoma raimondii*
Peru: 29, 30, 37.

Ochre-breasted Pipit *Anthus nattereri*
Argentina: 25.
Brazil: 098, 134, 138.
Paraguay: 18, 23.

Rufous-throated Dipper *Cinclus schulzi*
Argentina: 03, 04, 05, 07.
Bolivia: 26, 27.

Navas's Wren *Hylorchilus navai*
Mexico: 41, 53

Sumichrast's Wren *Hylorchilus sumichrasti*
Mexico: 40, 45.

Apolinar's Wren *Cistothorus apolinari*
Colombia: 38, 39, 40, 41, 42, 44, 45, 47.

Niceforo's Wren *Thryothorus nicefori*
Colombia: 34.

Socorro Mockingbird *Mimodes graysoni*
Mexico: 29.

Floreana Mockingbird *Nesomimus trifasciatus*
Ecuador: 50.

Unicoloured Thrush *Turdus haplochrous*
Bolivia: 02, 03, 04, 05.

Guadalupe Junco *Junco insularis*
Mexico: 01.

Sierra Madre Sparrow *Xenospiza baileyi*
Mexico: 36, 37, 38.

Slender-billed Finch *Xenospingus concolor*
Chile: 02, 03, 04, 05.
Peru: 63, 65, 87.

Grey-winged Inca-finch *Incaspiza ortizi*
Peru: 04, 09, 21.

Plain-tailed Warbling-finch *Poospiza alticola*
Peru: 21, 32, 35, 38, 44, 45.

Tucumán Mountain-finch *Poospiza baeri*
Argentina: 03, 04, 07, 09.

Cinereous Warbling-finch *Poospiza cinerea*
Brazil: 061, 062, 066, 069, 070.

Cochabamba Mountain-finch *Poospiza garleppi*
Bolivia: 18, 19, 20.

Rufous-breasted Warbling-finch *Poospiza rubecula*
Peru: 12, 40, 42, 44, 56, 57.

Temminck's Seedeater *Sporophila falcirostris*
Argentina: 15, 16.
Brazil: 053, 084, 086, 090, 102, 103, 111, 112, 113, 115, 116.
Paraguay: 05.

Buffy-throated Seedeater *Sporophila frontalis*
Argentina: 15, 22.
Brazil: 078, 084, 086, 088, 089, 092, 102, 103, 105, 106, 107, 109, 111, 112, 113, 114, 115, 116, 123, 129.

Rufous-rumped Seedeater *Sporophila hypochroma*
Argentina: 14, 24, 26, 28, 31, 32.
Bolivia: 02, 05, 06, 10.
Brazil: 019, 061.
Paraguay: 06, 19, 22.

Tumaco Seedeater *Sporophila insulata*
Colombia: 63.

Hooded Seedeater *Sporophila melanops*
No Key Areas identified.

Black-and-tawny Seedeater *Sporophila nigrorufa*
Bolivia: 05, 06, 08.
Brazil: 019

Marsh Seedeater *Sporophila palustris*
Argentina: 25, 26, 27, 28, 29, 31, 32.
Brazil: 017, 019, 061.
Paraguay: 06, 08, 22.
Uruguay: 04, 05.

Entre Ríos Seedeater *Sporophila zelichi*
Argentina: 28, 30, 31.

Mangrove Finch *Camarhynchus heliobates*
Ecuador: 50.

Yellow-headed Brush-finch *Atlapetes flaviceps*
Colombia: 28, 57.

Pale-headed Brush-finch *Atlapetes pallidiceps*
Ecuador: 28, 29, 41.

Tanager-finch *Oreothraupis arremonops*
Colombia: 11, 21, 52, 53, 66.
Ecuador: 07.

Yellow Cardinal *Gubernatrix cristata*
Argentina: 12, 14, 28, 29, 30, 31, 34, 37, 40.
Uruguay: 02,03.

Rufous-bellied Saltator *Saltator rufiventris*
Argentina: 03, 06.
Bolivia: 18, 19, 20, 24, 26, 27.

Cone-billed Tanager *Conothraupis mesoleuca*
No Key Areas identified.

Yellow-green Bush-tanager *Chlorospingus flavovirens*
Colombia: 52, 65.
Ecuador: 02.

Slaty-backed Hemispingus *Hemispingus goeringi*
Venezuela: 03, 05, 06.

Cherry-throated Tanager *Nemosia rourei*
Brazil: 079.

Gold-ringed Tanager *Buthraupis aureocincta*
Colombia: 19, 20, 21.

Golden-backed Mountain-tanager *Buthraupis aureodorsalis*
Peru: 32, 35, 49, 51.

Black-and-gold Tanager *Buthraupis melanochlamys*
Colombia: 10, 19, 20, 21.

Masked Mountain-tanager *Buthraupis wetmorei*
Colombia: 56.
Ecuador: 06, 12, 22, 24, 47.
Peru: 06.

Orange-throated Tanager *Wetmorethraupis sterrhopteron*
Ecuador: 49.
Peru: 13, 14.

Multicoloured Tanager *Chlorochrysa nitidissima*
Colombia: 20, 25, 51, 52, 53.

Azure-rumped Tanager *Tangara cabanisi*
Guatemala: 01.
Mexico: 56, 57, 58.

Seven-coloured Tanager *Tangara fastuosa*
Brazil: 031, 032, 034, 035, 036, 037.

Green-capped Tanager *Tangara meyerdeschauenseei*
Peru: 83, 84.

Black-backed Tanager *Tangara peruviana*
Brazil: 084, 091, 092, 093, 094, 105, 106, 107, 112, 113, 115, 116.

Sira Tanager *Tangara phillipsi*
Peru: 52.

Scarlet-breasted Dacnis *Dacnis berlepschi*
Colombia: 64, 65.
Ecuador: 01, 02, 11.

Turquoise Dacnis *Dacnis hartlaubi*
Colombia: 42, 43, 46, 51.

Black-legged Dacnis *Dacnis nigripes*
Brazil: 084, 086, 087, 088, 090, 092, 094, 102, 109, 110, 115, 116, 129.

Venezuelan Flowerpiercer *Diglossa venezuelensis*
Venezuela: 22, 23, 24, 25.

Golden-cheeked Warbler *Dendroica chrysoparia*
El Salvador: 01.
Guatemala: 05, 06.
Honduras: 04, 05.
Mexico: 52, 54, 55.

Black-polled Yellowthroat *Geothlypis speciosa*
Mexico: 31, 32, 33, 35.

Paria Redstart *Myioborus pariae*
Venezuela: 25.

Grey-headed Warbler *Basileuterus griseiceps*
Venezuela: 21, 22, 23, 24.

Tamarugo Conebill *Conirostrum tamarugense*
Chile: 02, 03, 04, 05, 06, 07.
Peru: 86, 88.

Black-capped Vireo *Vireo atricapillus*
Mexico: 09, 10, 11, 22, 25, 26, 27.

San Andrés Vireo *Vireo caribaeus*
Colombia: 01.

Baudó Oropendola *Psarocolius cassini*
Colombia: 16, 17, 18.

Selva Cacique *Cacicus koepckeae*
Peru: 54.

Saffron-cowled Blackbird *Xanthopsar flavus*
Argentina: 14, 24, 25, 27, 28, 31.
Brazil: 137, 138.
Paraguay: 09, 15.
Uruguay: 04, 05.

Pampas Meadowlark *Sturnella militaris*
Argentina: 34.
Brazil: 128.
Uruguay: 01,05.

Red-bellied Grackle *Hypopyrrhus pyrohypogaster*
Colombia: 08, 09, 11, 12, 13, 20, 24, 25, 26, 28, 59, 60.

Forbes's Blackbird *Curaeus forbesi*
Brazil: 035, 036, 068.

Red Siskin *Carduelis cucullatus*
Colombia: 29.
Venezuela: 13, 16, 23

Saffron Siskin *Carduelis siemiradskii*
Ecuador: 19, 20, 39.
Peru: 01.

Yellow-faced Siskin *Carduelis yarrellii*
Brazil: 024, 029, 031, 033, 034, 035, 036.
Venezuela: 14.

White-throated Jay *Cyanolyca mirabilis*
Mexico: 42, 43, 46, 51.

Dwarf Jay *Cyanolyca nana*
Mexico: 44, 47, 49.

APPENDIX 2: **Distribution of Key Areas within Endemic Bird Areas**

THE following list includes every Key Area found within each of the Neotropical Endemic Bird Areas (EBAs). The location of EBAs is shown in Figure 1 of each of the individual country treatments.

A02 Guadalupe Island
Mexico: 01.

A03 Baja California
No Key Areas.

A04 Sierra Madre Occidental
Mexico: 02, 03, 04, 05, 06, 07, 08, 24.

A05 North-west Mexican Pacific slope
Mexico: 21, 22, 23, 25, 26, 27.

A06 Sierra Madre Oriental
Mexico: 09, 10, 11, 12, 13, 14.

A07 North-east Mexican Gulf slope
Mexico: 15, 16, 17, 18.

A08 Central Mexican marshes
Mexico: 31, 32, 33, 35.

A09 Yucatán peninsula
No Key Areas

A10 Socorro Island
Mexico: 29.

A11 Central Mexican highlands
Mexico: 19, 20, 26, 27, 36, 37, 38, 39, 44, 47, 49.

A12 Sierra Madre del Sur
Mexico: 42, 43, 46, 50, 51.

A13 Isthmus de Tehuantepec
No Key Areas

A14 North Central American highlands
Guatemala: 01, 02, 03, 04, 05.
Honduras: 04, 05.
Mexico: 52, 54, 55, 56, 57, 58.

A15 North Central American Pacific slope
No Key Areas

A16 Central American Caribbean slope
Costa Rica: 04.

A17 South Central American Pacific slope
Costa Rica: 05, 06, 07, 08, 09, 10, 11, 12, 13, 14.
Panama: 01, 06.

A18 Costa Rica and Panama highlands
Costa Rica: 01, 02, 03, 04.
Panama: 02, 03, 04, 05.

A19 Darién and Urabá lowlands
Panama: 07, 08, 09, 10, 11.

A20 East Panama and Darién highlands
No Key Areas

A21 Cocos Island
No Key Areas.

A27 Balsas drainage
No Key Areas.

B01 Guiana shield
French Guiana: 01, 02.

B02 Tepuís
No Key Areas.

B03 Cordillera de Caripe and the Paria peninsula
Venezuela: 21, 22, 23, 24, 25.

B04 Cordillera de la Costa Central
Venezuela: 12, 13, 14, 15, 16, 17.

B05 Llanos
No Key Areas.

B06 Cordillera de Mérida
Venezuela: 03, 04, 05, 06, 09, 10.

B07 Caribbean dry zone
Colombia: 02, 03, 04.
Venezuela: 07, 08, 11, 12, 13, 18, 19, 20.

B08 Santa Marta mountains
No Key Areas.

B09 Nechí lowlands
Colombia: 05, 06, 07, 08, 09, 14, 15, 23.

B10 East Andes of Colombia
Colombia: 29, 30, 31, 32, 33, 34, 35, 36, 37, 38, 39, 40, 41, 42, 43, 44, 45, 46, 47, 48, 49, 50, 60.
Venezuela: 01, 02.

B11 Upper Río Negro and Orinoco white-sand forest
No Key Areas.

B12 Subtropical inter-Andean Colombia
Colombia: 09, 10, 12, 13, 24, 25, 26, 27, 28, 51, 53, 54, 55, 57, 58, 59.

B13 Dry inter-Andean valleys
No Key Areas.

B14 Chocó and Pacific-slope Andes
Colombia: 08, 11, 19, 20, 21, 52, 53, 64, 65, 66.
Ecuador: 01, 02, 04, 05, 07, 09, 11.

B16 Islas Galápagos
Ecuador: 50.

B17 North Central Andean forests
Colombia: 22, 24, 25, 26, 28, 56, 67, 68.
Ecuador: 13, 15.

B18 East Andes of Ecuador and northern Peru
Colombia: 69.
Ecuador: 12, 14, 22, 23, 45, 46, 47.

B19 Napo and upper Amazon lowlands
Peru: 53.

B20 Tumbesian western Ecuador and Peru
Ecuador: 03, 04, 05, 07, 09, 11, 16, 17, 18, 19, 20, 21, 25, 26, 28, 29, 31, 32, 34, 35, 36, 37, 38, 39, 40, 41, 42, 43, 44.
Peru: 01, 02, 03, 05, 10, 11, 24, 25, 26, 27, 28, 29, 30, 37.

B21 South Central Andean forests
Ecuador: 07, 22, 27, 33, 34, 44, 47.
Peru: 03, 05, 06, 10, 11, 28.

B22 Marañón valley
Peru: 04, 08, 15, 16, 17, 21, 28, 31, 33.

B24 Sub-Andean ridge-top forests of south-east Ecuador and northern Peru
Ecuador: 48, 49.
Peru: 07, 13, 14, 16, 17, 22, 23.

B25 North-east Peruvian cordilleras
Peru: 16, 17, 18, 19, 20, 32, 35, 36, 49, 50, 51.

B27 High Peruvian Andes
Peru: 09, 11, 12, 21, 32, 34, 35, 38, 39, 40, 41, 42, 43, 44, 45, 46, 47, 48, 55, 56, 57, 66, 68, 69, 70, 71, 72, 73, 74, 75, 76, 77, 85.

B28 Junín puna
Peru: 58, 59, 60, 61, 62, 67.

B29 Eastern Andean foothills of Peru
Peru: 78, 79.

B30 South-east Peruvian lowlands
Peru: 54, 79, 80, 81.

B32 South Peruvian and north Chilean Pacific slope
Chile: 02, 03, 04, 05, 06, 07.
Peru: 63, 65, 86, 87, 88, 89.

B33 Upper Bolivian yungas
No Key Areas.

B34 Lower Bolivian yungas
Bolivia: 04, 09, 12, 14, 16, 17.
Peru: 82, 83, 84.

B35 Bolivian Andes
Bolivia: 13, 15, 18, 19, 20, 21, 22, 23, 24.

B37 North Argentine Andes
Argentina: 03, 04, 05, 06, 07, 09.

B39 Argentine cordilleras
No Key Areas.

B40 Islas Juan Fernández
Chile: 17, 18.

B41 Central Chile
No Key Areas.

B42 Tierra del Fuego and the Falklands
No Key Areas.

B43 Central Amazonian Brazil
Brazil: 008, 012.

B45 Fernando de Noronha
No Key Areas.

B46 North-east Brazilian caatinga
Brazil: 028, 033, 039, 040.

B47 Alagoan Atlantic slope
Brazil: 030, 031, 032, 034, 035, 036, 037.

B48 Bahian deciduous forests
Brazil: 048, 050, 065.

B49 Minas Gerais deciduous forests
Brazil: 064.

B50 Serra do Espinaço
Brazil: 066.

B52 South-east Brazilian lowland and foothills
Argentina: 15, 16, 17, 18, 19, 20, 22, 23.
Brazil: 043, 045, 046, 051, 052, 053, 054, 055, 056, 067, 068, 074, 075, 076, 077, 078, 079, 084, 087, 088, 089, 090, 091, 092, 093, 094, 095, 099, 100, 102, 103, 104, 105, 106, 107, 108, 109, 110, 112, 113, 114, 115, 116, 120, 122, 123, 124, 125, 126, 128, 129, 130.
Paraguay: 05, 06, 07, 08, 09, 10, 11, 12, 13, 14, 16, 17, 22.

B53 South-east Brazilian mountains
Brazil: 072, 086, 088, 090, 100, 103, 108, 109.

B54 South-east Brazilian *Araucaria* forest
Argentina: 20, 21.
Brazil: 097, 099, 103, 131, 132, 133, 134, 135, 136, 137, 138, 140, 141, 142.

B55 Entre Ríos wet grasslands
Argentina: 27, 28, 29, 30, 31.

B56 Upper Rio Branco
Brazil: 001, 002, 003, 004.
Guyana: 01.

B57 Boliviano–Tucuman yungas
Argentina: 03, 04, 05, 07, 09.
Bolivia: 26, 27.

B58 Valdivian forests of central Chile and Argentina
Argentina: 25.

B60 Central Andean páramo
Colombia: 24, 56.
Ecuador: 06, 08, 10, 22, 24, 27, 30, 44, 47.
Peru: 06.

APPENDIX 3: Threatened birds of Central and South America, 1994

THE KEY AREA analysis takes as its basis the list of threatened species defined by Collar *et al.* (1992). Given below is the new list of threatened Neotropical species according to the re-evaluation of the world's threatened birds published in 1994 by Collar *et al.* (*Birds to watch 2: the world list of threatened birds*). For an explanation of the new IUCN threat categories used by this most recent analysis, see Appendix 4 (p. 305).

The species are listed by threat category; each is followed by its range-states (including those outside the Neotropics) and the following qualifiers.

[X] Thought to be extinct
[I] Introduced population
[V] Known only as a vagrant
[?] Records are questionable

EXTINCT

Atitlán Grebe *Podilymbus gigas*
Guatemala
Colombian Grebe *Podiceps andinus*
Colombia
Glaucous Macaw *Anodorhynchus glaucus*
Argentina, Brazil, Paraguay, Uruguay

EXTINCT IN THE WILD

Alagoas Curassow *Mitu mitu*
Brazil
Socorro Dove *Zenaida graysoni*
Mexico

CRITICALLY ENDANGERED

Magdalena Tinamou *Crypturellus saltuarius*
Colombia
Kalinowski's Tinamou *Nothoprocta kalinowskii*
Peru
Junín Grebe *Podiceps taczanowskii*
Peru
Galápagos Petrel *Pterodroma phaeopygia*
Ecuador
Guadalupe Storm-petrel *Oceanodroma macrodactyla*
Mexico
Brazilian Merganser *Mergus octosetaceus*
Argentina, Brazil, Paraguay
White-winged Guan *Penelope albipennis*
Peru
Blue-billed Curassow *Crax alberti*
Colombia
Red-billed Curassow *Crax blumenbachii*
Brazil
Bearded Wood-partridge *Dendrortyx barbatus*
Mexico
Austral Rail *Rallus antarcticus*
Argentina, Chile
Eskimo Curlew *Numenius borealis*
Argentina, Chile, Canada, U.S.A., Uruguay
Blue-eyed Ground-dove *Columbina cyanopis*
Brazil
Purple-winged Ground-dove *Claravis godefrida*
Argentina, Brazil, Paraguay
Lear's Macaw *Anodorhynchus leari*
Brazil
Spix's Macaw *Cyanopsitta spixii*
Brazil
Yellow-eared Parrot *Ognorhynchus icterotis*
Colombia, Ecuador
Fuertes's Parrot *Hapalopsittaca fuertesi*
Colombia
White-winged Nightjar *Caprimulgus candicans*
Bolivia, Brazil, Paraguay[X]
Hook-billed Hermit *Glaucis dohrnii*
Brazil
Sapphire-bellied Hummingbird *Lepidopyga lilliae*
Colombia
Honduran Emerald *Amazilia luciae*
Honduras
Scissor-tailed Hummingbird *Hylonympha macrocerca*
Venezuela
Juan Fernández Firecrown *Sephanoides fernandensis*
Chile
Bogotá Sunangel *Heliangelus zusii*
Colombia
Black-breasted Puffleg *Eriocnemis nigrivestis*
Ecuador
Turquoise-throated Puffleg *Eriocnemis godini*
Colombia[?], Ecuador
Imperial Woodpecker *Campephilus imperialis*
Mexico
Royal Cinclodes *Cinclodes aricomae*
Bolivia, Peru
White-browed Tit-spinetail *Leptasthenura xenothorax*
Peru
Alagoas Foliage-gleaner *Philydor novaesi*
Brazil
Alagoas Antwren *Myrmotherula snowi*
Brazil
Black-hooded Antwren *Formicivora erythronotos*
Brazil
Stresemann's Bristlefront *Merulaxis stresemanni*
Brazil
Kinglet Calyptura *Calyptura cristata*
Brazil
Peruvian Plantcutter *Phytotoma raimondii*
Peru
Niceforo's Wren *Thryothorus nicefori*
Colombia
Guadalupe Junco *Junco insularis*
Mexico
Pale-headed Brush-finch *Atlapetes pallidiceps*
Ecuador
Entre Ríos Seedeater *Sporophila zelichi*
Argentina
Tumaco Seedeater *Sporophila insulata*
Colombia
Cherry-throated Tanager *Nemosia rourei*
Brazil
Venezuelan Flowerpiercer *Diglossa venezuelensis*
Venezuela
Paria Whitestart *Myioborus pariae*
Venezuela
Grey-headed Warbler *Basileuterus griseiceps*
Venezuela
San Andrés Vireo *Vireo caribaeus*
Colombia
Forbes's Blackbird *Curaeus forbesi*
Brazil

ENDANGERED

Peruvian Diving-petrel *Pelecanoides garnotii*
Chile, Peru
Grey-backed Hawk *Leucopternis occidentalis*
Ecuador, Peru
Plumbeous Forest-falcon *Micrastur plumbeus*
Colombia, Ecuador
Cauca Guan *Penelope perspicax*
Colombia
Northern Helmeted Curassow *Pauxi pauxi*
Colombia, Venezuela
Southern Helmeted Curassow *Pauxi unicornis*
Bolivia, Peru
Gorgeted Wood-quail *Odontophorus strophium*
Colombia
Junín Rail *Laterallus tuerosi*
Peru
Plain-flanked Rail *Rallus wetmorei*
Venezuela
Bogotá Rail *Rallus semiplumbeus*
Colombia
Tolima Dove *Leptotila conoveri*
Colombia
Veracruz Quail-dove *Geotrygon carrikeri*
Mexico
Blue-throated Macaw *Ara glaucogularis*
Bolivia
Red-fronted Macaw *Ara rubrogenys*
Bolivia
Golden Parakeet *Guaruba guarouba*
Brazil
Thick-billed Parrot *Rhynchopsitta pachyrhyncha*
Mexico, U.S.A.[I]
Rufous-fronted Parakeet *Bolborhynchus ferrugineifrons*
Colombia
Brown-backed Parrotlet *Touit melanonota*
Brazil
Golden-tailed Parrotlet *Touit surda*
Brazil
Rusty-faced Parrot *Hapalopsittaca amazonina*
Colombia, Ecuador, Venezuela
Red-faced Parrot *Hapalopsittaca pyrrhops*
Ecuador, Peru
Red-spectacled Amazon *Amazona pretrei*
Argentina, Brazil, Paraguay[V]
Green-cheeked Amazon *Amazona viridigenalis*
Mexico, Puerto Rico[I], U.S.A.[I]
Red-browed Amazon *Amazona rhodocorytha*
Brazil
Red-tailed Amazon *Amazona brasiliensis*
Brazil
Yellow-headed Amazon *Amazona oratrix*
Belize, Guatemala, Mexico
Vinaceous Amazon *Amazona vinacea*
Argentina, Brazil, Paraguay
Blue-bellied Parrot *Triclaria malachitacea*
Argentina, Brazil
Banded Ground-cuckoo *Neomorphus radiolosus*
Colombia, Ecuador
Short-crested Coquette *Lophornis brachylopha*
Mexico
Táchira Emerald *Amazilia distans*
Venezuela
Chestnut-bellied Hummingbird *Amazilia castaneiventris*
Colombia
White-tailed Hummingbird *Eupherusa poliocerca*
Mexico
Oaxaca Hummingbird *Eupherusa cyanophrys*
Mexico
Little Woodstar *Acestrura bombus*
Ecuador, Peru
Esmeraldas Woodstar *Acestrura berlepschi*
Ecuador
Eared Quetzal *Euptilotis neoxenus*
Mexico, U.S.A.
Three-toed Jacamar *Jacamaralcyon tridactyla*
Brazil
White-mantled Barbet *Capito hypoleucus*
Colombia
Helmeted Woodpecker *Dryocopus galeatus*
Argentina, Brazil, Paraguay
Plain Spinetail *Synallaxis infuscata*
Brazil
Russet-bellied Spinetail *Synallaxis zimmeri*
Peru
Cipó Canastero *Asthenes luizae*
Brazil
White-throated Barbtail *Margarornis tatei*
Venezuela
Recurve-billed Bushbird *Clytoctantes alixii*
Colombia, Venezuela
Restinga Antwren *Formicivora littoralis*
Brazil
Fringe-backed Fire-eye *Pyriglena atra*
Brazil
Slender Antbird *Rhopornis ardesiaca*
Brazil
Grey-headed Antbird *Myrmeciza griseiceps*
Ecuador, Peru
Moustached Antpitta *Grallaria alleni*
Colombia
Bicoloured Antpitta *Grallaria rufocinerea*
Colombia
Brown-banded Antpitta *Grallaria milleri*
Colombia
Bahia Tapaculo *Scytalopus psychopompus*
Brazil
Banded Cotinga *Cotinga maculata*
Brazil
Kaempfer's Tody-tyrant *Hemitriccus kaempferi*
Brazil
Ash-breasted Tit-tyrant *Anairetes alpinus*
Bolivia, Peru
Antioquia Bristle-tyrant *Phylloscartes lanyoni*
Colombia
Minas Gerais Tyrannulet *Phylloscartes roquettei*
Brazil
Alagoas Tyrannulet *Phylloscartes ceciliae*
Brazil
Atlantic Royal Flycatcher *Onychorhynchus swainsoni*
Brazil
Ochre-breasted Pipit *Anthus nattereri*
Argentina, Brazil, Paraguay
Apolinar's Wren *Cistothorus apolinari*
Colombia
Floreana Mockingbird *Nesomimus trifasciatus*
Ecuador
Socorro Mockingbird *Mimodes graysoni*
Mexico
Sierra Madre Sparrow *Xenospiza baileyi*
Mexico
Worthen's Sparrow *Spizella wortheni*
Mexico, U.S.A.[X]
Yellow-headed Brush-finch *Atlapetes flaviceps*
Colombia
Yellow Cardinal *Gubernatrix cristata*
Argentina, Brazil, Paraguay, Uruguay
Plain-tailed Warbling-finch *Poospiza alticola*
Peru

Rufous-breasted Warbling-finch *Poospiza rubecula*
Peru
Cochabamba Mountain-finch *Poospiza garleppi*
Bolivia
Buffy-fronted Seedeater *Sporophila frontalis*
Argentina, Brazil, Paraguay
Temminck's Seedeater *Sporophila falcirostris*
Argentina, Brazil, Paraguay
Black-and-tawny Seedeater *Sporophila nigrorufa*
Bolivia, Brazil
Marsh Seedeater *Sporophila palustris*
Argentina, Brazil, Paraguay, Uruguay
Mangrove Finch *Camarhynchus heliobates*
Ecuador
Black-and-gold Tanager *Bangsia melanochlamys*
Colombia
Orange-throated Tanager *Wetmorethraupis sterrhopteron*
Ecuador, Peru
Azure-rumped Tanager *Tangara cabanisi*
Guatemala, Mexico
Seven-coloured Tanager *Tangara fastuosa*
Brazil
Black-backed Tanager *Tangara peruviana*
Brazil
Golden-cheeked Warbler *Dendroica chrysoparia*
El Salvador, Guatemala, Honduras, Mexico, Nicaragua, U.S.A.
Black-capped Vireo *Vireo atricapillus*
Mexico, U.S.A.
Baudó Oropendola *Psarocolius cassini*
Colombia
Saffron-cowled Blackbird *Xanthopsar flavus*
Argentina, Brazil, Paraguay, Uruguay
Pampas Meadowlark *Sturnella militaris*
Argentina, Brazil, Uruguay
Red-bellied Grackle *Hypopyrrhus pyrohypogaster*
Colombia
Red Siskin *Carduelis cucullata*
Colombia, Puerto Rico[I], Trinidad and Tobago[X], Venezuela
Dwarf Jay *Cyanolyca nana*
Mexico
White-throated Jay *Cyanolyca mirabilis*
Mexico

■ VULNERABLE

Tepuí Tinamou *Crypturellus ptaritepui*
Venezuela
Chocó Tinamou *Crypturellus kerriae*
Colombia, Panama
Taczanowski's Tinamou *Nothoprocta taczanowskii*
Peru
Lesser Nothura *Nothura minor*
Brazil
Dwarf Tinamou *Taoniscus nanus*
Argentina, Brazil
Galápagos Penguin *Spheniscus mendiculus*
Ecuador
Defilippe's Petrel *Pterodroma defilippiana*
Chile
Westland Petrel *Procellaria westlandica*
Chile
Pink-footed Shearwater *Puffinus creatopus*
Chile
Townsend's Shearwater *Puffinus auricularis*
Mexico
Black-vented Shearwater *Puffinus opisthomelas*
Mexico
Galápagos Cormorant *Phalacrocorax harrisi*
Ecuador
Andean Flamingo *Phoenicopterus andinus*
Argentina, Bolivia, Chile, Peru
Puna Flamingo *Phoenicopterus jamesi*
Argentina, Bolivia, Chile, Peru
White-necked Hawk *Leucopternis lacernulata*
Brazil
Crowned Eagle *Harpyhaliaetus coronatus*
Argentina, Bolivia, Brazil, Paraguay, Uruguay[X]
Galápagos Hawk *Buteo galapagoensis*
Ecuador
Rufous-headed Chachalaca *Ortalis erythroptera*
Ecuador, Peru
Bearded Guan *Penelope barbata*
Ecuador, Peru
Baudó Guan *Penelope ortoni*
Colombia, Ecuador
Chestnut-bellied Guan *Penelope ochrogaster*
Brazil
Black-fronted Piping-guan *Pipile jacutinga*
Argentina, Brazil, Paraguay
Horned Guan *Oreophasis derbianus*
Guatemala, Mexico
Piping Plover *Charadrius melodus*
Bahamas, Barbados, Canada, Cuba, Jamaica, Mexico, Puerto Rico, U.S.A., Virgin Islands (U.K.), Virgin Islands (U.S.A.)
Mountain Plover *Charadrius montanus*
Mexico, U.S.A.
Wattled Curassow *Crax globulosa*
Bolivia, Brazil, Colombia, Ecuador, Peru
Rusty-flanked Crake *Laterallus levraudi*
Venezuela
Rufous-faced Crake *Laterallus xenopterus*
Brazil, Paraguay
Brown Wood-rail *Aramides wolfi*
Colombia, Ecuador, Peru
Dot-winged Crake *Porzana spiloptera*
Argentina, Uruguay
Horned Coot *Fulica cornuta*
Argentina, Bolivia, Chile
Olrog's Gull *Larus atlanticus*
Argentina
Lava Gull *Larus fuliginosus*
Ecuador
Peruvian Pigeon *Columba oenops*
Peru
Ochre-bellied Dove *Leptotila ochraceiventris*
Ecuador, Peru
Hyacinth Macaw *Anodorhynchus hyacinthinus*
Bolivia, Brazil, Paraguay
Military Macaw *Ara militaris*
Argentina, Bolivia, Colombia, Costa Rica, Ecuador, Honduras, Mexico, Nicaragua, Panama, Peru, Venezuela
Blue-winged Macaw *Ara maracana*
Argentina, Brazil, Paraguay
Socorro Parakeet *Aratinga brevipes*
Mexico
Golden-capped Parakeet *Aratinga auricapilla*
Brazil, Paraguay[V]
Golden-plumed Parakeet *Leptosittaca branickii*
Colombia, Ecuador, Peru
Maroon-fronted Parrot *Rhynchopsitta terrisi*
Mexico
Blue-chested Parakeet *Pyrrhura cruentata*
Brazil
Santa Marta Parakeet *Pyrrhura viridicata*
Colombia
El Oro Parakeet *Pyrrhura orcesi*
Ecuador
White-necked Parakeet *Pyrrhura albipectus*
Ecuador
Flame-winged Parakeet *Pyrrhura calliptera*
Colombia

Yellow-faced Parrotlet *Forpus xanthops*
Peru
Spot-winged Parrotlet *Touit stictoptera*
Colombia, Ecuador, Peru
Yellow-faced Amazon *Amazona xanthops*
Bolivia, Brazil
Yellow-shouldered Amazon *Amazona barbadensis*
Aruba[X], Netherlands Antilles, Venezuela
Cocos Cuckoo *Coccyzus ferrugineus*
Costa Rica
White-chested Swift *Cypseloides lemosi*
Colombia, Ecuador
White-tailed Sabrewing *Campylopterus ensipennis*
Trinidad and Tobago, Venezuela
Mexican Woodnymph *Thalurania ridgwayi*
Mexico
Mangrove Hummingbird *Amazilia boucardi*
Costa Rica
Purple-backed Sunbeam *Aglaeactis aliciae*
Peru
Black Inca *Coeligena prunellei*
Colombia
Royal Sunangel *Heliangelus regalis*
Peru
Colourful Puffleg *Eriocnemis mirabilis*
Colombia
Violet-throated Metaltail *Metallura baroni*
Ecuador
Grey-bellied Comet *Taphrolesbia griseiventris*
Peru
Marvellous Spatuletail *Loddigesia mirabilis*
Peru
Chilean Woodstar *Eulidia yarrellii*
Chile, Peru
Glow-throated Hummingbird *Selasphorus ardens*
Panama
Coppery-chested Jacamar *Galbula pastazae*
Colombia, Ecuador
Five-coloured Barbet *Capito quinticolor*
Colombia
Tawny Piculet *Picumnus fulvescens*
Brazil
Ochraceous Piculet *Picumnus limae*
Brazil
Moustached Woodcreeper *Xiphocolaptes falcirostris*
Brazil
White-bellied Cinclodes *Cinclodes palliatus*
Peru
Más Afuera Rayadito *Aphrastura masafuerae*
Chile
Apurímac Spinetail *Synallaxis courseni*
Peru
Blackish-headed Spinetail *Synallaxis tithys*
Ecuador, Peru
Red-shouldered Spinetail *Gyalophylax hellmayri*
Brazil
Hoary-throated Spinetail *Synallaxis kollari*
Brazil, Guyana
Maquis Canastero *Asthenes heterura*
Bolivia
Pale-tailed Canastero *Asthenes huancavelicae*
Peru
Berlepsch's Canastero *Asthenes berlepschi*
Bolivia
Chestnut Canastero *Asthenes steinbachi*
Argentina
Austral Canastero *Asthenes anthoides*
Argentina, Chile
Orinoco Softtail *Thripophaga cherriei*
Venezuela
Striated Softtail *Thripophaga macroura*
Brazil
Rufous-necked Foliage-gleaner *Syndactyla ruficollis*
Ecuador, Peru
Bolivian Recurvebill *Simoxenops striatus*
Bolivia
Henna-hooded Foliage-gleaner
Hylocryptus erythrocephalus
Ecuador, Peru
Great Xenops *Megaxenops parnaguae*
Brazil
White-bearded Antshrike *Biatas nigropectus*
Argentina, Brazil
Speckled Antshrike *Xenornis setifrons*
Colombia, Panama
Plumbeous Antvireo *Dysithamnus plumbeus*
Brazil
Bicoloured Antvireo *Dysithamnus occidentalis*
Colombia, Ecuador
Rio de Janeiro Antwren *Myrmotherula fluminensis*
Brazil
Salvadori's Antwren *Myrmotherula minor*
Brazil
Ashy Antwren *Myrmotherula grisea*
Bolivia
Unicoloured Antwren *Myrmotherula unicolor*
Brazil
Band-tailed Antwren *Myrmotherula urosticta*
Brazil
Ash-throated Antwren *Herpsilochmus parkeri*
Peru
Pectoral Antwren *Herpsilochmus pectoralis*
Brazil
Narrow-billed Antwren *Formicivora iheringi*
Brazil
Orange-bellied Antwren *Terenura sicki*
Brazil
Yellow-rumped Antwren *Terenura sharpei*
Bolivia, Peru
Rio Branco Antbird *Cercomacra carbonaria*
Brazil, Guyana
Black-tailed Antbird *Myrmoborus melanurus*
Peru
Scalloped Antbird *Myrmeciza ruficauda*
Brazil
Rufous-fronted Antthrush *Formicarius rufifrons*
Peru
Giant Antpitta *Grallaria gigantea*
Colombia, Ecuador
Táchira Antpitta *Grallaria chthonia*
Venezuela
Cundinamarca Antpitta *Grallaria kaestneri*
Colombia
Hooded Antpitta *Grallaricula cucullata*
Colombia, Venezuela
Brasília Tapaculo *Scytalopus novacapitalis*
Brazil
Shrike-like Cotinga *Laniisoma elegans*
Brazil
Grey-winged Cotinga *Tijuca condita*
Brazil
Black-headed Berryeater *Carpornis melanocephalus*
Brazil
Chestnut-bellied Cotinga *Doliornis remseni*
Colombia, Ecuador, Peru
White-cheeked Cotinga *Zaratornis stresemanni*
Peru
Buff-throated Purpletuft *Iodopleura pipra*
Brazil
Cinnamon-vented Piha *Lipaugus lanioides*
Brazil

Turquoise Cotinga *Cotinga ridgwayi*
Costa Rica, Panama
White-winged Cotinga *Xipholena atropurpurea*
Brazil
Yellow-billed Cotinga *Carpodectes antoniae*
Costa Rica, Panama
Bare-necked Umbrellabird *Cephalopterus glabricollis*
Costa Rica, Panama
Long-wattled Umbrellabird *Cephalopterus penduliger*
Colombia, Ecuador
Three-wattled Bellbird *Procnias tricarunculata*
Costa Rica, Honduras, Nicaragua, Panama
Golden-crowned Manakin *Pipra vilasboasi*
Brazil
Black-capped Manakin *Piprites pileatus*
Argentina, Brazil
Buff-breasted Tody-tyrant *Hemitriccus mirandae*
Brazil
Fork-tailed Pygmy-tyrant *Hemitriccus furcatus*
Brazil
Cocos Flycatcher *Nesotriccus ridgwayi*
Costa Rica
Dinelli's Doradito *Pseudocolopteryx dinellianus*
Argentina
Rufous-sided Pygmy-tyrant *Euscarthmus rufomarginatus*
Bolivia, Brazil, Paraguay, Surinam
São Paulo Tyrannulet *Phylloscartes paulistus*
Argentina, Brazil, Paraguay
Restinga Tyrannulet *Phylloscartes kronei*
Brazil
Russet-winged Spadebill *Platyrinchus leucoryphus*
Argentina, Brazil, Paraguay
Pacific Royal Flycatcher *Onychorhynchus occidentalis*
Ecuador, Peru
Grey-breasted Flycatcher *Lathrotriccus griseipectus*
Ecuador, Peru
Santa Marta Bush-tyrant *Myiotheretes pernix*
Colombia
Black-and-white Monjita *Heteroxolmis dominicana*
Argentina, Brazil, Paraguay, Uruguay
White-tailed Shrike-tyrant *Agriornis andicola*
Argentina, Bolivia, Chile, Ecuador, Peru
Strange-tailed Tyrant *Alectrurus risora*
Argentina, Brazil, Paraguay, Uruguay
Ochraceous Attila *Attila torridus*
Colombia, Ecuador, Peru
Rufous-throated Dipper *Cinclus schulzi*
Argentina, Bolivia
Sumichrast's Wren *Hylorchilus sumichrasti*
Mexico
Nava's Wren *Hylorchilus navai*
Mexico
Cobb's Wren *Troglodytes cobbi*
Falkland Islands
Clarión Wren *Troglodytes tanneri*
Mexico
Yellow-green Finch *Pselliophorus luteoviridis*
Panama
Tanager-finch *Oreothraupis arremonops*
Colombia, Ecuador
Black-masked Finch *Coryphaspiza melanotis*
Argentina, Bolivia, Brazil, Paraguay, Peru
Slender-billed Finch *Xenospingus concolor*
Chile, Peru
Tucumán Mountain-finch *Poospiza baeri*
Argentina
Hooded Seedeater *Sporophila melanops*
Brazil
Cocos Finch *Pinaroloxias inornata*
Costa Rica
Rufous-bellied Saltator *Saltator rufiventris*
Argentina, Bolivia
Cone-billed Tanager *Conothraupis mesoleuca*
Brazil
Yellow-green Bush-tanager *Chlorospingus flavovirens*
Colombia, Ecuador
Slaty-backed Hemispingus *Hemispingus goeringi*
Venezuela
Black-cheeked Ant-tanager *Habia atrimaxillaris*
Costa Rica
Gold-ringed Tanager *Bangsia aureocincta*
Colombia
Golden-backed Mountain-tanager *Buthraupis aureodorsalis*
Peru
Masked Mountain-tanager *Buthraupis wetmorei*
Colombia, Ecuador, Peru
Multicoloured Tanager *Chlorochrysa nitidissima*
Colombia
Green-capped Tanager *Tangara meyerdeschauenseei*
Bolivia?, Peru
Turquoise Dacnis *Dacnis hartlaubi*
Colombia
Black-legged Dacnis *Dacnis nigripes*
Brazil
Scarlet-breasted Dacnis *Dacnis berlepschi*
Colombia, Ecuador
Belding's Yellowthroat *Geothlypis beldingi*
Mexico
Black-polled Yellowthroat *Geothlypis speciosa*
Mexico
Guaiquinima Whitestart *Myioborus cardonai*
Venezuela
Tamarugo Conebill *Conirostrum tamarugense*
Chile, Peru
Chocó Vireo *Vireo* sp.
Colombia
Noronha Vireo *Vireo gracilirostris*
Brazil
Selva Cacique *Cacicus koepckeae*
Peru
Yellow-faced Siskin *Carduelis yarrellii*
Brazil, Venezuela
Saffron Siskin *Carduelis siemiradzkii*
Ecuador, Peru

■ DATA DEFICIENT

Black Tinamou *Tinamus osgoodi*
Colombia, Peru
White-vented Storm-petrel *Oceanites gracilis*
Ecuador, Chile
Markham's Storm-petrel *Oceanodroma markhami*
Chile?, Peru
Ringed Storm-petrel *Oceanodroma hornbyi*
Chile, Peru
Speckled Crake *Coturnicops notatus*
Argentina, Brazil, Colombia, Guyana, Paraguay, Uruguay, Venezuela
Cayenne Nightjar *Caprimulgus maculosus*
French Guiana
White-fronted Swift *Cypseloides storeri*
Mexico
Coppery Thorntail *Popelairia letitiae*
Bolivia
Rondônia Bushbird *Clytoctantes atrogularis*
Brazil
White-masked Antbird *Pithys castanea*
Peru
Sinaloa Martin *Progne sinaloae*
Mexico

APPENDIX 4: **The new IUCN threat categories and criteria**

THE NEW IUCN criteria for assigning threat status and category (Mace and Stuart 1994, IUCN SSC 1995) are a less subjective and more accountable system than forms previously used. The criteria are numerically based and reflect stepwise increases in the risk of extinction as judged from measured or reasoned rates of decline, population levels and range size. The following box and table are adapted from those published in Collar *et al.* (1994).

Definitions of the new IUCN categories

EXTINCT (EX) A taxon is Extinct when there is no reasonable doubt that its last individual has died.

EXTINCT IN THE WILD (EW) A taxon is Extinct in the Wild when it is known only to survive in cultivation, in captivity, or as a naturalized population (or populations) well outside the past range. A taxon is presumed extinct in the wild when exhaustive surveys in known and/or expected habitat, at appropriate times (diurnal, seasonal, annual), throughout its historic range, have failed to record an individual. Surveys should be over a time frame appropriate to the taxon's life cycle and life form.

CRITICALLY ENDANGERED (CR) A taxon is Critically Endangered when it is facing an extremely high risk of extinction in the wild in the immediate future, as defined in any of the criteria (A–E)[1].

ENDANGERED (EN) A taxon is Endangered when it is not Critical but is facing a very high risk of extinction in the wild in the near future, as defined in any of the criteria (A–E)[1].

VULNERABLE (VU) A taxon is Vulnerable when it is not Critical or Endangered but is facing a high risk of extinction in the wild in the medium-term future, as defined in any of the criteria (A–E)[1].

CONSERVATION DEPENDENT (CD) Taxa which do not currently qualify as Critical, Endangered or Vulnerable may be classified as Conservation Dependent. To be considered Conservation Dependent, a taxon must be the focus of a continuing taxon-specific or habitat-specific conservation programme which directly affects the taxon in question. The cessation of this conservation programme would result in the taxon qualifying for one of the threatened categories above.

LOW RISK (LR) A taxon is Low Risk when it has been evaluated and does not qualify for any of the categories Critical, Endangered, Vulnerable, Conservation Dependent or Data Deficient. It is clear that a range of forms will be included in this category, including: (i) those that are close to qualifying for the threatened categories[2], (ii) those that are of less concern, and (iii) those that are presently abundant and unlikely to face extinction in the foreseeable future. It may be appropriate to indicate into which of these three classes taxa in Low Risk seem to fall. It is especially recommended to indicate an appropriate interval, or circumstance, before re-evaluation is necessary for taxa in the Low Risk class, especially for those indicated in (i) above.

DATA DEFICIENT (DD) A taxon is Data Deficient when there is inadequate information to make a direct, or indirect, assessment of its risk of extinction based on its distribution and/or population status. A taxon in this category may be well studied, and its biology well known, but appropriate data on abundance and/or distribution are lacking. DD is therefore not a category of threat or Low Risk. Listing of taxa in this category indicates that more information is required. Listing a taxon as DD acknowledges the possibility that future research will show that threatened classification is appropriate. It is important to make positive use of whatever data are available. In many cases great care should be exercised in choosing between DD and threatened status. If the range of a taxon is suspected to be relatively circumscribed, if a considerable period of time has elapsed since the last record of a taxon, or if there are reasonable chances of unreported surveys in which the taxon has not been found, or that habitat loss has had an unfavourable impact, threatened status may well be justified.

NOT EVALUATED (NE) A taxon is Not Evaluated when it has not yet been assessed against the criteria.

[1] See Table.

[2] This is the distinction which bears the title 'Near Threatened'.

The new IUCN threatened category thresholds at a glance

'Extent of occurrence' is the area contained within the shortest continuous imaginary boundary which encompasses all known, inferred or projected sites of present occurrence.

'Area of occupancy' is the area within the extent of occurrence which is occupied by a taxon (this measure is often applicable to species with highly specific habitats).

Criteria	Main numerical thresholds		
	Critical	**Endangered**	**Vulnerable**
A RAPID DECLINE	>80% over 10 years or 3 generations	>50% over 10 years or 3 generations	>50% over 20 years or 5 generations
B SMALL RANGE **fragmented, declining or fluctuating**	Extent of occurrence <100 km^2 or area of occupancy <10 km^2	Extent of occurrence <5,000 km^2 or area of occupancy <500 km^2	Extent of occurrence <20,000 km^2 or area of occupancy <2,000 km^2
C SMALL POPULATION **declining**	<250 mature individuals	<2,500 mature individuals	<10,000 mature individuals
D1 VERY SMALL POPULATION	<50 mature individuals	<250 mature individuals	<1,000 mature individuals
D2 VERY SMALL RANGE	—	—	<100 km^2 or <5 locations
E UNFAVOURABLE POPULATION VIABILITY ANALYSIS	Probability of extinction >50% within 5 years	Probability of extinction >20% within 20 years	Probability of extinction >10% within 100 years

APPENDIX 5: **Important Bird Areas: categories and criteria**

THE following categories and criteria are those which are now being used by the BirdLife Partnership in the process to identify Important Bird Areas throughout the world (see Box 3 in '*Key Areas in context*', p. 9).

CATEGORY	CRITERIA	NOTES
Globally threatened species	The site regularly holds significant numbers of a globally threatened species, or other species of global conservation concern.	The site qualifies if it is known, estimated or thought to hold a population of a species categorized as Critical or Endangered. Population-size thresholds for Vulnerable, Conservation Dependent, Data Deficient and Near Threatened species are set regionally, as appropriate, to help in site selection.
Restricted-range species	The site is known or thought to hold a significant component of the restricted-range species whose breeding distributions define an Endemic Bird Area (EBA) or Secondary Area (SA).	The site also has to form one of a set selected to ensure that, as far as possible, all restricted-range species of an EBA or SA are present in significant numbers in at least one site and, preferably, more.
Biome-restricted assemblage	The site is known or thought to hold a significant component of the group of species whose distributions are largely or wholly confined to one biome.	The site also has to form one of a set selected to ensure that, as far as possible, all species restricted to a biome are adequately represented.
Congregations	**(i)** The site is known or thought to hold, on a regular basis, 1% or more of a biogeographic population of a congregatory waterbird species. *or*	This applies to waterbird species as defined by Rose and Scott (1994). Thresholds are generated in some instances by combining flyway populations within a biogeographic region, but for others lacking quantitative data, thresholds are set regionally or inter-regionally, as appropriate. In such cases, thresholds will be taken as estimates of 1% of the biogeographic population.
	(ii) The site is known or thought to hold, on a regular basis, 1% or more of the global population of a congregatory seabird or terrestrial species. *or*	This includes those seabird species not covered by Rose and Scott (1994). Where quantitative data are lacking, numerical thresholds for each species are set regionally or inter-regionally, as appropriate. In such cases, thresholds will be taken as estimates of 1% of the global population.
	(iii) The site is known or thought to hold, on a regular basis, 20,000 or more waterbirds or 10,000 or more pairs of seabirds of one or more species. *or*	This is the Ramsar criterion for waterbirds, the use of which is discouraged wherever data are good enough to permit the use of (i) or (ii).
	(iv) The site is known or thought to exceed thresholds set for migratory species at bottleneck sites.	Thresholds are set regionally or inter-regionally, as appropriate.

INDEX OF KEY AREAS

Reference is made only to the main entries in the national inventories.